i_b	Borrowed (debt) fin...
i_c	Combined or nominal, interest rate (with inflation included)
i_r	Real interest rate (without inflation included)
I	Investment
I_t	Index of price change ($= I_n/I_k$) (inflation for period k to period n)
IRR	Internal rate of return
j	Percentage increase in amount each period (geometric series)
λ("Lamda")	Fraction of debt in total capitalization
M	Number of compounding periods per year
MACRS	Modified accelerated cost recovery system
MARR	Minimum attractive rate of return
MIRR	Modified internal rate of return
MV	Market value (also called salvage value, S)
N	Number of compounding periods (also, depreciable life)
N′	MACRS property class
NCF	Net cash flow
NOPAT	Net operating profit after taxes
NPV (or NPW)	Net present value (or worth)
Pr[·]	Probability of [·]
PW	Present worth (method, or equivalent)
PW-C	Present worth–cost (method, or equivalent)
r	Nominal interest rate per year (= annual percentage rate)
R_k	Revenues (cash inflows) for period k
R$	Real (uninflated, constant worth) dollars
RN	Random number
RND	Random normal deviate
RR	Rate of return
S	Number of standard deviations (for standardized normal distribution)
σ	Population (or estimated) standard deviation
SL	Straight-line method of depreciation
t	Effective (or marginal) income tax rate
TC_k	Total (marginal) cost for year k
TI	Taxable income
U[·]	Utility of [·]
V[·]	Variance of [·]
WACC	Weighted average cost of capital
WC	Working capital

PRENTICE HALL INTERNATIONAL SERIES
IN INDUSTRIAL AND SYSTEMS ENGINEERING
W. J. Fabrycky and J. H. Mize, Editors

AMOS AND SARCHET • *Management for Engineers*

AMRINE, RITCHEY, MOODIE, AND KMEC • *Manufacturing Organization and Management, 6/E*

ASFAHL • *Industrial Safety and Health Management, 5/E*

BABCOCK • *Managing Engineering and Technology, 3/E*

BADIRU • *Expert Systems Applications in Engineering and Manufacturing*

BANKS CARSON NELSON AND NICOL • *Discrete-Event System Simulation 4/E*

BLANCHARD • *Logistics Engineering and Management, 6/E*

BLANCHARD AND FABRYCKY • *Systems Engineering and Analysis, 2/E*

BUSSEY AND ESCHENBACH • *The Economics Analysis of Industrial Projects, 2/E*

BUZACOTT AND SHANTHIKUMAR • *Stochastic Models of Manufacturing Systems*

CANADA AND SULLIVAN • *Economics and Multi-Attribute Evaluation of Advanced Manufacturing* Systems

CANADA SULLIVAN AND WHITE • *Capital Investment Analysis for Engineering and Management 2e*

CHANG AND WYSK • *An Introduction to Automated Process Planning Systems*

CHANG, WYSK, AND WANG • *Computer Aided Manufacturing 2/E*

EBERTS • *User Interface Design*

ELSAYED AND BOUCHER • *Analysis and Control of Production Systems, 2/E*

FABRYCKY AND BLANCHARD • *Life-Cycle Cost and Economic Analysis*

FABRYCKY AND THUESEN• *Economic Decision Analysis, 3/E*

FISHWICK • *Simulation Model Design and Execution: Building Digital Worlds*

FRANCIS, MCGINNIS, AND WHITE • *Facility Layout and Location: An Analytical Approach, 2/E*

GIBSON • *Modern Management of the High-Technology Enterprise*

GRAEDEL AND ALLENBY • *Industrial Ecology, 2/E*

HALL • *Queuing Methods: For Services and Manufacturing*

HAMMER • *Occupational Safety Management and Engineering, 5/E*

HUTCHINSON • *An Integrated Approach to Logistics Management*

IGNIZIO AND CAVALIER • *Linear Programming*

KROEMER, KROEMER, AND KROEMER-ELBERT • *Ergonomics: How to Design for Ease and Efficiency2/E*

KUSIAK • *Intelligent Manufacturing Systems*

LANDERS, BROWN, FANT, MALSTROM, AND SCHMITT • *Electronics Manufacturing Processes*

LEEMIS • *Reliability: Probabalistic Models and Statistical Methods*

MUNDER AND DANNER • *Motion and Time Study: Improving Productivity, 7/E*

OSTWALD • *Engineering Cost Estimating, 3/E*

PINEDO • *Scheduling: Theory, Algorithms, and Systems 2/E*

PULAT • *Fundamentals of Industrial Ergonomics*

SHTUB, BARD, AND GLOBERSON • *Project Management: Engineering Technology and Implementation*

TAHA • *Simulation Modeling and SIMNET*

THUESEN AND FABRYCKY • *Engineering Economy, 9/E*

TURNER, MIZE, CASE, AND NAZEMETZ • *Introduction to Industrial and Systems Engineering, 3/E*

WOLFF • *Stochastic Modeling and the Theory of Queues*

THIRD EDITION

Capital Investment Analysis for Engineering and Management

John R. Canada
North Carolina State University, Emeritus

William G. Sullivan
Virginia Polytechnic Institute and State University, Emeritus
The University of Tennessee, Adjunct

Dennis J. Kulonda
Florida Institute of Technology

John A. White
University of Arkansas

PEARSON
Prentice
Hall

Upper Saddle River, NJ 07458

Library of Congress Cataloging-in-Publication Data

Canada, John R.
 Capital investment analysis for engineering and management / John R. Canada, William
G. Sullivan, Dennis J. Kulonda. -- [3rd ed.]
  p. cm.
 Includes bibliographical references and index.
 ISBN 0-13-143408-X
  1. Capital investments--Evaluation. I. Sullivan, William G., 1942– II. Kulonda, Dennis
J. III. Title.

 HG4028.C4C3 2004
 658.15'242--dc22   2004050537

Vice President and Editorial Director, ECS: *Marcia J. Horton*
Acquisitions Editor: *Dorothy Marrero*
Associate Editor: *Andrea Messineo*
Vice President and Director of Production and Manufacturing, ESM: *David W. Riccardi*
Executive Managing Editor: *Vince O'Brien*
Managing Editor: *David George*
Production Editor: *Daniel Sandin*
Director of Creative Services: *Paul Belfanti*
Art Director: *Jayne Conte*
Cover and Interior Design: *Bruce Kenselaar*
Art Editor: *Greg Dulles*
Manufacturing Manager: *Trudy Pisciotti*
Manufacturing Buyer: *Lisa McDowell*
Executive Marketing Manager: *Holly Stark*

© 2005, 1999 Pearson Education, Inc.
Pearson Prentice Hall
Pearson Education, Inc.
Upper Saddle River, NJ 07458

Pearson Prentice Hall® is a trademark of Pearson Education, Inc.

The author and publisher of this book have used their best efforts in preparing this book. These efforts include
the development, research, and testing of the theories and programs to determine their effectiveness. The author
and publisher make no warranty of any kind, expressed or implied, with regard to these programs or the documentation contained in this book. The author and publisher shall not be liable in any event for incidental or consequential damages in connection with, or arising out of, the furnishing, performance, or use of these programs.

Printed in the United States of America

10 9 8 7 6 5 4 3 2

ISBN 0-13-143408-X

Pearson Education Ltd., *London*
Pearson Education Australia Pty. Ltd., *Sydney*
Pearson Education Singapore, Pte. Ltd.
Pearson Education North Asia Ltd., *Hong Kong*
Pearson Education Canada, Inc., *Toronto*
Pearson Educación de Mexico, S.A. de C.V.
Pearson Education—Japan, *Tokyo*
Pearson Education Malaysia, Pte. Ltd.
Pearson Education, Inc., *Upper Saddle River, New Jersey*

TABLE OF CONTENTS

PREFACE

This book is intended primarily for graduate or advanced undergraduate study for students in engineering and business disciplines emphasizing capital investment decision methodologies. The scope and coverage of the book also make it suitable as a text and a reference in industry.

This book is an evolution of *Intermediate Economic Analysis for Management and Engineering* by John R. Canada, published by Prentice-Hall in 1971. That was revised with the capable coauthorship of John A. White and renamed *Capital Investment Decision Analysis for Management and Engineering,* published in 1980. The Second Edition, slightly renamed and published in 1996, incorporated the primary creative coauthorship talents of William G. Sullivan. This Third Edition benefits from the active coauthorship of all three of the preceding authors and is considerably enhanced by the addition of Dennis J. Kulonda, who provided accounting, finance, and managerial perspectives to broaden the usefulness of the book, particularly in the Engineering Management field. It continues as a text and reference that is more concise and yet more advanced than traditional applied works and is highly lucid—incorporating abundant example problems and solutions.

Part One places capital investment decisions within a systems analysis framework. Chapters 1–3 provide fundamental concepts and techniques important in accounting, cost measurement, and estimating. Extensive supplemental glossaries are provided in an Instructor's Manual.

Part Two begins with a rather succinct summary of basic time value-based evaluation techniques in Chapters 4–6. Chapter 7 covers currently important depreciation methodologies and federal income tax law provisions, with numerous examples of ways to make after-tax comparisons of alternatives. Chapter 8 shows how to consider inflation explicitly, if needed, and Chapter 9 addresses replacement analyses in moderate depth. Chapter 10 is all-new material on new product and expansion analyses. Chapter 11 summarizes capital budgeting management perspectives and includes appendices on adjusted present worth or value and bundled financing.

Part Three remains a solid treatment of formal approaches to evaluation of risk and uncertainty in capital investment analyses. Its Chapters 12–16 contain minimal modifications to such topics as risk, sensitivity, and decision tree analysis.

Part Four includes new material and specialized topics. Chapter 17 on real options analysis is based on an outstanding *Harvard Business Review* article by Timothy A. Luehrman. Chapter 18 contains considerably enhanced materials on activity-based costing and management. Finally, Chapter 19 provides a good overview of multiattribute decision techniques, with heavy emphasis on use of the analytic hierarchy process.

For Parts One and Two, no specialized mathematics are required; while, for much of Part Three, it is assumed that the student understands the basic analysis procedures of Part Two and has a fair knowledge of elementary probability and statistics.

Some probability concepts are explained in the text, but those who need further background will find that the first half of most probability and statistics texts will provide adequate reference material. Complete understanding of the application of some specialized quantitative techniques to capital investment analyses involving probabilities will be facilitated by prior exposure to the theory underlying those techniques.

For a course in economic evaluation of alternative projects at the advanced undergraduate or initial graduate level, Parts One and Two can be used for a broad overview as well as principles/techniques orientation and review purposes. Parts Three and Four provide the primary material not normally included in a first course. Throughout, ample student problem exercises are included, and even more comprehensive problems and cases are suggested in the Instructors Manual. Since the chapters in Parts Three and Four and largely independent of one another, one can include or delete chapters according to the needs of individual courses/interests.

The book is updated to reflect the increasing usefulness of Excel spreadsheets in lieu of tabled factors for computationally intricate decision problems. However, coverage of traditional interest factors and tables is retained, as it remains a desirable way to articulate concepts and enable quick feedback on the understanding of material.

Assumptions and Perspectives

We have made numerous assumptions throughout this book and have strived to notate them clearly, particularly if they differ from authoritative standards. By assumptions, we mean "educated guesses" and "plausible simplifications" to facilitate analyses—such as estimates of future cash inflows and outflows, interest rates, inflation rates, income tax regulations, and other matters affected by the future circumstances regarding the economy and the organization.

A key assumption in this book is that the viewpoint taken in an economy study is that of profit-seeking owners (i.e., shareholders) of an organization. For public organizations or regulated firms, the viewpoint typically is that of management endeavoring to maximize benefits or minimize costs. Consequently, we assume that managers who act on economic analysis results are rational persons making decisions objectively to take advantage of feasible investment opportunities available to them.

Another important assumption in much of this book is that most capital investment decisions are made independently of financing decisions. In this regard, we usually assume that engineers and others propose projects to be funded by a firm's overall pool of capital, and decisions regarding which projects to accept are separate from decisions concerning the sourcing of capital to fund these projects. This important assumption is often referred to as the "separation principle." We depart from this assumption briefly in Section 7.6 regarding "After-Tax Analyses that Include Specific Funding Arrangements," and in the Appendices to Chapter 11.

Finally, we assume through most of the book that engineering projects are irreversible and are to be undertaken in the near-term future if the return is acceptable. In reality, another possibly more useful comparison of alternatives would consider investing capital in a feasible project today or waiting one or more years to make the investment decision. The flexibility of deferring a project may add value to a prospective investment opportunity. Accordingly, Chapter 16, "Decision Tree Analysis," and Chapter 17, "Capital Investment Decisions as Real Options," include methods for qualifying the possible value of project postponement. Additionally, in Chapter 9 we recognize that equipment replacement decisions usually involve the option of waiting to invest at a more opportune time.

Numerical Precision

Most, but not all, of the numerical examples and problem exercises have been solved using computer-generated interest factors—sometimes contained within Excel-type programs and sometimes within handheld financial calculators. The interest factors tabulated in Appendices A and B are rounded to four decimal places and their use does not generally result in the precision of computer-generated answers. Either approach is quite close enough for usual economic analysis purposes. Indeed, given the inherent inaccuracy of many estimates in practice, it is common to round off answers to what are thought meaningful—such as to the nearest dollar or thousand dollars or, say, three significant digits. This edition contains tabled factors for some 15 different interest rates. Of course, factors for any other rates can be calculated using formulas in Table 4-1 or 4-3.

Acknowledgments

Innumerable people—friends, colleagues, and helpers—have contributed to the development of this work. We have acknowledged many of them in the prefaces of prior editions, and the legacy of help by them and others continues in this edition. Our wives, Wanda, Janet, Karen, and Mary Lib helped by providing encouragement and assistance. To all these, as well as the authors and publishers providing reprint permissions, and to many others unnamed, we wish to express our gratitude.

John R. Canada
William G. Sullivan
Dennis J. Kulonda
John A. White

PART ONE

Basic Concepts: Value,
Cost, and Capital
Investment Foundations

1

Value Creation and Financial Accounting

The prime purpose of engineers in our society is to harness physical elements, energy, and human creativity to create value for the organizations where they practice their craft. Intuitively, we equate the creation of value with being better off; but we cannot demonstrate value without defining and measuring it in both monetary and nonmonetary terms. Although nonmonetary items may dominate some issues, few decisions escape the scrutiny of assessment in terms of their monetary or economic impact. The reason is simple: Money is the only common denominator that we have available to aggregate the benefits of improved labor productivity, better quality, and faster response, and compare them with resources committed, such as more ambulances, an improved dispatching system, larger computers, or more capable operators. When faced with decision alternatives, assessment of the economic consequences of each alternative is a necessary prerequisite to any choice of action. To accomplish this, we must convert both the costs and benefits of each alternative into their monetary equivalents and choose the best if more than one is viable. Only then will we truly know whether we have created value. As readers will realize very shortly, such seemingly simple assessments can become intricate in a complex decision situation. Without careful assessments, ill-advised choices might be made. *The purpose of this text is to equip engineers and engineering managers with the concepts, skills, and tools necessary to cope with the wide range of capital investment situations that they may encounter, so that correct recommendations are made and value is created.*

The ideas we will discuss apply to people's personal lives, to the family, to their business, hobby, employer, and to society as a whole. To maintain some level of consistency in presentation, we will cast issues, problems, and approaches in the context

1

of the engineer as a member of an organization that is usually, but not always, driven by the need to generate profits or earnings in order to remain viable (survive) in our highly competitive society. That perspective will not inhibit us from delving into occasional problems, illustrations, and examples in which the concepts we teach apply dramatically to our lives as professionals, family business managers, and citizens. But, unless otherwise implied by the problem we are examining, our perspective is organizational, and our measure of effectiveness is the value we generate for the owners of the organization. Like it or not, our job is to create value by increasing the wealth of the owners. In a large public corporation, this is called shareholder's wealth; in a partnership, partner's wealth; in a small business, owner's equity (especially gratifying if you are the owner); and in a university, trustee's assets. Later chapters include methods for incorporating non-monetary factors into the analysis.

To begin this journey, we must develop an understanding of economic choices and common measures of economic performance.

1.1 Economic Choices and Decision Making

The starting point in any conscious attempt at rational decision making must be recognition that a problem or opportunity exists. Often, problem recognition is obvious and immediate. A broken machine or completely inadequate production capability, for example, causes awareness rather readily. But there are numerous significant opportunities for improvement or prevention of future problems that are not obvious without search and thought. It has long been acknowledged that economic analysis of complex alternatives should be most valuable when performed as an integral part of the "big picture" of relevant considerations facing the decision maker. This is often called the "systems analysis" approach. We shall develop our approach using this framework and, as we become more sophisticated in our abilities, expand upon it by adding more specific details.

1.2 The Systems Analysis Framework

Systems analysis is a coordinated set of procedures that address the fundamental issues of design and management: specifying how people, money, information, energy, and materials should be combined to achieve a larger purpose. It includes investigation of proper objectives; comparing quantitatively, where possible, the cost, effectiveness, and risks associated with the alternative policies or strategies for achieving them; and formulating additional alternatives if those examined are found wanting. The five primary elements of a systems analysis (consistent with the "scientific method") are as follows:

1. **Definition of Objectives**

     Knowing clearly and precisely what we wish to accomplish is the initial step in any choice process. A crisp statement of objectives sets the stage for a sound analysis. Skipping this step usually spells disaster for a project.

2. Generation of Alternatives

The need for imagination and creativity in the generation of alternatives cannot be overstated, for its lack is a common defect of many analyses. An enlightened statement of objectives can assist with the generation of alternatives. For example, the objective of repairing a broken machine excludes the alternatives of replacing the machine or subcontracting work. If the true objective is to complete a project that requires the use of the machine, then these added alternatives are potential solutions.

3. Formulation of Measures of Effectiveness

Decisions are always choices between alternatives. Sometimes the choices are simply between taking action and doing nothing. But the choice always involves a comparison. The measures that are compared should be specified in advance in order to avoid bias in the comparison.

4. Evaluation of Alternatives

The actual comparison should include a determination of the benefits and the costs that are incurred as a result of pursuing a particular alternative so that alternatives can be compared. Often there are multiple dimensions of comparison that complicate the comparison process. To cope with this complexity, most decision makers try to distill this information into simpler indicators such as monetary benefits and monetary costs.

5. Tentative Selection

Were all choices to be made on the basis of precise knowledge of all the costs and all the benefits of each alternative, selection would merely add arithmetic to the evaluation step above. Reality, however, is rarely that simple. Some data are known with precision; others, only within an order of magnitude. Not all costs and benefits are readily quantifiable. Some, like aesthetics, are intangible. Not all outcomes are certain, and each alternative may involve its own unique set of risks. Again, to cope with this complexity, we will initially analyze projects as if the information were available and known with certainty. Later, we will expand the analysis framework to explicitly consider variables such as risk, intangibles, and multiple attributes.

Since economic analyses are concerned with which alternative or alternatives are best for future use, they are, by nature, based on estimates of what is expected to happen in the future. A difficult part of an economic analysis is estimating relevant quantities for the future, for the analysis is no better than the estimates comprising it. Most estimates are based on past results, and the usual best source of information on past results is the accounting records of the enterprise. But accounting information is compiled under a very strict set of guidelines and conventions. Anyone attempting to use accounting information is highly likely to misinterpret it unless they understand how it was constructed. Further, accounting is the language spoken at the business level of the company. If we are to relate engineering projects to the economic measures used by top management, we must understand the fundamentals of financial reporting and the underlying accounting. To accomplish that end, we turn to the topic of financial accounting.

1.3 Financial Accounting

Every day, in the normal conduct of business, there are multiple concurrent financial transactions; products are sold, raw materials are ordered creating an obligation for future payment, payroll obligations are made, new employees are hired, perhaps new equipment is delivered, a license to use a designer name is purchased, or arrangements for a bank loan are made. Each of these transactions has important implications for the organization and impacts its ability to earn a profit for its owners. Depending on the size of the organization, there could easily be hundreds or thousands of such transactions. Obviously, there must be some framework to record and organize these transactions so that obligations are met and owners and other interested stakeholders can assess whether progress toward profitability is being made. That is the task of financial accounting: to record, summarize and communicate the net impacts of transactions that affect the profitability and financial status of the organization.

One expedient approach is to place receipts in a shoebox and withdraw money from the shoebox to pay expenses, noting periodically the amount of cash left in the shoebox. Of course, this "shoebox" accounting intermingles funds from all sources: cash sales, partners' contributions, loan proceeds, etc. Expenses are often handled on a first-come, first-served basis. A slow-paying customer can cause the cash balance in the shoebox to deplete quickly. Some order can be imposed on this chaos by developing categories of inflows and outflows, called a chart of accounts, and subsequently recording all transactions. This is called cash accounting and is used by some small businesses. Notice that only transactions involving a cash exchange are recognized, so that sales on credit, contracted loan repayments, and other obligations are not recognized until they occur. This lack of forward visibility in the accounting system inhibits ability to predict the cash balance at the end of the year.

Because of these limitations, monks in the Middle Ages developed a more sophisticated framework, called accrual accounting. It records transactions at the point that obligations are created and accrues their impact until the cash exchange occurs. It has been refined through the ages and has been developed to adhere to well-defined standards by regulatory groups in each nation. Accrual accounting has become the standard vehicle for the development of accounting information that will be provided to shareholders, creditors, government bodies, and even potential customers and prospective employees of the organization. Understanding how it is structured and its underlying concepts is essential knowledge for anyone making decisions that affect the financial status of the organization. This is not to say that it is essential to understand the bookkeeping details of transaction recording and processing. However, a solid understanding of the accounting basics, as presented in Appendix 1-A, will be very helpful in facilitating your understanding of the concepts presented in this chapter. Understanding is important because accounting information drives many important decisions in the company. Achieving that understanding depends upon mastery of a few basic concepts.

1.3.1 Underlying Concepts

The basic notion of dual entry accounting rests upon the concept of an accounting entity as an organization whose financial transactions are kept separate from the persons operating and/or owing it. A second premise is that the organization's productive resources or assets must be exactly equal to the financial resources or equity furnished by financiers such as owners, stockholders, creditors, and bondholders. It is immediately useful to separate these financiers into two categories:

> The first are the owners, partners, and shareholders who have furnished money to the firm in exchange for the prospects of a share of the profits. The portion of the equity that they provide is called *Owners' Equity, Shareholders' Equity* or, when the intent is clear, simply *equity*.

> The second category captures the resources of those who have lent money to the firm or have furnished goods and services on credit. These are called *Liabilities* since they represent the resources that must eventually be returned by specific agreement to their owners.

Since assets acquired by the organization must be equal to the resources furnished, the fundamental accounting equation is

$$\text{Assets} = \text{Liabilities} + \text{Owner's Equity} \qquad (1\text{-}1)$$

This equality simply asserts that the assets must come from somewhere. As the organization matures and earns profits, these flow to the owners and owner's equity increases. Of course, a corresponding increase in assets occurs: Cash balances may increase, new equipment may be purchased, or warehouses and inventory may be added to expand business. In fact, a major top management responsibility is to decide how new funds should be deployed within the organization to meet the best interests of the owners. That is also a key concern to engineering managers, who must make recommendations regarding the desirability of new capital investments and try to guide management in a specific direction. In a nutshell, the purpose of this book is to sharpen your understanding and capability to do exactly that.

To understand just how accounting information may contribute to that end, it is essential to understand the basic accounting concepts that provide the foundation for the reporting of accounting information. Once that is accomplished, we will move on to the accounting reports themselves. Key concepts that you must understand and accept as the foundation for accounting reports are discussed next. They are rule-of-thumb descriptions rather than the more accurate (but less useful at this stage) complete legal descriptions.

Business Entity.   Financial statements are developed for a business separate from its owners. What occurs in its owners' affairs is not considered. Personal assets of owners are not comingled with those of the business, as this would distort measurement of the business.

Going Concern.   Accounting statements for a business entity assume that the entity is a going concern that will continue operations into the foreseeable future, unless specifically stated otherwise. For example, accountants assume inventory is saleable in the future and hence value it at its manufacturing cost, unless there are specific reasons to believe that it deteriorates or will become obsolete.

Monetary Measurement.   Accounting measures only those transactions that have a monetary value. Note that there may be many very important factors which have value that are not considered in the accounting system; a talented workforce, a new revolutionary product, or an effective lobbyist all can create extraordinary opportunities. But their value to the organization is not reflected in the accounting records unless there is a verifiable financial transaction tied to their services.

Accounting Period.   Accounting statements reflect activity during or at the end of a standard time window, called the accounting period. Publicly held companies are required to report quarterly and annually—most often, but not always, matching the calendar year.

Historical Cost.   Business transactions are valued at the validated transaction cost or price at the point in time when the transaction occurred. Subsequent changes in market prices of assets are not normally recorded in the accounting statements; hence, the value of an enterprise is often different than its book value (the sum of the historical values of all its assets after allowance for depreciation).

Realization.   Revenue is realized in the accounting system whenever some standard objective indication of a sales transaction occurs. Often this occurs at the point of shipping, or at the point of invoicing. Exceptions to the delivery convention occur. For example, companies performing long-term projects often invoice progress payments to their customers.

Matching Costs and Revenues.   This important concept mandates the matching of revenues with the costs of resources consumed in providing the goods and services that generated the revenue. This is the basis for the accounting definition of profit or net income. It may require that some revenues be deferred to the future. For example, the revenue received for a three year magazine subscription may be held as an asset called "deferred income" until it is earned year by year and matched with the expenses of producing and delivering the magazine. Similarly, expense recognition may be delayed until intended revenues materialize. This is the case when raw materials are purchased for resale, but held in inventory until a sales occurs. The purchased lot of raw materials are said to be "capitalized," meaning that they are held as an asset. The amount used in subsequent sales are said to be "expensed," meaning that the asset has been sold and its value must be netted against the revenue earned to contribute toward the profit for the accounting period.

Consistency.   Consistency requires that the transaction processing conventions adopted to conform to accounting principles remain consistent from year to year. Auditors are required to certify that accounting statements have been prepared "on a basis consistent with that of the preceding year." When changes are made to accounting policies and practices, these should be disclosed in the annual reports as an "accounting change."

Disclosure.   Accounting reports must disclose enough so that they will not mislead careful readers. This disclosure can be made in the financial statements or in the notes to the statements.

Conservatism.   Conservatism must be applied in deciding which of several alternative treatments should be applied to account for a specific transaction. This means that preference should be given to those treatments which understate asset values or overstate liabilities.

Materiality.   Inconsequential transactions may be handled expediently. Material or significant transactions must conform to accounting standards.

Application of these concepts to specific transactions may result in conflicts regarding the proper way to record a transaction. The role of the accountant is to determine the most prudent and accurate way to report so that integrity is not compromised. That is the challenge in delivering fair and accurate financial statements. The next section explains the structure of commonly used accounting statements.

1.3.2 Accounting Statements

Standard accounting statements are the most widely accepted and most widely used vehicles to communicate information regarding the financial status of the enterprise. As indicated earlier, they are based upon the identification of income-producing assets that are held by the firm, but claimed either by its owners or its creditors.

The primary accounting statement, called the *Statement of Financial Condition*, or more commonly, the *Balance Sheet*, reports summarized financial information by subcategory for each term in the fundamental accounting equation. It is based upon accrual accounting conventions and the dual entry bookkeeping, explained in Appendix 1-A. Readers with no previous exposure to accounting should review that material before proceeding.

As observed from standard bookkeeping procedures, accrual accounting focuses upon changes in owner's equity as contrasted with cash accounting, which focuses upon changes in cash position. Both are important. The first relates to growth; the second to survivability. There are two major accounting statements that measure each during the course of an accounting period: the income statement and the cash flow statement.

The *Income Statement*, also called the *Profit and Loss Statement*, summarizes the net income earned during an accounting period. At the end of the period, the net income increases (or decreases) the retained earnings of the company. The retained earnings appear as one category of owner's equity on the balance sheet and reflect the cumulative profits earned by the firm since its inception. Other forms of owner's equity might include owner's contributions or the proceeds from the sale of shares of stock.

In contrast, the *Statement of Cash Flows* summarizes the impact of all transactions during the accounting period on the cash balance of the firm. It is useful in understanding how cash was generated and how it was used in supporting operations and changing the structure of the firm.

Notice that the latter two statements summarize the impact of transactions that occurred *during the accounting period*. In contrast, the balance sheet summarizes the status of all fund categories *at a point in time* corresponding to the end of an accounting period. With this background in hand, let us move to a deeper description of each statement.

1.3.2.1 The Balance Sheet.   *Assets.* Knowing the total dollar value of all the assets under the control of the firm is useful information in that it indicates the size of the enterprise. However, it provides no indication of the kind of assets, their useful life, or their ability to be liquidated or converted to cash. Common sense dictates that a more useful view can be constructed if assets were divided into meaningful asset categories and financial transactions were recorded by category. Each company develops its own preferred set of categories, called a chart of accounts, depending on the needs of its business and its investors. Every chart will be different, yet there are several common categories that extend across a wide variety of companies. Common ways to classify assets are explained below and illustrated in Table 1-1.

TABLE 1-1   Total Assets of Ajax Company as of December 31, 2006

Current Assets	($ in thousands)	
Cash	$ 10,000	
Marketable Short-Term Securities	8,000	
Accounts Receivable	1,000	
Inventory	10,000	
Prepaid Expenses	1,000	
Total Current Assets		$ 30,000
Long Term Assets		
Plant and Equipment	$20,000	
Less: Accumulated Depreciation	8,000	
Patent	17,000	
Less: Accumulated Amortization	1,000	
Total Fixed Assets		$28,000
Total Assets		$58,000

Current Assets include cash and short-lived assets readily convertible to cash, whereas fixed or long-term assets are not liquid. Current assets include cash on hand, near cash items such as short-term bonds, and accounts receivable. Accounts Receivable are records of the customer accounts wherein trade credit has been extended to the customer for items sold. When payment is received, the amount received is transferred to the cash account as a part of normal bookkeeping. Inventories of raw materials, work-in-process, and finished products are generally considered as current assets, since they exist only to be liquidated in product sales.

Long-term assets are productive entities held for the income that they will produce in future years. Productive plant and equipment are tangible assets that fit into this category. They are valued "on the books" at their initial installed cost. As they are "used up" in the production process, a portion of their value, called depreciation, is charged to expense to match with the revenue that was produced. Since there are somewhat complex IRS regulations that govern the various means by which this depreciation amount is determined—and since these amounts affect income tax payments and, hence, cash flows—detailed discussion of depreciation is deferred to Chapter 7, where its important effect on investment decisions will be considered. For examples used in this chapter, we will assume a simple straight-line depreciation where, for example, a $20,000 asset with a 10-year life is "written off" or depreciated at $2,000 annually. Again, the main point to understand here is that depreciation is an accounting transaction recorded to expense a capital asset over time. It arises from accrual accounting, which attempts to match expenses (use of asset value) with revenues (sales of products produced by assets) during an accounting period.

There may also be financial long-term assets and intangible assets. Intangible assets may include patents held by the company, or goodwill resulting from acquisition of another company, its expertise, and its customers. Following accounting principles, these assets are recognized only to the extent that a verifiable monetary amount can be associated with their acquisition. Like physical assets that are depreciated, these may be amortized over time. Financial assets include investments in other companies.

Liabilities and Net Worth.   The right-hand side of the balance sheet lists all of the claims on assets shown in the left-hand side. Claims arise from owners who have paid in capital to the enterprise (equity or net worth) and creditors who have lent money to the enterprise (liabilities). Because the latter have a priority claim, they are normally listed first on the balance sheet. As shown in Table 1-2, they are typically ordered by length of term with the shorter-term liabilities listed first.

Short-Term Liabilities are those claims which are due to mature in the near term—typically a year or less. These include such items as short-term bank loans, and accounts payable and accrued expenses that have been recognized under the matching principle but not yet paid. Short-Term Liabilities are essentially a list of financial obligations to the firm's suppliers for goods and services purchased, but not yet paid.

Long-term liabilities are usually extended financial obligations, such as bonds issued to acquire long-term capital assets. These are usually listed at the face value of the bonds or by the amount of outstanding principal.

TABLE 1-2   Total Liabilities and Net Worth of Ajax Company as of December 31, 2006

Liabilities and Net Worth		
Current Liabilities	($ in thousands)	
Accounts Payable	$ 4,000	
Notes Payable	3,000	
Accrued Expenses	2,000	
Total Current Liabilities		$ 9,000
Long-Term Debt	$15,000	
Total Long-Term Liabilities		$15,000
Total Liabilities		$24,000
Net Worth or Owner's Equity		
Common Stock at Par (4 million shares)	4,000	
Capital in Excess of Par	20,000	
Retained Earnings	10,000	
Total Net Worth		$34,000
Total Liabilities and Net Worth		$58,000

Owner's equity might include owners' contributions or the proceeds from the sale of shares of stock at their issue or par value. Since initial offerings of stock are usually sold at a premium above par value, the added capital received is shown as "additional paid-in capital" or "capital in excess of par." Retained earnings appear as one category of owner's equity on the balance sheet, and reflect the cumulative profits earned by the firm since its inception. The total owner's equity is often referred to as book value of the equity. This is to distinguish it from market value of the equity, obtained at any point in time by multiplying the quantity of outstanding shares by the then-current market price. Another frequent reference point is the total capitalization of the company, which refers to the sum of its long-term liability and equity. It is an indicator of enterprise size, and the equity is sometimes quoted at book value, other times at market value.

1.3.2.2 Income Statement.   The income statement reports the financial results of company operations during the accounting period. It is generally an intuitive and straightforward document summarizing the net results of the company's for-profit activities according to the obvious equation for net income,

$$\text{Net Income} = \text{Revenues} - \text{Expenses}. \tag{1-2}$$

As with the balance sheet, specific information categories are used in constructing the statement. There are only a few subtleties in its structure to master; however, some study is required to understand what conventions have been used in its construction for a specific firm. As with the balance sheet, we outline usual steps and categories of information presented, and advise that this is a starting point. Detailed interpretation will likely require a deeper understanding of the specific situation.

The sequence of steps in reporting income is logically plausible. First, note that revenues and cash inflows are not necessarily the same. Revenues result only from the conduct of the firm's operation of producing goods and services for its customers. The cash received from the sale of equipment or buildings, or other assets not specifically acquired for resale as a part of the firms business, is not revenue. It is merely an exchange between the long-term asset account and the cash account. Sometimes revenues are labeled as Sales, then relabeled as Net Sales after subtracting discounts and trade allowances or after subtracting returned items. A typical example for Ajax Company is shown in Table 1-3.

Division of Expenses into categories is even more diverse, but not any more complicated than defining points at which expenses are recorded and counted. We next discuss the most common, but not necessarily the only, conventions for manufacturing companies.

Cost of Sales (also called the Cost of Goods Sold) is the accounted cost of items sold during the period. It includes not only the material and direct labor costs in manufacturing, but also a "fair share" of the firm's overhead costs that have been allocated to the product according to guidelines adopted by the firm. This is an important topic

TABLE 1-3   Income Statement for Ajax Company for Year Ending December 31, 2006

	($ in thousands)	
Net Sales	$15,000	
Cost of Sales	5,000	
Gross Profit or Margin		$10,000
Selling, General and Administrative Expenses (SG&A)	800	
Depreciation and Amortization	2,000	
Earnings Before Interest and Taxes (EBIT)		7,200
Interest Income	90	
Interest Expense	1,800	
Taxable Income		5,490
Income Taxes	2,196	
Net Income After Taxes (NIAT)		3,294

and is the subject of Chapter 2 in this book. For now, you can simply accept that there is an available cost for each product or group of products sold.

An intermediate step is to subtract Cost of Sales from Net Sales and arrive at a Gross Margin, also called Gross Profit or Gross Income. But there are more expenses to deduct before arriving at Net Income. Some fixed expenses, dubbed Sales, General and Administrative or SG&A, are not allocated to products but are charged directly in the accounting period when they are incurred. If these are subtracted from the Gross Income, the result is often labeled Operating Income (also termed Earnings Before Interest and Taxes)[1].

Finally, there are non-operating revenues such as those from the interest earned on temporary investments of otherwise idle cash. There are financial charges as well: interest payments on borrowed money, financing charges, provision for income taxes, and extraordinary items not expected to recur. These are all subtracted from operating income to arrive at Net Income After Taxes. There are several other common rearrangements of these data that are useful in certain financial analyses. Sometimes, they are subtotaled directly in the income statement; sometimes, they must be reconstructed from data reported in the income statement. EBIT, Earnings Before Interest and Taxes, is net income − income taxes ± interest paid or received. It is ordinarily the same as operating income. Another useful aggregate is EBITDA, Earnings Before Interest, Taxes, Depreciation, and Amortization. It adds the noncash charges, depreciation, and amortization to EBIT, creating a pretax cash flow measure.

1.3.2.3 Change in Equity Statements.

The logical link between successive balance sheets is the *Income Statement*. However, tracing data between balance sheets using the income statement is incredibly complex. Some tools that explain the changes more easily are the *Change in Equity Statement* and the *Cash Flow Statement*. The *Change in Equity Statement* simply shows the reinvestment of net income in the firm via an increase in retained earnings in relation to dividends, which is the payment of some earnings to the stockholders. An example for Ajax is shown in Table 1-4. First, we obtain the retained earnings from the prior year's balance sheet. Suppose that amount is $ 7,106,000. Suppose also that management decides to pay dividends of $0.10 per share to each shareholder of the 4 million outstanding shares (See Table 1-2.) Then, the Statement of Retained Earnings would be shown in the statement in Table 1-4.

TABLE 1-4   Ajax Company Change in Equity Statement for Year Ending December 31, 2006

	($ in thousands)
Retained Earnings December 31, 2005	$ 7,106
Net Income January 1, 2006 to December 31, 2006	$ 3,294
Total Available to Stockholders	$10,400
Dividends Paid to Stockholders	$ 400
Retained Earnings December 31, 2006	$10,000

[1] Operating Expenses for the example in Table 1–3 would include Cost of Sales, SG&A Expenses, and Depreciation and Amortization.

1.3.2.4 Cash Flow Statements.   Just as the *Change in Equity Statement* links the retained earnings balances between successive statements, the *Cash Flow Statement* links the cash balances in successive statements. It adds information to the other statements by showing cash inflows and their disposition in the company. This provides evidence about the enterprise's current financial health, as well as its future plans, in that much can be inferred from the choices made regarding its uses of cash. In the United States, the accepted convention is to separate cash flows into three components: cash flows from operations, cash flows from investing, and cash flows from financing activities. The construction of such statements in any real setting is a task for well-trained accountants with an in-depth understanding of the prevailing accounting system. However, it is very valuable to develop a firm conceptual grasp of these tools and their importance in understanding a company's situation. Some useful exercises for accomplishing this are suggested in the problems at the end of this chapter. Cash Flow Statement analysis is not essential to understanding subsequent text chapters; however, understanding the concept of cash flow is critical. This is because analysis of any proposed capital investment hinges on the cash inflows generated by the investment, as well as the cash outflows associated with the investment. This concept is developed extensively in Part Two of the book. Before pursuing that goal, it is helpful to develop some additional concepts about enterprise health, growth, and value creation as they relate to financial accounting.

1.4 Financial Health and Accounting Statements

Audited and published financial statements are required of all publicly held companies in the United States by the Securities and Exchange Commission (SEC). They are presented on a comparative basis, showing at least one previous year's results, so that trends and changes can be observed. Their purpose is to provide information useful to potential investors, customers, creditors, and others who must make decisions regarding their relationship with the company. They are published in a company's annual report and are accompanied by a set of notes, which have been compiled by the auditor. Such notes should be considered an integral part of the statements because they explain interpretations, conventions, and various bases for judgmental items.

Analysts typically develop a series of measures or metrics that have been evolved over the years as tools that can be used to assess performance, liquidity, solvency, asset turnover, and other benchmarks. Although developing expertise in the creation and interpretation of these measures is beyond the scope of this text, it is professionally and personally helpful to develop some facility in the more commonly used measures. Such an opportunity is offered in the illustrative exercise, *Assessing A Company's Financial Health.*[2] The next section discusses some measures of economic performance.

---

[2] Available from Harvard Business School Publishing, Note 9-201-077, Rev. July 29, 2002.

1.5 Measures of Economic Performance

Measures of the economic performance of a plant, or any operating unit, are central to the concept of value creation, the chief purpose of capital investment projects.

1.5.1 Classical Indicators of Financial Performance

Common indicators of financial performance during an accounting period are simple ratios of net income to sales or net income to the amount of capital invested. Since these are not part of the material covered by accounting conventions, their exact form tends to follow preferences of the analyst. For example, it may make sense to watch trends in one of the following:

% net profit margin = net income / sales

% operating profit = operating profit / sales

Earnings per share (EPS) = net income / number of shares outstanding

The first is a bottom-line measure; the second may be more meaningful in assessing the true operating performance of a cash-rich company (such as Microsoft) that bolsters its operating profits with earnings from financial investments. The third is very commonly watched by existing and prospective shareholders. They are interested in value measures such as the P/E ratio, the price of the stock relative to its earnings, or growth measures such as the percent change in EPS. One shortcoming shared by these three measures is that they do not recognize the amount of capital employed in generating this performance. Other measures overcome this shortcoming by relating profitability to the amount invested. These Return on Investment (ROI) measures may take various specific forms:

Return on Assets (ROA) = Net Income / Total assets

Return on Equity (ROE) = Net Income / Owner's Equity

Return on Invested Capital (ROIC) = Net Income / Total capitalization

The first measure might be more appropriate for assessing plant managers; the second, for investors considering purchase of stock in the firm; the third, for creditors considering an extended line of credit. Of course, none of these decision makers would rely on a single measure. The point here is to illustrate why there are several very similar measures, and why an understanding of balance sheet basics is essential to financial performance measurement.

1.5.2 Economic Value Added

A potential difficulty with ROI measures is that a focus strictly on them might cause a company to overinvest or underinvest if there is a difference between their current ROI and their "cost of capital." To oversimplify a complex subject, think of the cost of capital as the interest rate that reflects the cost of obtaining new funds to invest in capital projects. As explained in Chapter 11, the cost of capital depends not just on the borrowing rate, but also upon the financial structure of the entire firm. Any over or under investment fails to maximize shareholder value, which is our prime objective in capital investment. To overcome this deficiency, Stern, Stewart & Co. popularized the concept of Economic Value Added[3] (EVA™) as a performance measure for ongoing operations.

EVA is calculated as

EVA = Net Sales

      − Operating Expenses

      = Earnings Before Interest and Taxes (EBIT)

      − Income Taxes

      = Net Operating Profit After Income Taxes (NOPAT)

      − Capital Charge (Invested Capital × Cost of Capital)

This approach levies a capital charge on profits and will be positive only if capital is employed wisely to generate profits that exceed its cost. Improving EVA improves apparent value to shareholders, at least for the period measured. Improving EVA can be accomplished by disposing of those assets that do not carry their weight, investing in new projects that produce a positive EVA, or by lowering the cost of capital. Of course, the focus of this book is the investment decision and its place in value creation. The methods that are developed in subsequent chapters will be shown to be consistent with EVA. In the process of that effort, more will be learned about asset disposition and financial structures that affect the cost of capital. EVA is discussed further in Chapter 7.

1.6 Other Measures

In addition to the financial performance measures just discussed, analysts use other metrics to assess business health and strength: liquidity measures and activity measures. Exact definitions of these measures can vary by company or industry, so it is important to understand the exact form being used in any context being studied in detail. Generic definitions are offered in the next sections. The previously cited exercise, *Assessing a Company's Financial Health*, is suggested as an efficient means to study these indicators in a specific context and to enhance understanding of the concepts, rather than the computations involved.

1.6.1 Liquidity Measures

Liquidity measures compare the short-term obligations (Current Liabilities) with the liquid assets (cash or near cash assets) available to pay them. For example, the Current Ratio, CR, is as follows

CR = Current Assets / Current Liabilities.

A high ratio provides comfort, while a low one (or a declining one over time) indicates some risk of financial failure. A more conservative measure, the Quick Ratio,

---

[3] Stewart, G.B., "EVA™ Fact and Fantasy," *Journal of Applied Corporate Finance*, Summer 1994.

QR, subtracts inventory from current assets before computing the ratio. The resulting formula,

$$QR = [\text{Current Assets} - \text{Inventory}] \ / \ \text{Current Liabilities},$$

is a more stringent measure that recognizes that some inventories may not be liquid. If the company is in the fashion business or the technology business, where product obsolescence occurs rapidly, this may be a safer measure for investors to use. Another commonly used measure is Net Working Capital, NWC, which is the difference between Current Assets and Current Liabilities. Negative values here correspond with a CR < 1 and are indicators of a need for immediate action. They are computed as

$$NWC = \text{Current Assets} - \text{Current Liabilities}.$$

For liquidity analysis, NWC overlaps the Current Ratio. Its real value is as a measure of the amount of working capital that must be supported by long-term investment funds. Long-term investment funds, often called Total Capitalization, are simply the sum of Long-term Liabilities (Bonds, Notes, etc.) and Equity (Stock, Retained Earnings). Total capitalization may be calculated at book value or market value, depending upon the purpose of the computation. It is usually most meaningful in terms of market values, which are often close to book value for debt but significantly different from book values for equity.

1.6.2 Activity Measures

Measuring the scale of activity of an enterprise is another way to gauge its vitality. Sales Volume in dollars and its growth rate are common measures, but there are other useful ones, such as

$$\text{Total Asset Turnover} = \text{Annual Sales} \ / \ \text{Total Assets, and}$$

$$\text{Inventory Turnover} = \text{Annual Cost of Goods Sold} \ / \ \text{Inventory.}$$

The preceding measures quantity the extent of use of Plant and Equipment and stocks of inventory, respectively. Improving either frees up money for investment purposes. Other measures convert current accounts, such as Accounts Receivable and Accounts Payable, into their number of days equivalent at current sales and spending rates, respectively.

1.7 Limitations of the Accounting System

The accounting system has both virtues and vices. Because it is based upon standards promulgated by The Financial Accounting Standards Board (FASB) and Generally Accepted Accounting Principles (GAAP), and records only historical facts, it does provide a factual baseline. The content of the financial statements is defined by the accounting conventions outlined earlier and by the interpretation of accountants in applying them to the business situation at hand. Information in the statements is always historical and fact based, with few exceptions. In any publicly held corporation, the accounting system must be certified by independent auditors. It is not intended to be a guide to future earning prospects, even though it is universally used as one indicator. The accounting system seldom contains all the information needed to make a decision about the next major steps that an entity should take, but it is an important reality check.

This accounting information may be embellished by performance measures and by liquidity/solvency measures that attempt to assess the financial health of the firm, but what remains is still an incomplete picture. The vitality of the firm depends not only on what it has accumulated and what it has done in the past; it also depends upon its capabilities to compete in a future that is rampant with new technologies, overseas competitors, new regulations, sophisticated customers, and imaginative marketers. Clearly, a company's health depends upon much more than its financial stability. It really depends upon a company's ability to evolve with a changing marketplace. To do so, the company must have a competitive strategy that defines what business it is in and how it expects to thrive in the business throughout the future. For that reason, capital investments must not only pass the tests that are developed for addition of economic value, but must also fit cohesively with the strategy of the enterprise.

While volumes can be and have been written about strategy formulation, we are barely able to scratch the surface herein. Two approaches are established, however. The first is to raise the notion of the need for a strategic fit as a part of the eventual assessment of capital investments. This will be addressed further in Chapter 10, where new product and capacity expansion opportunities are evaluated. The second is the concept of the Balanced Scorecard as a framework that both contains financial success measures, and incorporates measures tailored to a company's strategic endeavors.

1.8 Balanced Scorecard Concepts

As effective as EVA may be for measuring performance that guides value creation, it provides no driver that will ensure ongoing growth and development of competitive capabilities. Its perspective is purely financial and needs to be augmented by other measures if the idea is to guide the development of the enterprise toward some strategic position. Strategic positioning is a novel concept to many engineers. Most simply, Cook, Hunsaker, and Coffey[4] assert that strategy is the planned fit between an organization's capabilities and its evolving environment, crafted to achieve a favorable position within the competitive marketplace. Often, in practice, the strategic plan falls by the wayside in the heat of competitive battle, as the competitive market causes new pressures and as rival managers seek to solve problems through personal influence. Companies have suffered because day-to-day demands have a tendency to command the attention of employees, with attendant loss of focus on the long run. To

---

[4] Cook, Curtis W., Phillip L. Hunsaker, and Robert E. Coffey. *Management and Organizational Behavior*. New York: McGraw-Hill, 1997.

overcome that problem, the balanced scorecard has emerged as a vehicle for maintaining the focus of the entire organization upon achievement of the strategic plan for the enterprise.

The balanced scorecard (BSC) provides an orderly means to incorporate measures that guide ongoing operations and functional planning toward a company's strategic endeavors. Developed by Kaplan and Norton[5], the balanced scorecard has received widespread acceptance as a means to achieve strategic objectives. It supplements the financial perspective with customer, internal process, and learning and growth perspectives. It structures the development of objectives and performance measures for each, and links them in a strategy map to develop a cohesive whole. The notion, of course, is to introduce the approach as a replacement for the primarily financial approaches that have dominated the landscape in the past. Of course, a company's balanced scorecard would hardly appear in the literature. It exists as a tool for insiders to guide the destiny of their enterprise. The high level example pictured in Figure 1-1 illustrates the concept of the interlinked scorecard. The process starts from four perspectives: financial, customer, internal processes, and organizational learning. Each is linked as shown in Figure 1-1. Financial goals require customer-oriented measures and targets in order to meet sales objectives. Attaining sales objectives requires internal capabilities as developed by the internal processes used and by the skills and capabilities of a learning workforce. At each of the four perspectives, the BSC approach employs objectives, measures, targets, and initiatives to accomplish the objectives. The larger the enterprise, the more complex is the scorecard system required. Both the topic of formulating an effective competitive strategy and the topic of steering a large organization toward its attainment are broad and intricate. There are volumes written about each in the literature. Readers interested in developing a scorecard for their company are directed to the references at the end of this chapter. The message to others is straightforward. . . . Financial health is a necessary, but not sufficient, condition for overall corporate vitality and longevity.

1.9 Summary

Creation of value is a key responsibility of engineers. To accomplish that end, we employ a systems approach to define and resolve opportunities for improvement. A key component of that process is the selection of course(s) of action from set(s) of alternatives. Evaluation of each alternative should always include a financial assessment of the benefits gained versus the cost of the resources employed or the value of foregone opportunities associated with that alternative. Performing such benefit–cost analyses often involves projections of future cash flows that are not known with certainty. Often these are constructed from estimates based upon past information reported in the company's accounting system.

---

[5] Kaplan, Robert S. "The Balanced Scorecard—Measures That Drive Performance." *Harvard Business Review*, January–February 1992.

Responsible use of accounting information requires some familiarity with the conventions and rules associated with the collection and reporting of accounting data. Most companies use an accrual method of accounting. Until that concept is clearly understood, much of accounting appears to be unnecessarily arcane and somewhat confusing. An important task of this book is to overcome that barrier with a brief

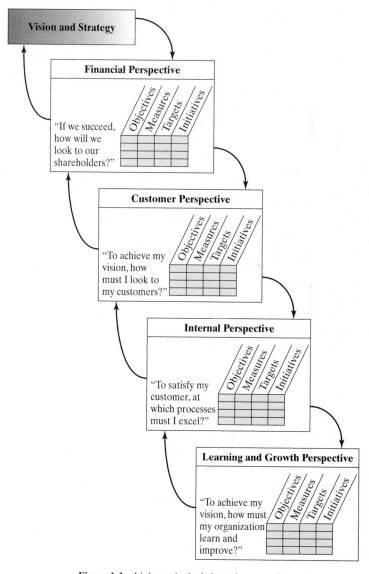

Figure 1-1   Linkages in the balanced scorecard.

Source: Kaplan, Robert S. and David P. Norton. *The Balanced Scorecard: Translating Strategy into Action.* Boston, Massachusetts: Harvard Business School Press, 1996. Reprinted by permission.

description of the main concepts and their relevance to engineering economics. The reader aspiring a deeper understanding for financial investment purposes is directed to the excellent Merrill Lynch publication cited in the references below. This chapter has focused upon the definition and limitations of information published in financial accounting and reporting systems. Chapter 2 extends that knowledge to help readers understand cost accounting systems in common use and the strengths and limitations of the information that they provide.

REFERENCES

Cook, Curtis W., Phillip L. Hunsaker, and Robert E. Coffey. *Management and Organizational Behavior*. New York: McGraw-Hill, 1997.

Frigo, Mark L. and Kip R. Krumweide. "The Balanced Scorecard: A Winning Performance Measurement System." *Strategic Finance*, January 2000.

Gallagher, Timothy J. and Joseph D. Andrew. *Financial Management Principles and Practice*. Upper Saddle River, NJ: Prentice Hall, 2003.

Horngren, Charles T., Srekant M. Datar, and George Foster. *Cost Accounting, A Managerial Emphasis*, Upper Saddle River, NJ: Prentice Hall, 2003.

Kaplan, Robert S. "The Balanced Scorecard—Measures that Drive Performance." *Harvard Business Review*, January–February 1992.

Kaplan, Robert S. and David P. Norton. *The Balanced Scorecard: Translating Strategy into Action*. Boston, Massachusetts: Harvard Business School Press, 1996.

Kaplan, Robert S. and David P. Norton. *The Strategy Focused Organization. How Balanced Scorecard Companies Thrive in the New Business Environment*. Boston, Massachusetts: Harvard Business School Press, 2001.

Merrill Lynch, *How to Read a Financial Report*. New Brunswick, NJ, 1997.

Morganstern, Julie H. and Sharon M. McKinnon. "Solving the Puzzle of the Cash Flow Statement." *Business Horizons*, January–February 1997.

Ostwald, Philip F. *Engineering Cost Estimating*, Third Edition. Englewood Cliffs, NJ: Prentice Hall, 1992.

Stewart, G.B. "EVA™ Fact and Fantasy." *Journal of Applied Corporate Finance*, Summer 1994.

Simon, William L. *Beyond the Numbers: How Leading Companies Measure and Drive Success*. Hoboken, NJ: Wiley, 1996.

PROBLEMS

1-1. Review the material in Appendix 1-A and comment on the following statement: "Engineers are not bookkeepers. There is no need for them to learn an intricate system of debits and credits."

1-2. Alan Rigg, managing partner of The Associates, a small management consulting firm, was meeting with the loan officer of a local bank. "Well, it looks like you have been losing money at the rate of $130,000 per month over the past three months," noted the banker. "Not really," said Alan. "We are still waiting for payment of a half-million on a job we just finished for a major corporation. They're not always sensitive to the needs of small companies." What can you conclude about the accounting system used at The Associates?

1-3. Suppose a new company starts on January 2005, with the issue of 10,000 shares of common stock for $100,000. The company buys $40,000 worth of raw materials on account and processing equipment worth $20,000 for cash. The equipment is expected to last for 10 years. During the course of the year, the company sells $90,000 worth and collects all but $10,000 in cash. In so doing, they consume $30,000 worth of the raw materials, pay wages of $20,000, and spend $10,000 for rent and miscellaneous administrative activities. They purchase an additional $30,000 worth of raw materials on credit. Later, they pay the material vendor a total of $50,000 before closing the books at year end.

Using the procedures developed in Appendix 1-A, work through the accounting cycle of posting entries, closing, and developing a balance sheet and an income statement.

1-4. (Adapted, Ostwald) Long Corporation started the year with the following balances:

Account	Balance as of January 1
Cash	$100,000
Inventory	100,000
New plant and equipment	400,000
Accounts payable	50,000
Owner's equity	550,000

Transactions during the year were limited to the following: paid $100,000 for labor; purchased $150,000 worth of materials; noted equipment depreciation of $50,000, added to inventory 300,000 units costing $1 to the manufacturer; sold 300,000 units for $2 each, cash; purchased new equipment costing $200,000. Accounts payable at the end of year were the same as at the beginning of the year. Neglect income taxes. Develop an income statement for the year just ended. Develop an end-of-year balance sheet.

1-5. (Adapted, Ostwald) Construct a balance sheet for Dynamics Corp. based on the following information:

Retained earnings	$610,000
Cash	150,000
Outstanding debt	450,000
Raw materials	100,000
Finished goods	50,000
Current liabilities	40,000
Stock outstanding	400,000
Fixed assets	1,100,000
In-process materials	100,000

What is the Net Working Capital? The Current Ratio?

1-6. (Adapted, Ostwald) Prepare a profit-and-loss statement by using the following account balances of EZ Machine Shop for the nine months ended September 30:

Sales	$700,000	Rent	$ 80,000
Sales returns	40,000	Salaries	120,000
Inventory, January 1	120,000	Interest earned	2,000
Purchases	270,000	Sales discounts	10,000
Purchase returns	20,000	Interest expense	5,000
Inventory, September 30	160,000	Income Taxes	47,000

1-7. (Adapted, Horngren) Suppose that a major corporation has two divisions. Balance sheets are kept at the corporate level; however, operating information is available for each division as tabled. The corporation estimates its income taxes at 40% of operating income and its cost of capital at 10% per year. Compare each division on the basis of Return on Assets (ROA), Return on Equity (ROE), Return on Invested Capital (ROIC), and Economic Value Added (EVA).

	East Coast	West Coast
Current Assets	$ 300,000	$ 2,000,000
Fixed Assets	700,000	3,000,000
Current Liabilities	250,000	1,500,000
Operating Income	200,000	750.000

Basic Accounting Mechanics

In Chapter 1, the results of accounting efforts are summarized in income statements and balance sheets with little concern for the accounting mechanics employed to generate them. This appendix presents a brief overview of the systematic procedures used by accountants to record and summarize transactions. These procedures are often called bookkeeping. The objective of this material is to help the reader without prior accounting training learn how accountants reduce, in a systematic manner, a complex set of business facts to the comprehensible set of relationships expressed in financial statements.

A. Accounts

Accountants use a series of accounts to record transactions. These accounts often correspond to the line items shown on the financial statements. When finer detail is useful, the records may be kept in subaccounts or natural breakdowns of the major line items. The hierarchy of accounts and subaccounts is called the chart of accounts. It will vary from company to company depending on their feelings about the most useful breakdowns. The transactions affecting each account are most easily visualized in a artificial format called a T-account, in which debits are shown on the left and credits on the right side of the "T." At the end of the accounting period, the entries on each side are summed and a net balance is displayed in the largest side. For example, the Cash account of a company might look like this:

	Cash	
(Increases)		(Decreases)
Balance at beginning of accounting period	$100,000	$ 3,000
	5,000	8,000
	20,000	
	10,000	
Total	$135,000	$11,000
New beginning balance at end of accounting period	$124,000	

All of the increases in cash are shown on one side. All of the decreases are recorded on the other. The new balance is determined by (1) adding all of the amounts listed on the increases and on the decreases side and (2) subtracting one total from the other.

B. Debit/Credit Mechanics

Each accounting transaction has two parts. This dual aspect convention is reflected in the statement "The payment by a customer of an account receivable increases cash and reduces accounts receivable." This statement uses layman's language to describe what occurred. The accountant would describe this transaction in terms of the debit/credit mechanism.

The accountant uses the term *debit* (Dr.) to describe that part of a transaction that

1. increases an asset account,
2. increases an expense account,
3. decreases a revenue account,
4. decreases an owner's equity account, or
5. decreases a liability account.

The term *credit* (Cr.) is used to describe that part of the transaction that

1. decreases an asset account,
2. decreases an expense account,
3. increases a revenue account,
4. increases an owner's equity account, or
5. increases a liability account.

Many people believe it is better to simply memorize these debit–credit rules than to try to determine their algebraic relationship to the basic accounting equation: Assets = Liabilities + Owners' Equity. They are easy to remember if you note that a debit increases a *use* of money. Money is *used* to purchase assets or to pay expenses. Note to the contrary that a credit increases a *source* of money: new capital, loan proceeds, or revenues from sales of products.

Here are some examples of the debit–credit terminology used to describe transactions:

1. A company buys $10,000 worth of raw materials from a supplier. The accounting effect of the transaction is

Dr. Raw Materials Inventory	$10,000	
Cr. Accounts Payable		$10,000

2. The company pays the supplier. The accountant would describe the transaction as

Dr. Accounts Payable	$10,000	
Cr. Cash		$10,000

3. The company releases $4,000 worth of raw materials to production. The accountant would describe the transaction as

Dr. In-Process Inventory	$4,000	
Cr. Raw Materials Inventory		$4,000

The words *debit* and *credit* have no meaning in accounting other than the following: Debit means the amount is entered on the left-hand side of the T-account; credit means the amount is entered on the right-hand side of the T-account. The words carry no value judgment. Depending on the account involved, they can be "desirable" or "undesirable" from the company's point of view. Notice in the three preceding examples that debits equal credits and the accounts remain in balance (i.e., Assets = Liabilities + Owner's Equity).

Figure 1-A-1 shows the relationship between T-accounts and the debit–credit mechanism, using the three example transactions. Each transaction is identified by the bracketed number. You should note that, because debit and credit are used to signify the left and right sides of the T-account, the side used to record an increase or a decrease depends on the account. For example, increases in assets are recorded on the left or debit side, whereas increases in liabilities (and owner's equity) are listed on the right or credit side.

C. Example—Your Firm

A series of such transactions is recorded as they occur until the accounting period ends. At that point, the books are closed, any adjusting entries required are made, and the income statements and balance sheets are developed from the accounting data. This is best demonstrated via an example with a limited number of transactions for

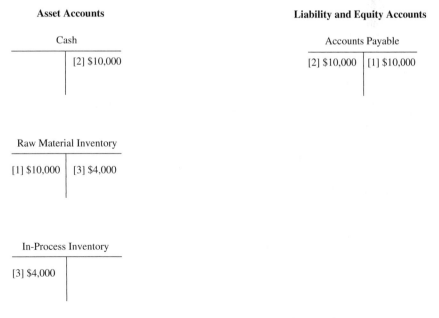

Figure 1-A-1   T-accounts showing example transactions.

simplicity. Suppose you decide to undertake an investment opportunity and that the following sequence of events occurs over a period of a year:

1. Organize a firm and invest $3,000 cash as capital.
2. Purchase equipment with a 4-year life for a total cost of $2,000 by paying cash.
3. Borrow $1,500 through note to bank at 5% interest annually.
4. Manufacture year's supply of finished goods inventory by purchasing $ 400 material on account and by paying $1,200 cash for labor.
5. Sell on credit all goods produced for year, 1,000 units at $3.00 each.
6. Collect $2,200 of account receivable.
7. Pay $400 account payable.
8. Pay $1,000 toward principal and $75 interest on bank note.

Your firm is a fairly simple organization. At this point, a quick scan of its history shows that a simple chart of accounts is adequate to capture all of its activity. Balance Sheet accounts are as follows:

D. Establishing the Chart of Accounts

Asset Accounts	Liability Accounts	Equity Accounts
Cash		Owners Equity
Equipment	Loans	Retained Earnings
Accounts Receivable	Accounts Payable	
Finished Goods Inventory		

Similarly, Revenue and Expense Accounts are as follows:

Revenue	Expense	Net Income
Sales	Materials Expense	Net Income
	Labor Expense	
	Depreciation Expense	
	Interest Expense	

Journalizing Transactions

The journal entries corresponding to the eight transaction steps are as follows:

1. Dr. Cash	$3,000		[Cash Increased]
Cr. Owners Equity		$3,000	[Ownership Increased]
2. Dr. Equipment	$2,000		[Equipment Increased]
Cr. Cash		$2,000	[Cash Decreased]

3. Dr. Cash	$1,500		[Cash Increased]
Cr. Loans		$1,500	[Loans Increased]

4. Dr. Inventory	$1,600		[Inventory Increased]
Cr. Accounts Payable		$400	[Payables Increased]
Cr. Cash		$1,200	[Cash Decreased]

5. Dr. Accounts Receivable	$3,000		[Receivables Increased]
Cr. Sales		$3,000	[Sales Increased]
Dr. Cost of Goods Sold	$1,600		[Expenses Increased]
Cr. Inventory		$1,600	[Inventory Decreased]

6. Dr. Cash	$2,200		[Cash Increased]
Cr. Accounts Receivable		$2,200	[Receivables Reduced]

7. Dr. Accounts Payable	$ 400		[Payables Increased]
Cr. Cash		$ 400	[Cash Decreased]

8. Dr. Loans	$1,000		[Loans Reduced]
Cr. Cash		$1,000	[Cash Decreased]
Dr. Interest Expense	$ 75		[Interest Expense Increased]
Cr. Cash		$ 75	[Cash Decreased]

The notes in the brackets explain the debits and credits. This example shows several of the kinds of transactions that normally occur. The dual entry system provides both an offsetting debit for every credit and also maintains the accounting identity as every transaction takes place. For example,

Transaction 1 shows the creation of an asset cash via an influx of owner's equity.

Transaction 3 shows the addition of cash assets via a liability, a bank loan.

Transaction 2 shows the exchange of one asset (cash) for another (equipment).

Transaction 4 shows the creation of an asset inventory through resource expenditures.

Transaction 5 shows a noncash sale in exchange for an asset (account receivable). A second accounting event matches the sale with the reduction in inventory (an asset) and the expensing of capitalized resources.

Transactions 6 and 7 show the exchange of cash for receivables and payables, respectively.

Transaction 8 shows the repayment of a loan and the charging of interest expense.

E. Posting to Account Ledgers

As already indicated, the process of posting transactions is conveniently described graphically in T-accounts. The posting process proceeds chronologically and would result in the T-accounts shown in Figure 1-A-2.

F. Adjusting Entries and Closing Entries

Once the period's transactions are completely posted, adjusting entries are added to the ledger. These are entries that do not necessarily reflect accounting transactions for cash, but involve maintaining the books in line with the accounting conventions and policies of the company. In this case, the only adjusting entry required is one that recognizes the gradual write-off of the equipment purchased for $2,000, which has a useful life of four years. A simple linear depreciation to a final value of zero would result in an annual depreciation charge of $500. We would add the transaction as follows:

9. Dr. Depreciation Expense	$ 500		[Matches pro rata equipment cost to revenue earned]
Cr. Equipment		$500	[Reduces asset value]

After adjusting costs are entered, the books can be closed. This involves "zeroing out" the revenue and expense accounts and developing the net income account; closing the net income account to retained earnings; and summarizing the balances in the balance sheet accounts. The results of this are shown in Figure 1-A-3, and the closing transactions are described as follows:

10. Dr. Sales	$3,000		[Zeros out Sales]
Cr. Income		$3,000	[Increases Income]
11., 12., 13. Dr. Income	$2,175		[Decreases Income]
Cr. Materials and Labor		$1,600	[Zeros Materials and Labor]
Depreciation		$ 500	[Zeros Depreciation Expense]
Interest Expense		$ 75	[Zeros Interest Expense]
14. Dr. Income	$ 825		[Zeros Income]
Cr. Retained Earnings		$ 825	[Increases Retained Earnings]

Financial statements are developed in conjunction with the closing process. Line items in the income and expense accounts are used to construct the income statement in Figure 1-A-4. Remaining balances in the asset, liability, and equity accounts are used to construct the balance sheet as shown in Figure 1-A-5. This completes the accounting cycle and readies the books for the next accounting period.

Your Firm's T-Accounts, Year Ending 20xx

ASSETS

EQUITY

PAID-IN CAPITAL

| | 3,000 [1] |

RETAINED EARNINGS

INCOME

INCOME

LIABILITIES

LOANS

| [8] 1,000 | 1,500 [3] |

ACCOUNTS PAYABLE

| [7] 400 | 400 [4] |

REVENUES

SALES

| | 3,000 [5] |

ACCOUNTS RECEIVABLE

| [5] 3,000 | 2,200 [6] |

INVENTORY

| [4] 400 | 1,600 [5] |
| [4] 1,200 | |

INTEREST

| [8] 75 | |

CASH

[1] 3,000	2,000 [2]
[3] 1,500	1,200 [4]
[6] 2,200	400 [7]
	1,075 [8]

EQUIPMENT

| [2] 2,000 | |

MAT & LABOR

| [5] 1,600 | |

EXPENSES

DEPRECIATION

Figure 1-A-2 T-accounts before closing

29

Your Firm's T-Accounts, Year Ending 20xx

ASSETS

CASH

[1] 3,000	2,000 [2]
[3] 1,500	1,200 [4]
[6] 2,200	400 [7]
	1,075 [8]
2,025	

ACCOUNTS RECEIVABLE

[5] 3,000	2,200 [6]
800	

EQUIPMENT

[2] 2,000	500 [9]
1,500	

INVENTORY

[4] 400	1,600 [5]
[4] 1,200	

LIABILITIES

LOANS

[8] 1,000	1,500 [3]
	500

ACCOUNTS PAYABLE

[7] 400	400 [4]

EQUITY

PAID-IN CAPITAL

	3,000 [1]

RETAINED EARNINGS

	825 [14]

REVENUES

SALES

[10] 3,000	3,000 [5]

EXPENSES

DEPRECIATION

[9] 500	500 [12]

INTEREST

[8] 75	75 [13]

MAT'L. & LABOR

[5] 1,600	1,600 [11]

INCOME

INCOME

[11] 1,600	3,000 [10]
[12] 500	
[13] 75	
[14] 825	
825	825

Figure 1-A-3 T-accounts after closing

30

Your Firm's
Income Statement for Year Ending 20xx

Operating revenues (Sales):		$3,000
Operating costs (Inventory depleted):		
Labor:	$1,200	
Material:	400	
Depreciation:	500	
Operating income		$ 900
Interest Expense		$ 75
Net Income		$ 825

Figure 1-A-4   Income Statement

Your Firm's
Balance Sheet as of End of Year 20xx

Assets		*Liabilities and Ownership*	
Cash	$2,025	Bank Loan:	$ 500
Accounts receivable:	800	Paid in capital	$3,000
Equipment:	1,500	Retained Earnings	$ 825
Total:	$4,325	Total:	$4,325

Figure 1-A-5   Balance Sheet

2
Cost Measurement and the Cost Accounting System

2.1 Introduction

Every engineer and manager has a keen interest in, and concern about, costs. Costs provide a common denominator for resource usage associated with building a product, delivering a service, or implementing an engineering project. Cost is the vehicle by which we can aggregate raw materials, labor hours, and expenditures on supplies and even expenditures on support services distantly related to the completion of the product, service, or project at hand. Comparisons of aggregated costs to the selling price of the products and services that we deliver, or to the value of the benefits of the project that we complete, are the conceptual ways in which we decide whether we have created value by our efforts.

In short, we must understand cost as an economic concept so that we can carefully decide how to measure costs and how to use cost information in the assessment of economic alternatives. As elementary as that may sound, it will soon become apparent that there are several issues and problems that must be overcome in order to develop, measure, and use cost information. *The purpose of this chapter is to provide the guidelines and concepts that you must understand in order to be able to use cost information properly.* A major source of the information that forms the starting point for cost analysis is found in the cost accounting systems used by most organizations. To be effective, you must thoroughly understand how the system you are working with is constructed and used. In reality, there is even more variety in managerial and cost accounting systems than there is in the financial accounting systems studied in Chapter 1. There is, however, a degree of commonality in the principles employed in the design and development of such systems. An understanding of these principles develops the skills needed to pursue the specifics. The goal of this chapter is to get you started on that journey. We begin with an examination of the origins of cost accounting.

2.2 Origins of Cost Accounting

The initial force driving the development of the practice of cost accounting is the need for a documented cost measure for inventory valuation. To see why this need arises, consider the flow of information in an accounting system for a manufacturer. This is illustrated in Figure 2-1 and explained in the following discussion. To put this into perspective, recall that, under accrual accounting, all costs are booked as assets (capitalized) until they are used (expensed) to generate revenues. For manufacturing, this means that not only the raw materials used in production but also the direct labor, the indirect labor (e.g., the foreman's salary), and even machine usage (i.e., depreciation) are held as assets on the balance sheet until the product is sold and generates revenue.

Thus, as shown in Figure 2-1, and explained next, incoming production materials are charged to raw materials inventory and valued at their cost. When a production order requiring some of the raw materials is issued, the value of the inventory affected (quantity required multiplied by unit cost, probably, but not necessarily) is credited to raw material inventory and debited to an account called work-in-process inventory. As effort is expended to make the product, the labor hours used (perhaps collected from time cards) are multiplied by the wage rate and are charged to the work-in-process account as well. Other associated costs, often called overhead costs, are incurred during the production process, and these must be charged to work-in-process as well, since there is not yet a sale to match them.

These overhead costs are problematic. They include a wide variety of costs, such as those for maintenance services, inspection, supervision, machine depreciation, engineering salaries, grinding wheels, shop towels, repair parts, factory heat and light, and more. They are incurred generically in the manufacture of the product, yet their amount is not easily traced to a specific production order flowing through the shop. Yet, matching requires that their cost be capitalized until the product is sold. In the generic situation, in which many products and production orders are underway simultaneously, questions arise on just how these overhead costs are assigned to specific jobs. This is critical because, as jobs are completed, the work-in-process account is relieved (credited) by the total dollar value of the job, and that value is charged (debited) to another asset account, finished goods inventory. Finally, when the products are sold, the revenue accounting entry credits sales and debits accounts receivable, while the cost entry relieves (credits) finished goods inventory and expenses (debits) an expense account, cost of goods sold. This achieves the desired matching of costs and revenues.

But the obstacle is determining how many dollars, or how much resource cost, to move from the work-in-process (WIP) to the finished goods (F/G) account. One way would be to track the costs associated with each production order. Raw materials often can simply be "charged" to each production order; labor expended directly on making the product can be recorded on time cards and "charged" to each production order. These are referred to as direct costs since they can be directly traced to the product with

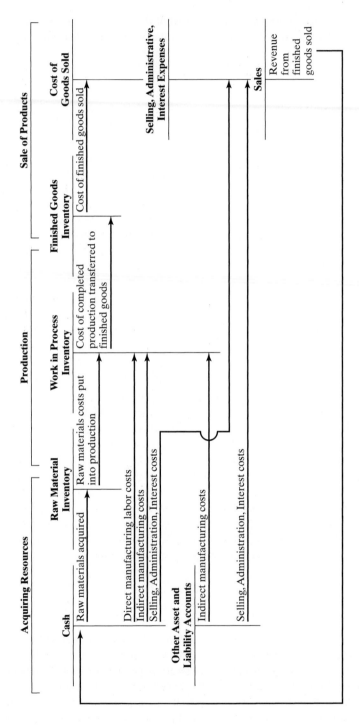

Figure 2-1 Flow of accounting information in manufacturing.

which they are associated. In contrast, the overhead costs are not traceable to specific products or specific production orders; hence, they are referred to as indirect costs. A way to assign these costs to production orders is needed for each order to bear its fair share of these indirect costs. One way to accomplish this is to estimate the total of these overhead or indirect costs for a period of time and prorate them to each production order issued during that time period. There are, of course, many ways to do this. Some balance must be struck between assigning information accurately and assigning it in a manner that is simple and understandable to all.

2.3 A Simple Solution: Production Order Costing

One solution is to develop a budget, which identifies all the manufacturing costs and their estimated annual amounts. That budget combines experience in the form of cost history with forward production plans in order to develop rules for assigning overhead costs to cost centers and to the products that flow through them. Developing the annual budget is no small task and is described in Section 2.5. Once this is accomplished, a common solution is to lump together all the estimated overhead expenditures for the budget year and divide them by the estimated direct labor for the year to determine an overhead rate, normally expressed as a percentage of direct labor cost or as overhead per direct labor hour. Once this is done, overhead costs can be charged to each production order on the basis of the direct labor charged. If the costs were forecast perfectly, then the total of the overhead allocations to the production orders will equal the planned overhead amount in the budget. In the (much) more usual case in which differences emerge, adjustments must be made where too much overhead was assigned to production orders (overabsorbed burden) or too little overhead was assigned (underabsorbed burden). When these adjustments are large, frequent, or both, profitability problems may emerge as the system is providing misleading information during the times between adjustments. This simplistic cost accounting system, called job-order or production-order costing, solves the issue of inventory valuation. How accurately it does so is open to question. Even if the company adheres closely to the overall production and cost forecast, distortions at the individual product level can occur. Additionally, it is based on budgeted costs but does not provide any mechanisms that would assure that actual expenditures will conform to the budget. To address this concern, we examine next an alternative to the job order cost system called a standard cost system.

2.4 The Concept of a Standard Cost

As previously outlined, inventory is valued in the accounting system by collecting direct costs for each production order and by subsequently assigning overhead costs to each. This assignment is often, but not necessarily, based on a simple formula such as multiplying the direct labor by the overall overhead rate. In some cases, cost collection by production order is time consuming; in others, products are so standardized that simple schedules are used to authorize production and no production order number exists. These nuances led to the development of standard cost systems in which inventory valuation is based upon the amount a product should cost (its standard cost) rather than its actual raw material and labor costs as collected during the workday.

Conceptually, the process of developing a standard cost is very simple. It is illustrated in Figure 2-2 and described next. Direct costs are simply gathered from engineering documents as follows:

> Raw materials can be traced to each product through an engineering document called the bill of material, which in its simplest form lists all the parts and raw materials required to make a unit of product together with their associated quantities.

> Standard direct labor hours are calculated by industrial engineers and recorded in another engineering document called the product routing. In its simplest form, the routing lists each labor step, its associated standard time, labor grade, and location.

> With this information, it is possible to calculate standard unit costs for direct materials and direct labor. Unit costs of burden or overhead are determined according to the rules of the standard cost system. For the simple allocation model already discussed, the unit labor cost is the direct labor standard hours per unit multiplied by the standard wage rate, while the unit overhead cost is simply the product of the labor cost and the overhead rate. As shown in Figure 2-2, summing these unit costs with the unit materials cost provides the total standard cost.

2.5 The Manufacturing Budgeting Process

As already observed, cost planning for manufacturing is complicated by the reality that the quantity of product manufactured, and hence the amount of production costs incurred, is seldom known in advance, except for very short time horizons for which customer orders are already available. Very often, customers expect delivery of products much more quickly than we can produce them from ground zero; therefore, we need to plan for resources in advance of customer orders. These resources include raw materials, direct labor to make the product, indirect labor to support manufacture, space, machinery, power, repair parts, and operating supplies, to name a few.

2.5.1 Budget Volume

A logical first step in establishing an annual budget is to establish a level of operating production output, or volume, to be used as a basis for budgetary planning. This so-called budget volume is certainly related to the sales forecast, but is not necessarily identical. It represents the level of output at which cost planning and budgeting is performed, whereas the sales forecast is a marketing target to be achieved.

In companies producing diverse products, a means of aggregating individual items to a common denominator is needed. For example, the sales forecast of annual unit sales by item is aggregated into total annual sales dollars by summing the product

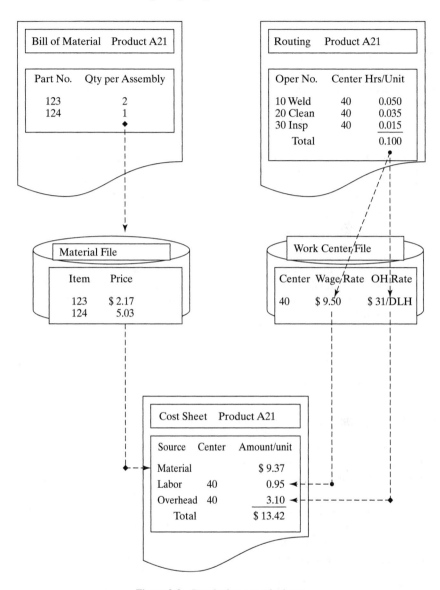

Figure 2-2   Developing a standard cost.

of all items and their respective prices. In manufacturing, a useful aggregate is annual budgeted direct labor hours, determined as the sum of item budget volume multiplied by standard labor hours per item. This is the annual budget for direct labor and also the budget volume of output measured in direct labor hours.

The budget volume becomes the reference point for a technique called flexible budgeting. That is, the budget "flexes" to provide a cost target that matches the actual volume experienced. If actual output is greater than the budget volume, then the spending allowed by the budget is higher and conversely, if volume is lower. This is a very sensible way to compensate for volume changes in cost planning, as many manufacturing costs vary with the level of output. Certainly, direct material costs and direct labor costs are expected to vary directly with output. Some overhead costs, such as operating supplies, may vary with output; others, such as heat and lighting expense, remain relatively fixed for a wide range of output variation. Establishing a budget for overhead costs, in particular, will require decisions about cost classification and assumptions about cost behavior.

2.5.2 Cost Classification

Budgeting and cost management efforts can be greatly simplified by segregating and classifying costs into groups that share common characteristics. The following are several useful dimensions that are often used:

Direct and Indirect Costs

Direct costs are those directly incurred with production of the product. They include direct materials and direct labor. Sometimes they are referred to as productive and nonproductive costs, to make the same distinction. Indirect costs are only indirectly related to production; but they are still considered inventoriable costs, and their accumulation is recorded as products progress from raw materials through finished goods inventories and eventually to cost of goods sold accounts.

Inventoriable and Noninventoriable Costs

Inventoriable costs pass through the accounts as just outlined so that they are matched to revenues when products are sold. Noninventoriable costs are those costs that are deemed to be unrelated to the production effort. They are sometimes called period costs because they are expensed during the accounting time period they are incurred. These typically include sales, general management and administrative expenses, often referred to as "SG&A" expenses. They are generally a separate line item in the overall company budget and are considered to be fixed with regard to production output or operating volume.

Fixed and Variable Costs

Costs are identified as fixed or variable solely on the presumption of their response to changes in output levels. Variable costs are presumed to increase directly with increases in output. Fixed costs are not immutable, but rather are presumed to remain fixed regardless of output. For example, heating costs are typically considered fixed, yet they change markedly with the intensity of winter weather. Direct costs are almost always considered as variable. SG&A costs are almost always considered as fixed. Between these two extremes is a variety of alternate assumptions or models of cost behavior as considered next.

Functional Cost Classifications

The classification of costs into direct and indirect is functional in nature; however, its usefulness in tracing or investigating cost overruns or in projecting future expenditures is limited. Further finer classifications, based on the nature or functions of the costs, can provide better information for decision making and analysis. If cost classifications are recorded at the point when the cost is incurred, then that information is available for use in the future. For example, separate raw material inventory accounts for castings, tubing, and flat stock can help management isolate the cause of an overage in steel inventory. Likewise, overhead expenses can be separated by functional category as in Table 2-1.

If finer granularity is useful, these accounts can be further subdivided. For example, operating supplies could be decomposed as shown in Table 2-2.

These kinds of functional classifications enable the company to collect information about the nature of overhead costs as items are issued to the shop floor. The issuing authorization might be a simple transaction slip signed by the shop foreman for a quantity of shop towels, a unit cost, and the account number. These are accumulated for all transactions for all accounts as a way to report actual expenditures by account and subaccount for a period of time; say, for example, a week. This can be compared to the budget for each subaccount to determine whether any corrective action is needed to maintain compliance with the budget. When compiled for the year, this provides an actual use basis for budget planning in the ensuing year, as well as a

TABLE 2-1   Example Overhead Chart of Accounts

Account	Overhead Cost Category
100	Indirect Labor
200	Operating Supplies
300	Tools and Grinding Wheels
400	Heat, Light, and Power
500	Maintenance
600	Maintenance Projects
700	Scrap and Rework
800	Depreciation

TABLE 2-2   Example Overhead Chart of Subaccounts

Sub Account	Supplies Cost Category
:	
20030	Coolant
20031	Shop Towels
20032	Speedy-dry
20040	Safety Glasses
20041	Ear Plugs
:	

means for identifying problem areas when actual costs exceed the volume-adjusted budget. A further refinement, cost center classification, enhances the usefulness of this budget versus actual spending comparison.

Cost Center Classification

The preceding approach implicitly described the collection of actual costs by subaccount over a time period for the entire organization. Further pinpointing of expenditures can be achieved by decomposing the organization into a collection of cost centers and by extending the transaction reporting to include the identification of the cost center incurring the expenditure. Budget performance can then be measured for each individual cost center. This is especially useful when the cost center includes a set of similar organizational functions under a common supervisor. In that case, the cost center supervisor can be measured and rewarded for cost management effectiveness. Cost centers can be grouped into a hierarchy to follow the next level of management called a responsibility center.

The amount of decomposition is limited only by the practicality of apportioning resources across cost centers that may share them. However, there is also a practical consideration of maintaining the discipline of diligently recording transactions. That will depend largely on the amount of attention that management devotes to using the information outputs of such a system. The important point to note here is that the cost system basics originally developed for inventory valuation have evolved into a framework for cost management. However, that framework comes at a cost of meticulous attention to detail. Further, that detail is historic in nature; it compiles what has already been spent and compares it with a budget allowance.

Cost Behavior Models

Thus far, we have used the notion of a budget allowance that is tied to output levels using what we have called flexible budgeting. We have introduced notions of fixed and variable costs, but have not discussed how these are used to construct a flexible budget and compute budget allowances for a specific output volume. This section addresses that issue.

The simplest and most commonly employed cost behavior model assumes a linear relationship between output volume, say in annual direct hours, and overhead costs in two components: total fixed dollars (FOH), and total variable dollars (VOH) at the budget volume, BV. A graphical model is shown in Figure 2-3.

The slope of the total overhead line is the Variable Overhead rate, R_V, in variable overhead dollars per direct labor hour or cost, as given by the following:

$$R_V = \text{Total Variable Overhead @ budget volume} \div \text{Budget Volume}$$
$$= \text{VOH/BV}$$

(2-1)

This model presumes that variable costs are directly proportional to output and can be calculated for any actual output volume (AV). The allowed spending on variable overhead items is determined by simply multiplying the variable overhead rate R_V by the actual volume AV. It further assumes that the fixed costs are uniform and invariant

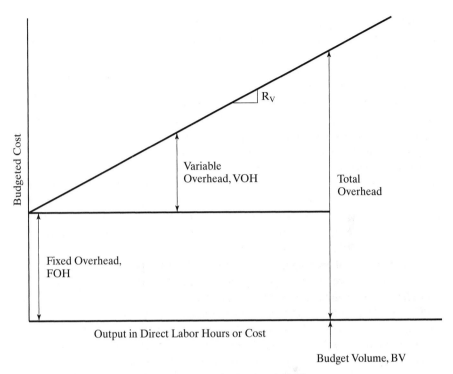

Figure 2-3   Flexible Budget with Fixed and Variable Components

for the budget year. For example, if we wished to calculate the allowed spending for month m, in which we produced 10% of the budgeted annual volume, we would perform the following calculations:

Actual Volume for month $m = AV = (0.10)(BV)$

Budgeted spending = Fixed Allowance / month + Variable Allowance / month
for month m

$$= (FOH/12) + (0.10) \, (BV) \times R_V$$

Thus, a simple fixed-variable model of cost behavior captures the essence of many cost patterns and provides a simple way to compute cost allowances or spending targets based upon the anticipated (before-the-fact) and actual (after-the-fact) levels of factory activity.

Some costs that behave as a mixture of fixed and variable costs are often described as semivariable costs. For example, items with increasing discounts based on level of usage result in costs increasing with volume, but at a slower rate as levels are reached where the discounts "kick in."

Some costs occur in step functions. Inspection costs and material handling costs increase in discrete jumps. When activity reaches the point at which another person must be added to the workforce, costs increase by the amount of that person's compensation. Then they remain fixed across a volume increase until another breakpoint is encountered. Similarly, storage space costs are fixed over a wide range of use until the point is reached at which no more space is available.

It is certainly possible to build overhead cost models with these ramifications included in the formulation. The issue is whether it is worth the time and effort required, as the cost system is pervasive and must be understood by many users at all levels of the organization. Most companies use a simple fixed-variable model as their basis for cost planning.

What we see in any instance is that the budgeting process developed for the construction of standard costs quite logically slides over to cost management by providing a flexible target for operations cost control based upon the realized level of output. This logic is very appealing, but it hinges on comparing after-the-fact dollar amounts with budget goals set at the beginning of the period (year). Hence, its value is limited in the sense that it measures spending that has already occurred. One partial remedy is to forecast the future period's volume and calculate the appropriate future allowances so that spending plans can be geared to meet them; however, this is no substitute for preventive controls such as target costing.

2.6 The Basic Product Costing Model

Recall our discussion of standard costs in Section 2.4. There, we advanced the notion of an overhead rate, which would be applied to the direct labor hours or cost associated with a product to compute the overhead to be applied to that product. Standard cost for the product would then simply equal the sum of direct material, direct labor, and overhead costs. The commonly used basic costing model uses the preceding simple budget model to develop the overhead rate, also called a costing rate R_C, to apply overhead costs to the direct labor hours or cost associated with a product in order to establish a unit overhead cost for the product in standard costing. Conceptually, Figure 2-4 shows the relationship between the budget model and the costing model.

Figure 2-4 is the same as Figure 2-3, except that an additional line extending from the origin to the total budgeted overhead point has been added to the graph. The slope of this new line is the costing rate R_C. That is,

$$R_C = \text{Total Overhead/BV} \tag{2-2}$$

This means that, if we produce exactly at the budgeted volume, the costs transferred to the cost of goods sold account will include exactly the amount of overhead that is budgeted for the producing departments. If we produce more than the budgeted volume, and maintain costs in line with budget, then we will overcost the items sold because our actual costs will amount only to A in Figure 2-4, but costs of goods sold will have costed production to a total of B. We have overabsorbed fixed costs, and hence our actual profitability will improve when the books are adjusted for this fact.

If we operate at less than the budget volume, we will underabsorb our fixed costs and need to restate profits downward. Obviously, from this perspective, it is prudent to set the budget volume at a level comfortably below the expected operating level to avoid end-of-year earnings surprises. However, dropping the budgeted volume too low will result in distorted standard costs by raising standard costs excessively. Thus,

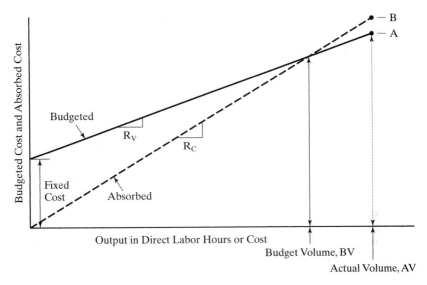

Figure 2-4   Relationship between R_V and R_C, showing overabsorbed cost (B–A) for AV > BV.

establishing the budgeted volume is a complex process based upon the company's situation and upon good managerial judgment.

2.6.1 Standard Costs in Inventory Valuation

The standard costing system, as just outlined, provides generally satisfactory solutions to the inventory valuation issues. It provides single unit costs for each product to move in-process work into finished goods at consistent values.

2.6.2 Standard Costs in Performance Measurement

Management control of spending depends upon measurement and comparisons to target amounts. Financial control of responsibility centers starts with a budget that has been compiled by the cost center. The cost center manager is then held responsible for adherence to the cost center budget. In a flexible budgeting arrangement, the amount he or she is allowed to spend depends upon the volume of product flowing through his or her cost center. In a standard costing system, this amount is routinely computed based upon production over a period of time.

Example 2-1
Mike Jones is foreman of the heat treating department at Nukem Company. His budget allowances for each overhead account, with variable rates expressed as a percentage of direct labor cost, are summarized in Table 2-3. The three products that he heat treats (A, B, and C) have standard labor contents of 30, 40, and 50 hours per hundred units, respectively. His operators' average wage is $10.17 per hour. His schedule for April 2006 is for 10,000 As, 6,000 Bs, and 15,000 Cs. What is his allowable spending under the budget?

    Solution   The standard labor amount that Mike will generate in April is shown in Table 2-4.

With 20 production days in April, each furnace attendant will work 160 hours, meaning that Mike needs a staff of 27 for each of the three shifts that the furnaces operate. His direct labor budget for April would be 12,900 hours × $10.17 = $131,193. Based upon that direct labor amount, his budget for each overhead account is shown in Table 2-5, resulting in a total overhead of $179,902.

TABLE 2-3   Heat Treat Department Overhead Budget

Account	Annual Fixed Budget	Variable Rate as % of Direct Labor Cost
Indirect Labor	$60,000	20.7%
Supplies		38.2%
Tooling		11.1%
Heat, Light, Power	$10,342	17.9%
Maintenance		36.6%
Depreciation	128,461	____
Total	$198,803	124.5%

TABLE 2-4   Standard Direct Labor Computation for April 2006

Product	Standard Hours (per hundred)	April Schedule (Units)	April Standard (Hours)
A	30	10,000	3,000
B	40	6,000	2,400
C	50	15,000	7,500
Total			12,900

TABLE 2-5   Overhead Budget Computation for April 2006

Account	Annual Fixed	Variable Rate as % of Direct Labor Cost	April Fixed	April Variable	April Total
Indirect Labor	$60,000	20.7%	$5,000	$27,157	$32,157
Supplies		38.2%	$0	$50,116	$50,116
Tooling		11.1%	$0	$14,562	$14,562
Heat, Light, Power	$10,342	17.9%	$862	$23,484	$24,345
Maintenance		36.6%	$0	$48,017	$48,017
Depreciation	$128,461	____	$10,705	$0	$10,705
Totals	$198,803	124.5%	$16,567	$163,335	$179,902

∎

The creation of a single "scientifically" determined standard cost creates a natural tendency to compare these item costs with the selling price to assess product profitability. This is a very dangerous comparison that can lead to poor economic decisions. The problem stems from the fact that the overhead costs are *allocations* of cost to the product. They are not necessarily traceable to the actual manufacture of the product. Consider the decision to drop one "unprofitable" product from a line of several products. The company would lose the revenue associated with the product and would save the costs of direct materials and labor. This is because they will no longer purchase the raw materials and will presumably reduce the direct labor workforce in accord with the production drop. They will probably save none of the fixed overhead associated with the product because there will likely be no saving in floorspace, machinery, heating bills, or top management that results from dropping this one product. If they save none of the fixed costs that had been allocated to the product, they will eventually be allocated to the remaining products in the line, thus raising their apparent costs. This problem is not restricted to the fixed costs. If some of the costs that are modeled as variable in the costing system are really only semivariable, then not all variable costs associated with the product will be saved either.

2.6.3 Direct and Incremental Costs in Pricing and Marketing Decisions

An alternative approach to comparing selling price to cost is to focus on direct and incremental costs only. Incremental costs are those costs that we are reasonably certain will actually change with an increase or decrease in production. Most of such costs will have been identified as variable costs in the budget modeling process. Yet it is not safe to assume that all costs budgeted as variable are in fact incremental. Some semivariable costs are likely to have been included in the variable cost budget. Further investigation on a case-by-case basis would be needed to sort this out.

It is also true that some costs modeled as fixed costs in the budget could become incremental with respect to a specific decision. For example, if costs really follow a step function, a large increment in output might require more expenditures on an item budgeted as fixed. Again, the only answer is to start analysis by using fixed and variable cost information, but to verify expected cost behavior by examining each cost category in the overhead accounts to ascertain whether the cost models in the budget are appropriate for the decision at hand.

2.6.4 The Contribution Margin Concept

Suppose, as in the preceding, we focus only upon direct and incremental costs. We can compare them to selling price and compute a contribution margin as follows:

$$M = P - DM - DL - IOC, \tag{2-3}$$

where

M is the contribution to the absorption of fixed overhead and attainment of profits per unit sold,

P is the selling price per unit,

DM is the Direct Material Cost per unit,

DL is the Direct Labor Cost per unit, and

IOC is the Incremental Cost per unit of any relevant *variable* overhead items.

Any product with $M<0$ is questionable, as the company does not recover even the direct and incremental costs of production. For any product with $M>0$, there is some contribution to the defrayment of fixed expenses, and continued production may be worthwhile if production capacity exists and no other alternative is available. The actual choices that should be made in a decision situation depend upon not only the value of M, but also upon the underlying situation and the range of alternatives.

Example 2-2

April was an especially busy month for Nukem Company, but business has tapered off somewhat as the economy has declined. Salesman Herb Tarlock has arrived on the scene with a new order for 5,000 of a special product D for rapid turnaround at a price of $ 5.00 each. The customer will furnish the material. Labor is estimated at 20 hours per hundred. The labor rate is $10.17 per hour. The cost accounting department has developed an overhead costing rate of 163% of labor cost to cover both fixed and variable overhead for the Heat Treat Department. Should Nukem accept the order?

   Solution   The labor cost for the product D is 20 hrs per hundred × $10.17 per hour = $2.03 each. With a costing rate of 163%, the overhead cost is $3.31 per unit and the total cost is $5.34 each. The accounting department wants to send poor Herb to Siberia but has the presence of mind to speak with Mike Jones first. Mike points out that only supplies and maintenance accounts (38.2% and 36.6% variable rates, respectively) would apply to the incremental business, as other overheads would remain the same. There would be $2,500 worth of special fixtures needed. These data suggest incremental overhead costs per unit of

Supplies	38.2% × $2.03	= $0.775
Maintenance	36.6% × $2.03	= 0.742
Special fixtures	$2500 / 5000	= 0.50
Total incremental Overhead/Unit		= $ 2.01

and $M = \$5.00 - 0 - 2.03 - 2.01 = \$0.97$, resulting in a decision to accept the contract and hopefully to buy Herb a steak dinner.   ■

2.7 Purposes of Cost Accounting—A Recapitulation

Starting with the origins of cost accounting, we have identified three distinct uses for cost accounting information and three corresponding purposes for cost accounting systems. These purposes are inventory valuation, cost management, and product costing.

   Inventory valuation is the least demanding requirement. As long as consistency is maintained between the yardstick used to measure costs flowing into work-in-process inventory and the valuations of finished items flowing into finished goods inventories, and as long as all inventoriable costs are captured in the system and reasonable allocations are made, this purpose is served.

Using this information to manage costs can evolve into an elaborate system of responsibility accounting and complex variance analysis that pinpoints the source of any deviation between budgeted costs and actual costs to operate the enterprise, or any portion of it. There are two difficulties which must be overcome. One is that the system design must be quite elaborate. It is most effective if responsibility center reports include and measure only those costs that are controllable at that responsibility level. While responsibility for cost management lies at all levels within the workforce, there needs to be a definition of boundaries that enables the foreman to communicate relevant, timely, and accurate information to his manager and vice versa. It makes no sense and serves no purpose for a department foreman to wrestle with a variance in a machine depreciation overhead account, for example. Furthermore, a high spending variance may force a foreman to dig through transactions to search for errors that might have given rise to an unfavorable variance that he must explain to his manager. Finally, the most important shortcoming is that these systems all focus on after-the-fact reporting. Managers are forced to explain history but with no specific tools to prevent recurrence.

This is one result of developing the cost control system as a by-product of the standard costing system. In contrast, systems designed *expressly* for cost management might base selected control on cost reduction targets on the basis of observable physical changes, rather than abstract dollar amounts.

Finally, cost accounting information often provides misleading information for product costing. Pricing studies, product profitability studies, and special analyses (e.g., make vs. buy studies) are clouded by simplistic overhead assignment and allocations, which do not reflect the realities of cash flow. For this reason, many companies have begun to look for more sophisticated systems. There is often a drive to make the same system serve all three purposes. As Kaplan[1] explains, such a goal may miss the point, which is simply to perform all three functions well. Attempts to develop more sophisticated systems to overcome the deficiencies of cost accounting, particularly in the area of product costing, are explained briefly next. They include multibase allocation systems and activity-based costing.

Multibase Costing Systems

So far, we have described a simple cost accounting model in which costs are collected by cost category (overhead account) and by cost center. Some allocation often takes place to obtain cost center allocations for costs that are shared plantwide. For example, the heating and lighting charges may be allocated to each cost center on the basis of the square footage it occupies. This may be inaccurate. For example, the heat treating department probably receives a higher than justifiable allocation according to the square footage scheme, but few precise scientific alternatives are available. There are other examples: Maintenance costs may be allocated based on machine hours, power charges based upon connected wattages, plant management salaries

---

[1] Kaplan, "One Cost System Isn't Enough." *Harvard Business Review*, vol 66, no.1 (January–February 1988) pp. 61–66.

based on volume of production, etc. Once all indirect costs are assigned to a cost center, the next task is to assign them to products passing through that center. This is a second stage allocation. In our simple costing model developed earlier, we allocated costs to products based upon their direct labor hours. In fact, accuracy might improve by using several bases and allocating costs on the basis of their best fit with a particular base to assign costs to products. For example, supervision costs may be assigned to products on the basis of their direct labor content; maintenance costs, on the basis of their machine hour content; material handling costs, on the basis of their weight; or material handling costs, on the basis of material costs or unit loads. These elaborate approaches attempt to remedy the shortcomings of their simpler counterparts by improving the precision with which cost assignments to products are made. The notion of flexibility in methods has great appeal as the cost of number crunching continues to decline in this digital era.

A further refinement, Activity Based Costing, has emerged and is the subject of Chapter 18. It provides for not only multiple bases, but also multiple drivers to further enhance the realism of standard cost information.

2.8 Relevant Costs in Economic Analyses

The focus of this book is upon concepts and methods for the analysis of the economic consequences of decisions involving choices between two or more alternative means to create value. We use cost to measure the value of resources sacrificed or consumed in the process.

For economic analyses, we should be concerned with the *marginal* costs of using resources for any alternative under consideration. What actually constitutes the appropriate marginal cost depends on both the scale and timing of the alternative. To make valid comparisons, marginal costs must be

- *incremental* with respect to the decision at hand,
- *relevant* to the decision,
- incurred in the *future* as a result of the decision,
- a *cash cost* resulting from the decision, or
- an *opportunity* foregone because of the decision.

This definition *avoids* unquestioning use of normal accounting information such as book values and standard costs and focuses on the change in the value of assets affected by a particular action or decision.

Chapter 3 elaborates and extends these principles to develop examples of appropriate and inappropriate uses of cost information. It also explores approaches for estimating the future costs that are relevant to a decision. The basis for estimating may include information from the accounting systems of a company or external sources of information or both.

REFERENCES

Horngren, Charles T., Srekant M. Datar, and George Foster. *Cost Accounting, a Managerial Emphasis*. Upper Saddle River, NJ: Prentice Hall, 2003.

Kaplan, Robert S. "One Cost System Isn't Enough." *Harvard Business Review* 66, no. 1 (January–February 1988): 61–66.

Ostwald, Philip F. *Engineering Cost Estimating*, 3rd Edition. Englewood Cliffs, NJ: Prentice Hall, 1992.

PROBLEMS

2-1. (Adapted, Ostwald) Nofziger Tool is developing their costing rates for the next budget year. Develop variable overhead and absorption costing overhead rates on the basis of direct labor dollars for the following production levels:

Production load	80%	100%	125%
Labor costs	$100,000	$125,000	$156,250
Number of units	20,000	25,000	31,250
Variable overhead cost	$60,000	$75,000	$93,750
Fixed overhead costs	$120,000	$120,000	$120,000

If 100% load is the best estimate of next year's volume, what costing rate should Nofziger use? Why?

2-2. (Adapted, Ostwald) An assembly area has floor space allocation as follows:

Drop area	300
Conveyor 1	100
Conveyor 2	150
Bench area	800
Total	1,350 square feet

Building service overhead costs are $8,000 for this area. Determine the overhead allocation ratio and overhead costs for each assembly production center.

2-3. (Adapted, Ostwald) A company is composed of four cost centers. Each month a budget is prepared anticipating the primary distribution of certain costs. The company expects to operate at 115% of budget volume in March, and this is reflected in the projected direct labor hours.

Cost center	March, 2006 Direct Labor Hours (DLH)	Floor Space Square Feet	Normal Maintenance	Normal Kilowatt Hours
Forging	3,580	5,000	$13,000	8,300
Machining	10,032	20,000	$23,500	43,000
Finishing	5,460	4,000	$ 3,450	3,050
Packing	2,133	2,100	$ 500	2,000
Total	21,205	31,100	$40,450	56,350

The referenced overhead items are listed at their normal (100% budget volume) levels. Maintenance and energy consumption are treated in the budget as purely variable expenses.

Suppose that the normal power bill is $17,800 and normal building services cost is $67,000. (a) Develop the March budget allowances for each cost center. (b) Develop the budgeted overhead costing rate for each cost center and a blanket overhead costing rate for the entire company.

2-4. (Adapted, Ostwald) Referring to Problem 2-3, assume that a job is routed through Machining and Finishing. Machining is heavily mechanized, with costly numerical control and other automatic equipment. Finishing has only a few simple tools. Obviously, overhead costs are high in Machining and low in Finishing. Job 1 takes 1 labor hour in Machining and 10 hours in Finishing. Job 2 takes 9 labor hours in Machining and only 2 in Finishing. If a single blanket rate based on labor hours is applied to both jobs, then the overhead allocation would be the same in both cases (11 hours for each job). What follows illustrates the previous discussion: Determine the overhead costs for jobs 1 and 2 by using department rates. Compare those with the overhead costs that would be computed using blanket rates. Discuss the importance of selecting the correct base.

2-5. (Adapted, Ostwald) Management is attempting to maintain an overhead rate of 175% for an assembled product. The "problem" area is the machining process. Here, the overhead rate is 225% on the basis of $23,000 direct labor. Total direct dollars, excluding machining, is $89,000. Overhead charges, excluding machining, are $150,000. What overhead rate will the company achieve on the basis of this information? Use direct labor dollars as a base.

2-6. The machining center at Mills Corp. generates 400,000 direct labor hours annually and incurs tool maintenance costs of $800,000 annually. Part A123 is broached at a standard rate of 0.4 direct labor hours per hundred parts. Overhead is simply assigned based on direct labor hours. An industrial engineer recommends that the tool maintenance expenditures be allocated by machine group before assigning them to products. His studies show that broaching incurs 20% of the tooling maintenance expenditures and generates 20,000 of the direct labor hours. How will the overhead rate and product cost of A123 change if the engineer's recommendations are followed?

2-7. In Example 2-1, Mike Jones plans to staff with 27 operators. If he produces *exactly* the April schedule, will he meet his direct labor budget? Should his direct labor budget allowance be based on that staff plan? Should his overhead budget allowances be based on that staff plan?

3
Relevant Costs and Revenues and Estimating

This chapter extends the cost concepts in Chapter 2 and introduces numerous techniques useful in estimating for economic analyses.

3.1 Cost Concepts

The word *cost* has many meanings in many different settings. The kinds of cost concepts that should be used depend on the decision problem at hand.

There are two facets to estimating for economic analyses: (1) determining the appropriate quantity to estimate, and (2) making the estimate itself. Most estimates to be made are costs and revenues (negative costs).

For economic analyses, the analyst should be concerned with the *marginal* costs of using resources for the alternative. What actually constitutes the appropriate marginal cost depends on both the scale and timing of the alternative. To determine this quantity, the term "cost" should be defined as follows:

> The cost of a resource is the decrease in wealth that results from committing this resource to a particular alternative; that is, before any of the benefits of the alternative are calculated.

This definition avoids unquestioning use of normal accounting information such as book values and one-time costs and focuses on what is the change in the value of assets effected by a particular action or decision. The following section discusses major cost concepts and issues useful for determining the marginal costs applicable in economic analyses.

3.1.1 Past Versus Future Costs

Costs and other financial events that previously have occurred or have been presumed are tabulated and summarized within the accounting function of an organization. Past costs can be no more than a guide or source of information for the prediction of future costs. Specifically, the analyst should resist the temptation to rely unquestioningly on what may be described as the "actual cost" data from the accounting function. Even if the data are accurate, they will be at best recorded historical costs for similar circumstances.

Future costs often differ significantly from the costs of similar activities in the past. The analyst should be prepared to adjust past costs to reflect probable changes by the time a proposed alternative is to be implemented, taking into account changes both in the prices of the resources and in the amount of each to be required. Prices of resources can be affected by innumerable factors such as climate, geography, and labor regulations.

In economic analyses, one is concerned with projecting what is expected to happen in the future as a result of alternative courses of action. Past costs often serve as a useful guide for such projections. It should be remembered that the viewpoints of the accountant and of the economic analyst are generally very opposite—one is an historian and the other is a fortuneteller.

3.1.2 Joint Costs

One of the difficulties in using accounting records is that some of the costs may be recorded in a single category even though they are, in fact, joint (or common) costs for many different activities. This may be true for labor and material as well as for overhead cost items. These joint costs are often allocated to different products, services, or projects by using more or less arbitrary formulas.

Any such formula should be scrutinized most carefully by the analyst interested in determining true marginal costs, that is, in the prediction of how total costs are affected by the alternatives under consideration. To determine applicable costs for any alternative, the analyst must both trace out each of the costs unique to the alternative and determine the portion of joint costs that is due to the alternative being examined. It is likely that the results will be very different from cost data readily available from the accounting function.

3.1.3 Usual Accounting Classification of Production Costs

As described in Chapters 1 and 2, the accounting function of an enterprise keeps records of happenings affecting the finances of the enterprise. Accounting records of production costs normally are separated into three main categories:

1. direct labor,
2. direct material, and
3. overhead.

Direct labor costs or *direct materials costs* are those labor or materials costs that can be conveniently and economically charged to products or jobs on which the costs are incurred. By contrast, indirect labor cost and indirect materials cost are those costs that cannot be conveniently and economically charged to particular products or jobs on which the costs are incurred. Indirect labor and indirect material costs are part of the third category, *overhead costs,* which includes all production costs other than the costs of direct labor and direct material. Examples of other types of overhead costs are power, maintenance, depreciation, insurance, etc. Overhead costs are often referred to as "indirect costs" or "burden."

3.1.4 Fixed and Incremental Costs

Fixed and incremental (also called *variable* or *marginal*) costs can best be defined with respect to changes in the alternatives (e.g., equipment, methods, production levels, etc.) under consideration. As the names imply, fixed costs remain constant, and incremental costs change (increment or vary) with respect to the change(s) under consideration.

One classical application of fixed and incremental (variable) cost concepts is in considering the effect of differing levels of production or facility utilization on total production cost. For example, space and equipment costs are often considered fixed costs, while direct labor and direct material are often considered variable costs of production. Such categorizations are made difficult because no cost is fixed for all time. What may be considered fixed over one period may be a variable cost for another period. The distinction between fixed and variable costs depends on the decision at hand and the time horizon relevant to them and should be reexamined for each analysis need.

To illustrate the problem, suppose a contractor has a backhoe for which the total costs for a year are the sum of depreciation, fuel, insurance, and operator wages. If the machine is idle during the year, then the fuel and wages are the only variable (marginal) costs, since the depreciation and insurance are fixed for that period of time regardless of how much the backhoe is used. Neglecting the effect of pricing during slack periods on prices and business that can be obtained during other periods, the decision maker should be willing to sell the services of the backhoe for any price greater than the cost of fuel and wages. If, however, the contractor were offered a yearly rental on the machine at a price that essentially covered only the cost of fuel and wages, he should not accept. For example, the relevant marginal costs over a year should also include the insurance costs that could be eliminated if the backhoe were not to be used for that time. Over an even longer term, even the depreciation expenses should be considered variable since the contractor may choose not to buy a new backhoe. This example illustrates the general principle that the proportion of total costs that may be considered variable decreases as the period of time for the decision under consideration increases.

If one is making an economic analysis of a proposed change, only the incremental (marginal) costs need be considered, since only prospective differences between alternatives need be taken into account.

3.1.5 Long- and Short-Run Costs

If an alternative system entails use of otherwise idle capacity, then the immediate opportunity costs and marginal costs will be low compared to the average costs. If, however, the system is already operating near capacity, the opportunity costs of additional output are likely to be high compared to the average costs. In either case, the long-run marginal costs will tend to be less extreme than the short-run marginal costs and will tend to somewhat approach, but not necessarily equal, the average costs.

Long-run marginal costs also depend on the changes in technology associated with shifts in level of production. In many industries, marginal costs decrease with increasing production over time as more efficient facilities are placed into operation. In many other industries, both the long-run and short-run marginal costs may increase with increasing output as added resources become more scarce.

Determination of long-run marginal costs rests on an adequate identification of the opportunities that will appear and disappear in the future. Although some industries may have been characterized by particular cost trends in the past, these may well not apply in any particular situation projected into the future. Indeed, past trends may well be reversed by future events.

3.1.6 Opportunity Costs

An opportunity (or alternative) cost is the value of that which is foregone (prior to the calculation of any benefits) because limited resources are used in a particular alternative, thereby causing one to give up the opportunity or chance to use the resources for other possible income-producing or expense-reducing alternatives. It is the same as the *shadow price* of a resource in classical economics. The opportunity cost is the usual appropriate measure of the marginal cost of a resource for economic analyses of alternatives. While it can be equal to the price paid for a resource, it is often very different from that actual outlay. Indeed, the use of a resource normally entails an opportunity cost even if the resource were obtained without cost.

As an example, suppose a particular project involves the use of firm-owned warehouse space that is presently vacant. The cost for the space that should be charged to the project in question should be the income or savings that the best perceived alternative use of the space would bring to the firm. This may be much more or much less than the average cost of the space that might be obtained from accounting records.

The opportunity cost of a resource is often fairly nebulous and hard to estimate. It may be the sacrifice of future earnings rather than present cash. In the most general sense, the cost of using a resource on one project is the cost of not having it available for the best alternative, whether that alternative is to sell the resource or to invest it productively in some other alternative that will bring future benefits. Extending the warehouse example, for instance, the opportunity cost of the space could be related to either the cash value that could be obtained from the outright sale of the space or the value associated with some other productive use of the space over time.

In estimating opportunity costs, it is useful to distinguish between resources that can be identically replaced, such as loads of sand or pieces of steel, and those that are somehow unique, such as a specific piece of real estate. For identically replaceable resources, for which there is a ready market, the opportunity cost for the resource is merely the market cost of the replacement or, equivalently, the salvage price of the resource if it is already possessed and will not be replaced.

For a resource that is somehow irreplaceable, the opportunity cost for the resource can be estimated as the cost of replacing the unique resource with the least undesirable substitute available. For example, suppose that the unique resource is an engineer who is especially talented at designing improved methods to reduce costs. The opportunity cost of using her on any project X is her particular value on some alternative project(s) Y where she could make the best improvements with the time available. If, however, the assignment of the especially talented engineer to this project X will cause one or more engineers to be hired to substitute for her on the project(s) she would have done otherwise, then the total cost of assigning her to project X is the net savings foregone plus the salaries of the substitute engineer(s).

As another example, consider a student who could earn $30,000 for working during a year and who chooses instead to go to school and spend $12,000 to do so. The total cost of going to school for that year is $42,000: $12,000 cash outlay and $30,000 for income foregone. (*Note:* This neglects the influence of income taxes and assumes that the student has no earning capability while in school.)

3.1.6.1 Opportunity cost in determination of interest rates for economic analyses.
A very important use of the opportunity cost principle is in the determination of the interest cost chargeable to a proposed capital investment project. The proper interest cost is not just the amount that would be paid for the use of borrowed money, but is rather the opportunity cost—the return foregone or expense incurred because the money is invested in this project rather than in other possible alternative projects. Even when internally owned funds rather than borrowed funds are used for investing, the interest cost chargeable is determined by the same opportunity cost principle. In classical economics terminology, the opportunity cost is a measure of the maximum benefit that, for any given situation, can be obtained from an extra unit of capital.

As an example, suppose a firm always has available certain investment opportunities, such as expansion or bonds purchases, that will earn a minimum of, say, $X\%$. This being the case, the firm would be unwise to invest in other alternative projects earning less than $X\%$. Thus, in computing the cost of various alternatives, the analyst may simply add in $X\%$ of the amount invested for each. Such a cost may be thought of as the opportunity cost of not investing in the readily available alternatives.

As another example, consider the interest cost for investment in a car, with money to be obtained from one of the following three financing alternatives:

(i)   borrow and pay 14% per year;

(ii) take out of savings account earning 5% per year;

(iii) cash in a "hot investment" that you are confident would earn 30% per year.

For financing alternative (i) the 14% is a cash cost rather than an opportunity cost and should present no conceptual difficulty. For (ii) the 5% is an opportunity cost because that amount will be given up if the money is not left in savings. Similarly, for (iii) the 30% is the opportunity cost if one invests in the car rather than keeping the "hotshot investment." The wide range between the 5% and the 30% is not highly unusual, for the opportunity cost of any resource (including money) is very much a function of what alternative(s) is foregone.

In economic studies, it is necessary to recognize the time value of money irrespective of how the money is obtained, whether it be through debt financing, through owners' capital supplied, or through reinvestment of earnings generated by the firm. Interest on project investments is a cost in the sense of an opportunity foregone, an economic sacrifice of a possible income that might have been obtained by investment of that same money elsewhere.

3.1.6.2 Opportunity cost in replacement analyses.

As another illustration of the opportunity cost principle, suppose a firm is considering replacing an existing piece of equipment that originally cost $50,000, presently has an accounting book value of $20,000, and can be salvaged now for $5,000. For purposes of an economic analysis of whether or not to replace the existing piece of equipment, the investment in that equipment should be considered as $5,000; for by keeping the equipment, the firm is giving up the *opportunity* to obtain $5,000 from its disposal. This principle is elaborated upon in Chapter 9.

3.1.7 Sunk Costs

Sunk costs are costs resulting from past decisions and are therefore irrelevant to the consideration of alternative courses of action. Thus, sunk costs should *not* be considered directly in economic analyses.

As an example, suppose Joe Student finds a used bike he likes on a Saturday and pays $50 as a "down payment," which will be applied toward the $1,000 purchase price but which will be forfeited if he decides not to take the bike. Over the weekend, Joe finds another bike that he considers equally desirable for a purchase price of $910. For purposes of deciding which bike to purchase, the $50 is a sunk cost and thus should not enter into the decision. The decision then boils down to paying $1,000 minus $50, or $950, for the first bike versus $910 for the second bike.

A classical example of a sunk cost occurs in the replacement of assets. Suppose that the piece of equipment examined in the last section, which originally cost $50,000, presently has an accounting book value of $20,000 and can be salvaged now for $5,000. For purposes of an economic analysis, the $50,000 is actually a sunk cost. However, the viewpoint is often taken that the sunk cost should be considered to be the difference between the accounting book value and the present realizable salvage value, which is called "book loss" or "capital loss." According to this viewpoint, the sunk cost is $20,000 minus $5,000, or $15,000. Neither the $50,000 nor the $15,000 should be considered in an economic analysis, except for the manner in which the $15,000 affects income taxes, as discussed in Chapter 7.

Often, sunk costs and opportunity cost considerations occur together. For another example, consider the plight of Joe and Joan Hapless, who bought a stock for $10,000 four years ago, only to see it decline to $1,000 in value as of one year ago. Since then, it has rebounded to be worth $3,000 now. Joan now says, "Sell the dog, we've lost $7,000!" but Joe says, "No, by keeping it for the last year we've made $2,000!" Who is right? The answer is neither, based on the respective rationales stated. Both are guilty of counting those sunk costs—the $10,000, the $1,000, and the $2,000 can all be considered "sunk" because they are all the result of the past. Only the $3,000 the stock is now worth is relevant. It is the opportunity cost if the stock is kept now; so the decision should boil down to what is the best use of the $3,000—leave it in the stock or sell the stock and use the $3,000 elsewhere.

3.1.8 Postponable Costs

A *postponable cost* is a cost that can be avoided or delayed for some period of time. As an example, the costs of certain types of maintenance or of personnel for certain planning functions may be postponable, while the cost of direct labor is unavoidable or not postponable if production is to continue.

3.1.9 Escapable Costs

When a reduction or elimination of business activity will result in certain costs being eliminated (with perhaps others increased), the net reduction in costs is considered the *escapable cost*. Escapable costs are related to declines in activity in a manner similar to the way variable costs are related to increases in business activity. The escapable cost when a business activity is decreased from X_2 to X_1 is frequently smaller than the variable cost that originally resulted when business level was expanded from X_1 to X_2. For example, it is usually a more difficult management task to reduce labor and other costs and commitments during a contraction than to increase them during an expansion. It is important in estimating net escapable costs that the amount of eliminated costs be reduced by the amount of any additional costs that would be incurred in related activities as a result of the change.

3.1.10 Replacement Costs

Replacement cost is, as the name implies, the cost of replacing an item. It is important to economic analyses because replacement cost rather than historical original cost is the relevant cost factor for most economic decisions. For example, if a storekeeper has been stocking an item costing $8, and selling that item for $12, and the price to the storekeeper for replacing the item is suddenly increased to $14, then the selling price should be raised to at least $14 before any additional units of that item are purchased.

3.1.11 Cash Costs Versus Book Costs

Costs that involve payments of cash or increases in liability are called *cash costs* to distinguish them from noncash (*book*) costs. Other common terms for cash costs are "out-of-pocket costs" or costs that are "cash flows." Book costs are costs that do not involve cash payments, but rather represent the amortization of past expenditures for items of lengthy durability. The most common examples of book costs are depreciation and depletion charges for the use of assets such as plant and equipment. In economic analyses, the only costs that need to be considered are cash flows or potential cash flows. Depreciation, for example, is not a cash flow and is important only in the way it affects income taxes, which are cash flows.

3.2 Cost Factors

In an economic analysis, a listing of main factors that may be relevant for projects under consideration is as follows:

First cost, installed and ready to run (or net realizable value)

Insurance and property tax

The life period of the machine until displaced from the proposed job

The salvage value at the date of displacement

The degree and the pattern of utilization; that is, the percent of capacity at which the machine will operate on the intended job with allowances for possible future changes in utilization

Routine maintenance and repair costs

Major repair items or periodic overhauls

Direct operating costs, including operating labor, fuel or power, scrap material, and rework

Indirect costs: indirect labor, tooling, supplies, floor space, inventory

Fringe benefits

Hazards and losses relative to equipment, material, and labor time

Changes in sales volume or price resulting from the choice

Changes in unit cost of labor, power, supplies, etc., resulting in changes in operating costs

3.3 Cost Savings and Revenues

The benefits of capital investment alternatives are often realized by savings in costs, which should reflect true incremental or marginal cost reductions. Whenever a single project or mutually exclusive alternatives involve revenues as well as costs, those incremental revenues and costs should be considered. Fortunately, when mutually exclusive alternatives are evaluated, many types of costs are often identical (fixed) for all alternatives and thus do not have to be estimated.

3.4 Introduction to Estimating

Probably the most difficult and expensive part of any economic analysis is to determine the estimates needed to complete the analysis. This chapter will attempt to provide a perspective and approaches to estimating, with emphasis on making single-valued estimates for the traditional assumed-certain analyses. Part Three will cover estimating in terms of probabilities and other measures of variability to reflect the risk inherent in predicting future outcomes. Chapter 18 discusses how activity-based cost data are used in the estimating process.

3.5 Estimating: Difficulty and Perspective

The basic difficulty in estimating for economic evaluations is that *forecasting* of critical elements associated with the manufacture of a product or the delivery of a service is unavoidable. Another difficulty is that most prospective projects for which estimations are to be made are unique; that is, substantially similar projects have not been undertaken in the past under conditions that are the same as expected for the future. Hence, outcome data that can be used in estimating directly and without modification often do not exist. It may be possible, however, to gather data on certain past outcomes that are related to the outcomes being estimated and to adjust and project that data based on expected future conditions. Techniques for collecting and projecting estimation data and also for making probabilistic estimates are founded in the field of statistics.

Whenever an economic analysis is for a major new product or process, the estimating for that analysis should be an integral part of comprehensive planning procedures. Such comprehensive planning would require the active participation of at least the marketing, design engineering, manufacturing, finance, and top management functions. It would generally include the following features:

1. a realistic master plan for product development, testing phase into production, and operation;
2. provision for working capital and facilities requirements;
3. integration with other company plans;
4. evaluation against company objectives for market position, sales volume, profit, and investment;
5. provision of a sound basis for operating controls if the project is adopted.

Obviously, such comprehensive planning is costly in time and effort, but when a new product or process has major implications for the future of a firm, it is generally a sound rule to devote a greater rather than a lesser amount of effort to complete planning, including estimates for the economic analysis that is a partial result of the planning. The application of this rule, of course, is bounded by constraints of limited time and talent; however, following the rule will tend to minimize the chance of poor decisions or lack of preparedness to implement projects once the decision to invest has been made.

3.6 Estimation Accuracy

Estimates or forecasts, by their nature, are evaluations of incomplete evidence indicating what the future may hold. They may be based on empirical observations of only somewhat similar or analogous situations, adjusted on the basis of the kind of personal hunch that grows out of the accumulation of the experiences. Or they may be inferences drawn from various kinds of available objective data, such as trade statistics, results experienced in analogous situations, or personal observations.

    Regardless of the estimate source, the estimate user should have specific recognition that the estimate will be in error to some extent. Even the use of formalized estimation techniques will not, in itself, eliminate error, although it will hopefully reduce error somewhat, or will at least provide specific recognition of the anticipated degree of error.

    The level of detail and accuracy of an estimate should depend on the following:

1. the estimability of that which is to be estimated;
2. methods or techniques employed;
3. qualifications of estimator(s);
4. time and effort available and justified by the importance of the study;
5. sensitivity of study results to the particular estimate.

    As estimates differ from conceptual to more detailed, the cost of preparing the estimate increases. As one might expect, the more expensive detailed estimates, which are usually associated with the final design of a system, are also more accurate. The general relationship between phase of design (conceptual, preliminary, and final) and its typical estimation accuracy versus cost of the estimate as a percentage of total project cost is shown in Fig. 3-1.

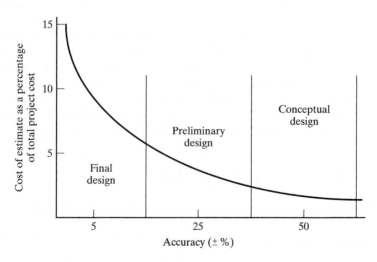

Figure 3-1   Estimation accuracy versus cost of the estimate.
(general relationship)

3.7 Sources of Data

The variety of sources from which estimating information can be obtained is too great for complete enumeration. The following four major sources, which are ordered roughly according to decreasing importance, are described in subsequent sections:

1. accounting records;
2. other sources within the firm;
3. sources outside the firm;
4. research and development.

3.7.1 Accounting Records

It should be emphasized again that, although data available from the records of the accounting function are a prime source of information for economic analyses, these data are very often not suitable for direct, unadjusted use. The data generated by the accounting function are often inherently misleading for economic analyses, not only because they are based on past results, but also because of the following limitations.

First, the accounting system is rigidly categorized. These categories for a given firm may be perfectly appropriate for operating decisions and financial summaries, but rarely are they fully appropriate to the needs of economic analyses for longer-term decisions.

Another limitation of accounting data for obtaining estimates is the misstatements imbedded by convention into accounting practice. These underestimates are based philosophically on the idea that management should avoid overstating the value of its assets and should therefore assess them very conservatively. This leads to such practices as not changing the stated value of one's resources as they appreciate due to rising market prices and depreciating assets over a much shorter life than actually expected. As a result of such accounting customs, the analyst should always be careful about treating such resources as cheaply (or, sometimes, as expensively!) as they might be represented.

The final limitations of accounting data are their illusory precision and implied authoritativeness. Although it is customary to present data to the nearest dollar or the nearest cent, the records are not nearly that accurate in general.

In summary, accounting records are a good source of historical data, but they have limitations when used in making estimates for economic analyses. Further, accounting records rarely directly contain the variable costs, especially opportunity costs, appropriate for economic analyses.

3.7.2 Other Sources within the Firm

The usual firm has a large number of people and records that may be excellent sources of estimates or information from which estimates can be made. Colleagues, supervisors, and workers can provide insights or suggest sources that can be obtained readily.

Examples of records that exist in most firms are sales, production, inventory, quality, purchasing, industrial engineering, and personnel. Table 3-1 provides a list of

TABLE 3-1   Types and Sources of Cost Estimating Data

Description of data	Sources
General design specifications	Product engineering and/or sales department
Quantity and rate of production	Request for estimate or sales department
Assembly or layout drawings	Product engineering or sales department or customer's contact person
General tooling plans and list of proposed subassemblies of product	Product engineering or manufacturing engineering
Detail drawings and bill of material	Product engineering or sales department
Test and inspection procedures and equipment	Quality control or product engineering or sales department
Machine tool and equipment requirements	Manufacturing engineering or vendors of materials
Packaging and/or transportation requirements	Sales department, shipping department, or product engineering (government specifications)
Manufacturing routings and operation sheets	Manufacturing engineering or methods engineering
Detail tool, gauge, machine, and equipment requirements	Manufacturing engineering or material vendors
Operation analysis and workplace studies	Methods engineering
Standard time data	Special charts, tables, time studies, and technical books and magazines
Material release data	Manufacturing engineering and/or purchasing department or materials vendors
Subcontractor cost and delivery data	Manufacturing engineering and/or purchasing department or customer
Area and building requirements	Manufacturing engineering, plant layout, or plant engineer
Historical records of previous cost estimates (for comparison purposes, etc.)	Manufacturing engineering, cost department, or sales department

the types of data that might be needed for cost-estimating purposes, together with typical sources (mostly intrafirm) for the data.

3.7.3 Sources Outside the Firm

There are innumerable sources outside the firm that can provide information helpful for estimating. The main problem is to determine which sources are potentially most fruitful for particular needs.

Published information such as technical directories, trade journals, U.S. government publications, and comprehensive reference books offer a wealth of information to the knowledgeable or persistent searcher.

Personal contacts are excellent potential sources. Vendors, salespersons, professional acquaintances, customers, banks, government agencies, chambers of commerce, and even competitors are often willing to furnish needed information if the request is serious and tactful.

Probably the most valuable estimating sources outside the firm, which are available and updated continuously, are cost indexes. Cost indexes provide a means for converting past costs to present costs through the use of dimensionless numbers, called *indexes,* to reflect relative costs for two or more points in time.

There are many cost indexes, and they cover almost every area of interest. Some are based on national averages; others are very specialized. Indicative values of indexes from the *Engineering News-Record* are shown in Table 3-2.

The Bureau of Labor Statistics of the U.S. Department of Labor publishes extensive data on price changes of many types of products and earnings of workers in almost every industry. Some of these data are components of many cost indexes, and others are useful in constructing highly specialized indexes.

Cost indexes are limited in their accuracy and, like all statistical devices, must be used with caution. Most indexes are based on data combined in more or less arbitrary fashion. A cost index, like cost data themselves, normally will reflect only average changes, and an average often has little meaning when applied to a specific case. Under favorable conditions a ±10% accuracy is the most that can be expected in projecting a cost index over a 4- or 5-year period.

TABLE 3-2   Typical Engineering Indexes*

Year	Materials price index	Skilled labor index	Building cost index	Construction cost index
1975	862	1921	1378	2128
1976	971	2061	1504	2322
1977	1077	2208	1620	2513
1978	1177	2350	1750	2693
1979	1303	2487	1826	2886
1980	1449	2670	1915	3159
1981	1480	2902	2014	3384
1982	1547	3244	2192	3721
1983	1641	3507	2352	4006
1984	1632	3691	2412	4118
1985	1604	3765	2406	4151
1986	1612	3808	2447	4231
1987	1648	3937	2518	4359
1988	1693	4061	2586	4484
1989	1677	4153	2612	4574
1990	1708	4283	2673	4691
1991	1693	4387	2715	4772
1992	1738	4536	2799	4927
1993	1846	4665	2915	5106
1994	2109	4764	3116	5381

*These index values are from the March 28, 1994 issue of *Engineering News-Record*, published by the McGraw-Hill Publishing Company. The base year is 1913, with an index value of 100.

3.7.4 Research and Development

If the information is not published and cannot be obtained by consulting some-one who knows, the only alternative may be to undertake research to generate it. Classic examples are developing a pilot plant and undertaking a test market program. These activities are usually expensive and may not always be successful; thus, this final step is taken only when there are very important decisions to be made and when the sources mentioned above are known to be inadequate.

3.8 Quantitative Estimating Techniques

An estimate, or forecast, is useful if it reduces the uncertainty surrounding a revenue or cost element. In doing this, a decision should result that creates increased value relative to the cost of making the estimate. This section describes three groups of estimating techniques that have proven to be very useful in preparing estimates for economic analysis. They are (1) time-series techniques, (2) subjective techniques, and (3) cost engineering techniques.

When revenue and/or cost elements are a function of time, such as unit sales per quarter, they are often referred to as a time series. Time-series data should be collected for the element under study and then carefully examined for underlying patterns. For example, a sudden increase in sales may be explained by increased government spending or a vendor filling its distribution pipeline. We will examine the use of *regression* for estimating causal relationships within time-series data and *exponential smoothing* for estimating future extensions to historical data patterns.

Frequently, the next stage of estimating is to apply expert judgment to the results of time-series techniques. Examples of subjective estimating approaches to be examined are the Delphi technique and technology forecasting. A highly effective estimating strategy is to couple a time-series technique, based on past data, with a subjective technique that introduces human judgment in attempting to discover how future revenue and cost elements will differ from those of the past.

Cost engineering techniques identify and utilize various revenue/cost drivers to compute estimates. They include models for estimating capital, material, labor, and many other factors of production. These models may utilize correlation and regression analysis, or they may be as simple as extensions of ratios of relevant cost indexes. For example, capital costs can often be accurately estimated by knowing the weight of a particular structure. Similarly, operating costs such as that for fuel can be computed from forecasts of kilowatt-hours generated by an electric power station.

3.8.1 Time-Series Techniques[1]

Two relatively simple, yet extremely useful, techniques for obtaining initial time-series forecasts of elements being estimated are described in this section: (1) linear regression analysis and (2) exponential smoothing.

---

[1] Section 3.8.1 is adapted from J. R. Canada and W. G. Sullivan, *Economic and Multiattribute Evaluation of Advanced Manufacturing Systems* (Englewood Cliffs, NJ: Prentice Hall, 1989). Reprinted by permission of the publisher.

3.8.1.1 Correlation and Regression Analysis.   Sometimes it is possible to correlate an element, such as revenue for a product line, with one or more economic indices, such as construction contracts awarded, disposable personal income, etc. Correlation concerns the explainable association between variables. When an index can be found to which an element to be estimated is highly correlated, but with a time lag, formal correlation analysis may be highly useful. In cases in which the lag is insufficient for longer-term forecasts, correlation of an element to be estimated with the available index still leaves the estimator with the need to predict the future value(s) of the index itself.

Regression is a statistical method of fitting a line through data to minimize squared error. It is exact; however, graphing might be used to provide a satisfactory approximation. With linear regression, approximated model coefficients can be used to obtain an estimate of a revenue/cost element.

In linear regression involving one independent variable x and one dependent variable y, the relationship that is used to fit n data points ($1 \le i \le n$) is

$$y = a + bx. \tag{3-1}$$

A mathematical statement of expressions used to estimate a and b in the simple linear regression equation 3-1 is as follows:

$$b = \frac{\sum_{i=1}^{n} x_i y_i - \bar{x}\sum_{i=1}^{n} y_i}{\sum_{i=1}^{n} x_i^2 - \bar{x}\sum_{i=1}^{n} x_i}, \tag{3-2}$$

$$a = \bar{y} - b\bar{x}. \tag{3-3}$$

Here, $\bar{x}$ and $\bar{y}$ are averages of the independent variable and dependent variable, respectively, for the n data points.

Example 3-1
A durable goods manufacturer has found personal disposable income in its market region in a given quarter to be strongly related to sales in the following quarter. These data are listed and summarized in Table 3-3. Because a plot of these data indicates an approximately linear relationship between the dependent variable (on the y-axis) and the independent variable (on the x-axis), linear regression is used to fit an equation to the data. The data and calculations summarized in Table 3-3 are utilized below to determine the linear regression equation:

$$b = \frac{\sum_{i=1}^{n} x_i y_i - \bar{x}\sum_{i=1}^{n} y_i}{\sum_{i=1}^{n} x_i^2 - \bar{x}\sum_{i=1}^{n} x_i} = \frac{2{,}626{,}817 - 250.6(9{,}788)}{1{,}416{,}926 - 250.6(5{,}012)}$$

$$= \frac{173{,}944.2}{160{,}918.8} = 1.081,$$

$$a = \bar{y} - b\bar{x} = 489.4 - 1.081(250.6) = 218.5.$$

Thus,

$$y = 218.5 + 1.081x.$$

Figure 3-2 shows the plotted data and the calculated regression line. As an example of how the regression equation is used, suppose that disposable income for the previous quarter is 310 (or 310×10^6). Then our forecast or estimate of sales (in thousands of dollars) for the current quarter, $\hat{y}$, is

$$\hat{y} = 218.5 + 1.081(310) = 553.6 \ (\text{or } \$553.6 \times 10^3).$$

Data point, i (period)	**TABLE 3-3** Calculations for Simple Linear Regression			
	y_i	x_i	$x_i y_i$	x_i^2
1	360	121	43,560	14,641
2	260	118	30,680	13,924
3	440	271	119,240	73,441
4	400	190	76,000	36,100
5	360	75	27,000	5,625
6	500	263	131,500	69,169
7	580	334	193,720	111,556
8	560	368	206,080	135,424
9	505	305	154,025	93,025
10	480	210	100,800	44,100
11	602	387	232,974	149,769
12	540	270	145,800	72,900
13	415	218	90,470	47,524
14	590	342	201,780	116,964
15	492	173	85,116	29,929
16	660	370	244,200	136,900
17	360	170	61,200	28,900
18	410	205	84,050	42,025
19	680	339	230,520	114,921
20	594	283	168,102	80,089
Totals	9,788	5,012	2,626,817	1,416,926

$$\sum_{i=1}^{n} x_i = 5,012 \qquad\qquad \sum_{i=1}^{n} y_i = 9,788$$

$$\bar{x} = \frac{\sum_{i=1}^{n} x_i}{n} = \frac{5,012}{20} = 250.6 \qquad\qquad \sum_{i=1}^{n} x_i^2 = 1,416,926$$

$$\bar{y} = \frac{\sum_{i=1}^{n} y_i}{n} = \frac{9,788}{20} = 489.4 \qquad\qquad \sum_{i=1}^{n} x_i y_i = 2,626,817$$

Key: y_i = actual quarterly sales ($\$10^3$) for period i
$\quad x_i$ = disposable income ($\$10^6$) in *preceding* period

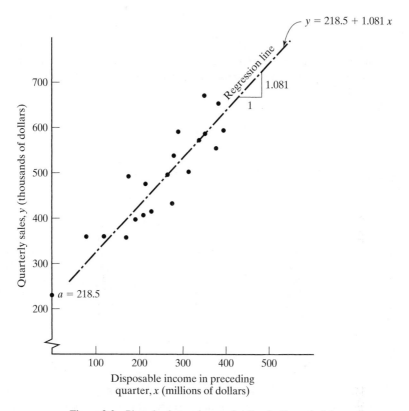

Figure 3-2   Plotted points and regression line for Example 3-1.

The correlation coefficient is a measure of the strength of the relationship between two variables only if the variables are linearly related. In Example 3-1, the correlation coefficient r, which measures the degree of strength, can be determined as follows:

$$r = \frac{S_{xy}}{\sqrt{S_{xx} \cdot S_{yy}}} \qquad (-1 \leq r \leq 1), \tag{3-4}$$

where

$$S_{xy} = \sum_{i=1}^{n} x_i y_i - \left(\sum_{i=1}^{n} x_i \right) \left(\sum_{i=1}^{n} y_i \right) / n \tag{3-5}$$

$$S_{xx} = \sum_{i=1}^{n} x_i^2 - \left(\sum_{i=1}^{n} x_i \right)^2 / n \tag{3-6}$$

$$S_{yy} = \sum_{i=1}^{n} y_i^2 - \left(\sum_{i=1}^{n} y_i \right)^2 / n. \tag{3-7}$$

If there is no relationship at all (a shotgun effect) between the dependent and independent variables, r will be zero, or nearly zero. A negative value of r indicates that one variable decreases as the other increases. When r is positive, the dependent and the independent variable both increase at the same time. The closer r is to -1 or $+1$, the more "perfect" is the correlation. Using Eqs. 3-4 through 3-7, the correlation coefficient for Example 3-1 is

$$S_{xy} = 2,626,817 - \frac{(5,012)(9,788)}{20} = 173,944$$

$$S_{xx} = 1,416,926 - \frac{(5,012)^2}{20} = 160,919$$

$$S_{yy} = 5,030,754 - \frac{(9,788)^2}{20} = 240,507$$

$$r = \frac{173,944}{\sqrt{(160,919)(240,507)}} = 0.88.$$

The positive value of r indicates that, as the independent variable (disposable income in the previous period) increases, the dependent variable (quarterly sales) will also tend to increase. This value of r indicates a good (but not great) relationship between the independent and dependent variables. One measure of the goodness of fit between x and y is called the coefficient of determination, which equals r^2. In our example, the value of r^2 is 0.77. The coefficient of determination measures the proportion of total variation that is explained by the regression line. Thus, in Example 3-1, the regression line, $y = 218.5 + 1.081x$, accounts for 77% of the variation in quarterly sales activity ($\$10^3$) in period i for the 20 observations of disposable income ($\$10^6$) in the preceding time period.   ■

3.8.1.2 Exponential Smoothing.

An advantage of the exponential smoothing method compared to simple linear regression for time-series estimates is that it permits the estimator to place relatively more weight on current data, rather than treating all prior data points with equal importance. Forecasting equations can quickly be revised with a relatively small number of calculations as each new data point is collected. Also, exponential smoothing does not assume linearity.

The main disadvantage of exponential smoothing is the basic assumption that trends and patterns of the past will continue into the future. However, it is more sensitive to changes than is linear regression. Because time-series analysis cannot predict turning points in the future, expert judgment and/or analysis of suspected causal factors should be used in interpreting results.

The basic exponential smoothing model that we shall discuss and illustrate is as follows:

$$S_t = \alpha' x_t + (1 - \alpha')S_{t-1} \qquad (0 \le \alpha' \le 1)$$

or

$$\begin{pmatrix} \text{Forecast for period } t + 1, \\ \text{made in period } t \end{pmatrix} = \alpha' \begin{pmatrix} \text{Actual data point} \\ \text{in period } t \end{pmatrix}$$

$$+ (1 - \alpha') \begin{pmatrix} \text{Forecast for period } t, \\ \text{made in period } t - 1 \end{pmatrix}. \qquad (3\text{-}8)$$

This term, α', the *smoothing constant*, merely provides a relative weighting for the new datum point compared to previous estimates. In general, α' should lie between 0.01 and 0.30, but the analyst should not hesitate to use a value outside this range if it gives better results with representative historical data.

An advantage of this technique of forecasting is its flexibility of weighting. If the weighting constant α' is 1, the mathematical model reduces to using the most recent period's outcome as the forecast. If α' is very close to 0, this is essentially equivalent to using a moving (arithmetic) average of actual outcomes over a large number of previous periods as the best estimate of the future outcome. Intermediate choices for α' between 0 and 1 provide forecasts that have more or less emphasis on long-run average outcomes versus current outcomes.

In estimating revenues, sales demand is an essential element for an economic analysis. Single exponential smoothing is illustrated with sales data listed in Table 3-4 that are graphed in Fig. 3-3. On the basis of Eq. 3-8, the following are sample calculations for S_t, which is termed a "smoothed statistic" (i.e., forecast) for period $t + 1$ (but made in period t):

$$S_t = \alpha'x_t + (1 - \alpha')S_{t-1},$$
$$S_1 = 0.3(50) + 0.7(50) = 50,$$
$$S_2 = 0.3(52) + 0.7(50) = 50.6,$$
$$S_3 = 0.3(47) + 0.7(50.6) = 49.52,$$
$$S_4 = 0.3(51) + 0.7(49.52) = 49.96.$$

TABLE 3-4   Exponential Smoothing Example

Period number, t	Demand, x_t (1,000 units)	S_t ($\alpha' = 0.30$)
0	—	50.00
1	50	50.00
2	52	50.60
3	47	49.52
4	51	49.96
5	49	49.67
6	48	49.17
7	51	49.72
8	40	46.80
9	48	47.16
10	52	48.61
11	51	49.33
12	59	52.23
13	57	53.66
14	64	56.76
15	68	60.13
16	67	62.19
17	69	64.23
18	76	67.76
19	75	69.93
20	80	72.95

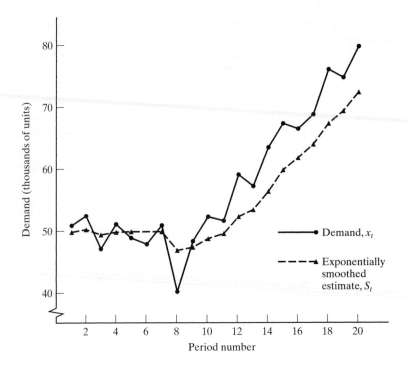

Figure 3-3   Exponential smoothing example.

In determining S_1, a value of S_0 must be estimated. Because no trend is assumed in the data for this model, an estimate based on the average of the first few data points is adequate. Here, S_0 was chosen to be 50.

To better understand the meaning of exponential smoothing, the following expression shows how demand data $x_t, x_{t-1}, x_{t-2}, \ldots$ are included in the forecast:

$$S_t = \alpha' x_t + (1 - \alpha') S_{t-1}. \tag{3-9}$$

Here,

$$S_{t-1} = \alpha' x_{t-1} + (1 - \alpha') S_{t-2} \tag{3-10}$$

and then

$$S_t = \alpha' x_t + \alpha'(1 - \alpha') x_{t-1} + (1 - \alpha')^2 S_{t-2}. \tag{3-11}$$

It is possible to continue substituting smoothed values in the same fashion until we get the following:

$$S_t = \alpha' x_t + \alpha'(1 - \alpha') x_{t-1} + \alpha'(1 - \alpha')^2 x_{t-2}$$
$$+ \alpha'(1 - \alpha')^3 x_{t-3} + \cdots + (1 - \alpha')^t S_0. \tag{3-12}$$

A forecast of T periods into the future is simply S_t because simple exponential smoothing assumes a constant pattern of data over time (i.e., no trend).

As you can see, every previous value of x is included in S_t. The x values are weighted so that the values more distant in time have successively smaller weighting factors. A large α' will place very little weight on remote data. The following calculations illustrate the weighting of the different data points included in S_4, as described by Eq. 3-12:

$$S_4 = 0.3(51) + 0.3(0.7)47 + 0.3(0.7)^2 52 + 0.3(0.7)^3 50 + (0.7)^4 50$$
$$= 0.30(51) + 0.21(47) + 0.147(52) + 0.1029(50) + 0.2401(50)$$
$$= 49.96.$$

This agrees with our calculation of S_4 earlier.

Table 3-4 shows a continuation of the preceding example for hypothetical demand data over 20 periods. It should be noted that the higher the value of α', the closer the new estimate will be to the most recent datum point.

3.8.2 Subjective Techniques

The time-series techniques discussed earlier are underpinned by the premise that the future is an extension of the past. However, the future will contain events that today are poorly understood or completely unanticipated. Hence, the aim of this section is to explain and illustrate two techniques for developing subjective information for purposes of making an estimate: (1) the Delphi method and (2) technology forecasting.

3.8.2.1 The Delphi Method.   Most analysts and decision makers draw on the advice of experts or other people thought knowledgable, as they form their judgments. Often the decision situation is highly complex and poorly understood, so that no single person can be expected to make an informed decision. The traditional approach to decision making in such cases is to obtain expert opinion through open discussions and to attempt to determine a consensus among the experts. However, results of panel discussions are sometimes unsatisfactory because group opinion is highly influenced by dominant individuals and/or because a majority opinion may be used to create the "bandwagon effect."

The Delphi method attempts to overcome these difficulties by forcing persons involved in the forecasting exercise to voice their opinions anonymously and through an intermediary. The intermediary acts as a control center in analyzing responses to each round of opinion gathering and in feeding back opinion to participants in subsequent rounds. By following such a procedure, it is hoped that the responses will converge on a consensus forecast that turns out to be a good estimator of the true outcome.

Two premises underlie the Delphi method. The first is that persons who are highly knowledgeable in a particular field make the most plausible forecasts. Second, it is believed that the combined knowledge of several persons is at least as good as that of one person.

Typically, the technique is initiated by writing an unambiguous description of the forecasting problem and sending this, along with relevant background information, to each participant in the study. Often, the participants are invited to list major areas of concern in their particular specialty as they may relate to the problem being addressed by the study. The first questionnaire sent out might request the opinion of each expert regarding likely dates for the occurrence of an event identified in the problem statement. Because responses to this type of question will normally reveal a spread of opinions, *interquartile* ranges are customarily computed and presented to the experts at the beginning of the second round. Interquartile ranges identify upper and lower quartile values in the continuum of responses such that 50% of the responses fall within that range.

In the second round of the Delphi technique, the participants are asked to review their response in the first round relative to interquartile ranges from that round. They then have the opportunity to revise their estimates in light of the group response. At this point, participants can request that additional information relevant to the forecasting problem be gathered and sent to them.

If an estimate departs appreciably from the group median, the respondent who furnished it is asked to give reasons for his or her position. Frequently, all panelists are urged to conceive statements that challenge or support estimates falling outside the interquartile or some other range of responses. These reasons, along with routine second-round estimates for the entire group, are again analyzed and statistically summarized (usually as interquartile ranges, although other measures capable of showing group convergence or divergence could be used).

In those cases in which a third-round questionnaire is felt necessary, participants receive a summary of second-round responses plus a request to reconsider and/or explain their estimate in view of group responses in the second round. They are again asked to reassess their earlier responses and possibly to explain why their estimates do not conform to the majority of group opinion.

An example of quantitative results of the Delphi technique is summarized in Fig. 3-4. The problem of concern to a manufacturer of large earth-moving equipment was to develop a forecast of total company sales during fiscal year 20XX. Six marketing and sales experts (A through F) were asked to consider historical company and industry data and anonymously prepare a forecast of bookings for a particular product. First-round questionnaire results are shown in the top part of Fig. 3-4. The median response in round 1 was 229, and the interquartile range of the responses was 85. Results of round 1 were fed back to each participant along with additional information pertaining to the forecast that each person requested. The second and third rounds were completed in a similar manner. Notice how forecasts in the three iterations of questioning tend to converge, with a final median forecast value of 260 units and an interquartile range of 47.

3.8.2.2 Technological Forecasting

Technological forecasting is a name given to a myriad of specialized forecasting techniques. It provides procedures for data collection and analysis to predict future technological developments and the impacts such developments will have on the environment and lifestyles of people. These techniques seek to make potential technological developments explicit, but more important, they force decision makers to try to anticipate future developments.

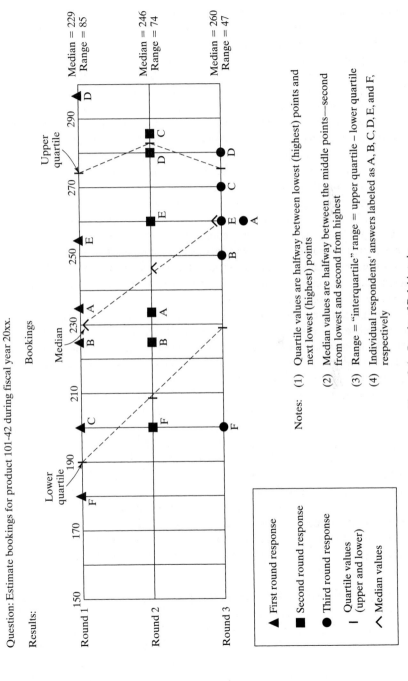

Figure 3-4 Results of Delphi study.

Technology forecasting is a method that can be used to estimate the growth and direction of a technology. A typical question that technology forecasting attempts to answer is What will be the machining tolerances of numerically controlled machine tools in the future? Or, what will be the operating characteristics of the next generation of integrated circuits for engineering workstations?

Trend extrapolation is often used to make technological forecasts. This technique is based on a historic time series for a selected technological parameter. It often is assumed that the factors influencing historical data are likely to remain constant rather than to change in the future. Usually, a single-function parameter such as speed, horsepower, or weight is extrapolated.[2] A good trend extrapolation depends on selection and prediction of key parameters of performance. The trend under study should be capable of quantification in order that it can be portrayed numerically, and an adequate database should exist on which to base a reliable trend line. An example of trend extrapolation is presented in Fig. 3-5. Notice that the *y*-axis is a logarithmic scale.

An advantage of trend extrapolation is that historical data are often readily obtainable. A straight-line or fitted-curve projection of the future is easily understood and used. A drawback to extrapolation stems from the assumption that factors that shaped the past will continue to hold basically unchanged in the future. Trend extrapolation techniques cannot predict unforeseen technology interactions such as unprecedented changes or inexplicable discoveries.

The *substitution curve* is based on the belief that a product or technology that exhibits a relative increase in performance over an older (i.e., established or conventional) product or technology will eventually substitute for the one having lesser performance. The relative increase in performance is the important factor in the substitution of one technology for another. A basic assumption with this method is that, once the substitution of one technology for another has begun, it will irreversibly continue to completion. Listed below are some common examples of the substitution effect:

Old technology	New technology
Petroleum lamps	Electric lamps
Horse-drawn carriages	Automobiles
Steam locomotive	Diesel locomotive
Cotton	Synthetic fibers
Leather	Vinyl
Soap	Detergent
Reciprocating engines	Turbojet engines
Hardwood floor	Vinyl flooring

The forecast starts with the observation that a new technology is starting to displace an older technology. A measurement term that best defines the fraction of total usage of each technology must be selected, and time-series data are gathered for both technologies. These data are used to establish the initial takeover rate and to predict

---

[2] Linear regression may be used for extrapolating data into the future or, more simply, a straight line can be "eyeballed."

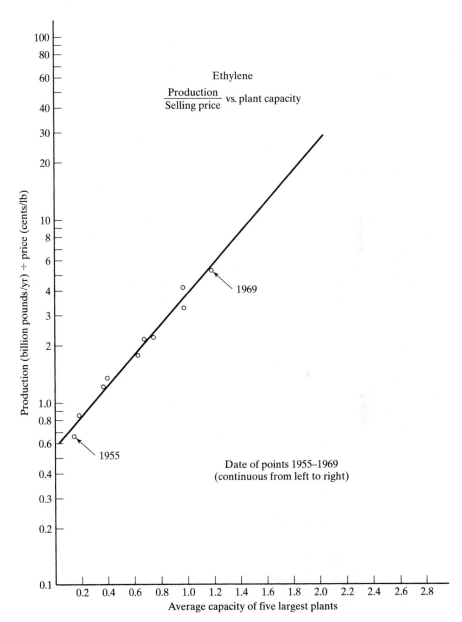

Figure 3-5   An example of linear trend extrapolation.

the year in which takeover will reach 50%. A typical substitution effect for two technologies is shown in Fig. 3-6.

Forecasting by analysis of *precursor events* uses the correlation of performance trends between two innovative technologies. Because technological advance
usually follows a pattern of continuous increase, situations frequently occur in which

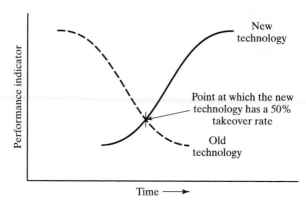

Typical technology life cycle

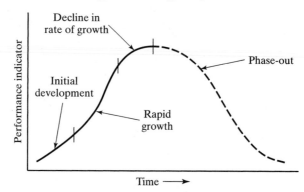

Figure 3-6   An example of substitution curves.

an indicator of technical progress lags another by a given period of time. It is thus possible to utilize the leading technology to predict the status of the lagging technology over a time period equal to the lag time. The frequently cited example of precursor events shown in Fig. 3-7 concerns the historical relationship between maximum speed of military aircraft to the maximum speed of commercial transport aircraft.

In this example, it was found that the speed of commercial aircraft followed the speed of military aircraft by 6 years in the 1920s and 11 years in the 1950s. As a result, it was predicted that commercial transport aircraft with speeds of Mach 2 would be expected no later than 1970, or if such aircraft were not introduced at this time, aircraft with speeds of Mach 3 would be introduced near 1976.

3.8.3 Cost Engineering Techniques

A wide array of techniques exists for estimating investment and working capital requirements associated with products, processes, and materials.[3] In this section, we

---

[3] An excellent reference is P. F. Ostwald, *Engineering Cost Estimating,* 3d ed. (Englewood Cliffs, NJ: Prentice Hall, 1992).

consider a small collection of useful techniques that permit these types of estimates to easily be made. Many of them utilize various kinds of revenue/cost indexes in the preparation of an estimate.

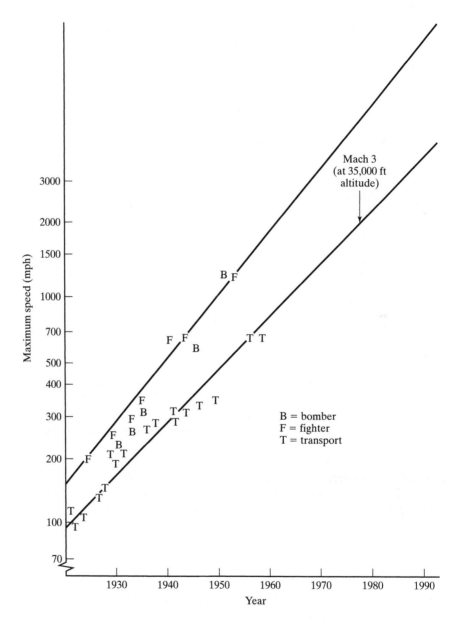

Figure 3-7   Comparative speed trends of combat and transport aircraft.

3.8.3.1 Unit Method.   A popular cost engineering technique is the *unit method*, which involves using an assumed or estimated "per unit" factor. Some examples are

- capital cost of plant per kilowatt of capacity,
- fuel cost per kilowatt-hour generated,
- capital cost per installed phone,
- revenue per long-distance call,
- operating cost per mile, and
- maintenance cost per day of use.

These factors may be multiplied by the appropriate unit to provide the total estimate. The following examples may be used for breaking quantities to be estimated into units that can be estimated readily:

1. in different units (*example*—dollars per week, to convert to dollars per year);
2. a proportion instead of a number (*example*—percent defective, to convert to number of defects);
3. a number instead of a proportion (*example*—number defective and number produced, to convert to percent defective);
4. a rate instead of a number (*example*—miles per gallon, to convert to gallons consumed);
5. a number instead of a rate (*example*—miles and hours traveled, to convert to average speed);
6. using an adjustment factor to increase or decrease a known or estimated number (*example*—defectives reported, to convert to total defectives).

Although the unit method is useful for preliminary estimating, the values can be very misleading because there is no consideration of the principle of economies of *scale* or economies of *scope*.

3.8.3.2 Factor Technique.   The *factor technique* is an extension of the unit method in which one sums the product of one or more quantities or components involving unit factors and adds these to any components estimated directly. That is,

$$C = \Sigma C_d + \Sigma f_i \times U_i, \qquad (3\text{-}13)$$

where

$$C = \text{value (cost, price, etc.) being estimated,}$$
$$C_d = \text{cost of selected components estimated directly,}$$
$$f_i = \text{cost per unit of component } i, \text{ and}$$
$$U_i = \text{number of units of component } i.$$

Example 3-2
Suppose that we need a slightly refined estimate of the cost of a house consisting of 1,500 ft^2, two porches, and a garage. Using unit factors of $80/ft^2, $2,000/porch, and $3,000/garage, we can calculate the estimate as

$$\$80 \times 1,500 + (\$2,000) \times 2 + \$3,000 = \$127,000.$$ ■

3.8.3.3 Exponential Costing.   *Exponential costing* may be used when the proposed plant has a different production capacity than the existing plant. According to de la Mare, "the principle of exponential costing states that, for many real-world production systems, proportionate increases in production capacity can be achieved by less than proportionate increases in capital cost. This principle is a special manifestation of the law of increasing returns to scale, and is known as the law of increasing technical returns to scale."[4]

The following general equation represents most types of equipment:

$$\frac{C_a}{C_b} = \left(\frac{Q_a}{Q_b}\right)^{\beta},\tag{3-14}$$

where

C_a = capital cost of the proposed facility,
C_b = known capital cost of an existing facility,
Q_a = production capacity of the proposed facility,
Q_b = production capacity of the existing plant, and
β = cost-exponent factor, which can range from 0.4 to greater than 1.00, but is usually in the range from 0.5 to 0.8.

Table 3-5 provides typical cost-exponent factors for selected types of industrial equipment. Equation 3-15 permits you to obtain an estimate of the capital cost of a proposed project by including factors to adjust for the effects of price increases as follows:

$$C_a = C_b \left(\frac{Q_a}{Q_b}\right)^{\beta} I_t,\tag{3-15}$$

where $I_t = I_n/I_k$ = index of price change (inflation) from period k to period n. A fairly detailed treatment of inflation, as it affects cost estimating and comparisons of alternatives, is provided in Chapter 8.

The accuracy of the exponential costing method depends largely on the similarity between the two projects, and on the accuracy of the cost-exponent factor β. Generally, error ranges from ±10 to ±30% of the actual final cost.

Example 3-3
A certain steam-generating boiler in the utility plant of a manufacturing complex produces 50,000 lb/hr of saturated steam. Eight years ago, this boiler was purchased for $250,000. If the

---

[4] R. F. de la Mare, *Manufacturing Systems Economics* (New York: Holt, Rinehart and Winston, 1982), p. 151.

TABLE 3-5 Typical Cost-Exponent Factors, β*

Process industrial equipment		Material-handling equipment		General industrial equipment	
Item	β	Item	β	Item	β
Agitators	0.3–0.5	Bagging machines	0.8	Air compressors	0.4
Centrifuges	0.7–1.3	Conveyors	0.7	Air driers	0.6
Evaporators	0.5–0.7	Conveyors (bucket)	0.6–0.8	Cranes	0.6
Heat exchangers	0.7–0.9	Conveyors (roller)	0.9	Driers (product)	0.4–0.5
Piping	0.7–0.9	Elevators	0.4	Electric motors	0.8
Pumps	0.5–0.9	Hoppers	0.7–0.9	Steam boilers	0.5
Tanks (rectangular)	0.5			Building	
				Single story	0.8
				Two story	0.7–0.8

*For other factors, see D. S. Remer and L. H. Chai, "Design Cost Factors for Scaling-up Engineering Equipment," *Chemical Engineering Progress* (August 1990):77–82.

price index for this type of boiler has increased at an average rate of 12% per year for the past 8 years, how much would a 150,000 lb/hr boiler cost now? The cost-exponent factor for this boiler is 0.50.

Solution   Utilizing Eq. 3-15,

$$C_a = C_b \left(\frac{Q_a}{Q_b} \right)^{\beta} I_t,$$

with $I_t = (1 + 0.12)^8$, the cost of the proposed boiler can be estimated as follows:[5]

$$C_a = \$250,000 \left[\left(\frac{150,000}{50,000} \right)^{0.5} (1.12)^8 \right]$$

$$= \$250,000[(3.00)^{0.5} (2.476)]$$

$$= \$1,072,140. \qquad \blacksquare$$

Example 3-4
Assume that 6 years ago an 80 kW diesel electric set cost $160,000. The plant engineering staff is considering a 120 kW unit of the same general design to power a small isolated plant. The cost-exponent factor is 0.6. The price index for this class of equipment 6 years ago was 187 and is now 194. Assume we want to add a precompressor, which, when isolated and estimated separately, now costs $18,000. Determine the total cost now of the 120 kW unit.

Solution   The required calculations are as follows:

The index of price change, $I_t = 194 / 187 = 1.0374.$

$$C_{now}(120 \text{ kW}) = \$160,000 \left[\left(\frac{120}{80} \right)^{0.6} (1.0374) \right] = \$211,700. \qquad \blacksquare$$

Total cost $= \$211,700 + \$18,000 = \$229,700.$

3.8.3.4 Learning Curves.   In repetitive operations involving direct labor, the average time to produce an item or provide a service is typically found to decrease over time because of learning that occurs. This phenomenon is observed for purchasing activities, assembly operations, food preparation, etc. As a result, cumulative average and unit times required to complete a task will drop dramatically as output increases. This can substantially reduce costs for large production runs, and failure to include this phenomenon can create large errors in cost estimates.

As a general rule, when cumulative production/service doubles, the total time required per output unit is reduced by x%. For example, if x = 4%, the time per unit required to go from, say, 4 to 8 units of output declines by 4% and we have a 96% (= 100% – 4%) learning curve. To illustrate this concept, suppose it takes 1 hour to produce the first unit of output. Assuming a 96% learning curve, it would take 0.96 hours

---

[5] See Section 8.1.1. The index of price change, $I_t = I_n/I_k$, equals $(F/P, 12\%, 8)$.

to produce the second unit, $(0.96)^2 = 0.9216$ hours to produce the fourth unit, $(0.96)^3 = 0.8847$ hours to produce the eighth unit, and so on.

A simple equation to estimate the time requirements of repetitive labor activities is

$$Y_i = Y_1 i^b,\tag{3-16}$$

where

Y_i = direct labor hours (or cost) for the ith production unit,
Y_1 = direct labor hours (or cost) for the initial (first) unit,
i = cumulative count of units of output, and
b = the learning curve exponent.

For instance, Y_{10} represents the time (or cost) required for the 10th unit of output. Furthermore, b equals the ratio of log(learning curve expressed as decimal) $\div$ log 2. If we have a 96% learning curve,

$$b = \log(0.96)/\log 2 = -0.017728/0.30103 = -0.0589,$$

and from Eq. 3-16, we have

$$Y_i = Y_1 i^{-0.0589}.$$

If $Y_1 = 1.0$ hour, the estimate of Y_{10} becomes

$$Y_{10} = 1.0(10)^{-0.0589} = 0.873 \text{ hrs.}$$

The cumulative average time for i production units ($i \geq 1,000$) can be approximated by

$$C_i = \frac{Y_1}{i(1 + b)}\left[(i + 0.5)^{1+b}\right], \text{ where } b < 0.\tag{3-17}$$

Hence, the value of C_{10} from Eq. 3-17 is approximately

$$C_{10} = \frac{1.0}{10(0.9411)}\left[(10.5)^{0.9411}\right] = 0.97 \text{ hrs.}$$

Example 3-5
As the number of manually assembled wire harnesses doubles from 20 to 40, there is a 20% reduction in labor hours per harness. If the first unit required 24 hours to assemble, how much time will it take for the fortieth harness?

Solution The required calculations are as follows:
The learning curve is 100% − 20% = 80% (= 0.80)

$$b = \log(0.8)/\log 2 = -0.322.$$

Using Eq. 3-16 as an approximation, we find that

$$Y_{40} = Y_1(40)^{-0.322} = 24(0.305) = 7.32 \text{ hrs.} \qquad \blacksquare$$

For convenience in making calculations, a range of learning curve exponents is provided in Table 3-6.

Example 3-6 further illustrates the use of a learning curve for estimating labor costs.

Example 3-6

One thousand units of a new product are to be produced. Direct labor has been estimated to average 8 hours per unit. Six months after production has begun, management has requested the following information:

    1. average labor hours per unit, to date;

    2. labor hours per unit, for the latest month's units;

    3. the % learning curve that production has followed.

The cumulative labor hours and units produced are listed in Table 3-7. Dividing the cumulative labor hours by the cumulative production shows that the average labor hours per unit to date is 15.75 labor hours. For the latest month's production, an average of $410/42 = 9.76$ labor hours per unit was reached.

In examining the data, it can be seen that the cumulative production at the end of month 6, (132 units) is double the cumulative production at the end of month 4 (66 units). The learning curve can be approximated by dividing the cumulative average labor hours per unit through

TABLE 3-6   Typical Learning Curve Exponents, b

Percent learning	b	Percent learning	b
100	0 (no learning)	70	−0.515
95	−0.074	65	−0.621
90	−0.152	60	−0.737
85	−0.234	55	−0.861
80	−0.322	50	−1.000
75	−0.415		

TABLE 3-7   Production Data

Month, i	Month's production	Cumulative production	Month's labor hours	Cumulative labor hours	Month's avg. labor hours/unit	Cumulative avg. labor hours/unit ($= Y_i$)
1	11	11	301	301	27.36	27.36
2	17	28	420	721	24.71	25.75
3	16	44	307	1028	19.19	23.36
4	22	66	356	1384	16.18	20.97
5	24	90	285	1669	11.87	18.54
6	42	132	410	2079	9.76	15.75

month 6 by the cumulative average labor hours per unit during month 4. The learning curve in effect is approximately 75%, as shown here:

$$\frac{Y_{132}}{Y_{66}} = \frac{15.75 \text{ labor hours/unit}}{20.97 \text{ labor hours/unit}} = 0.75$$

■

3.9 Additional Examples of Cost Engineering

Three examples are provided that further illustrate the techniques discussed in Section 3.5. Example 3-7 involves exponential costing in the process industry, and Example 3-8 demonstrates the use of linear regression in a highly automated drilling operation. Finally, Example 3-9 uses a learning curve and the factor technique for estimating the selling price of a manufactured item.

Example 3-7
The R Square Corporation is considering two plans, A and B, for expanding the capacity of its urea manufacturing plant. The relevant data and analysis for both plans are given next.

   Plan A
   R Square Corporation's office of agricultural and chemical development operates a urea plant that was built in 1993 at a cost of $4.3 million and capacity of 300,000 lb/yr. Due to increasing use of urea-based fertilizers and increasing maintenance costs on the existing unit, the feasibility of constructing a new plant in 2008 with a 750,000 lb/yr capacity is being studied. Time does not permit an in-depth cost estimation for the new plant, so engineers decide to obtain the cost of the new plant by scaling up the cost of the old plant. The cost-exponent factor is known to be 0.65, and the construction cost index is estimated to have increased an average of 10.5% per year for the past 15 years. What cost should the engineers report to the project review committee?

   Urea plant estimates are as follows:

   C_b = $4.3 million—cost of plant in 1993,

   Q_b = 300,000 lb/yr—capacity of existing plant,

   C_a = ?—cost of 750,000 lb/yr plant (but based on 1993 pricing),

   Q_a = 750,000 lb/yr—capacity of new plant,

   C_a' = cost reported to committee in 2008 dollars, and

   β = cost-exponent factor = 0.65.

   Solution   From Eq. 3-14,

$$C_a = C_b \left(\frac{Q_a}{Q_b} \right)^{\beta},$$

we find

$$C_a = \$4.3 \text{ million} \left(\frac{750,000 \text{ lb/yr}}{300,000 \text{ lb/yr}} \right)^{0.65} = \$7.8 \text{ million (in 1993)},$$

$$C_a' = C_a \, (F/P, 10.5\%, 15) = \$7.8 \text{ million} (1.105)^{15},$$

or

$$C_a' = \$34.88 \text{ million (estimated cost of new urea unit in 2008 dollars).}$$

Plan B

As an alternative plan, it is learned that a manufacturer of these units can prefabricate and install a unit on site for a total cost of $22 million. The R Square Corporation's constructed unit has an estimated life of 15 years, and the prefabricated unit has a life of 12 years. The operating costs for the company unit are $40,000 per year for years 1–10 and $30,000 per year for years 10–15. The operating costs for the prefabricated unit are $50,000 per year. The interest rate is assumed to be 10%. Which alternative should they select? What assumptions are involved?

Solution   *Note:* Assume negligible salvage value for both alternatives, and compare the alternatives by using the annual cost (AC) method to be discussed in Chapter 5.

	Prefabricated unit	Company unit
Initial cost:	$22,000,000	$34,880,000
Annual costs:	$50,000/yr	$40,000/yr, years 1–10
		$30,000/yr, years 10–15
Life:	12 yr	15 yr

Prefabricated unit:

$$AC = \$50,000 + \$22,000,000(A/P, 10\%, 12)$$
$$= \$3,279,600.$$

Company unit:

$$AC = \$30,000 + \$10,000(P/A, 10\%, 10)\,(A/P, 10\%, 15)$$
$$+\ \$34,880,000\,(A/P, 10\%, 15)$$
$$= \$4,893,884.$$

By a wide margin, the R Square Corporation should let the outside manufacturer construct and install the urea unit. However, numerous nonmonetary considerations could shift the decision to plan A. The monetary risks associated with expanding in an uncertain and highly competitive market may well cause neither plan to be acceptable. An indefinitely long study period has been assumed in the preceding analysis.   ∎

Example 3-8

For a certain drilling operation within a flexible machining cell, data regarding the time (in hours) to drill 1, 2, 3, and 4 holes in a $\frac{1}{4}$-inch sheet of carbon steel have been obtained. The data are as follows:

(a) Develop a linear regression equation for estimating the time (dependent variable) required for the number of holes drilled (independent variable).

(b) What is the correlation coefficient r for these data?

(c) Make an estimate for the time required to drill six holes, and discuss the danger in using the regression equation for this purpose.

Data point	x (No. of holes)	y (Hours)
1	2	0.0381
2	3	0.0720
3	4	0.1078
4	3	0.0815
5	1	0.0360
6	2	0.0605
7	1	0.0382
8	4	0.1318
9	3	0.0985
10	1	0.0468
11	2	0.0721
12	4	0.0950

Solution

(a) For the data given, various values must be calculated and substituted into Eqs. 3-2 and Eq. 3-3: $\Sigma x = 30$, $\Sigma y = 0.8783$, $\Sigma x^2 = 90.0$, $\Sigma y^2 = 0.0748$, $\Sigma xy = 2.5568$, and $\bar{x} = 2.5$. From these values, we find that

$$b = \frac{2.5568 - 2.5(0.8783)}{90.0 - 2.5(30.0)}$$

$$= 0.0241,$$

$$a = \frac{0.8783}{12} - 0.0241(2.5)$$

$$= 0.0129.$$

Thus, the linear regression equation is $y = 0.0129 + 0.0241x$. Estimates of time required to drill 1, 2, 3, and 4 holes are listed as follows:

No. of holes, x	Estimated time (hours), y
1	0.0370
2	0.0611
3	0.0852
4	0.1093

(b) The correlation coefficient for these data can be determined by using Eq. 3-4. It is $r = 0.909$, so the relationship between x and y is reasonably well estimated by a linear function.

(c) Finally, an estimate for drilling six holes is

$$y = 0.0129 + 0.0241(6) = 0.1575 \text{ hrs.}$$

Caution must be exercised in using linear regression to make estimates when the value of the independent variable lies outside the range of values (for holes drilled) that was utilized to develop coefficients (a and b) of the regression equation. In practice, it is not uncommon to observe this use of regression equations, but the practice cannot be recommended as a sound one. ■

Example 3-9

The ABC manufacturing company is trying to determine the unit selling price for a high-density, double-sided disk with 2 MB storage capacity. The disks are produced by installing a magnetic film into a plastic cartridge.

A total of three machining operations needs to be performed:

- Cut out disks from magnetic film.
- Apply disk control centerpiece.
- Insert into plastic cartridge.

The film, centerpiece, and cartridges are purchased from an outside manufacturer. A total of 10,000 disks is to be produced. Relevant information is as follows:

- The magnetic film is bought in rolls that cost $90 each. From each roll, 2,000 circular disks can be cut out.
- One person is needed to operate and supervise the cut-out machine. Installing a new roll takes 8 minutes, and cutting out 2,000 circular disks takes 25 minutes.
- No learning curve is applicable for the cut-out operation.
- The disk control centerpieces cost $0.12 per unit.
- One person is required to apply the centerpieces to the magnetic disks. Applying the first centerpiece takes 30 seconds, and for the remaining centerpieces an 80% learning curve is applicable.
- The plastic cartridges cost $0.15 per unit.
- One person is needed to supervise the disk-insertion operation. This operation is done automatically by a machine that can insert 1,500 disks per hour.
- No learning curve is applicable for inserting disks.
- The direct labor rate is $15.00/hour.
- Planning and liaison are 15% of factory labor.
- Quality control is 30% of factory labor.
- Factory overhead is 800% of total labor.
- General and administrative expense is 50% of total labor.
- Packing costs are 100% of total labor.
- The profit margin is 15% of total manufacturing cost.

Solution   The various components of total cost and required selling price are computed and summarized as follows:

Production Material Cost.   To produce 10,000 disks, 10,000/2,000 = 5 magnetic rolls are required. This costs 5 × $90 = $450.

10,000 centerpieces and cartridges cost 10,000($0.12 + $0.15) = $2,700. Thus, total production material cost = $450 + $2,700 = $3,150.

Direct Labor Hours.   Cutting out 2,000 disks from magnetic film takes: 8 + 25 = 33 minutes. Hence, producing 10,000 disks takes

$$5 \times 33 = 165 \text{ minutes.}$$

By summing Eq. 3-16, the cumulative time for applying 10,000 centerpieces (in minutes) with an 80% learning curve is

$$0.5 \text{ minutes} \times \sum_{i=1}^{10,000} i^{\log 0.8/\log 2} = 379.71 \text{ minutes}.$$

The disk-insertion operation requires

$$10,000/1,500 = 6.67 \text{ labor hours}.$$

Thus, a total of

$$(165 + 379.71)/60 + 6.67 = 15.75 \text{ labor hours}$$

is needed.

A popular cost-estimating template appears next.[6]

Customer:	Apex			
Model:	HDDS012X	Estimator:	Chas. Everyperson	
Part Name:	HD/DS Disk			
Part No:	012	Date:	February 20, 1994	
No. Parts Required:	10,000	Page:	1 of 1	
MANUFACTURING COST		HOURS	CHARGE RATE	DOLLARS
A. Factory Labor		15.75	$15.00	$ 236.25
B. Planning & Liaison Labor			15% of A	$ 35.44
C. Quality Control			30% of A	$ 70.88
D. TOTAL LABOR				$ 342.57
E. Factory Overhead			800% of D	$2,740.50
F. General & Admin. Expense			50% of D	$ 171.28
G. Production Material				$3,150.00
H. Outside Manufacture				$ 0.00
I. SUBTOTAL				$6,404.34
J. Packing Costs			100% of D	$ 342.57
K. TOTAL DIRECT CHARGE				$6,746.91
L. OTHER DIRECT CHARGE				$ 0.00
M. Facility Rental				$ 0.00
N. TOTAL MANUFACTURING				$6,746.91
O. Profit/Fee			15% of N	$1,012.04
TOTAL SELLING PRICE				$7,758.94
QUANTITY				10,000
UNIT SELLING PRICE				$ 0.78

---

[6] T. F. McNeill and D. S. Clark, *Cost Estimating and Contract Pricing* (New York: American Elsevier Publishing Co., 1966), p. 71.

3.10 Summary

In this chapter, important cost concepts and techniques for obtaining and/or developing data required in economic analyses have been discussed and illustrated. Specifically, we have concentrated on making single-valued estimates for analyses to be treated in Part Two of this book, which are commonly called "assumed certainty" analyses. Part Three, beginning with Chapter 12, will show methods that explicitly consider the variation in estimated quantities and incorporate many refinements for rational economic analyses.

   Regardless of who performs economic analyses or who makes the final investment decisions, the proper performance of these functions is critical to the economic health or even survival of the individual firm. Business decisions frequently involve investments that must be planned and executed many years before the expected returns will be realized. Moreover, the scale of the investments in research and capital assets required for our expanding economy grows as new technologies develop. Hence, knowledge of the principles and techniques underlying economic analyses is extremely important.

PROBLEMS

3-1. A supplier purchased an Ajax charger 5 years ago for $5,000, intending to sell it at its usual markup for $5,800.

   Before the supplier was able to obtain delivery, a competitor brought out a radically new charger for the same type of service, better in every way, but selling at a retail price of only $3,000. As a result, the Ajax charger has been a white elephant in the supplier's hands—it is a large piece of obsolete equipment that has been occupying valuable floor space that is now vitally needed.

   In discussing what to do, two members of the firm find themselves in disagreement. The president feels that the charger should be kept unless the $5,000 purchase price is realized on the sale. The accountant feels that the equipment should not be sold unless both the $5,000 cost and $500 cost of storage to date can be realized.

   Which course of action would you recommend? Why?

3-2. Smith purchased his house several years ago for $75,000 and was just offered $125,000 cash for it. Smith and his family had not been planning to sell and move, even though they are willing to do so. A neighborhood economist has correctly computed that the pretax annual rate of profit on the cash Smith has invested in the house would be 35%, and on this basis he recommends that Smith sell the house. What additional information does Smith need to make a decision? What irrelevant information was given?

3-3. A merchant has been attempting to maintain his stock of goods at a constant physical volume even though prices have been rising. His stock of one item was originally purchased for $10 per unit, which is the cost for accounting purposes. He sold these goods at $16 per unit (applying his usual markup) and immediately replaced them by identical ones purchased at the new wholesale price of $18 per unit. What do you think of the profitableness of this transaction? What should he have sold them for to make $0.01 per unit?

3-4. Certain factory space cost $20.00 per square foot to build and is estimated to have an economic life of 25 years and $0 salvage value. The minimum attractive rate of return on invested capital is 10%. The annual out-of-pocket cost of property taxes, heat, lights, and maintenance is $1.00 per square foot whether or not the space is being used. What should be the cost per square foot considered in an economic analysis of a new project A that entails proposed use of that space under each of the following conditions?

a. The space is now being used for another project B, which will have to be moved to new quarters costing $4.00 per square foot per year.

b. The space is idle and there is no alternative use of it expected for the entire period in which the project under consideration would exist.

c. The space is part of a large area that is used normally; hence, it is thought reasonable to charge only long-run average costs.

3-5. A firm has a manufacturing division with a normal manufacturing capacity of 1,000,000 units that sell for $120 each. The price consists of variable labor and material costs, $60; fixed costs, $40; and profit, $20. During a severe recession, only 200,000 units can be sold annually and only if the price is reduced to $100 each. The total fixed cost can be reduced 15% below normal if the plant remains open, and 30% below normal if it closes. The variable cost is directly proportional to output.

a. Disregard irreducible considerations. Should the plant remain open for the next year or two to produce 200,000 units a year, or should it shut down and reopen when business improves?

b. At the 200,000 unit production rate, to what level may the price be reduced during the recession before shutting down the plant becomes more economical than operating it?

3-6. Process A, designed to produce 10,000 units a year, has a fixed cost of $100,000 a year. Process B, with the same design capacity, has a fixed cost of $80,000 a year. Process A produces the initial 4,000 units at a variable cost of $10 and the next 6,000 units at a variable cost of $17. Process B produces the first 5,000 units at a variable cost of $9 each and produces the next 5,000 at $8 each. Show what load should be assigned to each plant if the demand for the product is varied from zero to 20,000 units. Assume that at no load the fixed costs will not be reduced.

3-7. A firm is considering whether to contract with vendors or use in-house crews for a type of maintenance work. The cost accounting system yields estimates per job as follows:

Direct labor	$5,000
Materials	2,200
Overhead	4,000
Total	$11,200

A vendor has offered to do the work for $10,000 per job. Discuss whether the firm should contract with the vendor under each of the following conditions:

a. The "in-house" crews would still be on the payroll and would be otherwise unoccupied if they were not doing this work.

b. The "in-house" crews on the payroll will always be productively occupied whether or not this maintenance work is done by them. That is, their labor time is worth $5,000 elsewhere if they do not work on this job. Of the standard overhead charge given, one-half is fixed and one-half is avoidable if the "in-house" crews do not do the job.

3-8. For the coming period, overhead costs for a firm are estimated to be $600,000 for production that is expected to sell for $2,800,000 and to require 200,000 direct labor hours costing $800,000 and direct materials costing $200.000. The firm allocates its overhead to various jobs on the basis of the following relation:

Overhead cost = $100,000 + 0.5(direct labor + direct materials).

The firm is considering changing its method of manufacturing a particular job with estimates as follows:

	Old Method	New Method
Direct labor	$100,000	$90,000
Direct material	8,000	16,000

Which method is more economical if actual overhead costs for the two methods are thought to

a. vary according to the allocation formula?

b. be the same for each method?

c. be $5,000 more for the new method than for the old method?

3-9. Explain by example the difference between book costs and sunk costs.

3-10. What is the difference between postponable costs and escapable costs?

3-11. Formulate a one-sentence rule for an engineering group indicating when they may eliminate consideration of overhead costs in the analysis of investment decisions.

3-12. List the costs associated with owning an automobile that you would classify as fixed costs and those that you would classify as variable costs according to miles driven during the next year. Assume that you plan to keep the car for many years.

3-13. Bad news—you have just wrecked your car! An automobile wholesaler offers you $2,000 for the car "as is." Also, your insurance company's claims adjuster estimates that there is $2,000 in damages to your car. Because you have collision insurance with a $1,000 deductibility provision, the insurance company mails you a check for $1,000. The odometer reading on the wrecked car is 58,000 miles.

What should you do? You need another car immediately. The best alternatives appear to be the following:

(1) Sell the wrecked car tomorrow to the wholesaler and invest your $7,000 life savings in a used $10,000 car (odometer reading = 28,000 miles);

(2) Spend $2,000 immediately to restore the wrecked car to its previous condition;

(3) Give the car to a "shade tree" body shop where they will repair it for $1,100 but will take an extra month to do so. During this repair time, you will rent a car at a monthly rate of $400.

In alternatives (2) and (3), you also have the option of selling the repaired car for $4,500 and then investing $5,500 of your savings in the used $10,000 car. If you elect to sell your refurbished car for $4,500, any money remaining in your life savings will be able to earn 5% annually in a certificate of deposit.

State your assumptions, and provide an analysis of what course of action you should take. State any nonmonetary considerations that may influence your decision.

3-14. Your firm is considering replacing its conventional trucks with turbine-powered vehicles. What information would you like to have in studying the decision? Where would you expect to get the information?

3-15. Explain how you would obtain information needed to study the economics of the following:

   a. Leasing versus purchasing a computer system.
   b. Maintaining equipment with in-house personnel or purchased services.
   c. Keeping versus replacing an old machine system.
   d. Coal-fired steam versus gas turbine generation plant.

3-16. Use the Delphi method to estimate the following:

   a. The ticket price for a 5,000-mile SST one-way trip.
   b. The cost of a year's college education in a state-supported university five years from now.
   c. The time to dig a trench $2' \times 4' \times 10'$ in soft clay using hand tools.
   d. The average annual growth rate for cellular telephone service over the next 10 years.
   e. The price of premium gasoline in 2020.
   f. The average length of time required to change an automobile tire.
   g. The percentage of football (or other) games to be won next year by the team of your choice.

3-17. Suppose that you own a small company that manufactures metal castings for several large automotive companies. Over the past several years, you have found that quarterly new-car sales tend to lag behind the prime interest rate by 3 months. You would like to make a forecast of next quarter's car sales so that the size of your work force can be anticipated. There is a direct relationship between car sales and demand for castings that your company produces. Use the data gathered and displayed on the accompanying chart to:

   a. Calculate a linear regression equation for these data, assuming that the interest rate is the independent variable.
   b. Calculate the correlation coefficient.
   c. Make a forecast of sales for the next quarter based on this quarter's prime interest rate of 7.50%.

Year	Quarter	Interest rate (%)	Next quarter sales ($M)	Year	Quarter	Interest rate (%)	Next quarter sales ($M)
1	1	8.00	$23	4	1	7.00	$25
	2	8.25	17		2	7.50	26
	3	8.50	18		3	7.50	17
	4	8.25	20		4	8.25	20
2	1	7.75	21	5	1	8.75	15
	2	7.25	25		2	8.50	18
	3	7.70	24		3	7.50	22
	4	7.25	29		4	7.00	23
3	1	7.50	24	6	1	7.50	?
	2	7.75	23				
	3	7.25	26				
	4	7.00	30				

3-18. Total operating costs and the corresponding production volumes for a particular process have been found to be as follows:

Operating costs ($M)	Production volume
800	1,000
1,000	1,100
700	900
600	850

a. Calculate the least-squares linear regression line to relate total operating costs as a function of production volume.

b. Estimate the operating costs for a production volume of 950 units.

c. Calculate the coefficient of correlation, and comment on whether this indicates a relatively good or poor fit of the regression line to the data.

3-19. In the packaging department of a large automotive parts distributor, a fairly reliable estimate of packaging and processing costs can be determined by knowing the weight of an order. Thus, weight is a cost driver that accounts for a sizable fraction of the packaging and processing costs at this company. Data for the past 10 orders are given on the accompanying chart.

a. Estimate the a and b coefficients, and write the linear regression equation to fit these data.

b. What is the correlation coefficient (r)?

c. If an order weighs 250 pounds, how much should it cost to package and process it?

Packaging and processing costs ($), y	Weight (lbs.), x
97	230
109	280
88	210
86	190
123	320
114	300
112	280
102	260
107	270
86	190

3-20. Suppose that comparison of Norcar Company sales with many economic indicators shows that sales correlate best with, say, the state's construction volume committed. The nature of the correlation is shown in the following table:

Norcar Company Sales and Construction
Volume Committed

Year	Sales ($ million)	Construction volume committed ($ million)
2005	3	40
2006	2	25
2007	5	50
2008	4	45

a. Determine the least-squares regression line that expresses the correlation between annual sales and construction volume.

b. Determine the correlation coefficient r between the two variables.

c. Obtain a linear extrapolation of sales for a committed construction index of $70 million.

d. What is the danger in using a construction volume outside the original range of $25–$50 million to forecast future sales?

3-21. This is an exercise to illustrate the use of various weighting constants for exponential smoothing forecasting. Suppose the actual sales of a firm were 500 units for year 1 and 600 for year 2. You forecasted it would be 550 units for year 2, and now you wish to forecast for year 3 and beyond.

a. What would be your forecast for year 3 if your smoothing constant, α', was, respectively, 0.1, 0.5, and 0.97?

b. Suppose actual sales turn out to be as follows:

Year	Actual sales (units)
3	700
4	800
5	700
6	600
7	600

What would have been the forecast for each year (4, 5, 6, and 7) using each of the three smoothing constants?

c. Distinguish between the actual results and the forecast for each year by using each of the three smoothing constants. What are your conclusions on the desirability and nondesirability of using a low value of α'?

3-22. Consider the following time-series data of demand for a certain company's product:

Month	Demand (booked orders)	Month	Demand (booked orders)
1	3,009	11	3,387
2	2,641	12	3,138
3	2,934	13	2,908
4	3,239	14	3,512
5	3,490	15	3,291
6	2,569	16	2,804
7	3,205	17	3,096
8	2,561	18	3,106
9	3,047	19	3,195
10	2,607	20	3,605

a. Plot these data on a piece of graph paper.

b. Apply single exponential smoothing to the data when $\alpha' = 0.20$, and make a forecast for $T = 1$ month into the future.

c. Repeat part (b) when $\alpha' = 0.05$ and $T = 3$ months.

3-23. Discuss the principal advantages and disadvantages of the Delphi method of forecasting.

3-24. In your class, attempt to run a Delphi study to determine the price of a compact disc player 3 years from now. Was group consensus affected by conducting two or three rounds of the procedure?

3-25. Try to think of some products that are presently in the early stages of a substitution curve effect. List them and try to estimate when the newer product will take more than half the market.

3-26. How could trend extrapolation be used to forecast future innovations in the aerospace industry? What performance characteristics do you believe are important here?

3-27. Use the factor technique to estimate the cost of installing a local area network in a factory environment having the following characteristics: One large building on a single level will require a total of 3,000 ft of coaxial (broadband) cable to network its six departments. Six network interface units (NIUs) will be required, and a total of 50 taps will have to be made to connect all the anticipated workstations and programmable devices. Two modems are needed in addition to one network manager/analyzer that costs $30,000. The information necessary to make the estimate may be obtained from the worksheet shown next. How accurate do you think such an estimate would be?

Component	Cost-estimating relationship	
1. Interbuilding connections	$100–$150 per foot	× ____ = ____
2. Intrabuilding connections	$20–$50 per foot	× ____ = ____
3. Cable installation	$20 per foot	× ____ = ____
4. Equipment		
a. Broadband		
CATV amplifier	$500–$1500	× ____ = ____
Taps	$17–$20 each	× ____ = ____
Splitters	$5–$15	× ____ = ____
NIUs	$500–$1,000 per port	× ____ = ____
Modems	$1,000 each	× ____ = ____
b. Baseband		
NIUs	$600 per port	× ____ = ____
Repeaters	$1,200–$1,500 each	× ____ = ____
Taps/transceivers	$200–$300 each	× ____ = ____
c. Network manager	$10,000–$30,000	____
Network analyzer	$30,000	____

3-28. A residential builder just finished constructing a 3,000 ft^2 home for $240,000. This cost did not include the lot or utility access fees. A detailed breakout of costs for this job is as follows:

Item	Fraction of finished cost
Lumber and carpentry	0.20
Electrical wiring	0.10
Plumbing	0.14
Concrete and masonry	0.09
Wallboard	0.04
Flooring	0.06
Foundation preparation	0.02
Accessories and appliances	0.05
Heating and air conditioning	0.09
Roofing	0.10
Painting	0.07
Miscellaneous	0.04
	1.00

  a. What is the unit cost for the just-finished home?

  b. If a 4,000-ft^2 home is to be built, estimate the total cost from the answer to part (a) and compare it with the total of estimated item costs based on the breakout given above.

3-29. A 100 kW diesel generator cost $140,000 seven years ago when a certain equipment cost index was arbitrarily set at 100. A similarly designed generator rated at 150 kW is now being proposed, and the cost index is 140. The cost-exponent factor β is 0.7 for this type of equipment.

  a. Determine the estimated cost of the proposed generator by using the appropriate cost-estimating relationship.

  b. Repeat part (a) when $\beta = 0.4$.

3-30. The Neptune Manufacturing Company is considering abandoning its old plant, built 23 years ago, and constructing a new plant that has 50% more square footage. The original cost of the old plant was $300,000, and its capacity, in terms of standardized production units, is 250,000 units per year. Capacity of the proposed plant is to be 500,000 units per year. During the past 23 years, costs of plant construction have risen by an average of 5% per year. If the cost exponent factor is 0.8, what is the estimated cost of the new plant?

3-31. A 60 kW diesel electric set, without a precompressor, cost $32,000 in 1974. A similar design, but using 140 kilowatts, is planned for an isolated installation. The cost-exponent factor $\beta = 0.7$, and the cost index in 1974 was 230. Now the cost index is 350. A precompressor is estimated separately at $1,900 now. Using the exponential costing model, find the estimated total equipment cost now.

3-32. A small plant has been constructed and the costs are known. A new plant is to be estimated using the exponential costing model. Major equipment, costs, and factors are as follows (note $mW = 10^6$ watts):

Equipment	Reference size	Unit reference cost	Cost-exponent factor	New design size
Two boilers	6 mW	$300,000	0.80	10 mW
Two generators	6 mW	400,000	0.60	9 mW
Tank	80,000 gal	106,000	0.66	91,500 gal

If ancillary equipment will cost an additional $200,000, find the cost for the proposed plant.

3-33. Your company is now making a product that has a raw material cost of exactly $0.53 per unit out of a total cost of $1.63 per unit. You are responsible for an analysis of the economics of tripling the present capacity. Should you automatically assume that the raw-materials cost for the added capacity will be $0.53 per unit? Why or why not?

3-34. A rule of thumb sometimes used is that when a unit is being operated at 50% capacity, the maintenance costs will be approximately 75% of the maintenance costs at 100% capacity.

a. Why would the maintenance costs not be 50% of the maintenance costs at 100% capacity?

b. Is it reasonable that maintenance costs at 120% of capacity could be less than 120% of the maintenance costs at 100% capacity? More than 120%? What might cause the difference?

3-35. The mechanical engineering department has a student team that is designing a formula car for national competition this coming spring.

The time required for the team to assemble the first car is 100 hours. Their improvement (or learning rate) is 0.8, which means that as output is doubled their time to assemble a car is reduced by 20%. For instance, $K = 100$ hours for unit 1, and so unit 2 will take 80 hours. Unit 4 will require 80 hours $(0.8) = 64$ hours, and so on.

a. How much time will it take the team to assemble the 10th car?

b. How much total time will be required to assemble the first 10 cars?

c. What is the estimated cumulative average assembly time for the first 10 cars?

3-36. The structural engineering design section within the engineering department of a regional electrical utility corporation has developed several standard designs for a group of similar transmission line towers. The detailed design for each tower is based on one of the standard designs. A transmission line project involving 50 towers has been approved. The estimated number of engineering hours to accomplish the first detailed tower design is 126. Assuming a 95% learning curve, what is your estimate of the number of engineering hours needed to design the eighth tower and to design the last tower in the project?

3-37. You have been asked to estimate the *per unit selling price* of a new line of widgets. Pertinent data are as follows:

Direct labor rate:	$15.00 per hour
Production material:	$375.00 per 100 widgets
Factory overhead:	125% of direct labor
Packing costs:	75% of direct labor
Desired profit:	20% of total manufacturing cost

Past experience has shown that an 80% learning curve applies to the labor required for producing widgets. The time to complete the first widget has been estimated to be 1.76 hours. Use the estimated time to complete the 50th widget as your standard time for the purpose of estimating the unit selling price.

3-38. If 846.2 labor hours are required for the third production unit, and 783.0 labor hours are required for the fifth production unit, determine the learning curve parameter (b).

3-39. You have been asked to estimate the *per unit selling price* of a new model of widgets. Pertinent data are as follows:

Direct labor rate:	$15.00 per hour
Production material:	$625.00 per 100 widgets
Factory overhead:	75% of factory labor cost
Packing costs:	40% of production material cost
Desired profit:	12% of total manufacturing cost

Initial studies have shown that an 85% learning curve applies to the labor required for producing the widgets. The time to complete the first widget has been estimated to be 1.5 hours. Use the estimated time to complete the 50th widget as your standard time for the purpose of estimating the unit selling price.

PART TWO
Basic Capital Investment Analysis: Applications

4
Computations Involving Interest

The key to understanding the material in this and several subsequent chapters is recognizing that *money has time value!* To understand what it means for money to have time value, suppose you are offered the opportunity to receive $1,000 today versus receiving it one year from today. Except in unusual circumstances,[1] you would opt to receive the $1,000 today. Having money sooner is generally preferred to having it later. Why? Because you can invest the money, earn a return on it, and accumulate a greater sum than you would otherwise. By deferring receipt of money, you defer the opportunity for investment.

Just as the spacing of forces is a primary consideration in mechanics, the spacing of cash flows (receipts and disbursements) is important in economic analyses. Often, the timing of the cash flows makes *the difference* in the profitability of an investment.

The mechanism used to express the time value of money is the *interest rate.* Also called the *discount rate* and *opportunity cost rate,* the interest rate expresses the change in the value of money as it moves forward or backward in time. As an illustration, if your time value of money is 10%, then you would be indifferent to receiving the $1,000 today and receiving $1,100 one year later. Stated differently, $1,000 today is *equivalent* to $1,100 one year in the future if the time value of money is 10% or if money *is worth* 10%.

From this we see that people's time values of money can differ. Money might be worth 10% to you, but to another person it might be worth 8% or 12%. What determines

---

[1] Some might be in a financial position such that deferring income is desirable due to income tax implications.

the time value of money? Many factors contribute to the determination of the time value of money. For example, suppose you can receive either $1,000 now or $X one year from today. For what value of X would you be indifferent? Subtracting $1,000 from X and dividing the result by $1,000 provides a decimal estimate of your time value of money.[2]

We have been given estimates of X ranging from $1,100 and $2,000. In fact, some individuals would choose to receive $1,000 today rather than receive $3,000 or $4,000 a year from today. Why? If their survival or that of their family depends on having money for food or medicine today, it might not matter to them what is offered a year from today. Or, perhaps, they don't trust the offeror to deliver the money a year from today. Regardless of the reasons, people have different time values of money— and so do corporations!

Another factor that often arises in considering what value to assign to the time value of money is the rate of inflation. While it is true that the rate of inflation can affect the time value of money, it is also the case that *money has time value in the absence of inflation.* Some have difficulty understanding why the latter is true. If you have difficulty accepting the claim that money has a time value in the absence of inflation, then consider the following. If you needed a place to live while at college, you would think nothing of paying rent for an apartment; indeed, you would expect to pay a rate that more than covers the costs of ownership of the apartment. You expect the owner of the apartment to make a profit on the transaction. Otherwise, why tie money up in apartments and forego the opportunity to invest it elsewhere and earn a return on it?

Likewise, if you do not own a car and need to use one for the weekend, you know that you can rent cars from an auto rental agency. Again, you expect to pay a rate that exceeds the cost of ownership of the car being rented. The same is true for chain saws, computers, offices, land, and a host of other rental items. Furthermore, even in the absence of inflation, you expect to pay rental rates that result in profits for the owners of the rental items.

Now, let's add one more to the list of rental items—money! If you need money, you can rent it from someone or some institution that owns money. In doing so, they will charge you rent for the money. However, they generally call it interest, not rent. And, as with other rental items, even in the absence of inflation, you expect to pay interest rates that will result in a profit for the owner of the money. Likewise, if you own the money, you expect others (including banks) to pay you interest on your money if you loan it to them, rent it to them, or invest it in them, *even in the absence of inflation.* The expectation of inflation normally increases the interest rate one expects to receive (or pay) for the use of the money. Chapter 8 focuses on the consideration of inflation.

4.1 Interest Calculations

Interest calculations may be based on interest rates that are either *simple* or *compound.*

---

[2] The purpose of the exercise is to illustrate the factors contributing to an individual's time value of money. Undoubtedly, different estimates of X would be provided under different circumstances of need, risk, and alternative investment opportunities.

4.1.1 Simple Interest

Whenever the interest charge for any period is based on the principal amount only and not also on any accumulated interest charges, the interest is said to be *simple*. Calculations involving simple interest may be performed utilizing the following formula:

$$I = P \times s \times N,$$

where

I = total amount of interest owed after N periods,

P = amount borrowed (invested),

s = simple interest rate, and

N = number of periods before repayment (withdrawal).

Example 4-1

An individual borrows $1,000 at a simple interest rate of 8% per year and wishes to repay the principal and interest at the end of 4 years. How much must be repaid?

Solution

$$I = P \times s \times N$$
$$= (\$1,000)(0.08)(4)$$
$$= \$320.$$

Therefore, $1,000 + $320 = $1,320 is repaid in 4 years.   ∎

4.1.2 Compound Interest

Whenever the interest charge for any interest period is based on the remaining principal amount plus any accumulated interest charges up to the beginning of that period, the interest is said to be *compound*. To illustrate the effect of compounding, the following example is given:

Example 4-2

An individual borrows $1,000 at a compound interest rate of 8% per year and wishes to repay the principal and interest in 4 years. How much must be repaid?

Solution

Year	Amount owed at beginning of year	Interest charge for year	Amount owed at end of year
1	$1,000.00	$1,000.00 × 0.08 = $ 80.00	$1,080.00
2	1,080.00	1,080.00 × 0.08 = 86.40	1,166.40
3	1,166.40	1,166.40 × 0.08 = 93.31	1,259.71
4	1,259.71	1,259.71 × 0.08 = 100.78	1,360.49

Thus, $1,360.49 is repaid. The difference between this and the $1,320.00 answer in the previous example utilizing simple interest is due to the effect of compounding of interest over the 4 years.   ∎

4.2 Equivalence

Before exploring various compound interest relationships, let's return to the notion of equivalence. Recall that we said that $1,000 is equivalent to $1,100 one year later if money is worth (has a time value of) 10%. Likewise, based on compound interest, $1,000 today is equivalent to $1,210 two years in the future if money is worth 10%. Generalizing this concept to a single sum or a series of money, it can have an infinite range of equivalent values over time, although it can have actual existence at only one point in time. Thus, to have precise meaning, an item of money must be identified in terms of timing as well as amount. For purposes of definition, two amounts of money or series of money at different points in time are said to be equivalent if they are equal to each other at some point in time at a given interest rate.

When money has a nonzero time value, three simple rules apply when performing arithmetic calculations with money:

Rule 1. money cannot be added or subtracted unless it occurs at the same point in time;

Rule 2. to move money forward in time by one time unit, multiply the magnitude of the money by $(1 + i)$, where i is the interest rate that reflects the time value of money; and

Rule 3. to move money backward in time by one time unit, divide the magnitude of the money by $(1 + i)$.

Since compound interest is encountered in practice much more often than simple interest, compound interest will be used throughout this book unless otherwise stated. The balance of the chapter deals with the use of compound interest formulas for equivalence conversions.

4.3 Compound Interest Formulas

A variety of compound interest formulas have been developed to facilitate the comparison of investment alternatives, the determination of loan payments, the determination of returns obtained from investments made, and the performance of other financial calculations. In general, the formulas were derived to facilitate calculations that would be laborious in the absence of computational hardware and software. However, given the widespread use of spreadsheets and sophisticated computers and hand-held calculators, there is less reliance today on compound interest formulas and their tabulated values than in the past.

Due to the existence of computation hardware and software to perform financial analyses, why have we chosen to include in this book compound interest formulas and tabulations of their values? The three principal reasons for doing so are as follows: not everyone has access to the technologies available; the assumptions underlying the software vary among products; and understanding how the formulas are obtained that underlie the software provides important insights into the implications of the values obtained from analysis. (The latter is the compelling reason, for even if everyone had access to the very same software, we would provide coverage of the compound interest formulas to ensure that readers understood what is behind the results obtained from the software.)

NOTATION AND CASH FLOW DIAGRAMS

The following notations are used throughout this book for compound interest calculations:

i = effective interest rate per interest period,

N = number of compounding periods,

P = present sum of money (the equivalent worth of one or more cash flows at a relative point in time called the present),

F = future sum of money (the equivalent worth of one or more cash flows at a relative point in time called the future),

A = end-of-period cash flows (or equivalent end-of-period values) in a uniform series continuing for a specified number of periods, and

G = uniform period-by-period increase or decrease in cash flows or amounts (the arithmetic gradient).

The use of time or cash flow diagrams is strongly recommended for most problems, at least whenever the analyst desires to visualize the cash flow situation. Whenever some distinction between types of cash flows seems desirable, it is recommended to use an upward arrow for a cash inflow and a downward arrow for a cash outflow.

4.3.1 Interest Formulas Relating Present and Future Sums

Figure 4-1 shows a time diagram involving a present single sum P and a future single sum F separated by N periods with interest at $i\%$ per period. Two formulas relative to those sums will be presented.

Find F When Given P

Applying Rule 2, if P dollars are deposited now in an account earning $i\%$ per period, the account will grow to $P(1 + i)$ by the end of one period; by the end of two periods, the account will be $P(1 + i)(1 + i) = P(1 + i)^2$; and by the end of N periods, the account will have grown to a future sum F, as given by

$$F = P(1 + i)^N, \qquad (4\text{-}1)$$

where the quantity $(1 + i)^N$, designated (F/P), is tabled in Appendix A for numerous values of i and N. Symbolically, we shall use the notation

$$F = P(F/P, i\%, N), \qquad (4\text{-}2)$$

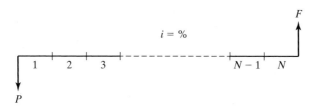

Figure 4-1   Time diagram for single sums.

where the symbol in parentheses denotes the unknown and known, the interest rate, and the number of periods, respectively.

Find P When Given F

Applying Rule 3, from page 102, the reciprocal of the relationship between P and F is given, mathematically as

$$P = F\left(\frac{1}{1+i}\right)^N,$$
(4-3)

where the quantity $1/(1+i)^N$ is tabled in Appendix A. Symbolically,

$$P = F(P/F, i\%, N).$$
(4-4)

4.3.2 Applying Interest Formulas to Cash Flow Series

Figure 4-2 depicts a time diagram involving a present single sum P and single sums $A_1, A_2, \ldots, A_N$ occurring at the end of periods 1, 2, ..., N, respectively. With interest at $i\%$ per period, the *present worth equivalent* of the cash flow series $\{A_1, A_2, \ldots, A_N\}$ can be obtained by summing the present worth equivalents of each individual cash flow in the cash flow series. Hence,

$$P = A_1(P/F, i\%, 1) + A_2(P/F, i\%, 2) + \cdots + A_N(P/F, i\%, N).$$
(4-5)

Similarly, the *future worth equivalent* of the cash flow series is given by the sum of the future worth equivalents of the individual cash flows,

$$F = A_N + A_{N-1}(F/P, i\%, 1) + \cdots + A_1(F/P, i\%, N-1).$$
(4-6)

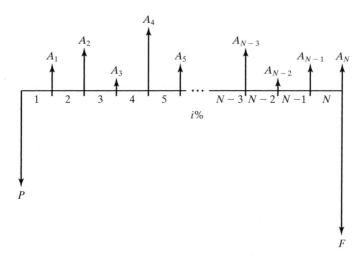

Figure 4-2   Time diagram for a series of cash flows.

In a number of instances, the present worth and future worth equivalents of cash flow series can be obtained in a closed mathematical form by using summation of series relations. In particular, if each cash flow in the series has the same value, A, then the series is referred to as a *uniform series*. If the value of a given cash flow differs from the value of the previous cash flow by a constant amount, G, then the series is referred to as a *gradient series*. When the value of a given cash flow differs from the value of the previous cash flow by a constant percentage, $j\%$, then the series is referred to as a *geometric series*. Closed-form expressions are available for P and F for uniform, gradient, and geometric series and will be presented in subsequent sections. Additional series for which closed-form series can be developed are explored in the problems at the end of this chapter.

4.3.3 Interest Formulas Relating Uniform Series of Payments to Their Present Worth and Future Worth

Figure 4-3 shows a time diagram involving a series of uniform cash flows of amount A occurring at the end of each period for N periods with interest at $i\%$ per period. As depicted in Fig. 4-3, the formulas and tables presented next are derived such that

1. P occurs one interest period before the first A; and
2. F occurs at the same point in time as the last A, and N periods after P.

Four formulas relating A to F and P are now given.

Find F When Given A

If A dollars are deposited at the end of each period for N periods in an account earning $i\%$ per period, the future sum F accrued at the end of the Nth period is

$$F = A[1 + (1 + i) + (1 + i)^2 + \cdots + (1 + i)^{N-1}].$$

It can be shown that this reduces to

$$F = A\left[\frac{(1 + i)^N - 1}{i}\right], \tag{4-7}$$

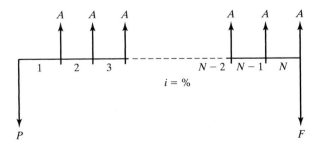

Figure 4-3   Standard time diagram for uniform series.

where the quantity $\{[(1 + i)^N - 1]/i\}$ is tabled in Appendix A. Symbolically,

$$F = A(F/A, i\%, N).\qquad(4-8)$$

Find A When Given F

The reciprocal of the relationship between A and F, from Eq. 4-7, is given mathematically as

$$A = F\left[\frac{i}{(1 + i)^N - 1}\right],\qquad(4-9)$$

where the quantity $\{i/[(1 + i)^N - 1]\}$ is tabled in Appendix A. Symbolically,

$$A = F(A/F, i\%, N).\qquad(4-10)$$

Find P When Given A

If we take the relation

$$F = A\left[\frac{(1 + i)^N - 1}{i}\right]$$

and substitute

$$F = P(1 + i)^N,$$

then we find

$$P = A\left[\frac{(1 + i)^N - 1}{i}\right]\left[\frac{1}{1 + i}\right]^N,$$

which simplifies to

$$P = A\left[\frac{(1 + i)^N - 1}{i(1 + i)^N}\right].\qquad(4-11)$$

The factor in the brackets is tabled in Appendix A. Symbolically,

$$P = A(P/A, i\%, N).\qquad(4-12)$$

Find A When Given P

The reciprocal of the relationship between A and P, from Eq. 4-11, is given mathematically as

$$A = P\left[\frac{i(1 + i)^N}{(1 + i)^N - 1}\right].\qquad(4-13)$$

Again, the factor in brackets is tabled in Appendix A. Symbolically,

$$A = P(A/P, i\%, N).\qquad(4-14)$$

A summary of the formulas and their symbols, together with example problems, is given in Table 4-1. It should be noted that, for all problems in this book involving uniform series, end-of-year payments are assumed unless stated otherwise.

4.3.4 Interest Factor Relationships

The following relationships exist among the six basic interest factors:

$$(P/F, i\%, N) = \frac{1}{(F/P, i\%, N)},\qquad(4-15)$$

$$(A/P, i\%, N) = \frac{1}{(P/A, i\%, N)},\qquad(4-16)$$

TABLE 4-1 Summarization of Discrete Compound Interest Factors and Symbols

To find	Given	Multiply "Given" by factor below	Factor name	Factor functional symbol	Example (answer for $i = 5\%$) (Note: All uniform series problems assume end-of-period payments.)
F	P	$(1+i)^N$	Single sum compound amount	$(F/P,i\%,N)$	A firm borrows $1,000 for 5 years. How much must it repay in a lump sum at the end of the fifth year? *Ans.*: $1,276.30
P	F	$\dfrac{1}{(1+i)^N}$	Single sum present worth	$(P/F,i\%,N)$	A company desires to have $1,000 8 years from now. What amount is needed now to provide for it? *Ans.*: $676.80
P	A	$\dfrac{(1+i)^N - 1}{i(1+i)^N}$	Uniform series present worth	$(P/A,i\%,N)$	How much should be deposited in a fund to provide for 5 annual withdrawals of $100 each? First withdrawal 1 year after deposit. *Ans.*: $432.95
A	P	$\dfrac{i(1+i)^N}{(1+i)^N - 1}$	Capital recovery	$(A/P,i\%,N)$	What is the size of 10 equal annual payments to repay a loan of $1,000? First payment 1 year after receiving loan. *Ans.*: $129.50
F	A	$\dfrac{(1+i)^N - 1}{i}$	Uniform series compound amount	$(F/A,i\%,N)$	If 4 annual deposits of $2,000 each are placed in an account, how much money has accumulated immediately after the last deposit? *Ans.*: $8,620.20
A	F	$\dfrac{i}{(1+i)^N - 1}$	Sinking fund	$(A/F,i\%,N)$	How much should be deposited each year in an account in order to accumulate $10,000 at the time of the fifth annual deposit? *Ans.*: $1,810

Key: i = Interest rate per interest period
N = Number of interest periods

A = Uniform series amount
F = Future worth

P = Present worth

$$(A/F, i\%, N) = \frac{1}{(F/A, i\%, N)},\tag{4-17}$$

$$(A/P, i\%, N) = i\% + (A/F, i\%, N),\tag{4-18}$$

$$(F/A, i\%, N) = (P/A, i\%, N)(F/P, i\%, N),\tag{4-19}$$

$$(P/A, i\%, N) = \sum_{k=1}^{N}(P/F, i\%, k),\tag{4-20}$$

and

$$(F/A, i\%, N) = \sum_{k=0}^{N-1}(F/P, i\%, k).\tag{4-21}$$

4.3.5 Interest Formulas for Uniform Gradient Series

Some economic analysis problems involve receipts or disbursements that are projected to change by a constant amount each period. For example, maintenance and repair expenses on specific equipment may increase by a relatively constant amount of change G each period.

Figure 4-4 is a cash flow diagram of a series of end-of-period disbursements increasing at the constant amount of change, G dollars per period. For convenience in derivation of the formulas, it is assumed that a series of uniform payments of amount G is started at the end of the second period, another series of amount G is started at the end of the third period, and so on. Each of these series terminates at the same time, the end of the Nth period. The future sum (at the of the Nth period) equivalent to the gradient series shown in Fig. 4-4 is

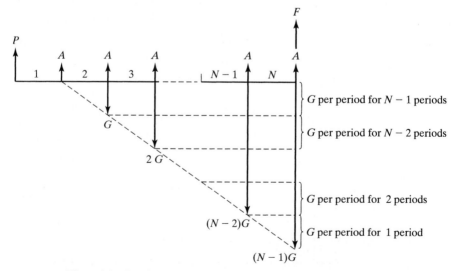

Figure 4-4 Cash flow diagram for uniform gradient of G dollars per period.

$$F = G\left[(F/A, i\%, N-1) + (F/A, i\%, N-2) + \cdots + (F/A, i\%, 2) + (F/A, i\%, 1)\right]$$

$$= \frac{G}{i}\left[(1+i)^{N-1} + (1+i)^{N-2} + \cdots + (1+i)^2 + (1+i) - (N-1)\right]$$

$$= \frac{G}{i}\left[(1+i)^{N-1} + (1+i)^{N-2} + \cdots + (1+i)^2 + (1+i) + 1\right] - \frac{NG}{i}.$$

The expression in the brackets reduces to

$$\frac{(1+i)^N - 1}{i} = (F/A, i\%, N).$$

Hence,

$$F = \frac{G}{i}\left[\frac{(1+i)^N - 1}{i} - N\right]. \tag{4-22}$$

The equivalent uniform annual worth of the gradient series may be found by multiplying the preceding sum of compound amounts by $(A/F, i\%, N)$. Hence,

$$A = F(A/F, i\%, N)$$

$$= \frac{G}{i}\left[\frac{(1+i)^N - 1}{i} - N\right]\left[\frac{i}{(1+i)^N - 1}\right]$$

$$= \frac{G}{i} - \frac{NG}{i}\left[\frac{i}{(1+i)^N - 1}\right] \tag{4.23a}$$

$$= G\left\{\frac{1}{i} - \left[\frac{N}{(1+i)^N - 1}\right]\right\}.$$

The factor in the braces is given in Table A-15 for a wide range of i and N. Symbolically, the relationship to find the uniform series equivalent to the gradient series is

$$A = G(A/G, i\%, N). \tag{4.23b}$$

To find the present worth of a uniform gradient series, by employing the relationship $P = A(P/A, i\%, N)$ and Eq. 4-23a, it can be shown that

$$P = G\left\{\frac{1}{i}\left[\frac{(1+i)^N - 1}{i(1+i)^N} - \frac{N}{(1+i)^N}\right]\right\}. \tag{4.24a}$$

The factor in braces is given in Table A-14 for a wide range of i and N. Symbolically, the relationship is

$$P = G(P/G, i\%, N). \tag{4.24b}$$

4.3.6 Interest Formulas for Geometric Series

Some economic analysis problems involve cash flows that are anticipated to increase over time by a constant percentage. Labor, energy, maintenance, and material costs are examples of items that may increase by a constant $j\%$ each period.

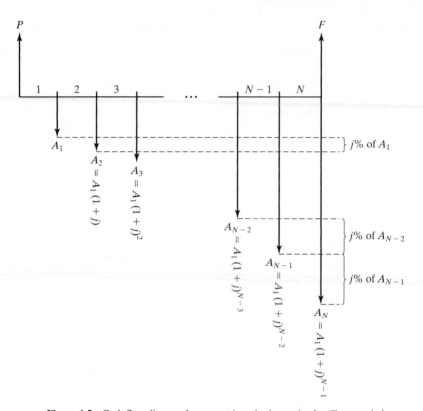

Figure 4-5   Cash flow diagram for geometric series increasing by $j\%$ per period.

Figure 4-5 gives a cash flow diagram of a series of end-of-period disbursements increasing at a constant rate of $j\%$ per period. If A_1 represents the size of the disbursement at the end of period 1, it can be seen that the size of the disbursement at the end of period k, A_k, is equal to $A_1(1 + j)^{k-1}$.

The present sum equivalent to the geometric series shown in Fig. 4-5 is

$$
\begin{aligned}
P &= A_1(P/F, i\%, 1) + A_2(P/F, i\%, 2) + A_3(P/F, i\%, 3) \\
&\quad + \cdots + A_N(P/F, i\%, N) \\
&= A_1(1 + i)^{-1} + A_2(1 + i)^{-2} + A_3(1 + i)^{-3} \\
&\quad + \cdots + A_N(1 + i)^{-N} \\
&= A_1(1 + i)^{-1} + A_1(1 + j)(1 + i)^{-2} + A_1(1 + j)^2(1 + i)^{-3} \\
&\quad + \cdots + A_1(1 + j)^{N-1}(1 + i)^{-N} \\
&= A_1(1 + i)^{-1}[1 + x + x^2 + \cdots + x^{N-1}],
\end{aligned}
\tag{4-25}
$$

where $x = (1 + j)/(1 + i)$. The expression in brackets in Eq. 4-25 reduces to $(1 - x^N)/(1 - x)$ when $x \neq 1$ or $j \neq i$. If $j = i$, then $x = 1$ and the expression in brackets reduces to N, the number of terms in the summation. Hence,

$$P = \begin{cases} A_1(1 + i)^{-1}(1 - x^N)/(1 - x) & j \neq i, \\ A_1 N(1 + i)^{-1} & j = i \end{cases} \tag{4-26}$$

which reduces to

$$P = \begin{cases} \dfrac{A_1[1 - (1 + i)^{-N}(1 + j)^N]}{i - j} & j \neq i, \\ A_1 N(1 + i)^{-1} & j = i \end{cases} \tag{4-27}$$

or

$$P = \begin{cases} \dfrac{A_1[1 - (P/F, i\%, N)(F/P, j\%, N)]}{i - j} & j \neq i. \\ A_1 N(P/F, i\%, 1) & j = i \end{cases} \tag{4-28}$$

By employing the appropriate interest formula from Table 4-1, we can convert the present sum equivalent of a geometric series into a future sum equivalent or a uniform series equivalent. In terms of the basic interest factors, the future sum equivalent is

$$F = \begin{cases} \dfrac{A_1[(F/P, i\%, N) - (F/P, j\%, N)]}{i - j} & j \neq i. \\ A_1 N(F/P, i\%, N - 1) & j = i \end{cases} \tag{4-29}$$

SOLVED PROBLEMS

1. Emma loans Austin $5,000 with interest compounded at a rate of 6% per year. How much will Austin owe Emma if he repays the loan at the end of 5 years?

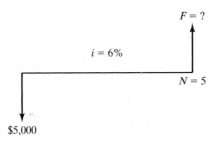

Since the problem is of the form "find F when given P," the formula to use is (4-2)

$$F = P(F/P, \ 6\%, \ 5)$$
$$= \$5,000(1.3382)$$
$$= \$6,691. \qquad \blacksquare$$

2. Charlotte wishes to accumulate $25,000 in a savings account in 10 years. If she wishes to make a single deposit today and the bank pays 6% compounded annually on deposits of this size, how much should Charlotte deposit in the account?

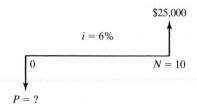

This problem is of the form "find P when given F," and the formula to use is (4-4)

$$P = F(P/F, 6\%, 10)$$
$$= \$25,000(0.5584)$$
$$= \$13,960.$$

∎

3. Sara Beth made annual deposits of $1,500 in an account that paid 5% compounded annually. How much money should have been in the account immediately after her 10th deposit?

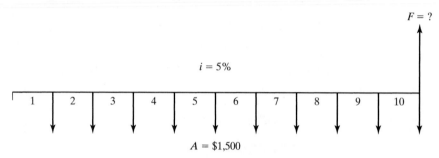

This problem is of the form "find F when given A," and the formula to use is (4-8)

$$F = A(F/A, 5\%, 10)$$
$$= \$1,500(12.5779)$$
$$= \$18,866.85.$$

∎

4. Jason has been making equal annual payments of $2,500 to repay a college loan. He wishes to pay off the loan immediately after having made an annual payment. He has four payments remaining. With an annual compound interest rate of 5%, how much should Jason pay?

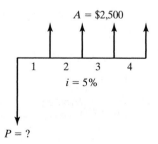

This problem is of the form "find P when given A," and the formula to use is (4-12)

$$P = A(P/A, \ 5\%, \ 4)$$
$$= \$2,500(3.5460)$$
$$= \$8,865.$$

5. Adriana wishes to accumulate $2,000,000 in 40 years. If 40 end-of-year deposits are made into an account that pays interest at a rate of 6% compounded annually, what size deposit is required each year to meet Adriana's stated objective?

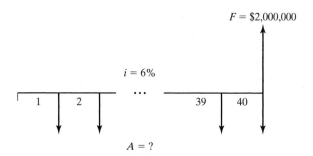

$F = \$2,000,000$

$i = 6\%$

1   2 $\cdots$ 39   40

$A = ?$

This problem is of the form "find A when given F," and the formula to use is (4-10)

$$A = F(A/F, \ 6\%, \ 40)$$
$$= \$2,000,000(0.0065)$$
$$= \$13,000.$$

6. Liu borrows $15,000 at 6% compounded annually. If the loan is repaid in 10 equal annual payments, what will be the size of Liu's payments if the first payment is made one year after borrowing the money?

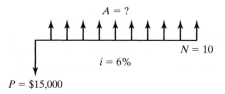

$A = ?$

$N = 10$

$i = 6\%$

$P = \$15,000$

This problem is of the form "find A when given P," and the formula to use is (4-14)

$$A = P(A/P, \ 6\%, \ 10)$$
$$= \$15,000(0.2374)$$
$$= \$3,561.$$

7. It is expected that a machine will incur operating costs of $4,000 the first year and that these costs will increase by $500 each year thereafter for the 10-year life of the machine. If money is worth 10% to the firm, what is the equivalent annual worth of the operating costs?

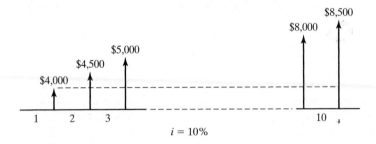

The overall cash flow series consists of the sum of two cash flow series: a uniform annual series of $4,000 and a gradient series of $500. Hence, the equivalent annual worth of the operating costs equals the uniform annual series amount of $4,000 plus the annual worth equivalent of the $500 gradient series, which is of the form "find A when given G" and is represented by formula (4-23b):

$$A = G(A/G,\ 10\%,\ 10)$$
$$= \$500(3.7255)$$
$$= \$1,862.75.$$

Therefore, the equivalent annual worth of the operating costs is

$$A = \$4,000 + \$1,862.75$$
$$= \$5,862.75. \qquad \blacksquare$$

8. Suppose the cash flows are reversed in Problem 7. What 10-year uniform annual series will be equivalent to the new cash flow series if money is worth 10% compounded annually?

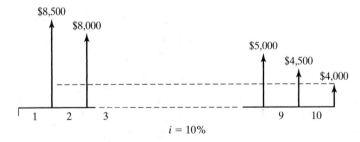

The overall cash flow series consists of the *difference* in a uniform annual series of $8,500 and a gradient series of $500. Hence, the equivalent annual worth of the operating costs equals the uniform annual series amount of $8,500 *minus* the annual worth equivalent of the $500 gradient series

$$A = \$8,500 - G(A/G, 10\%, 10)$$
$$= \$8,500 - \$500(3.7255)$$
$$= \$6,637.25. \qquad \blacksquare$$

9. In Problem 7, suppose the operating cost the first year is $4,000 and that each year thereafter, for the 10-year life of the machine, operating costs increase by 9% per year.

If money is worth 10% per year to the firm, what is the equivalent annual worth of the operating costs?

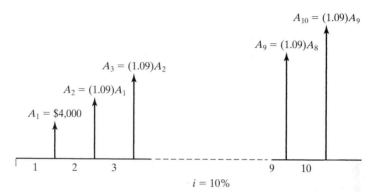

$i = 10\%$

This problem is of the form of a geometric series with $A_1 = \$4{,}000$, $i = 10\%$, $j = 9\%$, and $N = 10$ years. It can be characterized as "find A when given A_1." Formulas (4–29) and (4–10) will be used. Formula (4–29) gives

$$F = A_1[(F/P, 10\%, 10) - (F/P, 9\%, 10)]/(0.10 - 0.09)$$
$$= \$4{,}000[2.5937 - 2.3674]/(0.01)$$
$$= \$90{,}520.$$

Formula (4–10) converts the single sum future amount to a uniform annual series equivalent:

$$A = F(A/F, 10\%, 10)$$
$$= \$90{,}520(0.0627)$$
$$= \$5{,}675.60. \qquad\blacksquare$$

10. With interest at 6% compounded annually, how many years (integer-valued) must a certain amount be invested for it to double in magnitude?

$$(F/P, 6\%, N) = 2.00$$

By inspection of the 6% interest table (Table A-5), $N = 12$ years. $\blacksquare$

(Note: The *Rule of 72* is a popular rule-of-thumb for determining how long it takes for an investment to double in value. According to the *Rule of 72*, the number of years required for an investment to double in value equals the quotient of 72 and the interest rate; e.g., $72 \div 6\% = 12$ years. A quick scan of the $(F/P, i\%, N)$ values in Appendix A validates the accuracy of the *Rule of 72*.)

4.3.7 Deferred Uniform Payments

Frequently, uniform payments occur at points in time such that more than one interest formula must be applied in order to obtain the desired answer. For example, suppose a person borrows \$50,000 to purchase a small business and does not wish to begin repaying the loan until the end of the third year after purchasing the business. With an interest rate of 5% compounded annually, it is desired to determine the amount of the equal annual payment if 15 payments are made.

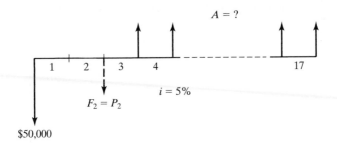

From the cash flow diagram, it is apparent that direct use of the formula "find A when given P" is not possible, since P does not occur one period prior to the first A. However, the problem can be solved logically in two steps. First, the amount of the loan, plus interest, after 2 years is

$$F_2 = P(F/P, 5\%, 2)$$
$$= \$50,000(1.1025) = \$55,125.$$

Note the use of the subscript 2 to denote the point in time. The direct use of $A = P(A/P, i\%, N)$ is now possible, since the amount to be repaid (\$55,125) occurs one period prior to the first loan payment. Therefore, the loan payment size will be

$$A = P_2(A/P, 5\%, 15)$$
$$= \$55,125(0.0963) = \$5,309.$$ ∎

4.4 Spreadsheets

A variety of software packages is available to facilitate the use of spreadsheets in performing economic analyses. Our purpose is not to endorse a particular spreadsheet program or provide sufficient details to ensure your proficiency in their use, but to demonstrate the power of such software in performing economic analyses. Arbitrarily, we have chosen to use Microsoft® Excel 2002.

Consider the spreadsheet given in Figure 4-6. The columns are identified alphabetically and the rows are identified numerically. As noted, we chose to enter in cell B1 the interest rate to be used. It is not necessary to designate a cell for storing the interest rate, but doing so facilitates sensitivity analyses. Either decimal or percentage formats may be used for the interest rate. Cash flow values are entered in column B, beginning with row 4. For illustrative purposes, a cash flow profile is shown that involves 11 equally spaced end-of-period (EOP) cash flows (B4:B14).

To determine the present worth equivalent for the cash flow profile, the NPV (net present value) financial function can be used. To do so, it is necessary to provide values for the interest rate and the cash flows to be incorporated in the calculation. For the example depicted in Figure 4-6, the cash flows are designated as the cells ranging from B5 to B14. The range begins with B5 because the present value obtained using the NPV function occurs one period before the first entry in the range. For the present value to occur at EOP 0, the first entry must occur at EOP 1. As a

	A	B
1	Interest Rate =	(enter interest rate)
2		
3	End of Period	Cash Flow
4	0	A_1
5	1	A_2
6	2	A_3
7	3	A_4
8	4	A_5
9	5	A_6
10	6	A_7
11	7	A_8
12	8	A_9
13	9	A_{10}
14	10	A_{11}
15	P	=npv(B1,B5:B14)+B4
16	A	=pmt(B1,A14,–B15)
17	F	=fv(B1,A14,–B16)

Figure 4-6   Sample spreadsheet.

result, B4, the cash flow occurring at EOP 0, is added to the NPV value obtained. Importantly, the NPV financial function does not interpret a blank cell to have zero value; hence, if the net cash flow for any period is zero, a value of zero must be entered in the appropriate cell.

To determine the annual worth equivalent for the cash flow profile, the PMT (payment) financial function can be used. The PMT function was designed to determine the payment size resulting from borrowing a specified amount at a given interest rate and for a given number of equal sized EOP payments. Hence, the parameters for the PMT function are the interest rate, number of payments, and amount borrowed. Having just calculated the net present value, it can serve as the "amount borrowed." From Figure 4-6, note that the PMT parameter values shown (B1, A14, and B15) correspond to the required parameters. Finally, since the PMT function is designed to determine the amount to be paid back for a loan amount, the loan amount is entered as a negative cash flow; hence, –B15 is entered as the final parameter value.

To determine the future worth equivalent for the cash flow profile, the FV (future value) financial function can be used. The FV function was designed to determine the balance in an account earning a specified interest rate, based on a specified number of equal EOP deposits of a specified amount being made. As shown in Figure 4-6, the FV parameter values shown (B1, A14, and B16) correspond to the required parameters. As with the PMT function, a negative value is assigned to the deposit; hence, –B16 is entered. (Alternatively, the future worth equivalent can be obtained as follows: =FV(B1,A14,–B15). In this case, the FV function determines the future value of a single deposit (B15) that remains in an account paying B1 interest for A14 periods.)

	A	B
1	i =	10.00%
2		
3	EOP	Cash Flow
4	0	−$10,000.00
5	1	$3,000.00
6	2	$3,000.00
7	3	$3,000.00
8	4	$3,000.00
9	5	$3,000.00
10	P	$1,372.36
11	A	$362.03
12	F	$2,210.20

Figure 4-7   Spreadsheet example with uniform series of annual returns.

To gain an appreciation of the power of spreadsheet software, consider Figures 4-7, 4-8, and 4-9. A $10,000 investment is considered, with returns occurring over a 5-year period. When an interest rate of 10% is used with annual returns of $3,000, a present worth equivalent of $1,372.36 results. (The annual worth and future worth equivalents are also computed, as shown.) When the annual return is reduced to $2,500, negative values occur for the present, annual, and future worth equivalents. Likewise, when the interest rate is increased to 16%, negative values occur. Through trial and error, you can determine the value of the annual return that would cause the investment to

	A	B
1	i =	10.00%
2		
3	EOP	Cash Flow
4	0	−$10,000.00
5	1	$2,500.00
6	2	$2,500.00
7	3	$2,500.00
8	4	$2,500.00
9	5	$2,500.00
10	P	−$523.03
11	A	−$137.97
12	F	−$842.35

Figure 4-8   Using a spreadsheet to analyze changes in cash flows.

	A	B
1	$i =$	16.00%
2		
3	EOP	Cash Flow
4	0	−$10,000.00
5	1	$3,000.00
6	2	$3,000.00
7	3	$3,000.00
8	4	$3,000.00
9	5	$3,000.00
10	P	−$177.12
11	A	−$54.09
12	F	−$372.01

Figure 4-9   Using a spreadsheet to analyze changes in the interest rate.

break even or produce present, annual, and future worth equivalents of zero. Likewise, you can determine the break-even interest rate. (As an exercise, determine the break-even annual return and break-even interest rate using a spreadsheet program. You will find that the break-even annual return is $2,637,975, and the break-even interest rate is 15.2382%. More attention will be given to the break-even interest rate in Chapters 5 and 6.)

The investment problem considered in Figures 4-7, 4-8, and 4-9 is quite simple and is easily solved using the equations provided in this chapter. For example, the present worth equivalent of five $3,000 cash flows at 10% interest equals $3,000(P/A, 10%, 5) = $3,000(3.7908) = $11,372.40. Hence, the net present worth for the $10,000 investment is $1,372.40. (The 4¢ difference compared to the P in Figure 4-7 is due to round-off error in the interest tables.) Similarly, the break-even annual return equals $10,000/(P/A, 10%, 5), which equals $2,637.97. (Again, the difference in results obtained by using the interest tables and by using Excel is from tabular round-off error.)

The power of spreadsheet software lies in its ability to handle large complex problems. Although neither very large nor complex, the investment depicted in Figure 4-10 is not easily solved by using conventional methods. However, spreadsheet software can solve it as quickly as the simplest examples.

Figures 4-11, 4-12, and 4-13 provide a framework for gradient series investments and depict increasing and decreasing gradient series, respectively. Figures 4-14, 4-15, and 4-16 provide a framework for geometric series investments and depict increasing and decreasing geometric series, respectively.

	A	B
1	$i =$	10.00%
2		
3	EOP	Cash Flow
4	0	–$50,000.00
5	1	–$5,550.00
6	2	$12,250.00
7	3	$13,500.00
8	4	$12,850.00
9	5	$11,100.00
10	6	$9,985.00
11	7	$8,250.00
12	8	$7,850.00
13	9	$7,250.00
14	10	$6,550.00
15	P	$22.14
16	A	$3.60
17	F	$57.42

Figure 4-10   Multiyear investment without a discernable pattern of annual returns.

	A	B
1	$i =$	(enter interest rate)
2	$G =$	(enter gradient size)
3	EOP	Cash Flow
4	0	A1
5	1	A2
6	2	=B2+B5
7	3	=B2+B6
8	4	=B2+B7
9	5	=B2+B8
10	6	=B2+B9
11	7	=B2+B10
12	8	=B2+B11
13	9	=B2+B12
14	10	=B2+B13
15	P	=npv(B1,B5:B14)+B4
16	A	=pmt(B1,A14,–B15)
17	F	=fv(B1,A14,–B16)

Figure 4-11   Spreadsheet framework for gradient series of annual returns.

	A	B
1	$i =$	10.00%
2	$G =$	$500.00
3	EOP	Cash Flow
4	0	–$10,000.00
5	1	$2,500.00
6	2	$3,000.00
7	3	$3,500.00
8	4	$4,000.00
9	5	$4,500.00
15	P	$2,907.87
16	A	$767.09
17	F	$4,683.15

Figure 4-12   Increasing gradient series spreadsheet example.

	A	B
1	$i =$	10.00%
2	$G =$	–$500.00
3	EOP	Cash Flow
4	0	–$10,000.00
5	1	$4,500.00
6	2	$4,000.00
7	3	$3,500.00
8	4	$3,000.00
9	5	$2,500.00
15	P	$3,627.64
16	A	$956.96
17	F	$5,842.35

Figure 4-13   Decreasing gradient series spreadsheet example.

	A	B
1	*i* =	(enter interest rate)
2	*j* =	(enter geometric rate)
3	EOP	Cash Flow
4	0	A1
5	1	A2
6	2	=(1+B2)*B5
7	3	=(1+B2)*B6
8	4	=(1+B2)*B7
9	5	=(1+B2)*B8
10	6	=(1+B2)*B9
11	7	=(1+B2)*B10
12	8	=(1+B2)*B11
13	9	=(1+B2)*B12
14	10	=(1+B2)*B13
15	*P*	=npv(B1,B5:B14)+B4
16	*A*	=pmt(B1,A14,-B15)
17	*F*	=fv(B1,A14,-B16)

Figure 4-14   Spreadsheet framework for geometric series of annual returns.

	A	B
1	*i* =	10.00%
2	*j* =	17.00%
3	EOP	Cash Flow
4	0	–$10,000.00
5	1	$2,500.00
6	2	$2,925.00
7	3	$3,422.25
8	4	$4,004.03
9	5	$4,684.72
15	*P*	$2,904.92
16	*A*	$766.31
17	*F*	$4,678.40

Figure 4-15   Increasing geometric series spreadsheet example.

	A	B
1	$i =$	10.00%
2	$j =$	−10.00%
3	EOP	Cash Flow
4	0	−$10,000.00
5	1	$4,500.00
6	2	$4,050.00
7	3	$3,645.00
8	4	$3,280.50
9	5	$2,952.45
15	P	$4,250.42
16	A	$1,121.25
17	F	$6,845.35

Figure 4-16   Decreasing geometric series spreadsheet example.

4.5 Compounding Frequency: Nominal and Effective Rates

In most economic studies, interest is accounted for as if compounding occurs once a year. In practice, the interest accumulation may take place more frequently, so it is important to note the effects of compounding frequency and to treat properly those problems where the assumption of annual compounding is not appropriate.

As an example, an interest rate may be stated as 12% compounded quarterly. In this case, the 12% is understood to be an annual rate, and is called the *nominal interest rate*. The number of quarterly compounding periods in a year is four. Hence, the interest rate per interest period is 12% ÷ 4 = 3% per quarter. The *effective interest rate* is the exact annual rate that takes into account the compounding that occurs within the year. The following formula may be used to calculate the effective interest rate:

$$\text{Effective rate} = (1 + r/M)^M - 1, \qquad (4\text{-}30)$$

where

$$M = \text{number of interest periods per year,}$$

and

$$r = \text{nominal interest rate.}$$

Recall that $(1 + r/M)^M$ is the single-sum compound amount factor; the effective interest rate may be determined directly from the interest tables using the relation

$$\text{Effective rate} = (F/P, r/M\%, M) - 1. \qquad (4\text{-}31)$$

Hence, for our example, the effective rate is $(F/P, 3\%, 4) - 1 = 12.55\%$.

The effective interest rate can be determined for the case of an infinite number of compounding periods from Eq. 4-30 by letting M, the number of interest periods per year, become infinitely large. Such a condition is termed *continuous compounding*. Thus, the effective rate of interest with continuous compounding is $e^r - 1$, where e is the base of the Naperian or natural logarithm and equals 2.7183. As an example, the effective rate of 12% compounded continuously is $e^{0.12} - 1 = 12.75\%$.

TABLE 4-2   Impact of Compounding Frequency upon Effective Interest Rate

		For a nominal rate of 12%	
Frequency of compounding	No. of compounding periods per year	Interest rate per period	Effective rate
Annual	1	12%	12.000%
Semiannual	2	6	12.360%
Quarterly	4	3	12.551%
Monthly	12	1	12.683%
Weekly	52	0.23077	12.734%
Daily	365	0.03288	12.747%
Continuously	$\rightarrow \infty$	$\rightarrow 0$	12.750%

The effect of compounding frequency on the effective interest rate for a nominal rate of 12% is given in Table 4-2.

4.6 Continuous Compounding Interest Formulas

In some instances, particularly when payments occur rather frequently within periods rather than at the beginning or end of periods, the additional theoretical accuracy of continuous compounding may be significant. Continuous compounding means that the interest or profit growth is proportional to the amount of total principal and interest at each instant.

To find the future worth of a present single sum under continuous compounding, we let N denote the number of years involved and substitute for i the effective interest rate for continuous compounding. Thus,

$$
\begin{aligned}
F &= P(1 + i_{\text{eff}})^N \\
  &= P[1 + (e^r - 1)]^N \\
  &= Pe^{rN}.
\end{aligned}
\tag{4-32}
$$

To find the present worth of a single sum under continuous compounding, the preceding equation can be transposed to

$$
P = Fe^{-rN}.
\tag{4-33}
$$

As an example, the future worth of a present amount of $10,000 at 10% nominal interest compounded continuously for 5 years is $10,000e^{0.10(5)} = \$16,487.21$. The effective rate of interest in this case can be calculated as $e^r - 1 = e^{0.10} - 1 = 10.517\%$. We can find that the future worth of the same $10,000 at effective interest of 10.517% compounded at the end of each year for 5 years is

$$
F = P(F/P, 10.517\%, 5) = \$10,000(1 + 0.10517)^5 = \$16,487.14.
$$

This illustrates the general principle that continuous compounding at a given nominal interest rate is equivalent to discrete annual compounding at the corresponding effective rate for continuous compounding. Hence, by letting i equal $e^r - 1$ and letting N be expressed in years, the discrete interest factors can be used to deal directly with continuous compounding. The continuous compounding interest factors for discrete payments are given by the first six entries in Table 4-3. Notice that the interest rate used in Table 4-3 is the nominal interest rate r. To distinguish between discrete and continuous compounding, the interest rate is underlined in the appropriate symbolic representation of the interest factor (e.g., $(F/P, \underline{r}\%, N)$).

Tables of continuous compounding interest factors are provided in Appendix B (e.g., $(F/P, \underline{10}\%, 5) = 1.6487$ in Table B-1).

TABLE 4-3   Summary of Continuous Compounding Interest Factors and Symbols

To find	Given	Multiply "given" by factor below	Factor name	Factor functional symbol
	Discrete payments			
F	P	e^{rN}	Continuous compounding compound amount factor (Discrete, single sum)	$(F/P,\underline{r}\%,N)$
P	F	e^{-rN}	Continuous compounding present worth factor (Discrete, single sum)	$(P/F,\underline{r}\%,N)$
P	A	$\dfrac{e^{rN} - 1}{e^{rN}(e^{r} - 1)}$	Continuous compounding present worth factor (Discrete, uniform series)	$(P/A,\underline{r}\%,N)$
A	P	$\dfrac{e^{rN}(e^{r} - 1)}{e^{rN} - 1}$	Continuous compounding capital recovery factor (Discrete, uniform series)	$(A/P,\underline{r}\%,N)$
F	A	$\dfrac{e^{rN} - 1}{e^{r} - 1}$	Continuous compounding compound amount factor (Discrete, uniform series)	$(F/A,\underline{r}\%,N)$
A	F	$\dfrac{e^{r} - 1}{e^{rN} - 1}$	Continuous compounding sinking fund factor (Discrete, uniform series)	$(A/F,\underline{r}\%,N)$
	Continuous payments			
P	$\bar{A}$	$\dfrac{e^{rN} - 1}{re^{rN}}$	Continuous compounding present worth factor (Continuous, uniform flow)	$(P/\bar{A},\underline{r}\%,N)$
$\bar{A}$	P	$\dfrac{re^{rN}}{e^{rN} - 1}$	Continuous compounding capital recovery factor (Continuous, uniform flow)	$(\bar{A}/P,\underline{r}\%,N)$
F	$\bar{A}$	$\dfrac{e^{rN} - 1}{r}$	Continuous compounding compound amount factor (Continuous, uniform flow)	$(F/\bar{A},\underline{r}\%,N)$
$\bar{A}$	F	$\dfrac{r}{e^{rN} - 1}$	Continuous compounding sinking fund factor (Continuous, uniform flow)	$(\bar{A}/F,\underline{r}\%,N)$

4.6.1 Continuous Payments throughout the Year

In some cases, money is disbursed uniformly throughout the year. Consequently, rather than depicting a cash flow of, say, $1,000 occurring at the end of the year, it is assumed that the $1,000 is spread uniformly over the year, as depicted in Fig. 4-17. To distinguish the discrete cash flow from the continuous cash flow, a bar above the symbol is used. Thus, $\bar{A}_k$ will denote the continuous cash flow spread uniformly over period k.

Suppose a total of $\bar{A}_1$ dollars flows uniformly and continuously throughout period 1. Divide $\bar{A}_1$ into M equal amounts to be deposited at equally spaced points in time during a year. The interest rate per period is r/M, and the future worth of the series of M equal amounts is

$$F = \frac{\bar{A}_1}{M} \, (F/A, r/M\%, M),$$

which reduces to

$$F = \frac{\bar{A}_1}{r} \left[\left(1 + \frac{r}{M} \right)^M - 1 \right].$$

Letting M approach infinity yields

$$F = \bar{A}_1 \frac{(e^r - 1)}{r}. \tag{4-34}$$

Hence, a continuous and uniform flow of $\bar{A}_k$ over year k is equivalent to a single discrete sum of $\bar{A}_k(e^r - 1)/r$ at the end of year k. Employing the previous convention of denoting by $\bar{A}_k$ the value of a single sum of money occurring at the end of year k, it is evident that a set of continuous cash flows $\{\bar{A}_1, \bar{A}_2, \ldots, \bar{A}_N\}$ spread uniformly over years 1 through N, respectively, is equivalent to a set of discrete cash flows $\{A_1, A_2, \ldots, A_N\}$ occurring at the end of years 1 through N, respectively, with $A_k = \bar{A}_k(e^r - 1)/r$.

In order to obtain the appropriate continuous compounding interest formulas for continuous flows of money, substitute $\bar{A}_k(e^r - 1)/r$ for A_k in the equation for continuous compounding of discretely spaced cash flows. The resulting interest formulas are summarized in Table 4-3. Tables in Appendix B provide factors for continuous compounding for $r = 10\%$ and 20%.

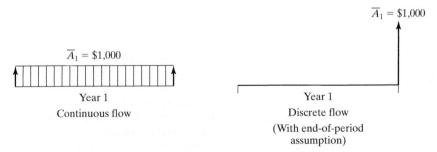

Year 1
Continuous flow

Year 1
Discrete flow
(With end-of-period assumption)

Figure 4-17 Comparison of continuous and discrete flows.

SOLVED PROBLEMS

1. A person makes six end-of-year deposits of $1,000 in an account paying 5% compounded annually. If the accumulated fund is withdrawn 4 years after the last deposit, how much money will be withdrawn?

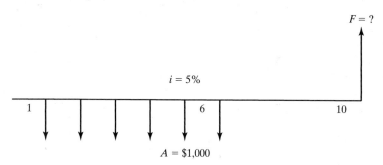

Since F does not occur at the time of the last A, it is necessary that the solution proceed in two steps. The amount of money in the account at the time of the last deposit may be computed as

$$F_6 = A(F/A, 5\%, 6)$$
$$= \$1,000(6.8019) = \$6,802.$$

The problem now is to find F_{10}, given $F_6 = P_6 = \$6,802$:

$$F_{10} = P_6(F/P, 5\%, 4)$$
$$= \$6,802(1.2155) = \$8,268 \qquad \blacksquare$$

2. Given the payments shown in the following cash flow diagram, what is the equivalent worth in 2010 with interest at 6%?

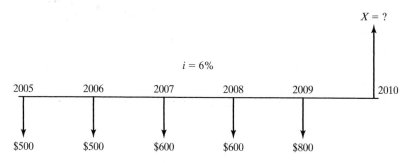

$$X = [\$500(F/A, 6\%, 5) + \$100(F/A, 6\%, 3) + \$200](F/P, 6\%, 1)$$
$$= [\$500(5.6371) + \$100(3.1836) + \$200](1.060) = \$3,537. \qquad \blacksquare$$

3. Money is to be invested for a child's college expenses. Annual deposits of $2,000 are made in a fund that pays 5% compounded annually. If the first deposit is made on the child's 5th birthday and the last on the child's 15th birthday, what is the size of four equal withdrawals on the child's 18th, 19th, 20th, and 21st birthdays that will just deplete the account?

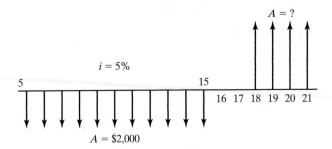

Amount in fund at $t = 15$ is

$$F_{15} = A(F/A, 5\%, 11)$$
$$= \$2,000(14.2068) = \$28,414.$$

Amount in fund at $t = 17$ is

$$F_{17} = P_{15}(F/P, 5\%, 2)$$
$$= \$28,414(1.1025) = \$31,326.$$

Amount of withdrawals is

$$A = P_{17}(A/P, 5\%, 4)$$
$$= \$31,326(0.2820) = \$8,834.$$ ∎

4. A college student borrows money in her senior year to buy a car. She defers payments for 6 months and makes 36 beginning-of-month payments thereafter. If the original note is for \$24,000 and interest is $\frac{1}{2}\%$ per month on the unpaid balance, how much will her payments be?

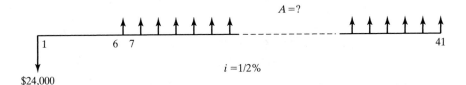

Amount owed at $t = 5$ is

$$F_5 = P_0 (F/P, \tfrac{1}{2}\%, 5)$$
$$= \$24,000(1.0253) = \$24,607.20.$$

Amount of monthly payments is

$$A = P_5(A/P, \tfrac{1}{2}\%, 36)$$
$$= \$24,607.20 \left[\frac{i(1+i)^N}{(1+i)^N - 1} \right]$$
$$= \$24,607.20 \left[\frac{(0.005)(1.005)^{36}}{(1.005)^{36} - 1} \right] = \$748.60$$

The problem can be solved by using Excel by making the following entry in a cell:

$$=\text{PMT}(0.5\%,36, - \text{FV}(0.5\%,5,, - 24000))$$

5. A \$15,000 investment is to be made with anticipated annual returns as shown in the next spreadsheet. If the investor's time value of money is 10%, what should be entered in cells B11, B12, and B13 to obtain present worth, annual worth, and future worth equivalents for the investment?

	A	B
1	EOP	Cash Flow
2	0	–\$15,000.00
3	1	\$2,000.00
4	2	\$2,500.00
5	3	\$3,000.00
6	4	\$3,500.00
7	5	\$4,000.00
8	6	\$4,000.00
9	7	\$4,000.00
10	8	\$4,000.00
11	P	
12	A	
13	F	

Notice that a cell is not designated for the interest rate. Therefore, the interest rate must be entered specifically in the NPV, PMT, and FV financial functions. The following entries in the designated cells will yield the results for *P, A,* and *F*:

B11=NPV(0.1,B3:B10) + B2

B12=PMT(0.1,A10, – B11) or =PMT(10%,8, – B11)

B13=FV(0.1,A10, – B12) or =FV(10%,8, – B11)

Making the designated entries yields P = \$2,189.02, A = \$410.32, and F = \$4,692.38. ∎

6. What is the effective interest rate for 4.75% compounded annually and 4.60% compounded quarterly?

$$\text{Effective rate} = (1 + r/M)^M - 1$$
$$= (1 + 0.0475)^1 - 1$$
$$= 4.75\%$$
$$\text{Effective rate} = (1 + r/M)^M - 1$$
$$= \left[1 + \frac{0.046}{4}\right]^4 - 1$$
$$= 4.68\%$$
∎

7. A loan company advertises that it will loan $1,000 to be repaid in 30 monthly install-ments of $44.60. What is the effective interest rate?

$$A = P(A/P, i\%, 30)$$

$$\frac{\$44.60}{\$1,000.00} = (A/P, i\%, 30) = 0.0446$$

By inspection (with interpolation in tables), $i = 2\%$, and

$$\text{Effective rate} = (F/P, 2\%, 12) - 1 = 0.2682 = 26.82\%. \qquad \blacksquare$$

8. Annual deposits of $1,000 are made in an account that pays 4% compounded quarterly. How much money should be in the account immediately after the fifth deposit?

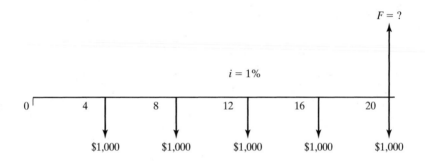

$$\text{Effective rate} = (F/P, 1\%, 4) - 1$$

$$= 4.06\%$$

$$F = A(F/A, 4.06\%, 5)$$

$$= \$1,000 \left[\frac{(1 + 0.0406)^5 - 1}{0.0406} \right] = \$5,423$$

or

$$A = \$1,000(A/F, 1\%, 4)$$

$$= \$1,000(0.2463) = \$246.30$$

$$F_{20} = \$246.30(F/A, 1\%, 20)$$

$$= 246.30(22.019) = \$5,423$$

An alternate solution method is to treat the five annual deposits as single sums of money. Therefore,

$$F = \$1,000[(F/P, 1\%, 16) + (F/P, 1\%, 12) + (F/P, 1\%, 8) + (F/P, 1\%, 4) + 1]$$

$$= \$1,000[1.1726 + 1.1268 + 1.0829 + 1.0406 + 1.000] = \$5,423. \qquad \blacksquare$$

9. An individual approaches the Loan Shark Agency for $1,000 to be repaid in 24 monthly installments. The agency advertises interest at $1\frac{1}{2}\%$ per month. They proceed to calculate the size of his payment in the following manner:

$$\begin{array}{lr}
\text{Amount requested:} & \$1,000 \\
\text{Credit investigation:} & 25 \\
\text{Credit risk insurance:} & \underline{5} \\
\text{Total:} & \$1,030
\end{array}$$

Interest: $(1030)(24)(0.015) = \$371$

Total owed: $\$1,030 + \$371 = \$1,401$

Payment: $\dfrac{\$1,401}{24} = \$58.50$

What effective interest rate is the individual paying?

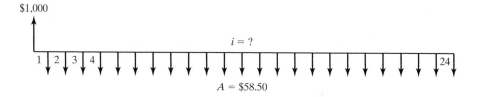

$\$1,000$

$i = ?$

$A = \$58.50$

$$A = P(A/P, i\%, 24)$$
$$\$58.50 = \$1,000(A/P, i\%, 24)$$

By interpolation in tables, $i \cong 2.9\%$ per month, and

Effective rate $\cong (F/P, 2.9\%, 12) - 1 \cong 1.41 - 1 \cong 41\%$ ■

10. What is the present worth of $100,000 in 10 years hence if interest is (a) 15% compounded annually? (b) 15% compounded continuously?

a. $P = F(P/F, 15\%, 10) = \$100,000(0.2472) = \$24,720$

b. $P = F(P/F, \underline{15}\%, 10) = \$100,000(0.2231) = \$22,310.$ ■

11. A firm spends $100,000 per year on materials, with the cost spread uniformly over each year. Annual rental payments for building and equipment total $300,000 per year, with the payments being made at the beginning of each year.

Using an interest rate of 10% compounded continuously, what is the present worth equivalent for 10 years of activity?

$$P = \overline{A}(P/\overline{A}, \underline{10}\%, 10) + [A(P/A, \underline{10}\%, 9) + A]$$
$$= \$100,000(6.3212) + \$300,000(5.6425) + \$300,000 = \$2,624,870$$

or

$$P = \frac{\$100{,}000(e^{0.10} - 1)}{0.10}(P/A, \underline{10}\%, 10) + (\$300{,}000(5.6425) + \$300{,}000)$$

$$= \$105{,}170.92(6.0104) + (\$300{,}000(5.6425) + \$300{,}000) = \$2{,}624{,}869 \quad \blacksquare$$

PROBLEMS[3]

4-1. If an investment earns annual compound interest of 6%, what is the smallest integer-valued number of years money must be invested to triple in value?

4-2. What is the smallest integer-valued annual compound interest that will result in an investment tripling in value in less than or equal to 20 years?

4-3. How much will a $5,000 investment today be worth in 10 years if it earns 8% simple interest? 8% annual compound interest?

4-4. What deposit today is required for it to be worth $50,000 in 25 years if it earns 8% annual compound interest?

4-5. What equal monthly investment is required over a period of 360 months to achieve a balance of $2,000,000 in an investment account that pays monthly interest of $1/2\%$?

4-6. How many monthly payments are necessary to repay a loan of $15,000 with an interest rate of 1% per month and end-of-month payments of $250?

4-7. What annual interest rate makes $5,000 today equivalent to $15,000 in 10 years?

4-8. $3,000 is invested annually at 5% annual compound interest. What is its value after 25 years? After 30 years? What is its value if the investment earns 6% interest? From this problem, what conclusions can be drawn regarding the impact of the interest earned versus the duration of the investment? (Hint: Duration is an exponent!)

4-9. $50,000 is borrowed at 6% annual compound interest. What size equal annual payment is required if the first of 10 payments is made one year after receiving the $50,000? What if the first payment is not made until 5 years after receipt?

4-10. Using an 8% annual compound interest rate, what investment today is needed in order to withdraw $2,000 annually for 10 years? Assume the first withdrawal occurs in one year. What if the first withdrawal does not occur for 5 years?

4-11. If money is worth more than 0% to you, would you rather receive $10,000/year for 5 years or receive $5,000/year for 10 years? What is your preference if you must pay these amounts, rather than receive them?

4-12. Suppose you deposit $5,000 in an account at the end of 2005, $4,500 at the end of 2006, $4,000 at the end of 2007, $3,500 at the end of 2008, $3,000 at the end of 2009, and $2,500 at the end of 2010. If the account pays interest of 6% compounded annually, how much will be in the account immediately after the last deposit?

4-13. In Problem 4-12, suppose the account pays interest of 5% annual interest during 2006, 5.5% annual interest during 2007, 6% annual interest during 2008, 6.5% interest during 2009, and 7% annual interest during 2010. How much will be in the account immediately after the last deposit?

4-14. As shown next, suppose you make annual deposits of $3,000 in an account that pays annual compound interest of 6% the first 2 years, 5% the next 2 years, 4% the next 2 years, and 3% the next 2 years. How much will be in the account after the ninth deposit?

---

[3] Unless stated otherwise, assume discrete compounding and end-of-period payments.

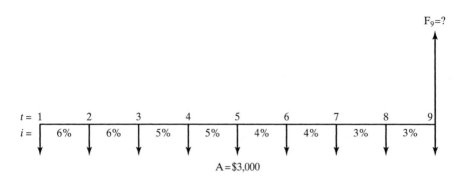

4-15. In Problem 4-14, suppose the annual compound interest rate is 4.5% over the investment period. How much will be in the account after the ninth deposit?

4-16. In 1970, first-class postage for a 1-ounce envelope was 6¢. In 2002, a first-class stamp for the same envelope cost 37¢. What annual compound increase in the cost of first-class postage was experienced during the 32-year period?

4-17. Julie borrowed $10,000 at 6% annual compound interest. She agreed to repay the loan with five equal annual payments. How much of each payment is interest, and how much reduces the unpaid principal balance on the loan?

4-18. Suppose you borrow $15,000 at 8% annual compound interest and agree to pay it off with 10 equal annual payments, with the first payment occurring 1 year after receiving the $15,000. What will be the size of the annual payments? How much of the second payment will go toward reducing the principal owed on the loan?

4-19. In Problem 4-18, suppose the first payment is not made until 5 years after receipt of the $15,000. What will be the size of the annual payments? How much of the second payment will reduce the principal?

4-20. When Juan was 25, he decided to begin planning for retirement in 40 years. Beginning with his 26th birthday, he invested $5,000 annually in an account that paid annual compound interest of 6%. How much was in the account immediately after his 40th deposit?

4-21. In Problem 4-20, how much would be in the account if every deposit was $1,000 larger than the previous deposit?

4-22. In Problem 4-20, how much would be in the account if every deposit was 8% larger than the previous deposit?

4-23. In Problem 4-20, if Juan decided to wait 10 years before investing for retirement, how much would he have to invest annually to have the same account balance on his 65th birthday?

4-24. In Problem 4-21, if Juan decided to wait 10 years before investing for retirement, how much would he have to invest on his 36th birthday to have the same account balance on his 65th birthday?

4-25. In Problem 4-22, if Juan decided to wait 10 years before investing for retirement, how much would he have to invest on his 36th birthday to have the same account balance on his 65th birthday?

4-26. What uniform annual investment is required in Problem 4-20 to achieve the same account balance as results in Problem 4-21?

4-27. What uniform annual investment is required in Problem 4-20 to achieve the same account balance as results in Problem 4-22?

4-28. What monthly compound interest rate would result in $1,000 monthly deposits having an accumulated value of $100,000 after 60 deposits? What would be the effective annual rate of return on the investment?

4-29. Maria purchased a refrigerator for $1,000. The store financed the refrigerator by charging 0.5% monthly interest on the unpaid balance. If the refrigerator is paid for with 30 equal end-of-month payments, what will be the size of the monthly payments? If the first payment is not made until 1 year after the purchase, what will be the size of the monthly payments?

4-30. A $100,000 investment is made over a 10-year period. A return of $23,000 occurs at the end of the first year. Each successive year yields a return that is $2,000 less than the previous year's return. If money is worth 8%, what is the equivalent present worth for the investment? What is the equivalent annual worth for the investment? What is the equivalent future worth for the investment? (Solve manually and then solve by using Excel or another spreadsheet software program.)

4-31. A $100,000 investment is made over a 10-year period. A return of $23,000 occurs at the end of the first year. Each successive year yields a return that is 12% less than the previous year's return. If money is worth 8%, what is the equivalent present worth for the investment? What is the equivalent annual worth for the investment? What is the equivalent future worth for the investment? (Solve manually and then solve by using Excel or another spreadsheet software program.)

4-32. A $75,000 investment is to be made with anticipated annual returns, as shown in the accompanying table. Money is worth 8% to the investor. What must be entered in cells B13, B14, and B15 to obtain present worth, annual worth, and future worth values?

	A	B
1	EOP	Cash Flow
2	0	-$75,000.00
3	1	$9,000.0
4	2	$10,000.00
5	3	$11,000.00
6	4	$12,000.00
7	5	$13,000.00
8	6	$13,000.00
9	7	$12,000.00
10	8	$11,000.00
11	9	$10,000.00
12	10	$9,000.00
13	P	
14	A	
15	F	

4-33. In Problem 4-30, if the annual returns are a geometric series, what annual percentage decrease is required to yield roughly the same present worth, annual worth, and future worth equivalents for the investment? Solve by using Excel or another spreadsheet software program. (Give your answer to the fifth decimal place (e.g., 8.35623%).)

4-34. What interest rate makes $2,500 today equivalent to five annual amounts of $1,000, with the first $1,000 cash flow occurring 1 year in the future?

4-35. A $60,000 investment is to be made with a return of $5,000 after 1 year. Thereafter, each year's annual return will be $1,000 greater than the previous year's return. Complete the spreadsheet and show the entries for cells B15, B16, and B17 to obtain present worth, annual worth, and future worth equivalents for the investment.

	A	B
1	i=	8.00%
2	G=	$1,000.00
3	EOP	Cash Flow
4	0	–$60,000.00
5	1	$5,000.00
6	2	
7	3	
8	4	
9	5	
10	6	
11	7	
12	8	
13	9	
14	10	
15	P	
16	A	
17	F	

4-36. What is the effective annual interest rate for 10% compounded (a) semiannually? (b) quarterly? (c) monthly? (d) weekly? (e) daily? (f) hourly? (g) every minute? (h) continuously? (Give your answer to the eighth decimal place (e.g., 10.51709181%).)

4-37. What amount of money today is equivalent to $100,000 in 25 years with an interest rate of 6% compounded (a) annually? (b) semiannually? (c) quarterly? (d) monthly? (e) continuously?

4-38. What is the effective interest rate for 8% compounded (a) annually? (b) semiannually? (c) quarterly? (d) monthly? (e) daily? (f) continuously?

4-39. What is the present worth of five equal annual expenditures of $2,500 based on an interest rate of 9.53% compounded continuously?

4-40. Over a 5-year period, a firm spends $35,000 per year for supplies. The supplies are consumed uniformly throughout the year. For purposes of analysis, they can be considered to flow continuously during the year. Based on an interest rate of 10% compounded continuously, what single sum of money at the beginning of the year is equivalent to the

continuous flow during the year? What single sum of money at the end of the year is
equivalent to the continuous flow during the year?

4-41. What is the present worth of operating expenditures of $150,000 occurring continu-
ously throughout a year if the firm's time value of money is equivalent to 10% compounded annually? What is the present worth equivalent if the firm uses an interest rate
of 10% compounded continuously?

4-42. Roberto borrows $10,000 and repays the loan with three equal annual payments. The
interest rate for the first year of the loan is 4% compounded annually; for the second
year of the loan the interest rate is 5% compounded annually; for the third year of the
loan the interest rate is 6% compounded annually.

   a. Determine the size of the equal annual payments.

   b. Compare the result from (a) with that which results from interchanging the interest
rates for the first and third years.

4-43. Marcida invests $50,000 in a fund that provides incentives for long-range investments.
During the first year, the fund pays interest at an annual compound rate of 4%. Thereafter, the interest rate earned on the cumulative investment balance increases by $1/4\%$
per year. Hence, the account will pay annual compound interest of 10% during the 25th
year of the deposit. How much will the investment be worth after 25 years? (This problem is easily solved by using a spreadsheet.)

4-44. What single sum of money today is equivalent to five equal annual future cash flows of
$5,000, with the first cash flow occurring 5 years from today? Money is worth 8% com-
pounded annually.

4-45. What single sum of money 10 years from today is equivalent to five equal annual future
cash flows of $5,000, with the first cash flow occurring 5 years from today? Money is
worth 8% compounded annually.

4-46. What single sum of money today is equivalent to five future cash flows, with the first
cash flow of $5,000 occurring 5 years from today? Subsequent cash flows are $500
larger than the previous cash flow. Money is worth 8% compounded annually.

4-47. What single sum of money today is equivalent to five future cash flows, with the first
cash flow of $5,000 occurring 5 years from today? Subsequent cash flows are 12%
larger than the previous cash flow. Money is worth 8% compounded annually.

4-48. What single sum of money today is equivalent to five future cash flows, with the first
cash flow of $5,000 occurring 5 years from today? Subsequent cash flows are $500
smaller than the previous cash flow. Money is worth 8% compounded annually.

4-49. What single sum of money today is equivalent to five future cash flows, with the first
cash flow of $5,000 occurring 5 years from today? Subsequent cash flows are 12%
smaller than the previous cash flow. Money is worth 8% compounded annually.

4-50. Given the cash flow diagram shown next and an interest rate of 8% per period, solve for
the value of an equivalent amount at the end of period (a) 5, (b) 9, (c) 12, and (d) 15.
(*Note:* Upward arrows denote cash inflows and downward arrows denote cash outflows.)

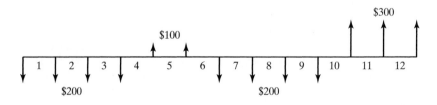

4-51. Consider the following cash flow profile:

EOY	Cash Flow	EOY	Cash Flow	EOY	Cash Flow
0	−$45,000	3	$9,000	6	$8,000
1	$10,000	4	$9,000	7	$7,000
2	$10,000	5	$8,000	8	$7,000

With an 8% annual interest rate, what single sum of money at the end of the fourth year will be equivalent to the cash flow series?

4-52. In Problem 4-51, what uniform annual series over [1,8] will be equivalent to the cash flow profile if money is worth 8% compounded annually?

4-53. In Problem 4-51, what uniform annual series over [2,7] will be equivalent to the cash flow profile if money is worth 8% compounded annually?

4-54. In Problem 4-51, suppose the positive-valued cash flows are replaced by a negative gradient series. If the cash flow at end-of-year 1 is $10,000, what negative gradient step is required for the cash flow profiles to be equivalent?

4-55. In Problem 4-51, suppose the positive-valued cash flows are replaced by a geometric series. If the cash flow at end-of-year 1 is $10,000, what geometric rate is required for the cash flow profiles to be equivalent?

5
Equivalent Worth Methods for Comparing Alternatives

In comparing investment alternatives, several different measures of economic effectiveness are often used. This and subsequent chapters consider various methods for studying the economic desirability of an individual project and for comparing the relative economic desirabilities of two or more projects. The methods are as follows:

1. present worth
2. annual worth
3. future worth
4. rate of return
5. benefit–cost ratio

The first three methods are treated in this chapter. Rate of return methods are presented in Chapter 6, and benefit–cost methods are included in Appendix 6–C. All of the methods are based on the concepts of the time value of money and equivalence, described in Chapter 4.

5.1 Measures of Economic Effectiveness

The various measures of economic effectiveness that will be used in this text are defined as follows:

Present worth: A determination of the present worth (PW) involves the conversion of each individual cash flow to its present worth equivalent and the summation of the individual present worths to obtain the net present worth.

Annual worth: The annual worth (AW) is determined by converting all cash flows to an equivalent uniform annual series of cash flows.

Future worth: The future worth (FW) is obtained by converting each individual cash flow to its future worth equivalent and determining the net future worth for the project.

Rate of return: Among the many definitions of rate of return, the most popular definition is the interest rate that yields a net present worth of zero; such a rate of return is referred to as the *internal rate of return* (IRR).

Benefit–cost ratio: There are several definitions of the benefit–cost ratio (B/C), but, in general, it can be defined as the ratio of the equivalent worth of benefits to the equivalent worth of costs.

5.2 Defining Investment Alternatives

In defining investment *alternatives,* it is important to ensure that the set of alternatives is mutually exclusive; the term "mutually exclusive" signifies that an "either–or, but not both" situation exists. That is, the choice of one excludes the choice of any other. Investment *opportunities* will denote projects, proposals, and other options available for investment. Distinct combinations of investment opportunities will be used to define the investment alternatives.

As an illustration, suppose two investment opportunities, A and B, are available. In this case, four mutually exclusive alternatives can be formed: neither A nor B, A only, B only, and both A and B. In general, if there are m investment opportunities, then 2^m investment alternatives can be formed. Of course, not all of the 2^m alternatives are necessarily feasible, because there are budget limitations, dependencies among investment opportunities, and other restrictions.

Example 5-1
Two industrial trucks, A and B, are being considered by the warehouse manager. In addition, two industrial truck attachments, C and D, are being considered. Opportunities A and B are mutually exclusive; opportunities C and D are mutually exclusive; and the purchase of an attachment is *contingent* upon the purchase of one of the trucks. The sixteen possible alternatives are

Do nothing	D only*	B and C	A, B, and D*
A only	A and B*	B and D	A, C, and D*
B only	A and C	C and D*	B, C, and D*
C only*	A and D	A, B, and C*	A, B, C, and D*

*Denotes an infeasible alternative.

Nine alternatives are infeasible because of the mutually exclusive and contingent dependencies among the investment opportunities. Thus, only seven alternatives (sometimes descriptively called *mutually exclusive combinations*) must be considered. ∎

5.3 A Systematic Procedure for Comparing Investment Alternatives

A systematic procedure for comparing investment alternatives can be stated as follows:

1. Define the objectives.
2. Define the alternatives.
3. Determine the study period.
4. Provide estimates of the cash flows for each alternative.
5. Specify the time value of money or interest rate.
6. Select the measure(s) of effectiveness.
7. Compare the alternatives.
8. Perform sensitivity analyses.
9. Select the preferred alternative.

As noted in the previous section, one may wish to aggregate investment opportunities in various combinations to arrive at the alternatives to be compared. As an illustration, suppose exactly one tractor-trailer combination is to be purchased. There might exist two tractor alternatives (A and B) and two trailer alternatives (C and D). Hence, there can be formed four tractor-trailer alternatives (AC, AD, BC, BD). Additionally, the do-nothing alternative (maintain the status quo) should not be overlooked.

The study period defines the period of time over which the analysis is to be performed. Cash flows that occur prior to and after the study period are not considered, except as they might influence cash flows during the study period. The study period may or may not be the same as the useful lives for equipment involved. If the study period is less than the useful life of an asset, then an estimate should be provided of its salvage value at the end of the study period; if the study period is longer than the useful life of an asset, then estimates of cash flows should be provided for subsequent replacements for the asset.

Estimates of the cash flows should be provided for each alternative by using the approaches previously described in Chapter 3. Even though an asset may continue to be used for a period of time beyond the study period, an estimate of its salvage value should be provided, assuming it will be disposed of at the end of the study period.

The interest rate or discount rate to be used is the minimum attractive rate for the firm. The *minimum attractive rate of return (MARR)*, treated in detail in Chapter 11, is defined to be the return that could be earned by investing elsewhere (opportunity cost concept).

As mentioned previously, a number of different measures of economic effectiveness can be selected. All of the methods we consider will yield the same choices. The recommended basis for selecting a particular measure of economic effectiveness is communication; namely, the method that is best understood by management should be used so long as it yields recommendations consistent with those made on the basis of correct time value of money analyses. We will have more to say about this matter at the end of Chapter 6.

When the measure of economic effectiveness is used, the alternatives should be compared and the most economic alternative identified. In the case of present worth, annual worth, and future worth comparisons, you can rank the alternatives on the basis of their worths. When rate of return and benefit–cost ratio comparisons are used, an incremental approach is necessary.

Because the comparison of investment alternatives involves the use of estimates of future economic conditions, errors can occur. Consequently, it is worthwhile to consider the consequences of such errors on the decision to be made. Sensitivity analyses can be performed to determine the effect of estimation errors on the economic performance of each alternative. Methods for performing sensitivity analyses are presented in Chapters 13 and 14.

The final selection of the preferred alternative is complicated by the presence of nonmonetary factors, multiple objectives, and risk and uncertainties concerning future outcomes. Methods for treating such conditions are presented in Parts Three and Four.

5.4 Judging the Economic Worth
of Investment Opportunities

The measures of economic effectiveness can be used to judge the economic worth of individual investment opportunities. In this section, it will be assumed that each investment opportunity is independent of other opportunities that may be under consideration. Furthermore, it is assumed that there are no restrictions on the number of such investment opportunities to be undertaken. The decision is not which of several opportunities is best, but rather, whether an individual opportunity is a worthwhile investment. We define an opportunity as being worthwhile if it has a nonnegative present worth, annual worth, or future worth, a rate of return at least equal to the minimum attractive rate of return, or a benefit–cost ratio at least equal to one.

Example 5-2
Given the following investment opportunities, and using each of the five methods given in Section 5.1, determine which of the following are worthwhile investment opportunities (a minimum attractive rate of return of 10% is to be used in the analysis):

Opportunity	Investment (P)	Study period (N)	Salvage value (S)	Net annual cash flow (A)
A	$10,000	5 yr	$10,000	+$2,000
B	12,000	5 yr	0	+ 3,000
C	15,000	5 yr	0	+ 4,167

Solution By using methods to be described subsequently, the following values are obtained:

Opportunity	Present worth	Annual worth	Future worth	Rate of return	Conventional benefit–cost ratio
A	+$3,790	+$1,000	+$6,106	20%	2.0000
B	− 630	− 170	− 1,015	8%	0.9475
C	+ 850	+ 217	+ 1,369	12%	1.0567

Thus, investments A and C (in that order) are deemed worthwhile.   ■

5.5 Present Worth (PW) Method

The term *present worth* (*PW*) means an amount at some beginning or base time that is equivalent to a particular schedule of receipts and/or disbursements under consideration. If disbursements only are considered, the term can be best expressed as *present worth-cost.*

5.5.1 Study Period in Comparisons of Alternative Projects

In comparing alternatives by the present worth (and future worth) method, it is essential that all alternatives be considered over the same length of time. If the alternatives all have the same expected life, there is no problem, for that life can be used. When the alternatives have different expected lives, it is common to use a study period equal to the lowest common multiple of the lives, or the length of time during which the services of the chosen alternative will be needed, whichever is less. For example, if two alternatives have expected lives of 3 and 4 years, respectively, then the lowest common multiple of the lives to use as a study period is 12 years. However, if the service for which the alternatives are being compared is expected to be needed for only 9 years, then 9 years should be the study period used.

5.5.2 Comparing Alternatives Using Present Worth Analysis When Receipts and Disbursements Are Known

When receipts (cash inflow) as well as disbursements (cash outflow) for more than one mutually exclusive project are known, the project with the highest net present worth should be chosen, as long as that present worth is greater than or equal to zero. As an example, consider two alternative lathes A and B, only one of which should be selected, if either. (That is, the "do nothing" alternative is feasible.)

Example 5-3

	Lathe	
	A	B
First cost:	$10,000	$15,000
Life:	5 yr	10 yr
Salvage value:	$2,000	$0
Annual receipts:	$5,000	$7,000
Annual disbursements:	$2,200	$4,000
Minimum attractive rate of return = 10%		
Study period = 10 yr		

Solution   The lowest common multiple of the lives is 10 years. Assuming that the service will be needed for at least that long and that what is estimated to happen in the first 5 years for project A will be repeated in the second 5 years, the solution is shown in the following table:

	Lathe*	
	A	B
Annual receipts = $5,000(P/A, 10%, 10):	$30,722.84	
$7,000(P/A, 10%, 10):		$43,011.97
Salvage value at Year 10 = $2,000(P/F, 10%, 10):	$771.09	
Total PW of cash inflow:	$31,493.92	$43,011.97
Annual disbursements = $2,200(P/A, 10%, 10):	−$13,518.05	
$4,000(P/A, 10%, 10):		−$24,578.27
First cost:	−$10,000.00	−$15,000.00
Replacement = ($10,000 − $2,000)(P/F, 10%, 5):	−$4,967.37	
Total PW of cash outflow:	−$28,485.42	−$39,578.27
Net PW:	$3,008.50	$3,433.70

*Values obtained from Excel—round-off errors account for differences in these results and what would be obtained by using the interest tables.

Thus, lathe B, having the higher net present worth greater than zero, is the better economic choice.   ■

5.5.3 Comparing Alternatives by Using Present Worth Analysis When Receipts Are Constant or Not Known

When alternatives that perform essentially identical services involve only known cash outflows, it is possible to compare the alternatives on the basis of present worth (PW). The method of study is the same as that illustrated for Example 5-3 except, of course, the alternative with the least negative present worth is best. As an example, consi der the following situation involving two compressors, each of which will do the desired job, but differ as shown:

Example 5-4

	Compressor	
	I	II
First cost:	$6,000	$8,000
Life:	6 yr	9 yr
Salvage value:	$1,000	$0
Annual operating disbursements:	$4,000	$3,200
Minimum return on investment = 15%		
Study period = 18 yr		

Solution   Again, assuming a study period equal to the lowest common multiple of lives (i.e., 18 years), we have the data in the following table:

	Compressor*	
	I	II
First cost:	-$6,000.00	-$8,000.00
First replacement = (-$6,000 + $1,000)($P/F$, 15%, 6):	-$2,161.64	
-$8,000($P/F$, 15%, 9):		-$2,274.10
Second replacement = (-$6,000 + $1,000)(P/F, 15%, 12):	-$934.54	
Operating disbursements = -$4,000($P/A$, 15%, 18):	-$24,511.86	
-$3,200($P/A$, 15%, 18):		-$19,609.49
Salvage value at Year 18 = $1,000(P/F, 15%, 18):	$80.81	
Net PW:	-$33,527.23	-$29,883.59

*Values obtained from Excel—round-off errors account for differences in these results and what would be obtained using the interest tables.

Since compressor II has the lower present worth of cost (least negative present worth), it is the better economic choice.   ■

5.6 Annual Worth (AW) Method

The term *annual worth* (*AW*) means a uniform annual series of money for a certain period of time that is equivalent in amount to a particular schedule of receipts and/or disbursements under consideration. If disbursements only are considered, the term is usually expressed as annual cost (AC) or equivalent uniform annual cost (EUAC).

5.6.1 Calculation of Capital Recovery Cost

The *capital recovery cost (CR)* for a project is the equivalent uniform annual cost of the capital invested. It is an annual amount that covers the following two items:

1. depreciation (loss in value of the asset);
2. interest (minimum attractive rate of return) on invested capital.

As an example, consider a machine or other asset having an investment cost (P) of $10,000, a life (N) of 5 years, and a salvage value (S) of $2,000. Further, the interest on invested capital i is 10%.

It can be shown that, no matter which method of calculating depreciation is used, the equivalent annual cost of the capital recovery is the same. For example, if straight-line depreciation is used, the equivalent annual cost of interest is calculated to be $710, as shown in Table 5-1. The annual depreciation cost (d) by the straight-line method is $d = (P-S)/N = (\$10,000 - \$2,000)/5 = \$1,600$. The $710 added to $1,600 results in a calculated capital recovery cost of $2,310.

There are several convenient formulas by which capital recovery cost can be calculated in order to obtain the preceding answer. The most apparent formula (using the same figures as for the machine we just discussed) is

$$CR = P(A/P, i\%, N) - S(A/F, i\%, N)$$
$$= \$10,000(A/P, 10\%, 5) - \$2,000(A/F, 10\%, 5)$$
$$= \$10,000(0.2638) - \$2,000(0.1638) = \$2,310.$$

Two other convenient formulas for calculating the capital recovery cost are

$$CR = P(i\%) + (P - S)(A/F, i\%, N)$$
$$= P(0.10) + (P - S)(A/F, 10\%, 5)$$
$$= \$10,000(0.10) + \$8,000(0.1638) = \$2,310$$

TABLE 5-1   Calculation of Equivalent Annual Cost of Interest Assuming Straight-Line Depreciation

Year	Investment at beginning of year	Interest on beginning-of-year investment @ 10%	Present worth of interest @ 10%	
1	$10,000	$1,000	$1,000(P/F, 10%, 1) =	$909
2	$8,400	$840	$840(P/F, 10%, 2) =	$694
3	$6,800	$680	$680(P/F, 10%, 3) =	$511
4	$5,200	$520	$520(P/F, 10%, 4) =	$355
5	$3,600	$360	$360(P/F, 10%, 5) =	$224
			Total:	$2,693

Annual equivalent cost of interest = $2,693(A/P, 10%, 5) = $710 and

$$CR = (P - S)(A/P, i\%, N) + S(i\%)$$
$$= (P - S)(A/P, 10\%, 5) + S(0.10)$$
$$= \$8,000(0.2638) + \$2,000(0.10) = \$2,310.$$

The last of the preceding formulas will be used primarily for the calculation of capital recovery cost throughout the rest of this book.

5.6.2 Comparing Alternatives by Using Annual Worth Analysis When Receipts and Disbursements Are Known

When receipts, as well as disbursements figures, are known for more than one mutually exclusive project, the project that has the highest net annual worth should be chosen, as long as that net annual worth is greater than or equal to zero.

As an example, consider the following two alternative lathes treated previously:

Example 5-5

	Lathe	
	A	B
First cost:	$10,000	$15,000
Life:	5 yr	10 yr
Salvage value:	$2,000	$0
Annual receipts:	$5,000	$7,000
Annual disbursements:	$2,200	$4,000
Minimum attractive rate of return = 10%		
Study period = 10 yr		

Solution

	Lathe*	
	A	B
Annual receipts:	$5,000.00	$7,000.00
Annual disbursements:	−$2,200.00	−$4,000.00
CR cost: = ($10,000 − $2,000)(A/P, 10%, 5) + $2,000(0.10):	−$2,310.38	
$15,000(A/P, 10%, 10):		−$2,441.18
Net AW:	$489.62	$558.82

*Values obtained from Excel—round-off errors account for differences in these results and what would be obtained using the interest tables.

Thus, lathe B, having the higher net annual worth, is the better economic choice. ∎

5.6.3 Comparing Alternatives by Using Annual Worth Analysis When Receipts Are Constant or Not Known

Very often, alternative projects are expected to perform almost identical functions so that each results in the same receipts, savings, or benefits. Sometimes, the savings or benefits are intangible or cannot be estimated; hence, the alternatives are judged on the basis of negative net annual worth, or annual cost, often called equivalent uniform annual cost and denoted EUAC. As an example, consider the same alternative compressors as were compared earlier using the PW method:

Example 5-6

	Compressor	
	I	II
First cost:	$6,000	$8,000
Life:	6 yr	9 yr
Salvage value:	$1,000	$0
Annual operating disbursements:	$4,000	$3,200
Minimum return on investment = 15%		
Study period = 18 yr		

Solution

	Compressor*	
	I	II
Annual operating disbursements:	−$4,000.00	−$3,200.00
CR cost: = ($6,000 −$1,000)(A/P, 15%, 6) + $1,000(0.15):	−$1,471.18	
$8,000($A/P$, 15%, 9):		−$1,676.59
Net AW:	−$5,471.18	−$4,876.59

*Values obtained from Excel—round-off errors account for differences in these results and what would be obtained using the interest tables.

Thus, compressor II, having the lower annual cost (least negative annual worth), is apparently the more economical choice. It should be noted that this analysis of competing alternatives with different lives makes certain assumptions that will be discussed subsequently.   ■

5.7 Future Worth (FW) Method

The term *future worth* (*FW*) means an amount at some ending or termination time that is equivalent to a particular schedule of receipts and/or disbursements under consideration. If disbursements only are considered, the term can best be expressed as *future worth-cost* (FW − cost) or *future cost*. The future worth is also referred to as the *terminal worth*.

5.7.1 Comparing Alternatives by Using Future Worth Analysis When Receipts and Disbursements Are Known

When receipt, as well as disbursement, figures are known for more than one mutually exclusive project, the project that has the highest net future worth should be chosen, as long as that net future worth is greater than or equal to zero.

As an example, consider the same two alternative lathes treated previously:

Example 5-7

	Lathe	
	A	B
First cost:	$10,000	$15,000
Life:	5 yr	10 yr
Salvage value:	$2,000	$0
Annual receipts:	$5,000	$7,000
Annual disbursements:	$2,200	$4,000
Minimum attractive rate of return = 10%		
Study period = 10 yr		

Solution

	Lathe*	
	A	B
Annual receipts = $5,000($F/A$,10%,10):	$79,687.12	
=$7,000($F/A$,10%,10):		$111,561.97
Salvage value at Year 10:	$2,000.00	
Total FW of cash inflow:	$81,687.12	$111,561.97
Annual disbursements = $2,200($F/A$,10%,10):	−$35,062.33	
=$4,000($F/A$,10%,10):		−$63,749.70
First cost = $10,000($F/P$,10%,10):	−$25,937.42	
=$15,000($F/P$,10%,10):		−$38,906.14
Replacement = ($10,000 - $2,000)($F/P$,10%,5):	−$12,884.08	
Total FW of cash outflow:	−$73,883.84	−$102,655.84
Net FW:	$7,803.28	$8,906.14

*Values obtained from Excel—round-off errors account for differences in these results and what would be obtained using the interest tables.

Thus lathe B, having the higher net future worth, is the better economic choice.   ■

5.7.2 Comparing Alternatives by Using Future Worth Analysis When Receipts are Constant or Not Known

As noted for present worth and annual worth analyses, alternative projects can be expected to perform identical functions such that only known cash outflows are given. It is possible to compare the alternatives on the basis of future worth-cost. The method of study is the same as illustrated for the preceding example; however, the alternative with the lowest future worth–cost is best. As an example, consider the same two compressors treated earlier:

Example 5-8

	Compressor	
	I	II
First cost:	$6,000	$8,000
Life:	6 yr	9 yr
Salvage value:	$1,000	$0
Annual operating disbursements:	$4,000	$3,200
Minimum attractive rate of return = 15%		
Study period = 18 yr		

Solution   It is assumed that replacements have identical cash flows.

	Compressor*	
	I	II
First cost = $6,000(F/P,15%,18):	−$74,252.72	
=$8,000(F/P,15%,18):		−$99,003.63
First replacement = ($6,000 – $1,000)(F/P,15%,12):	−$26,751.25	
=$8,000(F/P,15%,9):		−$28,143.01
Second replacement = ($6,000 – $1,000)(F/P,15%,6):	−$11,565.30	
Operating disbursements = $4,000(F/A,15%,18):	−$303,345.43	
=$3,200(F/A,15%,18):		−$242,676.34
Salvage value at Year 18:	$1,000.00	
Net FW:	−$414,914.71	−$369,822.98

*Values obtained from Excel—round-off errors account for differences in these results and what would be obtained using the interest tables.

Since compressor II has the least negative future worth, it is the better economic choice. ∎

5.8 Assumptions in Comparisons of Alternatives with Different Lives

In the comparison of projects A and B and also the comparison of compressors I and II, the alternatives compared had different expected lives. The solutions as shown are fully valid only if the following assumptions are reasonable:

1. The period of needed service for which the alternatives are being compared (study period) is either indefinitely long or a length of time equal to a common multiple of the lives of the alternatives.

   Note: The student should recognize that any point in time equal to a common multiple of lives would be a point at which each alternative would have just exhausted a life cycle.

2. What is estimated to happen in the first life cycle will happen in all succeeding life cycles, if any, for each alternative.

These assumptions are commonly made in economic analyses by default (i.e., they are made because there is no good basis for estimates to the contrary). The assumptions are sometimes referred to as the *repeatability* assumptions. They are implicitly contained in all examples and problems illustrating all methods of economic evaluations herein unless there is a statement to the contrary.

Applied to the earlier examples in which the compressors were compared, the assumptions mean the following:

1. The services of a compressor will be needed either 18 years, 36 years, 54 years, etc., or indefinitely.

2. When either compressor I or II is replaced at the end of a life cycle, it will be replaced with a compressor having characteristics affecting cost (i.e., first cost, life, salvage value, and annual operating disbursements) identical to the estimates used for the first life cycle.

Whenever alternatives to be compared have different lives and one or both of the conditions are not appropriate, then it is necessary to enumerate what receipts and what expenses are expected to occur at what points in time for each alternative for as long as service will be needed or the irregularity is expected to exist (i.e., the study period). This is sometimes referred to as the *coterminated* assumption. Various assumptions can be used to adjust the terminal value of each alternative at the cotermination time. For instance, if an alternative has a useful life shorter than the study period, the estimated annual cost of contracting for the services involved might be utilized during the remaining years. Similarly, if the useful life is longer than the study period, an estimated salvage value is customarily employed to serve as a terminal cash flow at the project's coterminated life. The enumerated information can then be converted into an equivalent PW, AW, FW, or other measure of merit by ordinary time value of money computations.

Example 5-9
Suppose that, for the earlier compressor illustration, it is expected that the standard assumptions are not met as follows: (a) a compressor is needed for only 12 years; and (b) the replacement for compressor II is expected to cost $14,000 rather than $8,000, and its salvage value after 3 years' service (end of the 12th year of study) is expected to be $400. Compare the two compressors by the annual cost method.

Solution   The annual cost (AC) for compressor I remains the same at $5,471.18. For compressor II, the cash flow diagram and solution are

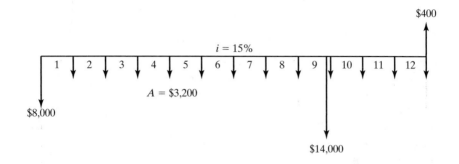

$$AW = -[\$8,000 - \$14,000(P/F, 15\%, 9) + \$400(P/F, 15\%, 12)](A/P, 15\%, 12)$$
$$ - \$3,200$$
$$ = -[\$8,000 + \$14,000(0.2843) - \$400(0.1869)](0.1845) - \$3,200$$
$$ = -\$5,396.$$

Thus, under the changed conditions, compressor II is still less costly, but by much less margin.

The easiest way to calculate the present worths (at time 0) and the future worths (at time 12) for comparison of the alternatives, given the preceding annual costs (worths), is as follows:

Alternative	PW = AW(P/A,15%,12)
I	−$5,471(5.4206) = −$29,656
II	−$5,396(5.4206) = −$29,250

Alternative	FW = AW(F/A,15%,12)
I	−$5,471(29.0017) = −$158,668
II	−$5,396(29.0017) = −$156,494

5.9 Relationship of Various Analysis Methods

All methods of economic evaluation considered to this point have the reassuring property of providing consistent results regarding the economic desirability or relative ranking of projects compared. In fact, it can be shown that the annual worths, present worths, and future worths for any projects under consideration are linearly proportional to each other.

Example 5-10

Show the consistency of economic comparison results for compressors I and II by the various methods given in this chapter.

Solution

$$\frac{PW_I}{PW_{II}} = \frac{AW_I}{AW_{II}} = \frac{FW_I}{FW_{II}}; \quad \frac{-\$33,527.23}{-\$29,883.59} = \frac{-\$5,471.18}{-\$4,876.59} = \frac{-\$414,914.71}{-\$369,822.98} = 1.1219. \quad \blacksquare$$

5.10 Using Excel to Determine Present Worth, Annual Worth, and Future Worth

As demonstrated in Chapter 4, spreadsheet software can prove to be powerful in performing economic analyses. Figures 5-1 and 5-2 provide spreadsheets for the lathe and compressor investment alternatives considered previously in this chapter. For example, Figure 5-2 shows the entries for cells B24–26 and cells C24–26 for PW, AW, and FW values, respectively, for the two compressors. Also shown are the corresponding Excel commands.

To illustrate the use of spreadsheet software in performing sensitivity analyses, Figure 5-3 depicts the effect of a 12.5% minimum attractive rate of return on the choice of lathes to purchase. Increasing the MARR from 10% to 12.5% causes the recommendation to change regarding the more economic lathe to purchase.

In Chapter 6, we will find that Lathe B is preferred for MARR values less than 12.13%; Lathe A is preferred for MARR values greater than 12.13% and less than 12.48%; and the "do nothing" alternative is preferred for MARR values in excess of 12.48%, because neither lathe has a positive discounted net worth. We will also find that Compressor B is preferred for MARR values less than 46.88%; otherwise Compressor A is preferred, since the "do nothing" alternative is not feasible.

	A	B	C
1	MARR = 10%		
2		*Lathe*	
3		A	B
4	EOY	Cash Flow	Cash Flow
5	0	−$10,000.00	−$15,000.00
6	1	$2,800.00	$3,000.00
7	2	$2,800.00	$3,000.00
8	3	$2,800.00	$3,000.00
9	4	$2,800.00	$3,000.00
10	5	−$5,200.00	$3,000.00
11	6	$2,800.00	$3,000.00
12	7	$2,800.00	$3,000.00
13	8	$2,800.00	$3,000.00
14	9	$2,800.00	$3,000.00
15	10	$4,800.00	$3,000.00
16	PW=	$3,008.50	$3,433.70
17	AW=	$489.62	$558.82
18	FW=	$7,803.28	$8,906.14

Figure 5-1   Spreadsheet for lathe example.

	A	B	C
1	MARR =	15%	
2			
3		Compressor	
		I	II
4	EOY	Cash Flow	Cash Flow
5	0	−$6,000.00	−$8,000.00
6	1	−$4,000.00	−$3,200.00
7	2	−$4,000.00	−$3,200.00
8	3	−$4,000.00	−$3,200.00
9	4	−$4,000.00	−$3,200.00
10	5	−$4,000.00	−$3,200.00
11	6	−$9,000.00	−$3,200.00
12	7	−$4,000.00	−$3,200.00
13	8	−$4,000.00	−$3,200.00
14	9	−$4,000.00	−$11,200.00
15	10	−$4,000.00	−$3,200.00
16	11	−$4,000.00	−$3,200.00
17	12	−$9,000.00	−$3,200.00
18	13	−$4,000.00	−$3,200.00
19	14	−$4,000.00	−$3,200.00
20	15	−$4,000.00	−$3,200.00
21	16	−$4,000.00	−$3,200.00
22	17	−$4,000.00	−$3,200.00
23	18	−$3,000.00	−$3,200.00
24	PW =	−$33,527.23	−$29,883.59
25	AW =	−$5,471.18	−$4,876.59
26	FW =	−$414,914.71	−$369,822.98

	A	B	C
24	PW =	=NPV(15%, B6:B23)+B5	=NPV(15%, C6:C23)+C5
25	AW =	=PMT(15%, 18, −B24)	=PMT(15%, 18, −C24)
26	FW =	=FV(15%, 18, −B25)	=FV(15%, 18, −C25)

Figure 5-2 Spreadsheet for compressor example.

153

	A	B	C
1	MARR =	12.5%	
2			*Lathe*
3		A	B
4	EOY	Cash Flow	Cash Flow
5	0	−$10,000.00	−$15,000.00
6	1	$2,800.00	$3,000.00
7	2	$2,800.00	$3,000.00
8	3	$2,800.00	$3,000.00
9	4	$2,800.00	$3,000.00
10	5	−$5,200.00	$3,000.00
11	6	$2,800.00	$3,000.00
12	7	$2,800.00	$3,000.00
13	8	$2,800.00	$3,000.00
14	9	$2,800.00	$3,000.00
15	10	$4,800.00	$3,000.00
16	PW=	$1,678.47	$1,609.29
17	AW=	$303.17	$290.67
18	FW=	$5,450.52	$5,225.89

Figure 5-3   Spreadsheet for lathe example with MARR changed to 12.5%.

PROBLEMS

5-1. On the basis of a MARR of 12%, recommend which of the five independent investment opportunities to undertake. Use the (a) PW method, (b) AW method, (c) FW method.

	Investment Opportunity				
	A	B	C	D	E
Initial investment:	$8,000	$10,000	$12,000	$15,000	$16,000
Annual net receipts:	2,000	3,000	3,200	3,800	4,400
Salvage value:	1,000	−1,000	1,000	2,000	500
Study period = 5 yr					

5-2. In Problem 5-1, suppose A through E constitute a set of feasible, mutually exclusive investment alternatives. Which would be chosen, using (a) the PW method? (b) the AW method? (c) the FW method?

5-3. Using a MARR of 10% and a study period of 10 years, determine which of the following independent (nonrepeating) investment opportunities are economically worthwhile. Make your determination by using (a) the PW method, (b) the AW method, (c) the FW method.

	Cash Flows (CF) for Investment Opportunity					
EOY	A	B	C	D	E	F
0	−$5,000	−$5,000	−$8,000	−$12,000	0	−$15,000
1	− 5,000	800	1,000	1,500	0	1,000
2	1,000	1,200	1,000	1,500	0	1,500
3	1,000	1,600	1,000	1,500	−$6,000	2,000
4	2,000	2,000	1,000	1,500	2,000	2,500
5	2,000	2,400	1,000	1,500	2,000	3,000
6	3,000	2,800	1,000	1,500	2,000	3,500
7	3,000	0	1,000	1,500	2,000	4,000
8	4,000	0	1,000	1,500	2,000	4,500
9	4,000	0	1,000	1,500	0	5,000
10	− 5,000	0	− 2,000	3,000	0	5,500

5-4. In Problem 5-3, suppose A through F constitute a set of feasible, mutually exclusive investment alternatives. Which would be chosen, using (a) the PW method? (b) the AW method? (c) the FW method?

5-5. If money is worth 8% to you, which (if any) of the following investment opportunities, each of which would require the same capital investment, would you favor? Why? Use the PW method in making your recommendation.

EOY	CF(A)	CF(B)	CF(C)
1	$1,000	$5,000	$3,000
2	2,000	4,000	3,000
3	3,000	3,000	3,000
4	4,000	2,000	3,000
5	5,000	1,000	3,000

5-6. In Problem 5-5, suppose the three cash flow profiles are for feasible, mutually exclusive investment alternatives and one must be chosen. Further, suppose your time value of money is greater than zero. Which would you choose?

5-7. Three investment proposals are being considered: A, B, and C. C is contingent on either A or B. How many feasible, mutually exclusive investment alternatives can be formed for this situation?

5-8. Four investment proposals are being considered: A, B, C, and D. A and C are mutually exclusive proposals; C is contingent on B; and D is contingent on A. How many feasible, mutually exclusive investment alternatives can be formed for this situation?

5-9. Five investment proposals are being considered: A, B, C, D, and E. A and B are mutually exclusive proposals; C is contingent on A; D is contingent on either B or C; E is contingent on either D or A; and the "do nothing" alternative is not feasible. How many feasible, mutually exclusive investment alternatives can be formed for this situation?

5-10. Five investment proposals are being considered: A, B, C, D, and E. A and B are mutually exclusive proposals. C is contingent on A; D is contingent on B; and E is contingent on either A or B. How many feasible, mutually exclusive investment alternatives can be formed?

5-11. A choice between two machines must be made. It is estimated that Machine Q will have an initial cost of $1,500,000; first-year annual operating and maintenance costs of $135,000, increasing by 10% annually over the 5-year planning horizon; and a negligible salvage value at the end of the planning horizon. Machine R is estimated to have an initial cost of $1,750,000; $75,000 first-year operating and maintenance costs, increasing by 15% per year over the 5-year planning horizon; and a negligible salvage value at the end of the planning horizon. With a MARR of 10%, compare the two machines, using (a) the PW method, (b) the AW method, (c) the FW method.

5-12. Three mutually exclusive investment alternatives including the "do nothing" alternative are illustrated in the accompanying table. The cash flows (CF) are shown for a 5-year planning horizon. The MARR is 12%. Compare the alternatives, using (a) the PW method, (b) the AW method, (c) the FW method.

EOY	CF(A)	CF(B)	CF(C)
0	$0	−$100,000	−$150,000
1	0	+ 30,000	+ 50,000
2	0	+ 55,000	+ 75,000
3	0	+ 80,000	+ 100,000
4	0	+ 105,000	+ 125,000
5	0	+ 130,000	+ 150,000

5-13. Four mutually exclusive investment alternatives are illustrated in the accompanying table; the "do nothing" alternative is not feasible. The cash flows (CF) are shown for a 10-year planning horizon. The MARR is 12%. Determine which is best, using (a) the PW method, (b) the AW method, (c) the FW method.

EOY	CF(A)	CF(B)	CF(C)	CF(D)
0	−$ 50,000	−$125,000	−$200,000	−$100,000
1	− 100,000	− 75,000	+ 50,000	− 150,000
2–5	+ 50,000	+ 70,000	+ 50,000	+ 75,000
6–9	+ 50,000	+ 70,000	+ 75,000	+ 85,000
10	+ 75,000	+ 70,000	+ 75,000	+ 90,000

5-14. Four mutually exclusive investment alternatives including the "do nothing" alternative are illustrated in the accompanying table. The cash flows (CF) are shown below over a 7-year planning horizon. The MARR is 10% per year. Determine which is best, using (a) the PW method, (b) the AW method, (c) the FW method.

EOY	CF(A)	CF(B)	CF(C)	CF(D)
0	$0	−$100,000	−$180,000	−$125,000
1	0	− 100,000	− 20,000	− 75,000
2	0	+ 0	+ 100,000	+ 60,000
3	0	+ 25,000	+ 75,000	+ 60,000
4	0	+ 50,000	+ 50,000	+ 60,000
5	0	+ 75,000	+ 25,000	+ 60,000
6	0	+ 100,000	+ 0	+ 60,000
7	0	+ 125,000	+ 100,000	+ 60,000

5-15. Four mutually exclusive investment alternatives are illustrated in the accompanying table; the "do nothing" alternative is not feasible. The cash flows (CF) are shown for a 7-year planning horizon. Based on a MARR of 15%, determine which is best, using (a) the PW method, (b) the AW method, (c) the FW method.

EOY	(CF)A	CF(B)	CF(C)	CF(D)
0	−$ 50,000	−$100,000	−$250,000	−$225,000
1	− 100,000	− 100,000	+ 75,000	− 75,000
2–7	+ 50,000	+ 70,000	+ 75,000	+ 100,000

5-16. Three order-picking systems are being considered for use in a catalog distribution center. Alternative A involves picking items manually from bin shelving. With alternative B, items are picked manually from a horizontal carousel conveyor; hence, essentially no walking is required by the order picker. With alternative C, items are picked automatically by an automatic item retrieval machine. The alternatives have different space and labor requirements. Likewise, they have different acquisition, maintenance, and operating costs.

For the study period of 10 years, the estimates for the alternatives are given as follows:

	A	B	C
First cost:	$1,150,000	$1,250,000	$2,000,000
Salvage value:	750,000	750,000	875,000
Annual disbursements:	425,000	400,000	275,000

Based on a MARR of 12%, determine which alternative is preferred, using (a) the PW method, (b) the AW method, (c) the FW method.

5-17. In Problem 5-16, suppose there is uncertainty regarding the annual disbursements required for alternative C; what is the upper limit on annual disbursements for it to be the preferred alternative?

5-18. In Problem 5-16, how sensitive is the selection to the MARR being (a) 0%? (b) 11%? (c) 22%?

5-19. A machine can be repaired today for $7,500. If repairs are not made, operating expenses are expected to increase by $750 each year for the next 5 years. If repairs are made, the increases in operating expenses will not occur. The MARR is 10% per year. Assuming the machine will have a negligible salvage value at the end of the 5-year period, and by using the PW method, determine whether or not the machine should be repaired?

5-20. Three alternatives are available to fill a given need that is expected to last for at least 12 years. Each is expected to have a negligible salvage value at the end of the life cycle.

	Plan A	Plan B	Plan C
First cost:	$4,000	$12,000	$24,000
Life cycle:	6 yr	3 yr	4 yr
Annual disbursements:	$7,000	$2,000	$500

Using a MARR of 12%, compare the alternatives, using (a) the PW method, (b) the AW method, (c) the FW method.

5-21. Two manufacturing methods have been proposed for a new production requirement. One method involves two general-purpose machines that cost $17,500 each, installed. Each will produce 10 pieces per hr and will require an operator costing $10.00 per hr during operation. The other method requires a special-purpose machine costing $55,000 that will produce 20 pieces per hr and will require an operator costing $12.00 per hr during operation. Both types of machines are expected to last 10 years and have negligible salvage values. Other relevant data are as follows:

	General-purpose machine (each)	Special-purpose machine
Power cost per hr:	$ 0.50	$ 0.75
Fixed maintenance per yr:	750.00	1,000.00
Variable maintenance per hr:	0.25	0.20
Insurance and floor space per yr:	4,900.00	5,500.00

a. If the expected output is 20,000 pieces per yr and the minimum before-tax rate of return is 15%, which method has the lower total annual cost?

b. At what annual output rate would you be indifferent between the two methods?

5-22. It is desired to determine the most economical thickness of insulation for a large cold-storage room. Insulation is expected to cost $250 per 1,000 sq ft of wall area per in. of thickness installed. The insulation is expected to have a useful life of 15 years, at which time it will have to be removed at a cost of $10 per 1,000 sq ft. of wall area per in. of thickness installed. Annual property taxes and insurance are expected to be 3% of the first cost of the insulation.

Heat loss per 1,000 sq ft of wall area is a function of insulation thickness. For thicknesses between 3 in. and 8 in., heat loss (measured in Btu/hr) is estimated to equal $4,500(0.82)^{t-3}$, where t is the thickness of insulation, measured in inches. Heat loss occurs continuously throughout the 365-day year. The cost of heat removal is estimated at $0.01 per 1,000 Btu/hr. A minimum attractive rate of return of 10% applies. (For purposes of analysis, end-of-year discrete cash flows for heat loss may be assumed.)

Determine the optimum thickness of insulation in the interval [3″, 8″], on the basis of insulation being available in (a) 1 in. increments of thickness, (b) $\frac{1}{4}$ in. increments of thickness, and (c) 0.1 in. increments of thickness.

5-23. Alternative methods I and II are proposed for a plant operation. The following is comparative information:

	Method I	Method II
Initial investment:	$10,000	$40,000
Life:	5 yr	10 yr
Salvage value:	$1,000	$5,000
Annual disbursements:		
Labor:	$12,000	$4,000
Power:	$250	$300
Rent:	$1,000	$500
Maintenance:	$500	$200
Property taxes and insurance:	$400	$2,000

All other expenses are equal for the two methods, and the income from the operation is not affected by the choice. If the minimum attractive rate of return is 12% and the study period is 10 years, which is the better choice using the annual worth (cost) method?

5-24. Compare the net future worths of two temporary structures that will be retired at the end of 10 years. Assume the minimum attractive rate of return = 10% and that estimates are as follows:

	Structure A	Structure B
First cost:	$20,000	$25,000
Net salvage value:	5,000	− 1,000
Annual maintenance and		
property taxes:	5,000	1,500

5-25. A proposed material for covering the roof of a building will have an estimated life of 8 years and will cost $5,000. A heavier grade of this roofing material will cost $1,400 more but will have an estimated life of 12 years. Installation costs for either material will be $1,800. Compare the annual costs, using a minimum attractive return of 12% and the "repeatability" assumptions.

5-26. A small tractor is required for snow removal. It can be purchased for $4,250 and is expected to have a $500 salvage value at the end of its economic life of 5 years. Its annual operating cost is $1,000, and maintenance will be $300 the first year and increase by $100 per year. The minimum attractive rate of return is 10%. A contractor will provide this service for $2,300 per year, payable at the beginning of the year. Which alternative has lower total present worth-cost?

5-27. Compare the annual costs of pumps A and B for a 15-year service life, using an interest rate of 12%.

	Pump A	Pump B
First cost:	$3,500	$5,500
Estimated salvage value:	0	1,500
Annual pumping cost:	450	300
Annual repair cost:	150	80

5-28. It is desired to determine the optimal height for a proposed building that is expected to last 30 years and then be demolished at zero salvage. The following are pertinent data:

	Number of floors			
	2	3	4	5
Building first cost:	$200,000	$250,000	$300,000	$350,000
Annual revenue:	45,000	60,000	75,000	90,000
Annual cash disbursements:	15,000	25,000	35,000	45,000

In addition to the building first cost, the land requires an investment of $500,000 and is expected to retain that value throughout the life period. If the minimum required rate of return is 12%, show which height, if any, should be built on the basis of annual worth comparisons.

5-29. A manufacturing process can be designed for varying degrees of automation. The minimum required rate of return is 12%. Which degree should be selected if the economic life is 10 years and the salvage values are negligible? Use a present worth analysis.

	Machine A	Machine B	Machine C
First cost:	$ 20,000	$ 25,000	$ 35,000
Annual receipts:	150,000	180,000	200,000
Annual disbursements:	138,000	166,000	184,000

5-30. An individual is faced with two mutually exclusive investment alternatives. By investing $10,000, a single sum of $15,000 will be received 4 years after the investment; alternatively, by investing $15,000, a single sum of $18,500 will be received 2 years after the investment. Throughout, it cannot be assumed that either investment is repeatable. For a MARR of 10% per year, determine the preferred alternative, using the following method(s): (a) PW, (b) AW, (c) FW. What are the implications of changing the MARR to 8%?

5-31. In the design of a special-use structure, two mutually exclusive alternatives (D1 and D2), having the following values for the economic parameters, are under consideration:

	D1	D2
Capital investment	$50,000	$120,000
Annual expenses	$9,000	$5,000
Useful life (years)	20	50
Market value (at end of useful life)	$10,000	$20,000

If the repeatability assumption is applicable, which design alternative do you recommend (a) using annual worth analysis? (b) using present worth analysis? (c) using future worth analysis? Assume a MARR of 10% per year.

5-32. A new highway is to be constructed. Design A calls for a *concrete* pavement costing $90 per foot with a 20-year life; two paved ditches costing $3 per foot each; and three box culverts every mile, each costing $9,000 and having a 20-year life. Annual maintenance will cost $1,800 per mile; the culverts must be cleaned every 5 years at a cost of $450 each per mile.

Design B calls for a *bituminous* pavement costing $45 per foot with a 10-year life; two sodded ditches costing $1.50 per foot each; and three pipe culverts every mile, each costing $2,250 and having a 10-year life. The replacement culverts will cost $2,400 each. Annual maintenance will cost $2,700 per mile; the culverts must be cleaned yearly at a cost of $225 each per mile; and the annual ditch maintenance will cost $1.50 per foot per ditch.

Compare the two designs on the basis of equivalent cost per mile for a 20-year period. Find the most economical design on the basis of equivalent annual worth and present worth with a MARR of 6% per year.

5-33. Assume you have $80,000 available for investment, and it is currently earning interest at an annual compound rate of 10%. Two investment alternatives have become available. One requires investing $80,000 now; the other requires investing $80,000 in 2 years, but a commitment is required now if it is to be pursued. In both cases, the investments will terminate 5 years from now. The cash flows for each alternative are shown in the accompanying table. Using present worth analysis, which do you recommend?

	Mutually Exclusive Alternative	
Year	A	B
0	−$80,000	
1	$21,750	
2	$21,750	−$80,000
3	$21,750	$33,333
4	$21,750	$33,333
5	$21,750	$33,333

5-34. A firm is considering purchasing a new machine to increase output of an existing production process. If each of these machines provides the same service over the useful
lives and the MARR is 12%, which machine should be selected on the basis of an annual
worth comparison and applying the repeatability assumption?

	Mutually Exclusive Alternative			
	A	B	C	D
Initial investment	$75,000	$75,000	$100,000	$50,000
Net Annual Revenue	$20,435	$16,212	$22,675	$17,879
Useful life (years)	5	6	10	4
Market value (at end of useful life)	$15,000	$12,000	$25,000	$12,000

5-35. Consider the following end-of-year cash flows for two mutually exclusive alternatives,
one of which must be chosen. A 12% MARR is to be used.

	Machine	
	X	Y
Initial investment	$6,000	$14,000
Annual expenses	$2,500	$2,400
Useful life (years)	12	18
Market value (at end of useful life)	$0	$2,800

a. Determine the preferred alternative if the repeatability assumption holds.

b. Determine the preferred alternative if the repeatability assumption does NOT
hold, the study period is 18 years, and a machine can be leased for $8,000 per year
(including all annual expenses) after the useful life ends for either machine.
Assume beginning-of-year lease payments.

5-36. Two electric motors (A and B) are being considered to drive a centrifugal pump. One
of the motors must be selected. Each motor is capable of delivering 60 horsepower
(output) to the pumping operation. It is expected that the motors will be in use 800
hours per year. At the end of their useful lives, both motors will have negligible salvage
values. An 8% annual MARR and the repeatability assumption are applicable. Electricity costs $0.07 per kilowatt-hour. (Recall, 1 hp = 0.746 kW.)

	Motor	
	A	B
Initial cost	$1,200	$1,000
Annual maintenance	$160	$175
Useful life (years)	3	5
Electrical efficiency	90%	80%

a. On the basis of the data shown and using an annual worth comparison, which motor should be selected?

b. What basic tradeoff is being made in this problem?

5-37. Two electric motors (C and D) are being considered to drive a bulk conveyor. One of the motors must be selected. Each motor is capable of delivering 50 horsepower (output) to the conveyor operation. It is expected that the motors will be used 1,000 hours per year. At the end of their useful lives, both motors will have negligible salvage values. An 8% MARR and the repeatability assumptions hold. Electricity costs $0.085 per kilowatt-hour. (Recall, 1 hp = 0.746 kW.)

	Motor	
	C	D
Initial cost	$1,100	$950
Annual maintenance	$150	$100
Useful life (years)	5	15
Electrical efficiency	82%	62%

a. On the basis of the data shown in the accompanying table and using an annual worth comparison, which motor should be selected?

b. What basic tradeoff is being made in this problem?

5-38. Two investment alternatives are under consideration. Alternative A requires an initial investment of $100,000 and will generate net annual returns of $20,000 per year for 6 years; it will have a $30,000 salvage value at the end of its useful life of 6 years. Alternative B also requires an initial investment of $100,000; it will produce a $10,000 annual gradient series of annual returns, with the last annual return equaling $60,000 at the end of its 7-year useful life. If the two alternatives are considered, using an annual worth comparison with a MARR of 10% per year, which would be preferred, assuming the repeatability assumptions hold?

5-39. Two mutually exclusive alternatives are being considered (see the accompanying table) for the environmental protection equipment at a petroleum refinery. One of the alternatives must be selected. The repeatability assumptions hold and the MARR is 20% per year. Which alternative should be recommended?

	Alternative	
	A	B
Initial investment	$20,000	$38,000
Annual expenses	$5,500	$4,000
Useful life (years)	5	10
Market value (at end of useful life)	$1,000	$4,200

a. Which environmental protection equipment alternative should be selected?

b. Suppose the study period is 5 years and the market value for alternative B is $15,000 after 5 years. Which should be selected?

5-40. Basing your answer on the repeatability assumptions, a MARR of 8% per year, and an annual worth comparison for the data shown next, which machine should be chosen?

	Machine	
	A	B
Investment investment	$15,000	$22,500
Annual revenue	$6,250	$7,300
Annual expenses	$2,750	$3,000
Annual insurance (% of initial investment)	7%	4%
Useful life (years)	10	15
Market value (at end of useful life)	$2,500	$3,000

6
Rate of Return Methods for Comparing Alternatives

In the previous chapter, present worth, annual worth, and future worth analyses were considered in the comparison of investment alternatives. In this chapter, rate of return methods are presented. The use of rate of return methods in measuring the acceptability of individual investment *opportunities* was mentioned at the beginning of Chapter 5. It was noted that an individual investment opportunity is worthwhile if its rate of return is not less than the minimum attractive rate of return (MARR).

Any *rate of return* (*RR*) method of economic comparison involves the calculation of a rate or rates of return and comparison against a minimum standard of desirability (i.e., the MARR).

Two common techniques for calculating rates of return can be said to be theoretically sound, because they directly take into account the effects of any particular timing of cash flows throughout the study period considered. These methods, which can lead to slightly different calculated results, will be referred to as follows:

1. the internal rate of return (IRR) method, and
2. the external rate of return (ERR) method.

The most common method of calculation of the internal rate of return for a single project involves finding the interest rate at which the present worth of the cash inflow (receipts or cash savings) equals the present worth of the cash outflow (disbursements or cash savings foregone). That is, one finds the interest rate at which the PW of cash inflow equals the PW of cash outflow; or, at which the PW of net cash flows equals zero. Because the PW, AW, and FW differ only by a constant, the IRR can also be calculated by finding the interest rate that equates to zero the AW and/or FW of net cash flows.

To solve for the IRR, you can use trial-and-error methods and manually compute the PW, AW, or FW—attempting to find an interest rate that yields a value of zero for the particular measure of worth. Generally, the process concludes by interpolating between interest rates that yield positive and negative values for the measure of worth. Alternatively, you can use one of the financial calculators or computer software packages available.

To demonstrate the errors inherent in using linear interpolation to determine the IRR, several investment scenarios will be considered. For our purposes, cash inflows are denoted by a positive sign and cash outflows are denoted by a negative sign.

6.1 Computation of Internal Rate of Return (IRR) for a Single Investment Opportunity

Example 6-1

Capital investment:	$10,000
Project life:	5 yr
Salvage value:	$2,000
Annual receipts:	$5,000
Annual disbursements:	$2,200

Solution   Expressing PW of net cash flow, we have

$$- \$10,000 + (\$5,000 - \$2,200)(P/A, i\%, 5) + \$2,000(P/F, i\%, 5) = 0,$$

$$@i = 15\%: -\$10,000 + \$2,800(P/A, 15\%, 5) + \$2,000(P/F, 15\%, 5) = 0$$

$$\$380.39 \neq 0,$$

and

$$@i = 20\%: -\$10,000 + \$2,800(P/A, 20\%, 5) + \$2,000(P/F, 20\%, 5) = 0$$

$$-\$822.53 \neq 0.$$

Since we have both a positive and a negative PW of net cash flow, the answer is bracketed. Linear interpolation for the answer can be set up as follows:

i	PW of net cash flow
15%	$380.39
$x\%$	0
20%	−$822.53

The answer can be found by solving for $x\%$ in the following:

$$\frac{15\% - x\%}{15\% - 20\%} = \frac{\$380.39 - \$0}{\$380.39 - (-\$822.53)}$$

or

$$x\% = 15\% + \frac{\$380.39}{\$380.39 + \$822.53}(20\% - 15\%).$$

Solving, $x\% = 16.58\%$.

Using Excel or a handheld financial calculator, a more accurate IRR value was obtained, 16.4763%. For linear interpolation to provide an accurate estimate of IRR, a very small interval is needed. As examples, using interest rates of 16% and 17% yield present worths of $120.25 and –$129.61, respectively. By using linear interpolation, a value of 16.4813% (rounded to 16.5%) is obtained—this is closer to the value obtained by using Excel and the financial calculator than that obtained by interpolating between 15% and 20%.   ∎

6.1.1 Principles in Comparing Alternatives by a Rate of Return Method

When comparing alternatives by a RR method, when at most one alternative will be chosen, there are two main principles to keep in mind:

1. Each increment of investment capital must justify itself (by sufficient RR on that increment).
2. Compare a higher investment alternative against a lower investment alternative only if that lower investment alternative is justified.

The usual criterion for choice when using a RR method is, "Choose the alternative requiring the highest investment for which each increment of investment capital is justified."

This choice criterion assumes that the firm wants to invest any capital needed as long as the capital is justified by earning a sufficient RR on each increment of capital. In general, a sufficient RR is any RR greater than or equal to the MARR.

6.1.2 Alternative Ways to Find the Internal Rate of Return on Incremental Investment

The internal rate of return on the incremental investment for any two alternatives can also be found by two approaches.

1. finding the rate at which the PW (or AW or FW) of the net cash flow for the difference between the two alternatives is equal to zero, or
2. finding the rate at which the PWs (or AWs or FWs) of the two alternatives are equal.

6.1.3 Comparing Alternatives When Receipts and Disbursements Are Known

Consider the same two alternative lathes A and B compared in the last chapter, and determine which is better by the internal rate of return method, using the first approach previously cited.

Example 6-2

	Lathe	
	A	B
First cost:	$10,000	$15,000
Life:	5 yr	10 yr
Salvage value:	$2,000	$0
Annual receipts:	$5,000	$7,000
Annual disbursements:	$2,200	$4,000
Minimum attractive rate of return = 10%		
Study period = 10 yr		

   Solution   The first increment of investment to be studied is the $10,000 for lathe A. This project is the same as illustrated in the "single project" solution in Example 6-1. The IRR for the lathe, and hence the first increment of investment, was shown to be approximately 16.5%. Since 16.5% is greater than the minimum required rate of return of 10%, the increment of investment in lathe A is justified.
   The next step is to determine weather the second increment of investment (i.e., increasing the investment from $10,000 in lathe A to $15,000 in lathe B) is justified. An easy way to obtain the solution is to calculate the year-by-year difference in net cash flow for the two projects and then to find the IRR on the difference. In order for this year-by-year difference in net cash flow to be computed, the cash flows for each project must be shown for the same number of years (length of study period). The study period should be a common multiple of the lives of the projects under consideration, or the length of time during which the services of the chosen alternatives will be needed, whichever is less. For the example lathes, a study period of 10 years will be used as illustrated here:

Year	Lathe A	Lathe B	Difference Lathe B − Lathe A
0	−$10,000	−$15,000	−$5,000
1	+ 2,800	+ 3,000	+ 200
2	+ 2,800	+ 3,000	+ 200
3			
4			
5	−$8,000		+$8,000
6			
7			
8			
9			
10	+ 2,800 + $2,000	+ 3,000	+ 200 − $2,000

The equation expressing the present worth of the net cash flow for the difference between the two lathes is

$$-\$5,000 + \$200(P/A,i\%,10) + \$8,000(P/F,i\%,5) - \$2,000(P/F,i\%,10) = 0.$$

By using a financial calculator or computer software, a value of 12.1337% (~12.1%) yields a present worth of zero for the increment of investment. Thus, the IRR on the incremental investment is approximately 12.1%. Because the MARR is 10% and the IRR > the MARR, the increment of investment is justified. Hence, lathe B is preferred to lathe A.

The IRR for the individual alternatives can be obtained by finding the interest rate that equates their individual present worths to zero. For lathe A (the same "single investment opportunity" given in Example 6-1), the IRR is approximately 16.5%. For lathe B, a value of approximately 15.1% is obtained for the IRR by finding the interest rate that satisfies the following equality:

$$-\$15,000 + \$3,000(P/A, i\%, 10) = 0.$$

On the basis of the results obtained, it is clear that B would be chosen so long as the MARR is not greater than approximately 12.1%. Even though the IRR for B is slightly greater than 15%, for a MARR of 15%, lathe A should be chosen, not lathe B. In effect, lathe B earns 16.5% on the first $10,000 of investment and 12.1% on the additional $5,000 increment; overall, its IRR is 15.1%. If the MARR is 15%, then we would be better off financially to invest $10,000 in lathe A and earn 16.5%; the balance of $5,000 should remain in the "opportunity fund" where it is expected to earn at least the MARR of 15%.   ■

The Excel spreadsheet results for the lathe example are provided in Figure 6-1. To determine the IRR for a series of cash flows, the IRR function is used. As shown in Figure 6-1, IRR values are obtained by including within the parentheses the cash flows for which the IRR is to be determined. Entering = IRR(B5:B15) in a cell results in the rate of return that equates to zero the discounted present worth of cash flows B5 through B15.

6.1.4 Comparing Numerous Alternatives

The next example is given to further illustrate the principle that the return on each increment of investment capital should be justified. To make the computations easier, each alternative in this example has a uniform series of annual savings or returns and a salvage value equal to the investment. In such cases, the IRR can be calculated directly by dividing the annual net cash, inflow, or savings by the investment amount. In the tabulated solution shown, the symbol Δ is used to mean "incremental" or "change in." The letters on each end of the arrows designate the projects for which the increment is considered.

Example 6-3

	Alternative investment					
	A	B	C	D	E	F
Investment:	$1,000	$1,500	$2,500	$4,000	$5,000	$7,000
Annual savings in						
cash disbursements:	150	375	500	925	1,125	1,425
Salvage value:	1,000	1,500	2,500	4,000	5,000	7,000

If the company is willing to invest any capital that will earn at least 18%, find which alternative, if any, should be chosen using the IRR method.

Solution   It should be noted that the alternatives are arranged in order of increasing investment amount and that calculations regarding an increment must be completed before one knows which increment to consider next. The symbol ΔIRR means internal rate of return (IRR)

	A	B	C	D
1	MARR=	10%		
2				
3			Lathe	Increment
		A	B	B – A
4	EOY	Cash Flow	Cash Flow	Cash Flow
5	0	–$10,000.00	–$15,000.00	–$5,000.00
6	1	$2,800.00	$3,000.00	$200.00
7	2	$2,800.00	$3,000.00	$200.00
8	3	$2,800.00	$3,000.00	$200.00
9	4	$2,800.00	$3,000.00	$200.00
10	5	–$5,200.00	$3,000.00	$8,200.00
11	6	$2,800.00	$3,000.00	$200.00
12	7	$2,800.00	$3,000.00	$200.00
13	8	$2,800.00	$3,000.00	$200.00
14	9	$2,800.00	$3,000.00	$200.00
15	10	$4,800.00	$3,000.00	–$1,800.00
16	PW =	$3,008.50	$3,433.70	$425.20
17	AW =	$489.62	$558.82	$69.20
18	FW =	$7,803.28	$8,906.14	$1,102.85
19	IRR =	16.4763%	15.0984%	12.1337%

	A	B	C	D
19	IRR =	=IRR(B5:B15)	=IRR(C5:C15)	=IRR(D5:D15)

Figure 6-1 Spreadsheet for (PW, AW, FW, and IRR values for the lathes) in Example 6-2.

on incremental investment.

Increment considered	A	B	B → C	B → D	D → E	E → F
ΔInvestment:	$1,000	$1,500	$1,000	$2,500	$1,000	$2,000
ΔAnnual savings:	$150	$375	$125	$550	$200	$300
ΔIRR:	15%	25%	12.5%	22%	20%	15%
Is increment justified?	No	Yes	No	Yes	Yes	No

By the preceding analysis, alternative E would be chosen, because it is the alternative requiring the highest investment for which each increment of investment capital is justified. Note that the analysis was performed without even considering the IRR on the total investment for each of the alternatives.

In choosing alternative E, several increments of investment were justified, as shown in the following table:

Increment	Investment	Internal rate of return on increment (ΔIRR)
B	$1,500	25%
B → D	2,500	22
D → E	1,000	20
Total:	$5,000	

∎

Finally, although the special structure of the example does not require knowledge of the durations of the investments, for demonstration purposes, assume that each investment has a duration of two years. The Excel spreadsheet for the example is provided in Figure 6-2. Notice that, by the PW, AW, and FW methods, alternative E is preferred. Also, notice that alternative B (in column C) has the highest overall IRR, and that the alternative with the greatest investment (alternative F in column G) has an overall IRR greater than the minimum of 18%. Nevertheless, alternative E would be chosen by using the IRR method on the rationale that the company wants to invest any increment of capital when and only when that increment will earn at least the minimum attractive rate of return.

6.1.5 Comparing Alternatives When Disbursements Only Are Known

When disbursements only are known, IRRs can be calculated for incremental investments only and not for the investment in any one alternative. Thus, the lowest investment has to be assumed to be justified (or necessary) without being able to calculate the IRR on that alternative. As an example, consider the same alternative compressors compared in the last chapter, and determine which is the better alternative.

	A	B	C	D	E	F	G
1	MARR =	18%					
2	EOY	CF[A]	CF[B]	CF[C]	CF[D]	CF[E]	CF[F]
3	0	−$1,000	−$1,500	−$2,500	−$4,000	−$5,000	−$7,000
4	1	$150	$375	$500	$925	$1,125	$1,425
5	2	$1,150	$1,875	$3,000	$4,925	$6,125	$8,425
6	PW =	−$46.97	$164.39	$78.28	$320.96	$352.27	$258.33
7	AW =	−$30.00	$105.00	$50.00	$205.00	$225.00	$165.00
8	FW =	−$65.40	$228.90	$109.00	$446.90	$490.50	$359.70
9	IRR =	15.0%	25.0%	20.0%	23.1%	22.5%	20.4%

Figure 6-2   PW, AW, FW, and IRR values for the multiple investment alternatives (Example 6-3).

Example 6-4

	Compressor	
	I	II
First cost:	$6,000	$8,000
Life:	6 yr	9 yr
Salvage value:	$1,000	$0
Annual operating disbursements:	$4,000	$3,200
MARR = 15%		
Study period = 18 yr		

Solution   Listing the cash flows for the lowest common multiple of lives yields the values given in Figure 6-3. Note that, for compressor I, the net cash flow in year 6 is obtained by summing the annual disbursement in year 6 (−$4,000), the salvage value at the end of year 6 ($1,000), and the first cost for the replacement compressor (−$6,000), yielding a value of −$9,000.

To use the IRR method, rank the alternatives in increasing order of investment (compressor I first and then compressor II). Next, determine the incremental investment required to invest in the alternative with the greatest initial investment (compressor II) by subtracting the cash flows for compressor I from those required for compressor II. The IRR for the difference between the compressors (i.e., on the incremental investment) is obtained by solving the following equation for the difference in the net cash flows:

$$-\$2,000 - \$8,000(P/F, i\%, 9) - \$1,000(P/F, i\%, 18) + \$800(P/A, i\%, 18)$$
$$+\$5,000(P/F, i\%, 6) + \$5,000(P/F, i\%, 12) = 0.$$

By trial and error, the IRR can be found to be approximately 47%. Since this return on the increment of investment is greater than the MARR of 15%, compressor II is the economical choice. (As shown in Figure 6-3, with Excel, the IRR obtained for the incremental investment is 46.878%.) It should be noted that the IRR method can be used even though we are unable to compute the IRR for either compressor I or compressor II.

	A	B	C	D
1	MARR=	15%		
2			*Compressor*	Increment
3		I	II	II – I
4	EOY	Cash Flow	Cash Flow	Cash Flow
5	0	–$6,000	–$8,000	–$2,000
6	1	–$4,000	–$3,200	$800
7	2	–$4,000	–$3,200	$800
8	3	–$4,000	–$3,200	$800
9	4	–$4,000	–$3,200	$800
10	5	–$4,000	–$3,200	$800
11	6	–$9,000	–$3,200	$5,800
12	7	–$4,000	–$3,200	$800
13	8	–$4,000	–$3,200	$800
14	9	–$4,000	–$11,200	–$7,200
15	10	–$4,000	–$3,200	$800
16	11	–$4,000	–$3,200	$800
17	12	–$9,000	–$3,200	$5,800
18	13	–$4,000	–$3,200	$800
19	14	–$4,000	–$3,200	$800
20	15	–$4,000	–$3,200	$800
21	16	–$4,000	–$3,200	$800
22	17	–$4,000	–$3,200	$800
23	18	–$3,000	–$3,200	–$200
24	IRR =			46.878%

Figure 6-3   IRR comparison of the two compressors (Example 6-4).

The use of AW calculations in determining the IRR on an incremental investment provides a convenient shortcut for problems in which the two "repeatability" assumptions stated in Chapter 5 (regarding the period of needed service and the repeatability of cash flows) hold true. For the preceding compressor problem, the IRR may be found by solving the following:

$$(\$6,000 - \$1,000)(A/P, i\%, 6) + \$1,000(i\%) + \$4,000$$
$$= \$8,000(A/P, i\%, 9) + \$3,200.$$

As with the PW approach, the equation is satisfied for an interest rate of approximately 47%. ■

6.1.6 Differences in Ranking of Investment Opportunities

It was pointed out previously that the IRR method will always give results that are consistent (regarding project acceptance or rejection) with results, using the PW, AW, or FW method. However, the IRR method can give a different *ranking* of the order of desirability of individual investment opportunities than the PW, AW, or FW method. As an example, consider Fig. 6-4 depicting the relation of IRR to net present worth for two investment opportunities, projects X and Y.

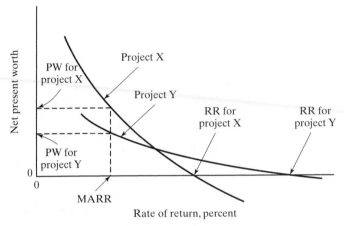

Figure 6-4   Relation of IRR to PW for projects X and Y.

The IRR for each project is the rate at which the net present worth for that project is zero. The net present worth for each project is shown for a typical interest rate. For the hypothetical, but quite feasible, relationship shown in Fig. 6-4, project Y has the higher IRR while project X has the higher net present worth for all IRRs less than the rate at which the net present worths are equal. This illustrates a case in which the IRR method does result in a different ranking of alternatives as compared with the PW (AW or FW) method. However, since both projects had a net present worth greater than zero, and the IRR for both projects is greater than the minimum attractive rate of return, the determination of acceptance of both projects is consistently shown by either method.

6.1.7 Problems in Which either No Solution or Several Solutions for Rate of Return Exist

It is possible, but not commonly experienced in practice, to have situations in which there is no single IRR solution by the discounted cash flow method. Descartes' rule of signs indicates that multiple solutions can occur whenever the cash flow series reverses sign (from net outflow to net inflow or the opposite) more than once over the period of study. As an example, consider the following project for which the IRR is desired:

Example 6-5
Find the IRR(s) for these data:

Year	Net cash flow
−1	+$ 500
0	− 1,000
1	0
2	+ 250
3	+ 250
4	+ 250

Solution

Year	Net cash flow	PW @ 35%		PW @ 63%	
		Factor	Amount	Factor	Amount
−1	+$ 500	1.35	+$ 676	1.63	+$ 813
0	− 1,000	1.00	− 1,000	1.00	− 1,000
1	0				
2	+ 250	0.55	+ 137	0.38	95
3	+ 250	0.41	+ 102	0.23	57
4	+ 250	0.30	+ 75	0.14	35
		Net PW:	Σ = 0		Σ = 0

Thus, the present worth of the net cash flows equals 0 for interest rates of 35% and 63%. Whenever multiple answers such as these exist, it is likely that neither is correct.   ■

   An effective way to overcome this difficulty and obtain a "correct" answer is to manipulate cash flows as little as necessary so that there is only one reversal of the cumulative net cash flow. This can be done by using the minimum attractive rate of return to manipulate the funds, and then solving by the discounted cash flow method. For Example 6-5, if the minimum attractive rate of return is 10%, the + $500 at year − 1 can be compounded to year 0 to be $500($F/P$,10%,1) = + $550. This, added to the − $1,000 at year 0, equals −$450. The − $450, together with the remaining cash flows, which are all positive, now fits the condition of only one reversal in the cumulative net cash flow. The interest rate at which the present worth of the net cash flows equals 0 can now be shown to be 19% as shown in the following table:

Timing	Net cash flow	PW @ 19%	
		Factor	Amount
0	−$450	1.00	−$450
2	+ 250	0.70	+ 177
3	+ 250	0.59	+ 150
4	+ 250	0.48	+ 123
	Net PW:		Σ = 0

It should be noted that, whenever a manipulation of net cash flows is performed, the calculated IRR will vary depending on what cash flows are manipulated and at what interest rate. The less manipulation and the closer the minimum rate of return is to the calculated rate of return, the less is the variation in the final calculated rate of return.

   Appendix 6-A gives another example of the preceding type of problem and shows how the ERR method can be used to obtain a solution.

6.2 External Rate of Return (ERR) Method

Implicit in the internal rate of return method is the assumption that recovered funds are reinvested at a rate equal to the internal rate of return. Since it is often the case that opportunities do not exist for investing recovered funds and earning such a rate, the notion of an explicit reinvestment rate of return method has appeal. Additionally, the possibility of multiple solutions using the internal rate of return method can result in misinterpretations and misunderstandings of the rate of return figure obtained.

Because of the desire to incorporate explicitly the reinvestment rate in rate of return calculations and the desire to use a method that will yield a unique solution, an external rate of return method was developed. If complicated manipulations are to be avoided, the explicit reinvestment rate of return method is limited to a single invest-ment (negative cash flow) and a uniform series of returns (positive cash flows). A more general approach is to define the external rate of return as the interest rate that equates the future worth of investments to the accumulation of reinvested returns. Recovered funds are assumed to be reinvested at the minimum attractive rate of return, based on the opportunity cost concept. To illustrate, consider the same single project for which rates of return were calculated at the beginning of the chapter.

Example 6-6

Determine the ERR of these cash flows:

Capital investment:	$10,000
Project life:	5 yr
Salvage value:	$2,000
Annual receipts:	$5,000
Annual disbursements:	$2,200

Solution The following solution by the external rate of return (ERR) method uses a reinvestment rate or minimum attractive rate of return (MARR) of 10%:

Year	Net annual cash flow
0	−$10,000
1–5	2,800
5	2,000

Future accumulation of recovered monies = $2,800(F/A, 10%, 5) + $2,000 = $19,094$. Future worth of investments = $10,000(F/P, i\%, 5)$. Thus,

$$10,000(F/P, i\%, 5) = $19,094$$

or

$$(F/P, i\%, 5) = 1.9094.$$

From the interest tables it is seen that

i	$(F/P, i\%, 5)$
12%	1.7623
15%	2.0114

By using linear interpolation, a value of 1.9094 is obtained for the $(F/P, i\%, 5)$ factor when $i\%$ equals, approximately, 13.77%. The exact value of $i\%$ can be obtained by using logarithms and recalling the mathematical relation represented by the $(F/P, i\%, N)$ factor, namely,

$$(1 + i)^N = (F/P, i\%, N).$$

Hence, for the example,

$$(1+i)^5 = 1.9094,$$
$$5\ln(1+i) = \ln 1.9094,$$
$$\ln(1+i) = 0.1293,$$
$$1+i = \ln^{-1}(0.1293),$$

and

$$i = 0.1381.$$

Thus, an external rate of return of 13.81% will result from the project. Since 13.81% is greater than the MARR of 10%, the project is justified economically.

   In general, a project justified economically from using the internal rate of return method will also be justified by using the external rate of return method. A unique solution always occurs using the ERR method.   ■

   The ERR can be obtained relatively easily using Excel. As shown in Figure 6-5, the original set of cash flows {CF} is represented as the union of two sets of cash flows: {CF(–)}, containing the negative-valued cash flows, and {CF(+)}, containing the positive-valued cash flows. The FV function is used to determine the future worth of {CF(+)}. The FV obtained is entered in the last cell in {CF(–)} to generate a new cash flow profile {CF}. Since the external rate of return is the interest rate that makes the future worth of {CF(–)} equal to the FV, the IRR function can be used to determine the interest rate that yields a value of zero for the FW, AW, or PW of {CF}. As shown, the ERR value obtained in cell E10 is 13.81%.

   For cash flow profiles involving a single negative cash flow, the Modified Internal Rate of Return (MIRR) function can be used to obtain ERR values. Its parameters are MIRR(values,finance_rate,reinvest_rate). The finance rate is the interest rate paid if the investment is borrowed; the reinvestment rate is the MARR. As shown in Figure 6-5, by entering =MIRR(B3:B8,,B1) the MIRR function calculated the interest rate that made the future value of the negative-valued cash flows equal to the future value of the positive-valued returns, assuming the latter are reinvested at the MARR. As shown in cell B11, the MIRR value obtained is also 13.81%.

   If, however, the cash flow profile does not fit the assumptions of the MIRR function, erroneous solutions can result. For instance, in the situation just considered, suppose the $10,000 investment is spread over a 2-year period as shown in Figure 6-6. The ERR is 12.21%. However, the MIRR value obtained is 11.38%.

6.2.1 Comparing Alternatives by Using the External Rate of Return Method

   An incremental approach is required when comparing investment alternatives by using any RR method. Specifically, the external rate of return on incremental investment is defined to be the interest rate that equates the future worth of the incremental investment and the future accumulation of incremental positive-valued cash flows. To illustrate, consider the lathe example presented previously.

	A	B	C	D	E
1	MARR =	10%			
2	EOY	CF	CF(−)	CF(+)	{CF}
3	0	−$10,000	$10,000	$0	−$10,000
4	1	$2,800	$0	$2,800	$0
5	2	$2,800	$0	$2,800	$0
6	3	$2,800	$0	$2,800	$0
7	4	$2,800	$0	$2,800	$0
8	5	$4,800	$0	$4,800	$19,094
9	FW =			$19,094	
10	ERR =				13.81%
11	MIRR =	13.81%			

	A	B	C	D	E
8	5	$4,800	$0	$4,800	=D9
9	FW =			=FV(B1,5,−2800)+2000	
10	ERR =				=IRR(E3:E8)
11	MIRR =	=MIRR(B3:B8,,B1)			

Figure 6-5 Spreadsheet solution for the external rate of return (ERR).

	A	B	C	D	E
1	MARR =	10%			
2	EOY	CF	CF(−)	CF(+)	{CF}
3	0	−$6,000	$6,000	$0	−$6,000
4	1	−$4,000	$4,000	$0	−$4,000
5	2	$2,800	$0	$2,800	$0
6	3	$2,800	$0	$2,800	$0
7	4	$2,800	$0	$2,800	$0
8	5	$2,800	$0	$2,800	$0
9	6	$4,800	$0	$4,800	$19,094
10	FW =			$19,094	
11	ERR =				12.21%
12	MIRR =	11.38%			

Figure 6-6   Example of incorrect calculation of the external rate of return (ERR), using the MIRR function.

Example 6-7

Use the ERR method to select either lathe A or lathe B.

	Lathe	
	A	B
First cost:	$10,000	$15,000
Life:	5 yr	10 yr
Salvage value:	$2,000	$0
Annual receipts:	$5,000	$7,000
Annual disbursements:	$2,200	$4,000
Minimum attractive rate of return = 10%		
Study period = 10 yr		

Solution   It was shown previously that lathe A has an ERR of 13.81% and is economically justified. It remains to determine if the incremental investment required to purchase lathe B is justified. Over the 10-year study period, the incremental cash flows have the following profile:

End of year	CF(B) − CF(A)
0	−$5,000
1–4	+ 200
5	+ 8,200
6–9	+ 200
10	− 1,800

Future accumulation of positive-valued cash flows equals

$$\$8,000(F/P,10\%,5) + \$200(F/A,10\%,9)(F/P,10\%,1) = \$15,871.$$

Future worth of negative-valued cash flows equals

$$\$5,000(F/P,i\%,10) + \$1,800.$$

Hence,

$$\$5,000(F/P,i\%,10) + \$1,800 = \$15,871$$

or

$$(F/P, i\%, 10) = 2.8142.$$

Thus,

$$\log(1 + i\%)^{10} = \log 2.8142.$$

Solving for the ERR gives a value of 10.9%, which is greater than the MARR; hence, the incremental investment required for B is justified. ■

Because of the negative incremental cash flow in the tenth year, the MIRR function cannot be used to determine the ERR. However, as shown in Figure 6-7, Excel and the IRR function can be used to determine the ERR value by subdividing the set of cash flows as before. The ERR on the incremental investment required to purchase lathe B is 10.9%, as previously determined.

	A	B	C	D	E	F	G
1	MARR =	10%					
2			*Lathe*		Increment		
4	EOY	CF(A)	CF(B)	CF(B–A)	CF(–)	CF(+)	{CF}
5	0	–$10,000	–$15,000	–$5,000	$5,000	$0	–$5,000
6	1	$2,800	$3,000	$200	$0	$200	$0
7	2	$2,800	$3,000	$200	$0	$200	$0
8	3	$2,800	$3,000	$200	$0	$200	$0
9	4	$2,800	$3,000	$200	$0	$200	$0
10	5	–$5,200	$3,000	$8,200	$0	$8,200	$0
11	6	$2,800	$3,000	$200	$0	$200	$0
12	7	$2,800	$3,000	$200	$0	$200	$0
13	8	$2,800	$3,000	$200	$0	$200	$0
14	9	$2,800	$3,000	$200	$0	$200	$0
15	10	$4,800	$3,000	–$1,800	$1,800	$0	$14,071.56
16	FW =	$7,803.28	$8,906.14	$1,102.85		$15,871.56	
17	ERR =						10.90%
18	MIRR =			8.85%			

Figure 6-7   Correct and incorrect calculations of the ERR using the IRR and MIRR functions.

6.2.2 Comparing Alternatives When Disbursements Only Are Known

When disbursements only are known, the external rate of return method is applied to incremental cash flows in the same manner as other rate of return methods. Thus, the alternative having the lowest investment is assumed to be justified (or necessary). As an illustration, consider again the compressor example.

Example 6-8

	Compressor	
	I	II
First cost:	$6,000	$8,000
Life:	6 yr	9 yr
Salvage value:	$1,000	$0
Annual operating disbursements:	$4,000	$3,200
Minimum attractive rate of return = 15%		
Study period = 18 yr		

Solution

End of year	II– I
0	−$2,000
1	800
2	800
3	800
4	800
5	800
6	800 + $5,000 = $5,800
7	800
8	800
9	800 − $8,000 = −$7,200
10	800
11	800
12	800 + $5,000 = $5,800
13	800
14	800
15	800
16	800
17	800
18	800 − $1,000 = −$200

Future accumulation of positive-valued cash flows equals

$$\$800(F/A,15\%,18) - \$8000[(F/P,15\%,9)] + \$5,000[(F/P,15\%,12)+(F/P,15\%,6)]$$
$$= \$95,371.34.$$

Future worth of negative-valued cash flows equals

$$\$2,000(F/P,i\%,18) + \$7,200(F/P,i\%,9) + \$200.$$

$$@i = 20\% \qquad \$90,597.09 < \$95,371.34$$
$$@i = 25\% \qquad \$164,866.48 > \$95,371.34$$

Interpolating for the $i\%$ value that equates the two cash flows yields an ERR estimate of 20.32%, which is greater than the MARR of 15%; hence, compressor II is recommended, using the ERR method. An Excel solution is depicted in Figure 6-8, with a more accurate ERR estimate of 20.43%.

The ERR solution for the compressor investment constructed the two sets using the net cash incremental cash flows. Alternately, one could have constructed the two sets as shown in Figure 6-9, which includes in the set $\{CF(-)\}$ each replacement investment and the loss of a salvage value at the end of the study period. In such a case, the resulting ERR is reduced to 18.69%, which is still greater than the MARR.

Recall, the external rate of return is intended to represent the *return on invested capital* when *recovered capital* is reinvested at the MARR. Therefore, in setting up the problem, it is important to include the *invested capital* in the set $\{CF(-)\}$ and the

	A	B	C	D	E	F	G
1	MARR =	15%					
2		*Compressor*		Increment			
4	EOY	CF(I)	CF(II)	CF(II–I)	CF(–)	CF(+)	{CF}
5	0	–$6,000	–$8,000	–$2,000	$2,000	$0	–$2,000
6	1	–$4,000	–$3,200	$800	$0	$800	$0
7	2	–$4,000	–$3,200	$800	$0	$800	$0
8	3	–$4,000	–$3,200	$800	$0	$800	$0
9	4	–$4,000	–$3,200	$800	$0	$800	$0
10	5	–$4,000	–$3,200	$800	$0	$800	$0
11	6	–$9,000	–$3,200	$5,800	$0	$5,800	$0
12	7	–$4,000	–$3,200	$800	$0	$800	$0
13	8	–$4,000	–$3,200	$800	$0	$800	$0
14	9	–$4,000	–$11,200	–$7,200	$7,200	$0	–$7,200
15	10	–$4,000	–$3,200	$800	$0	$800	$0
16	11	–$4,000	–$3,200	$800	$0	$800	$0
17	12	–$9,000	–$3,200	$5,800	$0	$5,800	$0
18	13	–$4,000	–$3,200	$800	$0	$800	$0
19	14	–$4,000	–$3,200	$800	$0	$800	$0
20	15	–$4,000	–$3,200	$800	$0	$800	$0
21	16	–$4,000	–$3,200	$800	$0	$800	$0
22	17	–$4,000	–$3,200	$800	$0	$800	$0
23	18	–$3,000	–$3,200	–$200	$200	$0	$95,171.34
26	FW =					$95,371.34	
27	ERR =						20.43%

Figure 6-8   ERR comparison of the two compressors.

	A	B	C	D	E	F	G
1	MARR =	15%					
2		*Compressor*		Increment			
4	EOY	CF(I)	CF(II)	CF(II-I)	CF(–)	CF(+)	{CF}
5	0	–$6,000	–$8,000	–$2,000	$2,000	$0	–$2,000
6	1	–$4,000	–$3,200	$800	$0	$800	$0
7	2	–$4,000	–$3,200	$800	$0	$800	$0
8	3	–$4,000	–$3,200	$800	$0	$800	$0
9	4	–$4,000	–$3,200	$800	$0	$800	$0
10	5	–$4,000	–$3,200	$800	$0	$800	$0
11	6	–$9,000	–$3,200	$5,800	$0	$11,800	–$6,000
12	7	–$4,000	–$3,200	$800	$0	$800	$0
13	8	–$4,000	–$3,200	$800	$0	$800	$0
14	9	–$4,000	–$11,200	–$7,200	$7,200	$800	–$8,000
15	10	–$4,000	–$3,200	$800	$0	$800	$0
16	11	–$4,000	–$3,200	$800	$0	$800	$0
17	12	–$9,000	–$3,200	$5,800	$0	$11,800	–$6,000
18	13	–$4,000	–$3,200	$800	$0	$800	$0
19	14	–$4,000	–$3,200	$800	$0	$800	$0
20	15	–$4,000	–$3,200	$800	$0	$800	$0
21	16	–$4,000	–$3,200	$800	$0	$800	$0
22	17	–$4,000	–$3,200	$800	$0	$800	$0
23	18	–$3,000	–$3,200	–$200	$200	$800	$144,765.51
26	FW =					$144,965.51	
27	ERR =						18.69%

Figure 6-9   Alternate ERR comparison of the two compressors.

recovered capital in the set {CF(+)}. For individual investments, that is generally easy to determine. However, when the cash flow profile represents incrementalinvestments and incremental recovered capital, sometimes how the comparison should be performed is not as obvious. For that reason, we recommend that incremental analysis be based on net incremental cash flows, rather than creating entries for the {CF(–)} set.

It is important to remember that the resulting recommendation of the preferred alternative, using the ERR method will be consistent with the recommendation made by using PW, AW, FW, or IRR methods of comparison. Also, while the ERR method offers the possibility of ambiguity when dealing with incremental cash flows, incremental cash flows are more likely to produce situations in which the IRR method produces multiple values. It was precisely because of the complexity introduced by having multiple IRR values that the ERR method was developed.

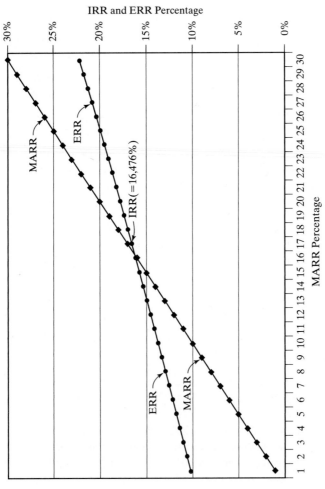

Figure 6-10 ERR vs. MARR for Lathe A.

Finally, when using the ERR method, it is useful to remember its relationship with the IRR and MARR. Namely, the following ordering will occur: either IRR > ERR > MARR, or IRR < ERR < MARR, or IRR = ERR = MARR. For the investment required to purchase Lathe A, Figure 6-10 plots the ERR values that would result from various MARR values; the intersection point for the two curves occurs at MARR = 16.4763%, which is the IRR for the investment.

In addition to the internal rate of return and external rate of return methods of calculating rates of return, other so-called rate of return and investment recovery measures of merit are used in practice. Three such measures are described briefly in Appendix 6–B.

PROBLEMS

6-1. Determine which of the independent investment opportunities shown in the accompanying table should be undertaken. Use a MARR of 12% and (a) the IRR method, (b) the ERR method.

	Investment opportunity				
	A	B	C	D	E
Initial investment:	$8,000	$10,000	$12,000	$15,000	$16,000
Annual net receipts:	2,000	3,000	3,200	3,800	4,400
Salvage value:	1,000	−1,000	1,000	2,000	500
Study period = 5 yr					

6-2. In Problem 6-1, suppose A through E constitute a set of feasible, mutually exclusive investment alternatives. Determine which should be chosen using (a) the IRR method, (b) the ERR method.

6-3. Determine which of the independent (nonrepeating) investment opportunities featured in the accompanying table are economically worthwhile. Use a MARR of 10%, a study period of 10 years, and (a) the IRR method, (b) the ERR method.

EOY	Cash flows for investment opportunity					
	A	B	C	D	E	F
0	−$5,000	−$5,000	−$8,000	−$12,000	$ 0	−$15,000
1	− 5,000	800	1,000	1,500	0	1,000
2	1,000	1,200	1,000	1,500	0	1,500
3	1,000	1,600	1,000	1,500	− 6,000	2,000
4	2,000	2,000	1,000	1,500	2,000	2,500
5	2,000	2,400	1,000	1,500	2,000	3,000
6	3,000	2,800	1,000	1,500	2,000	3,500
7	3,000	0	1,000	1,500	2,000	4,000
8	4,000	0	1,000	1,500	2,000	4,500
9	4,000	0	1,000	1,500	0	5,000
10	− 5,000	0	− 2,000	3,000	0	5,500

6-4. In Problem 6-3, suppose A through F in the previous table constitute a set of feasible, mutually exclusive investment alternatives. Which should be chosen based on (a) the IRR method? (b) the ERR method?

6-5. Determine which of the independent (nonrepeating) investment opportunities shown in the accompanying table are economically worthwhile. Use a MARR of 10%, a study period of 5 years, and (a) the IRR method, (b) the ERR method.

	Cash flows for investment opportunity					
EOY	A	B	C	D	E	F
0	−$10,000	−$5,000	−$8,000	−$12,000	−$6,000	−$15,000
1	2,000	1,000	3,000	5,000	1,000	4,500
2	2,000	1,500	3,000	5,000	1,500	4,500
3	2,000	2,000	5,000	6,000	2,000	4,500
4	3,000	2,500	0	0	2,500	4,500
5	3,000	0	0	0	3,000	0

6-6. In Problem 6-5, in the previous table suppose A through F constitute a set of feasible, mutually exclusive investment alternatives. Determine the economic choice using (a) the IRR method, (b) the ERR method.

6-7. If money is worth 8% to you, use the ERR method to determine which (if any) of the following investment opportunities you would favor:

EOY	CF(A)	CF(B)	CF(C)
0	−$12,500	−$12,500	−$12,500
1	$1,000	$5,000	$3,000
2	2,000	4,000	3,000
3	3,000	3,000	3,000
4	4,000	2,000	3,000
5	5,000	1,000	3,000

6-8. In Problem 6-7, suppose the three cash flow profiles are for feasible, mutually exclusive investment alternatives and one must be chosen. Further, suppose your time value of money is 5%. Using the ERR method, which would you choose?

6-9. Suppose you compute the internal rate of return for an investment alternative and find that it equals 13%. Will the external rate of return be equal to, less than, or greater than 13% if the MARR is equal to (a) 5%, (b) 10%, (c) 13%, (d) 15%?

6-10. Four mutually exclusive investment alternatives are illustrated in the accompanying table; the "do-nothing" alternative is not feasible. The cash flows are shown for a 7-year planning horizon. The MARR is 15%. Determine the preferred alternative using (a) the IRR method, (b) the ERR method.

EOY	CF(A)	CF(B)	CF(C)	CF(D)
0	−$ 50,000	−$100,000	−$250,000	−$225,000
1	− 100,000	− 100,000	+ 75,000	− 75,000
2–7	+ 50,000	+ 70,000	+ 75,000	+ 100,000

6-11. Four mutually exclusive investment alternatives are illustrated in the accompanying table; the "do-nothing" alternative is not feasible. The cash flows are shown for a 10-year planning horizon. The MARR is 12%. Determine the preferred alternative, using (a) the IRR method, (b) the ERR method.

EOY	CF(A)	CF(B)	CF(C)	CF(D)
0	−$ 50,000	−$125,000	−$200,000	−$100,000
1	− 100,000	− 75,000	+ 50,000	− 150,000
2–5	+ 50,000	+ 70,000	+ 50,000	75,000
6–9	+ 50,000	+ 70,000	+ 75,000	85,000
10	+ 75,000	+ 70,000	+ 75,000	90,000

6-12. A construction firm is considering leasing a crane needed on a project for 5 years for $150,000 payable at the beginning of each year. The alternative is to buy a crane now for $500,000 and sell it at the end of 5 years for $100,000. Annual maintenance costs for ownership only are expected to be $15,000 per year for the first 2 years and $20,000 per year for the last 3 years. At what interest rate (IRR) are the two alternatives equivalent?

6-13. An industrial machine costing $5,000 will produce net savings of $2,000 per year. The machine has a 5-year economic life, but must be returned to the factory for major repairs after 3 years of operation. These repairs will cost $2,500. The company's MARR is 12%. What IRR will be earned on the purchase of this machine? Do you recommend it?

6-14. An improved facility costing $250,000 has been proposed. Construction time will be 2 years with expenditures of $100,000 the first year and $150,000 the second year. Savings beginning the first year after construction completion are as follows:

Year	Savings
1	$ 30,000
2	50,000
3	70,000
4	90,000
5	110,000

The facility will not be required after 5 years and will have a salvage value of $25,000. Determine the (a) IRR and (b) ERR using a reinvestment rate of 10%.

6-15. A distillery is considering the erection of a bottle-making plant. The number of bottles needed annually is estimated at 600,000. The initial cost of the facility is $150,000, with an estimated life of 10 years. Annual operation and maintenance costs are expected to be $40,000, and annual taxes and insurance $15,000. Should the distillery erect the bottle-producing facility or buy the bottles from another company at $0.15 each? Use the IRR and ERR methods with a minimum attractive rate of 12%.

6-16. What is the internal rate of return on an investment of $2,000 that will yield $528 per year for 5 years?

6-17. There are five alternative machines to do a given job. Each is expected to have a salvage value of 100% of the investment amount at the end of its life of 4 years. If the firm's minimum attractive rate of return is 10%, which machine is the best choice based on the data shown? Use (a) an internal rate of return comparison, and (b) an external rate of return comparison.

	Machine				
	A	B	C	D	E
Investment:	$1,000	$1,400	$2,100	$2,700	$3,400
Net cash flow per yr:	110	180	280	340	445

6-18. A firm is considering purchasing a new machine to increase output of an existing production process. If each of these machines provides the same service over its useful life and the MARR is 12%, which machine should be selected, based on an internal rate of return comparison and applying the repeatability assumption?

	Mutually Exclusive Alternative			
	A	B	C	D
Initial investment	$75,000	$75,000	$100,000	$50,000
Net annual revenue	$20,435	$16,212	$22,675	$17,879
Useful life (years)	5	6	10	4
Market value (at end of useful life)	$15,000	$12,000	$25,000	$12,000

6-19. Two investment alternatives are under consideration. Alternative A requires an initial investment of $100,000 and will generate net annual returns of $20,000 per year for 6 years; it will have a $30,000 salvage value at the end of its useful life of 6 years. Alternative B also requires an initial investment of $100,000; it will produce a $10,000 annual gradient series of annual returns, with the last annual return equaling $60,000 at the end of its 7-year useful life. If the two alternatives are considered by using (a) an internal rate of return comparison and (b) an external rate of return comparison, with a MARR of 10% per year, which would be preferred, assuming the repeatability assumption holds?

6-20. Consider the two mutually exclusive alternatives shown in the accompanying table, both of which have 10-year useful lives and negligible salvage values. Based on (a) internal rate of return and (b) external rate of return comparisons, with a 10% MARR, which should be selected?

	Alternative	
	A	B
Initial investment	$780,000	$1,840,000
Net annual receipts	$138,060	$311,000

6-21. Three mutually exclusive investment alternatives are under consideration. The estimated cash flows for each alternative are given in table that follows. Based on the repeatability assumption, a 15% MARR, and an internal rate of return comparison, which alternative, if any, should be selected?

	Mutually Exclusive Alternative		
	A	B	C
Capital investment	$30,000	$60,000	$40,000
Annual costs	$16,000	$30,000	$25,000
Annual revenues	$28,000	$53,500	$38,000
Useful life (years)	5	5	6
Market value (at end of useful life)	$10,000	$10,000	$10,000

6-22. Given the data shown, use the same conditions as in Problem 6-21 to determine which alternative, if any, should be chosen.

	Mutually Exclusive Alternative		
	D	E	F
Capital investment	$22,000	$26,200	$17,000
Annual costs	$7,000	$7,500	$5,800
Annual revenues	$14,000	$15,000	$12,000
Useful life (years)	4	10	5
Market value (at end of useful life)	$4,000	$5,000	$3,500

6-23. Three mutually exclusive designs are illustrated in the accompanying table. Based on 10-year useful lives for all three designs, which should be selected, using (a) internal rate of return and (b) external rate of return comparisons, with a 10% MARR and negligible market values at the end of the 10-year study period?

	Mutually Exclusive Design		
	K	L	M
Initial investment	$170,000	$330,000	$300,000
Annual expenses	$70,000	$79,000	$64,000
Annual revenues	$114,000	$147,000	$130,000

6-24. Assume you have $80,000 available for investment and it is currently earning interest at an annual compound rate of 10%. Two investment alternatives have become available. One requires investing $80,000 now; the other requires investing $80,000 in 2 years from now, but a commitment is required now if it is to be pursued. In both cases, the investments will terminate 5 years from now. The cash flows for each alternative are shown in the table that follows. Which do you recommend, using (a) internal rate of return and (b) external rate of return comparisons?

	Mutually Exclusive Alternative	
Year	A	B
0	−$80,000	
1	$21,750	
2	$21,750	−$80,000
3	$21,750	$33,333
4	$21,750	$33,333
5	$21,750	$33,333

6-25. Consider two investment alternatives having cash flow profiles as shown in the table that follows. Use both an internal and an external rate of return comparison to determine the recommended alternative. Base your analysis on a study period of 5 years and a minimum attractive rate of return of (a) 10%, (b) 20%, (c) 40%.

End of year	A	B
0	$ 0	−$10,000
1	− 20,000	5,000
2	30,000	3,500
3	250	40,000
4	20,000	2,576
5	3,864	30,000

6-26. Two alternative machines will produce the same product, but one machine will produce higher quality items that can be expected to return greater revenue. Given the following data, determine which machine is better, using the IRR and ERR methods, a MARR of 10%, and a study period of 10 years:

	Machine A	Machine B
First cost:	$ 20,000	$ 35,000
Salvage value:	2,000	0
Annual receipts:	150,000	180,000
Annual disbursements:	138,000	163,000

6-27. An individual is faced with two mutually exclusive investment alternatives. By investing $15,000, a single sum of $18,500 will be received 2 years after the investment; alternatively, by investing $20,000, a single sum of $30,000 will be received 4 years after the investment. Using IRR and ERR methods, determine the preferred alternative based on a MARR of (a) 5%, (b) 10%, (c) 20%.

6-28. Using a 10-year study period, determine, from the data given in the accompanying table, the preferred incinerator alternative by using the following method(s): (a) IRR, (b) ERR.

	Incinerator				
	A	B	C	D	E
First cost:	$3,000	$3,800	$4,500	$5,000	$6,000
Salvage value:	0	0	0	0	0
Annual operating disbursements:	1,800	1,770	1,470	1,320	1,000
MARR = 10%					

6-29. Four mutually exclusive investment alternatives including the "do-nothing" alternative are illustrated in the table that follows. The cash flows are shown over a 7-year planning horizon. The MARR is 10%. Using the ERR method, determine the preferred alternative.

EOY	CF(A)	CF(B)	CF(C)	CF(D)
0	$0	−$100,000	−$180,000	−$125,000
1	0	− 100,000	− 20,000	− 75,000
2	0	0	+ 100,000	+ 60,000
3	0	+ 25,000	+ 75,000	+ 60,000
4	0	+ 50,000	+ 50,000	+ 60,000
5	0	+ 75,000	+ 25,000	+ 60,000
6	0	+ 100,000	0	+ 60,000
7	0	+ 125,000	+ 100,000	+ 60,000

6-30. Three order-picking systems are illustrated in the table that follows for use in a catalog distribution center. Alternative A involves picking items manually from bin shelving. With alternative B, items are picked manually from a horizontal carousel conveyor; hence, essentially no walking is required by the order picker. With alternative C, items are picked automatically by an automatic item retrieval machine. The alternatives have different space and labor requirements. Likewise, they have different acquisition, maintenance, and operating costs.

A study period of 10 years and a MARR of 12% are to be used. For the study period, the cash flow estimates for the alternatives are given.

	A	B	C
First cost:	$1,150,000	$1,250,000	$2,000,000
Salvage value:	750,000	750,000	875,000
Annual disbursements:	425,000	400,000	275,000

a. Using the IRR method, which alternative is preferred?

b. Would your preference change if the MARR were (i) 10%? (ii) 20%? (iii) 30%?

6-31. Given the mutually exclusive projects shown in the next table and assuming $N = \infty$, which investment should be selected if the company's MARR is 25%? Use the IRR method.

	A	B	C	D	E	F
Investment:	$30,000	$60,000	$75,000	$50,000	$55,000	$70,000
Annual savings:	10,000	18,000	21,500	14,000	16,000	20,500

6-32.[1] Consider the data provided in Problem 6-7. If they are mutually exclusive investment alternatives, determine the preferred alternative, using the net cash flow B/C method. For the preferred investment, determine the discounted payback period.

6-33.[1] Solve Problem 6-10, using the net cash flow B/C method. For the preferred investment, determine the discounted payback period.

6-34.[1] Solve Problem 6-11, using the conventional B/C method. For the preferred investment, determine the discounted payback period.

6-35.[1] Solve Problem 6-23, using the conventional B/C method. For the preferred investment, determine the discounted payback period.

---

[1] This problem is based on material presented in Appendix 6-B and Appendix 6-C.

Use of the ERR Method to Overcome Multiple-Solutions Difficulty with the IRR Method

In order to illustrate the differences in approach, using the IRR and ERR methods, consider Solomon's classic pump problem.[2]

Example

A firm is considering installing a new pump that will move oil faster than the present pump. The new pump will finish the job in one year; the present pump will require two years. The total value of the oil to be pumped is $20,000. The new pump costs $1,600, and neither pump will have any salvage value.

The cash flow profiles for each pump and the incremental cash flow of the new pump over the old are presented in Table 6-A-1. If the firm's MARR is 25%, determine which pump is more economical, using the IRR and ERR methods.

TABLE 6-A-1 Cash Flow Profiles

End of year	Old pump	New pump	Increment New pump–Old pump
0	–$ 0	–$ 1,600	–$ 1,600
1	10,000	20,000	10,000
2	10,000	0	–10,000

Solution Using FW calculations to solve for the internal rate of return on the incremental net cash flows gives

$$-1,600(1+i)^2 + 10,000(1+i) - 10,000 = 0.$$

This equation has two roots yielding internal rates of return of $i = 0.25$ (25%) and $i = 4.0$ (400%). With two (or more) roots, there may be some confusion as to which should be compared to the MARR in order to select the correct alternative. Actually, neither may be correct.

Using the same incremental net cash flows to compute the external rate of return results in

$$1,600(1+i)^2 + 10,000 = 10,000(1+r).$$

Solving for i as a function of the reinvestment rate r,

$$i = 2.5\sqrt{r} - 1.$$

Table 6-A-2 gives different values of r, the corresponding external rate of return i, and the recommended alternative to select.

Note that for those ERR values in excess of the reinvestment rate, the decision is to purchase the new pump. Also, note the correspondence between the internal rates of return (25%

---

[2] Solomon, Ezra, "The Arithmetic of Capital Budgeting Decisions," *The Journal of Business* Vol. 29, No. 2(1956):124–129.

TABLE 6-A-2   Values of *r*, *i*, and Recommended Alternative

r Reinvestment rate	*i* = ERR External rate of return	Recommended alternative
0.10	−0.209	Old pump
0.15	−0.032	Old pump
0.20	0.118	Old pump
0.25	0.250	Either pump
0.30	0.369	New pump
0.50	0.768	New pump
1.00	1.500	New pump
2.00	2.536	New pump
3.00	3.330	New pump
4.00	4.000	Either pump
5.00	4.590	Old pump

and 400%) and the values of the external rate of return at which one is indifferent about the alternatives. Naturally, for ERR values less than *r*, it is preferable to invest elsewhere and earn the MARR *r* rather than invest in the new pump. ■

Other Rate of Return and Investment Recovery Measures of Merit for Financial and Management Purposes

Chapter 6 demonstrated two theoretically sound methods for calculating rates of return over a project life or study period. Several additional methods that often have significant weaknesses, but are sometimes used for computational simplicity or to relate to commonly understood accounting figures, are briefly explained here.

Consider the Lathe B (in Example 6-2) with the following given data:

First cost: (Capital Investment):	$15,000
Project life:	10 yr
Salvage value:	$0
Annual receipts:	$7,000
Annual disbursements:	$4,000

The IRR for the project is 15.098%, and the ERR for the project is 12.291% (for an assumed reinvestment rate of 10%). Both of these methods are theoretically correct (keeping in mind that the IRR method assumes reinvestment at the computed rate—15.098% in this case).

A. Other Rates of Return (Based on Accounting Averages over Life)

We will now demonstrate calculation of other (nontheoretically correct) rates of return, with minimal explanations and assuming straight-line depreciation:

$$\text{Depreciation/yr} = \frac{\text{First cost} - \text{Salvage value}}{\text{Project life}}$$

$$= \frac{\$15,000 - \$0}{10} = \$1,500/\text{yr}.$$

1. RR: Average Return/Yr on Original Investment

$$= \frac{\text{Net profit/yr}}{\text{Original investment}} = \frac{\text{Net cash flow/yr} - \text{Depreciation/yr}}{\text{Original investment}}$$

$$= \frac{\$7,000 - \$4,000 - \$1,500}{\$15,000} = \frac{\$1,500}{\$15,000} = 10\%.$$

2. RR: Average Return/Yr on Average Investment

$$= \frac{\text{Net profit/yr}}{\frac{\text{Original investment} + \text{Salvage value}}{2}}$$

$$= \frac{\$7,000 - \$4,000 - \$1,500}{\$15,000/2} = \frac{\$1,500}{\$7,500} = 20\%.$$

B. Rate of Return for a Given Period (Year)

This measure is commonly and correctly applied by using accounting statement results for a business or firm, where

$$RR = \frac{\text{Book profit for period (year)}}{\text{Book value (i.e., Undepreciated investment)}}.$$

However, when applied to a particular project, such as previously given, its results are very dependent on the period for which the measure is made and do not reflect the RR over the life of the project. For example, if the RR were calculated at the end of the 5th year when the book value had been depreciated down to $7,500, the RR would be $1,500/$7,500 = 20%.

C. Measures of Investment Recovery

Analysts and managers often use quick screening measures of merit based on how quickly the investment will be recovered, with or without consideration of the time value of money. In essence, the payback period is the point in time when the investment "breaks even." Two common measures are the payback period method (which does not consider interest) and the discounted payback period method.

     1. *Payback (Payout) Period Method*
         Ignoring the time value of money, or assuming MARR = 0%, the payback period equals the ratio of the investment to the net cash flow per year. Hence, for our earlier example,

$$Payback\ Period = \frac{\$15,000}{\$7,000 - \$4,000} = 5.0\ yr.$$

         When the annual cash flow is not a uniform series, the cumulative sum of the cash flows is determined and the point in time is noted when the cumulative sum equals zero. The payback period is an indicator of how soon the investment is expected to be recovered (without interest).

     2. *Discounted Payback Period Method*
         The discounted payback period is the point in time when the cumulative net present worth equals zero. Since the example involves a uniform series of annual cash flows, its discounted payback period is the value of N that satisfies one of the following relations:

$$\$15,000 + (\$7,000 - \$4,000)(P/A,10\%,N) = 0$$

     or

$$(P/A,10\%,N) = 5.0.$$

From the interest tables and using linear interpolation, $N \cong 7.28$ yr. This is an indicator of how soon the investment (and interest) are expected to be recovered.

When the annual cash flows are not a uniform series, Excel can prove helpful in determining the discounted payback period for an investment. As shown in Figure 6-B-1, for Solved Problem 5, presented near the end of Chapter 4, using linear interpolation, the discounted payback period is 6.843 yr (7 year, $$$).

Figure 6-B-1 also contains alternative methods of computing the cumulative present worth. Depending on your familiarity with Excel, and if you know how to "anchor points with dollar signs before dragging down a column," you might prefer to forego using the NPV function and use the direct method shown in the right-most column.

	A	B	C
1	MARR =	10%	
2	EOP	Cash Flow	Cum PW
3	0	-$15,000.00	-$15,000.00
4	1	$2,000.00	-$13,181.82
5	2	$2,500.00	-$11,115.70
6	3	$3,000.00	-$8,861.76
7	4	$3,500.00	-$6,471.21
8	5	$4,000.00	-$3,987.53
9	6	$4,000.00	-$1,729.63
10	7	$4,000.00	$323.00
11	8	$4,000.00	$2,189.03
12	PW =	$2,189.03	

	C
1	
2	Cum PW
3	=B3
4	=NPV($B$1, $B$4:B4)+$B$3
5	=NPV($B$1, $B$4:B5)+$B$3
6	=NPV($B$1, $B$4:B6)+$B$3
7	=NPV($B$1, $B$4:B7)+$B$3
8	=NPV($B$1, $B$4:B8)+$B$3
9	=NPV($B$1, $B$4:B9)+$B$3
10	=NPV($B$1, $B$4:B10)+$B$3
11	=NPV($B$1, $B$4:B11)+$B$3
12	

	C
1	
2	Cum PW
3	=B3
4	=C3+B1/(1+B1)^A4
5	=C4+B2/(1+B1)^A5
6	=C5+B3/((1+B1)^A6
7	=C6+B4/(1+B1)^A7
8	=C7+B5/(1+B1)^A8
9	=C8+B6/(1+B1)^A9
10	=C9+B7/(1+B1)^A10
11	=C10+B8/(1+B1)^A11
12	

Figure 6-B-1 Discounted payback period solution for example problem, showing alternate setups, using Excel.

APPENDIX 6-C

Benefit–Cost Ratio Measure of Merit

In addition to present worth, annual worth, future worth, internal rate of return, and external rate of return, the benefit–cost ratio method has experienced considerable usage in the public sector. Many government agencies and departments, a number of public utilities, and other public service organizations use benefit–cost ratio methods in performing some types of economic analyses.

The *benefit–cost ratio (B/C)* can be defined as the ratio of the equivalent worth (present, annual, or future) of benefits or net benefits to the equivalent worth of costs. As such, the B/C method is similar to the rate of return method.

When independent investment opportunities are evaluated by using B/C ratios, they should be rank ordered by decreasing ratios, and the highest (with B/C ratios > 1.0) chosen within funds limitations. An example (not including calculation details) is given as part of Example 5–2.

When mutually exclusive alternatives are compared by using B/C ratios, incremental principles must be used. As with the rate of return, you should choose the highest cost alternative for which each increment of investment is justified by $\Delta B/\Delta C > 1.0$.

The B/C ratio is also referred to as the *savings-investment ratio (SIR)* by some government agencies and departments.

Three commonly used formulations of the benefit–cost ratio are as follows:

1. Net Cash Flow B/C Method is shown as

$$B/C = \frac{PW\{CF(+)\}}{PW\{CF(-)\}},$$

where CF(+) and CF(−) denote positive and negative cash flows, respectively.

2. Conventional B/C Method is shown as

$$B/C = \frac{PW(\text{benefit to user})}{PW(\text{total cost to supplier})} = \frac{PW[B]}{PW[CR+(O+M)]}$$

or

$$B/C = \frac{AW(\text{benefit to user})}{AW(\text{total cost to supplier})} = \frac{B}{CR+(O+M)},$$

where

$$B \ = \ \text{annual worth of benefits to user,}$$

CR = capital recovery cost or the equivalent
annual cost of the initial investment,
considering salvage value and interest,

O = uniform annual operating cost, and

M = uniform annual maintenance cost.

3. Modified B/C Method is shown as

$$B/C = \frac{PW[B-(O+M)]}{PW(CR)}$$

or

$$B/C = \frac{B-(O+M)}{CR}.$$

The numerator of the modified B/C expresses the present (or annual) worth of the net benefits and after operating and maintenance costs; the denominator includes only the investment costs, expressed on a present or annual basis.

Example

To illustrate the use of the net cash flow B/C method in comparing investment alternatives, recall the choice between the two alternative lathes. The data are as follows:

	Lathe	
	A	B
First cost:	$10,000	$15,000
Life:	5 yr	10 yr
Salvage value:	$2,000	$0
Annual receipts:	$5,000	$7,000
Annual disbursements:	$2,200	$4,000
Minimum attractive rate of return = 10%		
Study period = 10 yr		

1. Net Cash Flow B/C Solution

As with IRR and ERR, the alternatives are ranked in increasing order of the initial investment. Next, lathe A is analyzed to determine if it is economically viable. Using formulation 1, from Figure 6-C-1, it is seen that B/C = 1.186 > 1.0. Therefore, lathe A is justified.

Next, the increment of investment required to purchase lathe B is ana-lyzed. As shown in Figure 6-C-2, $\Delta B/\Delta C = 1.075 > 1.0$. Therefore, the increment of investment is justified and lathe B is recommended.

	A	B	C	D
1	MARR =	10%		
2	EOY	CF	CF(+)	CF(−)
3	0	−$10,000	$0	$10,000
4	1	$2,800	$2,800	$0
5	2	$2,800	$2,800	$0
6	3	$2,800	$2,800	$0
7	4	$2,800	$2,800	$0
8	5	$4,800	$4,800	$0
9	PW =		$11,856	$10,000
11	B/C =	1.186		

Figure 6-C-1   Net cash flow B/C solution for Lathe A.

	A	B	C	D	E	F
1	MARR =	10%				
2			Lathe		Increment (Δ)		
4	EOY	CF(A)	CF(B)	CF(B–A)	CF(+)	CF(−)
5	0	−$10,000	−$15,000	−$5,000	$0	$5,000
6	1	$2,800	$3,000	$200	$200	$0
7	2	$2,800	$3,000	$200	$200	$0
8	3	$2,800	$3,000	$200	$200	$0
9	4	$2,800	$3,000	$200	$200	$0
10	5	−$5,200	$3,000	$8,200	$8,200	$0
11	6	$2,800	$3,000	$200	$200	$0
12	7	$2,800	$3,000	$200	$200	$0
13	8	$2,800	$3,000	$200	$200	$0
14	9	$2,800	$3,000	$200	$200	$0
15	10	$4,800	$3,000	−$1,800	$0	$1,800
16	PW =				$6,119.18	$5,693.98
17	B/C =			1.075		

Figure 6-C-2   Net cash flow B/C solution for incremental investment in Lathe B.

2. Conventional B/C Solution

In using the conventional B/C method for the lathe example, recall that we previously assumed that lathe A will be replaced after 5 years with another lathe having an identical cash flow profile. Hence, annual worths for individual life cycles can be used in computing the capital recovery cost B/C. Recall, from Example 5–5, that the CR cost of lathe A equals $2,310. Annual disbursements total $2,200. Therefore,

$$B/C(A) = \frac{\$5,000}{\$4,510} = 1.109.$$

Since B/C(A) > 1.0, the equivalent worth of the benefits exceeds the equivalent worth of the costs and lathe A is justified economically. An Excel solution to the example is provided in Figure 6-C-3.

Next, the increment of investment required to purchase lathe B is considered. Recall that the CR cost for lathe B equals $2,441. Therefore, its annual cost, including annual disbursements, totals $6,441. The incremental B/C calculation reduces to

$$\Delta B/\Delta C = \frac{\$7,000-\$5,000}{\$6,441-\$4,510} = \frac{\$2,000}{\$1,931} = 1.036.$$

Again, since B/C > 1.0, the increment of investment is justified. Therefore, lathe B is recommended.

3. Modified B/C Solution

From the conventional B/C solution,

$$B/C(A) = \frac{\$2,800}{\$2,310} = 1.212.$$

Since B/C > 1.0, lathe A is justified.

For the incremental investment required to acquire lathe B,

$$\Delta B/\Delta C = \frac{\$3,000-\$2,800}{\$2,441-\$2,310} = \frac{\$200}{\$131} = 1.527.$$

Since $\Delta B/\Delta C$ > 1.0, the incremental investment is justified and lathe B is recommended.

	A	B	C	D
1	MARR =	10%		
2	EOY	CF	CF(+)	CF(−)
3	0	−$10,000	$0	$10,000
4	1	$2,800	$5,000	$2,200
5	2	$2,800	$5,000	$2,200
6	3	$2,800	$5,000	$2,200
7	4	$2,800	$5,000	$2,200
8	5	$4,800	$5,000	$200
9	PW =		$18,954	$17,098
11	B/C =	1.109		

Figure 6-C-3   Conventional B/C solution for Lathe A.

It should be obvious that all the above formulations of the B/C method yield recommendations that are consistent with those obtained using by PW, AW, FW, IRR, and ERR methods.

The application of the B/C method is relatively straightforward when disbursements only are known. In this situation, differences in cash flows for pairs of mutually exclusive alternatives must be evaluated as with the IRR method. For this reason, we limit treatment of such alternatives to the problems at the end of the chapter.

7
Consideration of Depreciation and Income Taxes

This chapter provides a brief overview of the principal depreciation considerations and a general technique for including the effect of income taxes in economic studies.

Only cash flows need be considered in determining the economic desirability of an alternative in an economic analysis. Income taxes are relevant cash flows and should be considered whenever their omission may cause the selection of an uneconomical alternative.[1] Even though depreciation write-offs are not, in themselves, cash flows, they do affect income taxes, and hence they affect cash flows.

7.1 Introduction to Depreciation

The primary purpose of depreciation accounting is to provide for the recovery of capital invested in property that is expected to decline in value as a result of time and/or use. This is done through the mechanism of *depreciation charges,* or write-offs, which are allocations made periodically for the purpose of distributing the cost of capital assets over their useful lives. Thus, depreciation is the decrease in value of physical properties with the passage of time. More specifically, depreciation is an accounting concept that establishes an annual deduction against before-tax income such that the effect of use and time on an asset's value can be reflected in a firm's financial statements. Although depreciation does occur and is easily recognized, the determination of its magnitude in advance is not easy. The actual amount of depreciation can never be established until the asset is retired from service. Because

---

[1] This is demonstrated in Example 7-20.

depreciation is a *noncash cost* that affects income taxes, we must learn to consider it properly when making after-tax engineering economy studies.

Depreciable property is property used in a business to produce income. The cost of depreciable property can be deducted from business income for income tax purposes over a future period of time. Property is depreciable if it meets these requirements:

1. It must be used in business or held for the production of income.

2. It must have determinable life, and the life must be longer than one year.

3. It must be something that wears out, decays, gets used up, becomes obsolete, or loses value from natural causes.

In general, if property does not meet all three of these conditions, it is not depreciable.

Depreciable property may be classified as *tangible* or *intangible*. Tangible property is any property that can be seen or touched, and intangible property is property, such as a copyright or franchise, that is not tangible. Additionally, depreciable property may be classified as *real* or *personal*. Personal property is property that can be transported, such as machinery or equipment, which is not real estate. Real property is land and generally anything that is erected on, growing on, or attached to land. However, land itself is never depreciable.

The purpose of the following sections is to acquaint the student with several depreciation methods that have been employed for many decades. These methods are of interest to us because they form the foundation for modern depreciation models in widespread use today.

7.1.1 Straight-Line Method

Straight-line depreciation is the simplest depreciation method. It assumes that a constant amount is depreciated each year over the life of the asset. The following definitions are used in the equations presented. If we define

N = depreciable life (write-off period) of the asset in years,

B = cost basis,

d_k = annual depreciation deduction in year k ($1 \le k \le N$),

BV_k = book value at end of year k,

BV_N = estimated book (salvage) value in year N, and

d_k^* = cumulative depreciation through year k,

then

$$d_k = (B - BV_N)/N ; \tag{7-1}$$

$$d_k^* = k d_k \quad \text{for } 1 \le k \le N ; \tag{7-2}$$

$$BV_k = B - d_k^*. \tag{7-3}$$

Notice that, for this method, you must have an estimate of the final BV, which will also be the final book value at the end of year N.

Example 7-1

A machine costs $15,000 installed. The allowable write-off period is 12 years, at which time the salvage value is estimated to be $1,500. What will be the annual depreciation charge and what will be the book value at the end of the third year?

    Solution

$$d = \frac{(B - BV_N)}{N} = \frac{\$15,000 - \$1,500}{12} = \$1,125;$$

$$BV_3 = B - 3d = \$15,000 - 3\left[\frac{\$15,000 - \$1,500}{12}\right]$$

$$= \$11,625.$$

■

7.1.2 Declining Balance Method

In the declining balance method, sometimes called the *constant percentage method,* it is assumed that the cost of depreciation for any year is a fixed percentage (designated R) of the book value at the *beginning* of that year. In this method, $R = 2/N$ when a 200% declining balance is being used (i.e., twice the straight-line rate of $1/N$), and N equals the allowable life of an asset. If the 150% declining balance method is specified, then $R = 1.5/N$. The following relationships hold true for the declining balance method:

$$d_1 = B(R); \tag{7-4}$$

$$d_k = B(1 - R)^{k-1}(R); \tag{7-5}$$

$$d_k^* = B[1 - (1 - R)^k]; \tag{7-6}$$

$$BV_k = B(1 - R)^k; \tag{7-7}$$

$$BV_N = B(1 - R)^N. \tag{7-8}$$

In order for the book value to equal the estimated salvage value at the end of the write-off period, N years, R should be calculated as

$$R = 1 - \sqrt[N]{\frac{BV_N}{B}}. \tag{7-9}$$

Example 7-2

For the previous example using straight-line depreciation, determine the book value at the end of the third year and the depreciation charge for the fourth year by using the 200% declining balance method of depreciation.

    Solution

$$R = 2/12 = 0.1667;$$

$$BV_3 = B(1 - R)^3 = \$15,000(0.8333)^3 = \$8,679.51;$$

$$d_4 = B(1 - R)^3(R) = \$1,446.88.$$

■

Example 7-3

Repeat Example 7-2 when it is desired that the terminal book value in year 12 equal exactly $1,500.

Solution

$$R = 1 - \sqrt[N]{BV_N/B} = 1 - \sqrt[12]{1{,}500/15{,}000}$$
$$= 1 - 0.826 = 0.174;$$
$$BV_3 = \$15,000(1 - 0.174)^3 = \$8,460;$$
$$d_4 = \$8,460(0.174) = \$1,470. \qquad\blacksquare$$

7.1.3 Comparison of Depreciation Methods

To provide a common basis for comparing the two methods of depreciation examined here so far, Table 7-1 shows year-by-year depreciation charges for a typical asset that costs $16,000, is expected to last 5 years, and then has a terminal book value of $1,000.

7.1.4 Units-of-Production Depreciation

All the depreciation methods discussed to this point are based on elapsed time, on the theory that the decrease in value of property is mainly a function of time. When the decrease in value is mostly a function of use, depreciation may be based on the *units-of-production method.*

This method results in the total depreciable investment being allocated equally over the units produced and requires an estimate of the total lifetime of productive use. The depreciation rate is calculated as

$$\text{Depreciation per unit of production} = \frac{B - BV_N}{\text{Estimated lifetime production}}. \qquad (7\text{-}10)$$

Example 7-4

A vehicle used in a business has a first cost of $50,000 and is expected to have a $10,000 salvage value when traded after 100,000 miles of use. It is desired to find its depreciation rate based on functional use and to find its book value after 20,000 miles of use.

TABLE 7-1 Comparison of Depreciation Charges by Using Two Methods for a Machine Having a $16,000 Investment, Five-Year Life, and $1,000 Terminal Book Value

End of year	Straight line	Double declining balance
1	$ 3,000	$ 6,400
2	3,000	3,840
3	3,000	2,304
4	3,000	1,382
5	3,000	1,074*
Totals	$15,000	$15,000

*Year 5 depreciation set equal to $15,000 less cumulative depreciation through the fourth year ($13,926) so that BV5 = $1,000.

Solution

$$\text{Depreciation per unit of production} = \frac{\$50,000 - \$10,000}{100,000 \text{ miles}} = \$0.40 \text{ per mile;}$$

$$\text{After 20,000 miles,} \quad BV = \$50,000 - \frac{\$0.40}{\text{mile}}(20,000 \text{ miles}), \text{ or } BV = \$42,000.$$

■

7.2 The Tax Reform Act of 1986 and Its Depreciation Provisions

The Modified Accelerated Cost Recovery System (MACRS) method was created by the Tax Reform Act of 1986 (TRA 86) and is now the principal means for computing depreciation expenses (termed *recovery allowances* under TRA 86). MACRS is mandatory for most tangible depreciable assets *placed in service* after July 31, 1986. Under MACRS, $BV_{N'+1} = 0$, and useful life estimates are not directly utilized in calculating depreciation. Here, N' is the MACRS property class of an asset.

MACRS consists of two methods for depreciating property. The main method is called the General Depreciation System (GDS), while the second is called the Alternate Depreciation System (ADS). Unless required by law or specifically elected, GDS is normally used to determine the appropriate depreciation deduction.

MACRS allows a business to recoup the *cost basis* of recovery property over a *recovery period.* The cost basis is normally the cash purchase price of a property plus the cost of making the asset serviceable, thus including shipping and handling, insurance, installation, and training expenses.

Example 7-5

In 2005, your firm purchased a used machine for $10,500 for use in producing income. An additional $1,000 was spent to recondition and install the machine. What is the cost basis?

Solution

The cost basis of the machine is $11,500 (total costs of $10,500 plus $1,000).　　　　■

The procedure for computing MACRS depreciation deductions in any given year of an asset's useful life is given in Fig. 7-1. Step 1 is to obtain the property's asset depreciation range (ADR) value from Table 7-2. The ADR value, which depends on the type of property and industry involved, is used in Step 2 to establish the asset's MACRS property class in Table 7-3.

Based on the MACRS property class, a set of GDS recovery rates is located in Table 7-4 (Step 3) and is then used in Step 4 to determine MACRS depreciation deductions in year k ($1 \le k \le N' + 1$),

$$d_k(p) = r_k(p) \cdot B,$$

where　　　　　　　　　　　　　　　　　　　　　　　　　　　　　(7-11)

$d_k(p)$ = depreciation deduction in year k for recovery property class p,

$r_k(p)$ = MACRS rate (a decimal) for year k in recovery property class p, and

B = cost basis of the recovery property.

MACRS does not allow deductions for the cost of property, such as land, which has no determinable life.

TABLE 7-2   Selected Asset Depreciation Range (ADR) Guideline Periods

Description of depreciable assets	ADR Guideline period (years)
Transportation	
Automobile, taxis	3
Buses	9
General-purpose trucks	
Light	4
Heavy	6
Air transport (commercial)	12
Petroleum	
Exploration and drilling assets	14
Refining and marketing assets	16
Manufacturing	
Sugar and sugar products	18
Tobacco and tobacco products	15
Carpets and apparel	9
Lumber, wood products, and furniture	10
Chemical and allied products	9.5
Cement	20
Fabricated metal products	12
Electronic components	6
Rubber products	14
Communication	
Telephone	
Central office buildings	45
Distribution poles, cables, etc.	34
Radio and television broadcasting	6
Electric utility	
Hydraulic plant	50
Nuclear	20
Transmission and distribution	30
Services	
Office furniture and equipment	10
Computers and peripheral equipment	6
Recreation—bowling alleys, theater, etc.	10

SOURCE: *Depreciation*, IRS Publication 534, Washington, D.C.:
U.S. Government Printing Office, Dec. 2001 (rev.).

Notice that in Table 7-4 there are $N' + 1$ rates shown for a MACRS property class of N' years. If a depreciable asset is disposed of *after* year $N' + 1$, the final BV of the item will be zero. Also, from Eq. 7-11, it is apparent that MACRS rates in Table 7-4 are applied to the cost basis (B) only, regardless of an asset's expected market value when disposal occurs.

Example 7-6

In May 2004, your company traded in a computer, used in its business, that had a book value at that time of $25,000. A new, faster computer system having a fair selling price of $400,000 was acquired. Because the vendor accepted the older computer as a trade-in, a deal was agreed to whereby your company would pay $325,000 cash for the new computer system.

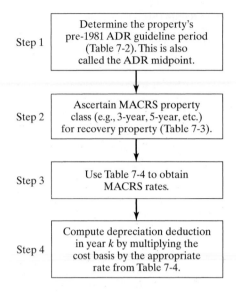

| Step 1 | Determine the property's pre-1981 ADR guideline period (Table 7-2). This is also called the ADR midpoint. |

| Step 2 | Ascertain MACRS property class (e.g., 3-year, 5-year, etc.) for recovery property (Table 7-3). |

| Step 3 | Use Table 7-4 to obtain MACRS rates. |

| Step 4 | Compute depreciation deduction in year k by multiplying the cost basis by the appropriate rate from Table 7-4. |

Figure 7-1   Flow diagram for computing depreciation GDS deductions under MACRS per TRA 86.

(a) What is the MACRS property class of the new computer?

(b) How much depreciation can be deducted each year on the basis of this class life? (Refer to Fig. 7-1).

Solution

(a) The new computer has an ADR guideline period of 6 years (see Table 7-2). Hence, its MACRS recovery period is 5 years (see Table 7-3).

(b) The cost basis for this property is $350,000, which is the sum of the $325,000 cash price of the computer and the $25,000 book value remaining on the trade-in (in this case, the trade-in was treated as a nontaxable transaction).   ∎

MACRS rates that apply to the $350,000 cost basis are found in Table 7-4. An allowance (half-year) is built into the year 1 rate, so it does not matter that the computer was purchased in May 2004, instead of, say, November 2004. The GDS depreciation deductions for 2004 through 2009 can be computed with Eq. 7-11 and the 5-year MACRS rates as follows:

Property	Date placed in service	Cost basis	ADR guideline period	MACRS property class
Computer system	May 2004	$350,000	6 years	5 years

Year	Depreciation deductions
2004	$0.20 \times \$350,000 = \$\ 70,000$
2005	$0.32 \ \times \ 350,000 = \ 112,000$
2006	$0.192 \times \ 350,000 = \ \ 67,200$
2007	$0.1152 \times \ 350,000 = \ \ 40,320$
2008	$0.1152 \times \ 350,000 = \ \ 40,320$
2009	$0.0576 \times \ 350,000 = \ \ \underline{20,160}$
	Total $350,000

∎

TABLE 7-3   MACRS Property Classes and Permissible Methods for
Calculating Depreciation Rates

MACRS property class and depreciation method	ADR guideline period	Special rules
3-year, 200% declining balance	4 years or less	Includes some racehorses. Excludes cars and light trucks.
5-year, 200% declining balance	More than 4 years to less than 10 years	Includes cars and light trucks, semiconductor manufacturing equipment, qualified technological equipment, computer-based central office switching equipment, some renewable and biomass power facilities, and research and development property.
7-year, 200% declining balance	10 years to less than 16 years	Includes single-purpose agricultural and horticultural structures and railroad track. Includes property with no ADR midpoint.
10-year, 200% declining balance	16 years to less than 20 years	None.
15-year, 150% declining balance	20 years to less than 25 years	Includes sewage treatment plants, telephone distribution plants, and equipment for two-way voice and data communication.
20-year, 150% declining balance	25 years or more	Excludes real property with ADR midpoint of 27.5 years or more. Includes municipal sewers.
27.5-year, straight line	N/A	Residential rental property.
31.5-year, straight line	N/A	Nonresidential real property.

SOURCE: *Depreciation*, IRS Publication 534, Washington, D.C.: U.S. Government Printing Office, Dec. 2001 (rev.).

From Example 7-6 we can conclude that the following relationship is true:

$$\text{Cost basis} = \text{BV of the trade-in (if any)} \\ + \text{cash price of new equipment after trade-in.} \tag{7-12}$$

7.2.1 Half-Year Convention

We must be careful to observe that MACRS uses a half-year convention. The IRS assumes that assets are purchased halfway through the year, no matter when the asset is actually purchased. This means the rate for year 1 in Table 7-4 is for only a half-year of depreciation. This is built into the calculations used to determine the rates, so there is no need to adjust the rate. Thus, the cost of 3-year property is recovered by using the half-year convention for the first year; then the full allowance for the second and third years is deducted, and the remaining balance is deducted for the

TABLE 7-4 Modified ACRS Rates per TRA 86

| | Class life (i.e., property class) | | | | | |
Year	3-year[a]	5-year[a]	7-year[a]	10-year[a]	15-year[b]	20-year[b]
1	0.3333	0.2000	0.1429	0.1000	0.0500	0.0375
2	0.4445	0.3200	0.2449	0.1800	0.0950	0.0722
3	0.1481	0.1920	0.1749	0.1440	0.0855	0.0668
4	0.0741	0.1152	0.1249	0.1152	0.0770	0.0618
5		0.1152	0.0893	0.0922	0.0693	0.0571
6		0.0576	0.0892	0.0737	0.0623	0.0528
7			0.0893	0.0655	0.0590	0.0489
8			0.0446	0.0655	0.0590	0.0452
9				0.0656	0.0591	0.0447
10				0.0655	0.0590	0.0447
11				0.0328	0.0591	0.0446
12					0.0590	0.0446
13					0.0591	0.0446
14					0.0590	0.0446
15					0.0591	0.0446
16					0.0295	0.0446
17						0.0446
18						0.0446
19						0.0446
20						0.0446
21						0.0223

[a] These rates are determined by applying the 200% declining balance method to the appropriate property class with the half-year convention applied to the first and last years. Rates for each class life must sum to 1.0000.

[b] These rates are determined with the 150% declining balance method instead of the 200% declining balance method and are rounded off to four significant digits.

SOURCE: *Depreciation*, IRS Publication 534, Washington, D.C.: U.S. Government Printing Office, Dec. 2001 (rev.).

fourth year. The "recovery period" is still 3 years, but the deductions are spread over 4 years if the asset is kept in service for 4 years or more.

When an asset is disposed of, the half-year convention is also used. If disposal occurs at the end of year $N' + 1$ of the recovery period or later, nothing is changed. *If the asset is disposed of before this period, then only half of the normal deduction can be taken for that year. This means that the rate in Table 7-4 should be divided by 2 to compute depreciation in the year of disposal if the asset is sold before year $N' + 1$.* In the case of early disposal, the BV in the year of disposal is

$$BV_k = B - d_k^* \text{ through } k \text{ years of depreciation.}$$

Example 7-7

A firm purchased a new piece of semiconductor manufacturing equipment in July 2004. The cost basis for the equipment is $100,000. Determine (a) the depreciation charge permissible in the fourth year (2007), (b) the BV at the end of 2007, (c) the cumulative depreciation through 2006, and (d) the BV at the end of 2006 if the equipment is disposed of (sold) in 2006.

Solution

From Table 7-2, it can be seen that the semiconductor (electronic) manufacturing equipment has an ADR guideline period of 6 years, and from Table 7-3 it has a 5-year MACRS class life. The rates that apply are given in Table 7-4.

(a)   The depreciation deduction, or cost recovery allowance, that is allowable in 2007 (d_4) is 0.1152($100,000) = $11,520.

(b)   The BV at the end of 2007 (BV_4) is the cost basis less depreciation charges in years 1–4:

$$BV_4 = \$100,000 - \$100,000(0.20 + 0.32 + 0.192 + 0.1152)$$
$$= \$17,280.$$

(c)   Accumulated depreciation through 2006, d_3^*, is the sum of depreciation in 2004, 2005, and 2006, or

$$d_3^* = d_1 + d_2 + d_3$$
$$= \$100,000(0.20 + 0.32 + 0.192)$$
$$= \$71,200.$$

(d)   The depreciation deduction in 2008 (year 5) can be only (0.5)(0.1152)($100,000) = $5,760 when the equipment is disposed of (sold) prior to 2009 (year 6). Thus, the BV at the end of year 5 is $BV_4 - \$5,760 = \$11,520.$ ∎

7.2.2 The Alternate MACRS Method

TRA 86 provides for the use of an alternate method (the ADS). Election to adopt the Alternate Depreciation System for a class of property applies to all property in that class placed in service during the tax year. The decision to use the alternative MACRS method, once made, is irrevocable for the assets involved. Under this method, depreciation is calculated using (1) the straight-line method of depreciation with $BV_{N'+1} = 0$, and (2) a half-year convention for the depreciable (ADR) life of the property. (See Table 7-2.)

Example 7-8

A large manufacturer of sheet metal products purchased a new, modern, computer-controlled flexible manufacturing system in October 2005 for $3.0 million. Because this company would not be profitable until the new technology had been in place for several years, it elected to utilize the ADS in computing its depreciation deductions. Thus, the company could slow down its depreciation allowances in hopes of postponing income tax advantages until it became a profitable concern. What depreciation allowances can be claimed for the new system?

Solution

From Table 7-2, the ADR guideline period for a manufacturer of fabricated metal products is 12 years. This flexible manufacturing system would normally be depreciated, using 7-year MACRS rates (Table 7-3). However, under the ADS, the straight-line method with no terminal book value is applied to the 12-year depreciation period, using the half-year convention. Consequently, depreciation in year 1 (2005) would be

$$\frac{1}{2}\left(\frac{\$3,000,000}{12}\right) = \$125,000.$$

Depreciation deductions in years 2–12 (2006–2016) would be $250,000 each year, and depreciation in year 13 (2017) would be $125,000. Notice how the half-year convention extends depreciation deductions over 13 years. ∎

7.3 A Comprehensive Example

We now consider an asset for which depreciation is computed by most methods discussed to this point. Be careful to observe the differences in the mechanics of each method, as well as the differences in the annual depreciation amounts themselves.

Example 7-9

The La Salle Bus Company has decided to purchase a new bus for $85,000, which includes a trade-in of its old bus. The old bus has a book value of $10,000 at the time of the trade-in. The new bus will be kept for 10 years before being sold. Its estimated book (salvage) value at that time is expected to be $5,000.

    First, we must calculate the cost basis. The basis is the original purchase price of the bus plus the book value of the old bus that was traded in. Thus, the cost basis is $85,000 + $10,000, or $95,000. We will also need to determine the ADR guideline period. Hence, we need to look at Table 7-2 and find "buses" under the Transportation category. We find that buses have a 9-year ADR guideline period. This will be used as the number of years over which we depreciate the bus with historical methods discussed earlier.

Solution: Straight-Line Method

For the straight-line method, we use the ADR guideline period of 9 years even though the bus will be kept for 10 years. By using Eqs. 7-1 and 7-3, we obtain the following information:

$$d_k = \frac{\$95,000 - \$5,000}{9 \text{ years}} = \$10,000 \quad \text{for } k = 1 \text{ to } 9.$$

Straight-Line Method

EOY k	d_k	BV_k
0	—	$95,000
1	$10,000	85,000
2	10,000	75,000
3	10,000	65,000
4	10,000	55,000
5	10,000	45,000
6	10,000	35,000
7	10,000	25,000
8	10,000	15,000
9	10,000	5,000

    Notice that no depreciation was taken after year 9 because the ADR life was only 9 years. Also, the BV will remain at $5,000 from the last ADR year until the bus is sold.

Solution: Declining Balance Method

To demonstrate this method, we will use the 200% declining balance equations. By using Eqs. 7-4, 7-5, and 7-7, we derive these sample calculations:

$$R = 2/9 = 0.2222;$$
$$d_1 = \$95,000(0.2222) = \$21,111;$$
$$d_5 = \$95,000(1 - 0.2222)^{5-1}(0.2222) = \$7,726;$$
$$BV_5 = \$95,000(1 - 0.2222)^5 = \$27,040.$$

200% Declining Balance Method

EOY k	d_k	BV_k
0	—	$95,000
1	$21,111	73,889
2	16,420	57,469
3	12,771	44,698
4	9,932	34,765
5	7,726	27,040
6	6,009	21,031
7	4,674	16,357
8	3,635	12,722
9	2,827	9,895

Solution: MACRS Method

For this method, we will follow the steps given in Fig. 7-1.

1. Determine the ADR guideline period from Table 7-2. We have already found it to be 9 years.

2. Ascertain the MACRS property class for recovery property. By Table 7-3, an ADR guideline period of 9 years corresponds to a 5-year MACRS property class.

3. Now, use Table 7-4 to determine d_k.

EOY k	MACRS rate	d_k	BV_k
0	—	—	$95,000
1	0.2000	$19,000	76,000
2	0.3200	30,400	45,600
3	0.1920	18,240	27,360
4	0.1152	10,944	16,416
5	0.1152	10,944	5,472
6	0.0576	5,472	0
7	—	—	0
8	—	—	0
9	—	—	0
10	—	—	0

Notice that the depreciation life is much shorter than for the previous methods, and the BV at the end of year $N' + 1$ is zero. The estimated salvage value of $5,000 does not affect MACRS depreciation amounts.

Solution: MACRS Half-Year Convention

To demonstrate the half-year convention, we will change the La Salle bus problem so that the bus is now sold in year 5 in part (a) and in year 6 for part (b).

(a) Selling bus in year 5

EOY	Factor	d_k	BV_k
0	—	—	$95,000
1	0.2000	$19,000	76,000
2	0.3200	30,400	45,600
3	0.1920	18,240	27,360
4	0.1152	10,944	16,416
5	0.0576	5,472	10,944

(b) Selling bus in year 6

EOY	Factor	d_k	BV_k
0	—	—	$95,000
1	0.2000	$19,000	76,000
2	0.3200	30,400	45,600
3	0.1920	18,240	27,360
4	0.1152	10,944	16,416
5	0.1152	10,944	5,472
6	0.0576	5,472	0

Notice that when we sold the bus in year 5 before the recovery period had ended, we claimed only half of the normal depreciation (MACRS rate = 0.1152/2). The other years (years 1–4) were not changed. When the bus was sold in year 6 (at the end of the recovery period), we did not divide the last year amount by 2.

Selected methods of depreciation, illustrated in Example 7-9, are compared in Fig. 7-2. ∎

7.4 Introduction to Income Taxes

The main types of taxes important to economic analyses are as follows:

1. *Property taxes* are based on the valuation of property owned, such as land, equipment, buildings, inventory, etc., and the established tax rates. They do not vary with income and are usually much lower in amount than income taxes.

2. *Sales taxes* are taxes imposed on product sales, usually at the retail level. They are relevant in engineering economy studies only to the extent that they add to the cost of items purchased.

3. *Excise taxes* are taxes imposed upon the manufacture of certain products, such as alcohol and tobacco.

4. *Income taxes* are taxes on pretax income of an organization in the course of regular business. Income taxes are also levied on gains on the disposal of capital property. They are usually the most significant type of tax to consider in economic analyses and are the subject of the remainder of this chapter.

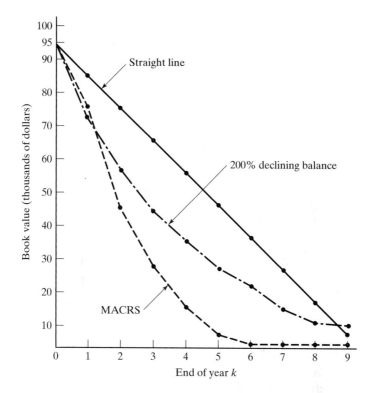

Figure 7-2   BV comparisons for selected methods of depreciation (Example 7–9).

7.4.1 The Discount Rate to Use in After-Tax Studies

Any practical definition of a minimum attractive rate of return (MARR) should serve as a guide for investment policy to attain the goals toward which a firm is, or will be, striving. One obvious goal is to meet shareholder expectations. Clearly, long-term goals and capital structure will be changing as a firm matures. In this regard, it is widely accepted that in after-tax engineering economy studies, the MARR should be *at least* the tax-adjusted weighted average cost of capital (WACC),

$$\text{WACC} = \lambda(1-t)i_b + (1-\lambda)e_a, \tag{7-13}$$

where

λ = fraction of a firm's pool of capital that is borrowed from lenders,

t = effective income tax rate as a decimal,

i_b = before-tax interest paid on borrowed capital, and

e_a = after-tax cost of equity capital.

The adjustment on the borrowed capital component of WACC, $(1-t)i_b$, accounts for the fact that interest on borrowed capital is tax deductible.

Often, a value higher than the WACC is assigned to the after-tax MARR to reflect the opportunity cost of capital, perceived risk/uncertainty of projects being evaluated, and policy issues such as organizational growth and shareholder satisfaction. We are now in a position to understand how a firm's after-tax MARR is determined and why its value may change substantially over time. Moreover, it is possible to approximate a value of the before-tax MARR that has been used for discounting cash flows in previous chapters:

$$\text{Before-tax MARR} \approx \frac{\text{After-tax MARR}}{1-t} \qquad (7\text{-}14)$$

To summarize, suppose a firm's WACC is 10% per year. The after-tax MARR may be set at 15% per year to better reflect the business opportunities that are available for capital investment. If the firm's effective income tax rate is 40%, the approximate value of the before-tax MARR is $15\%/(1 - 0.4) = 25\%$ per year.

7.4.2 When Income Taxes Should Be Considered

In the preceding chapters, we have treated income taxes as if they are either not applicable, or taken into account by using a before-tax rate of return that is larger than the after-tax rate of return. The intention in this second case is to adjust resulting cash flows so they will be sufficient to provide for both the after-tax rate of return and the income tax.

As previously noted, an approximation of the before-tax rate of return (RR) requirement to include the effect of income taxes in studies using before-tax cash flows can be determined from the following relationship:

$$[\text{Before-tax RR}] = \frac{[\text{After-tax RR}]}{[1 - \{\text{Effective income tax rate}\}]}.$$

If the property is nondepreciable and there is no capital gain or loss or investment tax credit, then the relationship in Eq. 7-14 is exact, not an approximation. If the salvage value is less than 100% of first cost and if the life of the property is finite, then the depreciation method selected for income tax purposes affects the timing of income tax payments. Therefore usually the relationship in Eq. 7-14 is an approximate only.

In practice, it is important to make after-tax analyses in any income-tax-paying organization. After-tax analyses can be performed by exactly the same methods (PW, IRR, etc.) as before-tax analyses. *The only difference is that after-tax cash flows must be used in place of before-tax cash flows, and the calculation of a measure of merit such as PW is based on an after-tax MARR.*

The mystery behind the sometimes complex computation of income taxes is reduced when you recognize that income taxes paid are just another type of expense, while income taxes saved (through business deductions, expenses, or direct tax credits) are identical to other kinds of reduced expenses (e.g., savings).

The basic concepts underlying federal and state income tax regulations that apply to most economic analyses of capital investments generally can be understood and applied without difficulty. This chapter is not intended to be a comprehensive treatment of federal tax law. Rather, we utilize some of the more important provisions of the Tax Reform Act of 1986 (TRA 86), followed by illustrations of a general procedure for computing after-tax cash flows (ATCFs) and conducting after-tax economic analyses. Where appropriate, important changes to income tax provisions enacted by the Jobs and Growth Relief Reconciliation Act of 2003 (JGRRA 03) are also included in this chapter.

7.4.3 Taxable Income of Business Firms

At the end of each tax year, a corporation must calculate its net (i.e., taxable) before-tax income or loss. Several steps are involved in this process, beginning with the calculation of *gross income*. Gross income represents the gross profits from operations (revenues from sales minus the cost of goods sold) plus income from dividends, interest, rent, royalties, and gains (or losses) on the exchange of capital assets. The corporation may deduct from gross income all ordinary and necessary operating expenses, including interest, to conduct the business. Deductions for depreciation are permitted each tax period as a means of consistently and systematically recovering capital investment. Consequently, allowable expenses and depreciation deductions may be used to determine *taxable income* for a firm, as shown in Eq. 7-15:

$$\text{Taxable income} = \text{gross income}$$
$$- \text{ all expenses (except capital expenditures)} \qquad (7\text{-}15)$$
$$- \text{ depreciation deductions.}$$

Taxable income is often referred to as *net income before taxes,* and when income taxes are subtracted from it, the remainder is called *net income after taxes.* There are two types of income for tax computation purposes: ordinary income (and losses) and capital gains (and losses). These types of income are explained in subsequent sections.

Example 7-10
In 2006, a company generates $1,500,000 of gross income and incurs operating expenses of $800,000. The total depreciation deductions in 2006 equal $114,000. (a) What is the taxable income of this firm? (b) If operating expenses had been $1,590,000, what would the net operating loss (NOL) have been?

Solution

(a) On the basis of Eq. 7-15, this company's taxable income in 2006 would be

$$\$1,500,000 - \$800,000 - \$114,000 = \$586,000.$$

(b)   NOL $= |\$1,500,000 - \$1,590,000 - \$114,000| = \$204,000.$ ∎

7.4.4 Calculation of Effective Income Tax Rates

Although the regulations of most of the states with income taxes have the same basic features as the federal regulations, there is great variation in income tax rates. State income taxes are in most cases much lower than federal taxes and can often be closely approximated as a constant percentage of federal taxes, typically ranging from

6% to 12% of taxable income. No attempt will be made here to discuss the details of state income taxes. An understanding of the applicable federal income tax regulations usually will enable the analyst to apply the proper procedures if state income taxes must also be considered.

To illustrate the calculation of an effective (combined federal and state) income tax rate for a very large corporation, suppose that the federal income tax rate is 35% and the state income tax rate is 8%. Further assume the common case in which taxable income is computed the same way for both types of taxes, except that state taxes are deductible from taxable income for federal tax purposes, but federal taxes are *not* deductible from taxable income for state income tax purposes. The general expression for the effective income tax rate (t) is

$$t = \text{federal rate} + \text{state rate} - (\text{federal rate})(\text{state rate}). \qquad (7\text{-}16)$$

In this example the effective income tax rate would be

$$t = 0.35 + 0.08 - 0.35(0.08) = 0.402, \text{ or about } 40\%.$$

It is the effective income tax rate on increments of taxable income that is of importance in engineering economy studies. In this chapter, we will use an effective income tax rate of 40% in most of the examples that follow.

7.4.5 Income Taxes on Ordinary Income (and Losses)

Ordinary (taxable) income is the net income before taxes that results from the routine business operations (such as the sale of products or services) performed by a corporation or individual. For federal income tax purposes, virtually all ordinary income adds to taxable income and is subject to a graduated rate scale.

The corporate federal income tax rates as of 2003 are given in Table 7-5. For example, suppose that a firm in 2003 has a gross income of $5,270,000, expenses (excluding capital) of $2,927,500, and depreciation of $1,874,300. Its taxable income and federal income tax would be determined with Eq. 7-15 and Table 7-5 as follows:

$$
\begin{aligned}
\text{Taxable income} &= \text{gross income} - \text{expenses} - \text{depreciation} \\
&= \$5,270,000 - \$2,927,500 - \$1,874,300 \\
&= \$468,200.
\end{aligned}
$$

	Taxable income	Income tax
Income tax = 15% of first	$ 50,000	$ 7,500
+ 25% of next	25,000	6,250
+ 34% of next	25,000	8,500
+ 39% of next	235,000	91,650
+ 34% of remaining	133,200	45,288
Total	$468,200	Total $159,188

TABLE 7-5   Corporate Federal Income Tax as of 2003

If taxable income is:		The tax is:	of the amount
over	but not over		over
$ 0	$ 50,000	15%	$ 0
50,000	75,000	$ 7,500 + 25%	50,000
75,000	100,000	13,750 + 34%	75,000
100,000	335,000	22,250 + 39%	100,000
335,000	10,000,000	113,900 + 34%	335,000
10,000,000	15,000,000	3,400,000 + 35%	10,000,000
15,000,000	18,333,333	5,150,000 + 38%	15,000,000
18,333,333	....	6,416,667 + 35%	18,333,333

The total income tax liability in this case is $159,188. Because engineering economy studies are concerned with incremental differences among alternatives, we are normally concerned with what (incremental) income tax rates apply to those differences

Example 7-11

A corporation is expecting an annual taxable income of $45,000. It is considering investing an additional $100,000, which is expected to create an added annual net cash flow (receipts minus expenses) of $35,000 and an added annual depreciation charge of $20,000. What is the corporation's federal income tax liability, based on rates in effect in 2003 (a) without the added investment, and (b) with the added investment?

   Solution

(a)

Income taxes	Rate	Amount
On first $45,000	15%	$ 6,750
	Total	$ 6,750

(b)

Taxable income		
Before added investment		$45,000
+ added net cash flow		35,000
− depreciation		(20,000)
	Total	$60,000

Income taxes on $60,000	Rate	Amount
On first $50,000	15%	$ 7,500
On next $10,000	25%	2,500
	Total	$10,000

The increased income tax liability from the investment is $10,000 − $6,750 = $3,250.

As an added note, the change in tax liability can usually be determined more readily by an incremental approach. For instance, this example involved changing the taxable income from $45,000 to $60,000 as a result of the new investment. Thus, the change in income taxes per year (for 2003) could be calculated as follows:

$$\text{First, } \$50,000 - \$45,000 = \$5,000 \text{ at } 15\% = \$\ 750;$$
$$\text{Next, } \$60,000 - \$50,000 = \$10,000 \text{ at } 25\% = \underline{\ 2,500}$$
$$\text{Total} \quad \$3,250. \qquad \blacksquare$$

7.4.6 Income Taxes on Capital Gains (and Losses)

When a capital asset[2] is disposed of for more (or less) than its book value, the resulting capital gain (or capital loss) historically had been taxed (or saved taxes) at a rate different from that for ordinary income. However, for corporations, TRA 86 eliminated the differential between income taxes on capital gains and ordinary income. Thus, for firms with federal taxable income greater than $18,333,333, capital gains are taxed at a rate of 35% and capital losses create tax savings amounting to 35% of the loss.

In equation form, the determination of capital gains and losses is straightforward:

$$\text{Capital gain (loss)} = \text{net selling price (i.e., market value)} - \text{book value} \qquad (7\text{-}17)$$
$$= MV - BV.$$

Or,

$$\text{Capital gain (loss)} = \text{net selling price (i.e., MV)} - \text{original first cost}$$
$$\text{(i.e., cost basis)} + \text{accumulated depreciation deductions}$$
$$= MV - B + d_k^*. \qquad (7\text{-}18)$$

In many cases, personal property used in a trade or business is sold for an amount that is greater than its BV but less than its cost basis at the time of disposal. Such gains are technically *not* capital gains, but instead are referred to as *depreciation recapture*. TRA 86 specifies that this difference between the disposal price (i.e., MV) of a depreciable asset and its BV at the time of disposal is to be taxed as ordinary income. In this chapter, capital gains are generally taxed at an effective income tax rate of 40%. When an asset's selling price is less than its BV, the resulting loss on disposal creates an income tax credit savings that is 40% of the loss.

Example 7-12

A large, profitable company has just made two transactions in which a gain and a loss have been recorded. The first transaction involved the sale of long-term bonds having a face value of $100,000. Because interest rates in the economy are substantially higher than those in force when the bond was purchased five years ago, the company was able to obtain only $95,800 from the sale of these bonds. The second transaction was the sale of a piece of earth-moving equipment for $23,100. This depreciable asset has a current BV of $18,000.

    (a)   Classify each transaction and determine its *federal* income tax consequence.

    (b)   What is the net income tax liability (or credit)? Let $t = 0.40$

---

[2] Capital assets include all property owned by a taxpayer except (1) property held mainly for sale to customers, (2) most accounts or notes receivable, (3) depreciable property utilized to carry out the production process, (4) real property used by the business, and (5) copyrights and certain types of short-term, non-interest-bearing state/federal notes. Stock owned in another company is a familiar example of a capital asset.

Solution

(a) The first transaction involves a capital asset, and the *capital loss* is $100,000 − $95.800 = $4,200. The resultant tax *credit* (savings or reduction in income taxes) is ($4,200)(0.40) = $1,680. The second transaction is an illustration of depreciation recapture because the selling price is higher than the BV of the equipment. Here, a *gain on disposal* is $23,100 − $18,000 = $5,100, and the income tax *liability* is $5,100(0.40) = $2,040.

(b) The net consequence of these transactions is a tax liability of $2,040 − $1,680 = $360. This figure could also be obtained by offsetting the capital loss against the gain from depreciation recapture and taxing the difference:

Taxes owed = 0.40[$5,100 (gain) − $4,200 (loss)] = $360. ∎

7.4.7 Investment Tax Credit

A special provision of the federal income tax law, *when in force,* is the investment tax credit (ITC). Originally enacted in 1962, it permits businesses to subtract from their overall tax liability a stated percentage of the cost basis in qualifying property purchased during a particular tax year; thus, it encourages capital investment. If the ITC is 10% of the value of a qualifying property, for instance, the net impact is to reduce the after-tax cost of the asset to 90% of its cost to a company. Qualifying property is depreciable, tangible property used in a trade or business.

TRA 86 *repealed the ITC* in the interest of broadening the corporate tax base and lowering income tax rates on ordinary income. Because the ITC has been repealed and reinstated several times since 1962, it is highly likely that Congress will restore the credit in some form in the future. A simple example is provided here to illustrate the mechanics of the investment tax credit.

Example 7-13

A firm purchased a computerized machine tool for $250,000. Suppose that Congress has restored a 10% investment tax credit that applies to domestically manufactured machine tools. (a) What is the after-tax cost of acquiring this equipment? (b) What is the MACRS depreciation deduction in the first year of ownership if this property's class life is 7 years?

Solution

(a) The 10% investment tax credit is $25,000, assuming that the entire investment is qualifying property. This financial incentive applies dollar for dollar against tax liabilities that the firm incurs, and it effectively reduces the machine tool's first cost (after taxes) to $225,000.

(b) The basis in the machine tool is unaffected by the 10% ITC. This represents another financial incentive of ITCs. Hence, using Table 7-4, the first-year MACRS depreciation deduction is 0.1429($250,000) = $35,725. ∎

Only minimal coverage of some of the main provisions of TRA 86 and JGRRA 2003 that are important to engineering economy studies has been provided here. It is by no means complete, but it is intended to establish a basis for illustrating after-tax (i.e., after income tax) economic analyses. In general, the analyst should either search out specific provisions of the federal and/or state income tax law affecting projects being studied or seek information from persons qualified in income tax law.[3]

The remainder of the chapter illustrates various types of after-tax problems by using a tabular form for computing after-tax cash flows (ATCFs). We also explain after-tax treatment of specific equity-debt financing plans, and we discuss "economic value added."

7.5 General Procedure for Making After-Tax Economic Analyses for Projects

After-tax economic analyses can be performed by using exactly the same methods as before-tax analyses. The only difference is that after-tax cash flows (ATCFs) are used in place of before-tax cash flows (BTCFs) by including expenses (or savings) due to income taxes and then making equivalent worth calculations using an after-tax MARR. The income tax rates and governing regulations may be complex and subject to changes, but once those rates and regulations have been translated into their effect on ATCFs, the remainder of the after-tax analysis is relatively straightforward. We use ATCF as a principal metric for determining a project's profitability.

A note of caution concerning the definition of BTCFs (and ATCFs) for projects is in order at this point. BTCF is defined to be annual cash revenue (or savings) attributable to a project minus its annual cash expenses. These expenses should *not* include interest and other financial cash flows. The reason is that *project* cash flows should be analyzed separately from financial cash flows. Including interest expense with project cash flows is incorrect when the firm's pool of capital is being used to undertake an engineering project. Why? A firm's pool of capital consists of debt capital and equity capital. Because the MARR typically uses the weighted average cost of capital as its lower limit, discounting at the MARR for investments from the pool of capital takes account of the cost of debt capital (interest). Thus, there is no need to subtract interest expense in determining BTCFs—to do so amounts to double counting the interest expense associated with debt capital.

However, sometimes it is appropriate to analyze the equity cash flows associated with an engineering project. In this special case, all the financial cash flows (e.g., interest cash flow on debt capital) must be explicitly considered. This situation is discussed in Section 7.6 and in Appendix 11-A.

To formalize the procedure for determining ATCFs, the following notation and equations are provided: For any given year k of the study period, $k = 0, 1, 2, \ldots, N$, let

---

[3] Some applicable publications are as follows:

a. *Tax Guide for Small Business,* IRS Publication 334, Washington, D.C.: U.S. Government Printing Office, published annually.

b. *Your Federal Income Tax,* IRS Publication 17, Washington, D.C.: U.S. Government Printing Office, published annually.

c. J. K. Lasser, *Your Income Tax* (New York: Simon and Schuster, published annually).

R_k = revenues from the project; this is the positive cash flow from the project during period k,

E_k = cash outflows during year k for deductible expenses (not including, interest),

d_k = sum of all noncash, or book, costs during year k, such as depreciation and depletion,

t = effective income tax rate on *ordinary* income (federal, state, and other); t is assumed to remain constant during the study period,

T_k = income taxes paid during year k, and

$ATCF_k$ = ATCF from the project during year k.

Because the taxable income for a project is $(R_k - E_k - d_k)$, the *ordinary income tax lia-bility* when $R_k > (E_k + d_k)$ is computed with Eq. 7-19:

$$T_k = -t(R_k - E_k - d_k). \tag{7-19}$$

The ATCF associated with a project is

$$ATCF_k = (R_k - E_k - d_k)(1 - t) + d_k \tag{7-20}$$

or

$$ATCF_k = (R_k - E_k)(1 - t) + td_k. \tag{7-21}$$

In many economic analyses of engineering and business projects, ATCFs in year k are computed in terms of $BTCF_k$ (i.e., year k before-tax cash flows):

$$BTCF_k = R_k - E_k. \tag{7-22}$$

Thus,[4]

$$ATCF_k = BTCF_k + T_k \tag{7-23}$$

$$= (R_k - E_k) + -t(R_k - E_k - d_k)$$

$$= (1 - t)(R_k - E_k) + td_k \tag{7-24}$$

Tabular headings to facilitate the computation of after-tax cash flows with Eqs. 7-19 and 7-24 are as follows:

Year	(A) Before-tax cash flow	(B) Depreciation	(C) = (A) − (B) Taxable income	(D) = −t(C) Cash flow for income taxes	(E) = (A) + (D) After-tax cash flow
k	$R_k - E_k$	d_k	$R_k - E_k - d_k$	$-t(R_k - E_k - d_k)$	$(1 - t)(R_k - E_k) + td_k$

Column A consists of the same information used in before-tax analyses, namely, the cash revenues (or savings) less the deductible expenses. Column B contains depreciation that can be claimed for tax purposes. Column C is the taxable income, or amount subject to income taxes. Column D contains the income taxes paid

---

[4] In Eq. 7-19, we use $-t$ in column D, so subtraction of income taxes in Eq. 7-23 is accomplished.

(or saved). Finally, column E shows the ATCFs to be used directly in after-tax economic analyses.

The column headings of Fig. 7-3 indicate the arithmetic operations for computing columns C, D, and E when year $k = 1, 2, \ldots, N$. When $k = 0$ and $k = N$, capital expenditures are usually involved and their tax treatment (if any) is illustrated in the examples that follow. The table should be used with the conventions of + for cash inflow or savings and − for cash outflow or opportunity foregone.

Example 7-14

If the revenue from a project is $10,000 in 2004, out-of-pocket expenses are $4,000, and depreciation claimed for income tax purposes is $2,000, what is the ATCF when $t = 0.40$?

Solution From Eq. 7-20, we have

$$\text{ATCF}_{2004} = (\$10,000 - \$4,000 - \$2,000)(1 - 0.4) + \$2,000 = \$4,400.$$

The same result can be obtained with Eq. 7-21 (or 7-24):

$$\text{ATCF}_{2004} = (\$10,000 - \$4,000)(1 - 0.4) + 0.4(\$2,000) = \$4,400.$$

Equation 7-21 clearly shows that depreciation contributes a credit of td_k to the after-tax cash flow in operating year k. ∎

> The ATCF attributable to depreciation (a tax savings) is td_k in year k. This is often referred to as a "depreciation tax shield." After income taxes, an expense becomes $(1 - t)E_k$ and a revenue becomes $(1 - t)R_k$.

Example 7-15

Suppose that an asset, with a cost basis of $100,000 and a depreciable life of five years, is being depreciated with *alternate MACRS* as follows:

Year	2004	2005	2006	2007	2008	2009
Depreciation deduction	$10,000	$20,000	$20,000	$20,000	$20,000	$10,000

If the firm's effective income tax rate remains constant at 40% during 2004–2009, what is the PW of after-tax savings resulting from depreciation when MARR = 10% per year (after income taxes)?

Solution The PW of tax credits (savings) due to this depreciation schedule is

$$\text{PW}(1090) = \sum_{k=2004}^{2009} 0.4 d_k (1.10)^{-k} = \$4,000(0.9091) + \$8,000(0.8264) + \ldots$$

$$+ \$4,000(0.5645) = \$28,948. \quad ∎$$

Example 7-16

The asset in Example 7-15 is expected to produce net cash inflows (net revenues) of $30,000 per year during the 2004–2009 period, and its terminal market value is negligible. If the effective income tax rate is 40%, how much can a firm afford to spend for this asset and still earn the

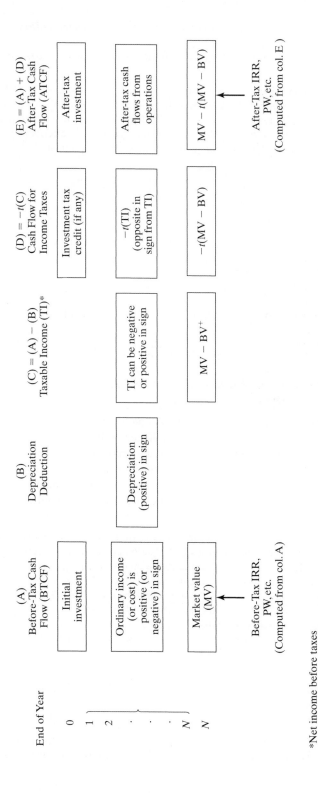

Figure 7-3 General format for determining ATCF.

*Net income before taxes
⁺BV = book value at end of year N

MARR? What is the meaning of any excess in affordable amount over the $100,000 cost basis given in Example 7-15?

Solution After income taxes, the PW of net revenues is $(1 - 0.4)(\$30,000)$ $(P/A, 10\%, 6) = \$18,000(4.3553) = \$78,395$. After adding to this the PW of tax savings computed in Example 7–15, the affordable amount is $107,343. Because the cost of the investment is $100,000, the net PW equals $7,343 (which is the excess in affordable amount over $100,000). This same result can be obtained by using the general format of Fig. 7-3:

EOY	(A) BTCF	(B) Depreciation deduction	(C) = (A) – (B) Taxable income	(D) = –0.4(C) Income taxes	(E) = (A) + (D) ATCF
0	–$100,000	—	—	—	–$100,000
1	30,000	$ 10,000	$20,000	–$8,000	22,000
2	30,000	20,000	10,000	– 4,000	26,000
3	30,000	20,000	10,000	– 4,000	26,000
4	30,000	20,000	10,000	– 4,000	26,000
5	30,000	20,000	10,000	– 4,000	26,000
6	30,000	10,000	20,000	– 8,000	22,000
Totals	$ 80,000	$100,000	$80,000		$ 48,000
				PW(10%) = $	7,343

■

7.5.1 Illustration of Computations of ATCFs

Examples 7-17 and 7-18 illustrate the computation of ATCFs, as well as many common situations that affect income taxes. All problems include the assumption that income tax expenses (or savings) occur at the same time (year) as the income or expense that gives rise to the taxes. For purposes of comparing the effects of various situations, the after-tax IRR and PW are computed for each example. One can observe from the results of Example 7-17 that the faster (i.e., earlier) the depreciation deduction is effective, the more favorable the after-tax IRR and PW will become.

Example 7-17

Certain new machinery placed in service in 2004 is estimated to cost $180,000. It is expected to *reduce* net annual operating expenses by $36,000 per year for 10 years and to have a $30,000 market value (MV) at the end of the 10th year. (a) Develop the before-tax and after-tax cash flows, and (b) calculate the before-tax and after-tax IRR using MACRS depreciation with a federal income tax rate of 35% plus a state income tax rate of 5%. State income taxes are deductible from federal taxable income. This machinery has a MACRS class life of 5 years. (c) Calculate the after-tax PW when the after-tax MARR = WACC = 12%.[5] In this example, the study period is 10 years but the tax life is 6 years (which includes the carryover effect of the half-year convention).

Solution

(a) Table 7-6 applies the format illustrated in Fig. 7-3 to calculate the BTCF and ATCF for this example. In column D the effective income tax rate is very close to 0.38 (from Eq. 7–16) on the basis of the information just provided.

[5] We use Equation 7-13 to find the WACC. The annual after-tax cost of equity is 18% and $(1-\lambda) = 0.50$. Borrowed capital, before income taxes, costs 10% per year and $t \approx 0.38$ from Equation 7-16.

(b) The before-tax IRR is computed from column A:

$$0 = -\$180,000 + \$36,000(P/A, i'\%, 10) + \$30,000(P/F, i'\%, 10).$$

By trial and error, $i' = 16.1\%$.

The entry in the last year is shown to be \$30,000, since the machinery will have this estimated MV. However, the asset was depreciated to zero with the MACRS method. Therefore, when the machine is sold at the end of year 10, there will be \$30,000 of *recaptured depreciation,* which is taxed at the effective income tax rate of 38%. This tax entry is shown in column D (EOY 10).

By trial and error, the after-tax IRR for Example 7-17 is found to be 12.4%.

(c) When MARR = 12% is inserted into the PW equation at the bottom of Table 7-6, it can be determined that the after-tax PW of this investment is \$2,588.   ∎

If the machinery in Example 7-17 has been classified as 10-year MACRS property instead of 5-year property, depreciation deductions would be slowed down in the early years of the study period and shifted into later years, as shown in Table 7-7. Compared with Table 7-6, entries in columns C, D, and E of Table 7-7 are less favorable in the sense that a fair amount of ATCF is deferred until later years, producing a lower after-tax IRR and PW. For instance, the PW is reduced from \$2,588 in Table 7-6 to −\$6,207 in Table 7-7. The only difference between Table 7-6 and Table 7-7 is the *timing of the ATCF,* which is a function of the timing and magnitude of the depreciation deductions. In fact, the curious reader can confirm that the sums of entries in columns A through E of Tables 7-6 and 7-7 are identical (except for round-off differences) in both tables—timing of cash flows does, of course, make a difference!

It is also of interest to note the impact that an ITC would have on the after-tax IRR in Example 7-17. If a 10% ITC could have been claimed for the investment, it would have reduced other income taxes payable by the firm, thereby reducing the out-of-pocket investment by \$18,000 to \$162,000. As a result, the after-tax PW is increased by \$18,000 and the after-tax IRR is increased to 15.3% (the depreciation schedule is assumed not to change). The affected portion of the cash flow table would be as follows:

Year, k	BTCF	Depreciation deduction	Taxable income	Cash flow for income taxes	ATCF
0	−\$180,000			+\$18,000	−\$162,000
1	36,000	\$36,000	\$ 0	0	36,000
2	36,000	\$57,600	− 21,600	− 8,202	44,208
•	•	•	•	•	•
•	•	•	•	•	•
•	•	•	•	•	•

The investment tax credit is normally treated as a "year 0" cash flow because of the quarterly income tax payments required of corporations. The end of the first

TABLE 7-6 ATCF Analysis of Example 7-17 (5-year MACRS Property Class)

End of year, k	(A) BTCF	(B) Basis	Depreciation deduction MACRS rate		Deduction	(C) = (A) − (B) Taxable income	(D) = − 0.38(C) Cash flow for income taxes	(E) = (A) + (D) ATCF
0	−$180,000	—	—		—	—	—	−$180,000
1	36,000	$180,000	× 0.2000	=	$36,000	$ 0	$ 0	36,000
2	36,000	180,000	× 0.3200	=	57,600	− 21,600	+ 8,208	44,208
3	36,000	180,000	× 0.1920	=	34,560	1,440	− 547	35,453
4	36,000	180,000	× 0.1152	=	20,736	15,264	− 5,800	30,200
5	36,000	180,000	× 0.1152	=	20,736	15,264	− 5,800	30,200
6	36,000	180,000	× 0.0576		10,368	25,632	− 9,740	26,260
7–10	36,000	0	—		0	36,000	− 13,680	22,320
10	30,000	—	—		—	30,000[a]	− 11,400[b]	18,600
Total	$210,000							$130,201

PW(12%) = $ 2,588

[a] Depreciation recapture = MV − BV = $30,000 − 0 = $30,000.

[b] Tax on depreciation recapture = $30,000(0.38) = $11,400.

After-tax IRR: Set PW of Column E = 0 and solve for i' in the following equation:

$0 = -\$180,000 + 36,000(P/F,i',1) + 44,208(P/F,i',2) + 35,453(P/F,i',3) + 30,200(P/F,i',4) + 30,200(P/F,i',5) + 26,260(P/F,i',6) + 22,320(P/A,i',4)(P/F,i',6)$
$+ 18,600(P/F,i',10);$ IRR = 12.4%.

TABLE 7-7 Reworked Example 7-17 with a 10-year MACRS Property Class

End of year, k	(A) BTCF	(B) Basis	Depreciation deduction × MACRS rate	=	Deduction	(C) = (A) = (B) Taxable income	(D) = −0.38(C) Cash flow for income taxes	(E) = (A) + (D) ATCF
0	−$180,000	—	—		—			−$180,000
1	36,000	$180,000	× 0.1000	=	$18,000	$18,000	−$ 6,840	29,160
2	36,000	180,000	× 0.1800	=	32,400	3,600	− 1,368	34,632
3	36,000	180,000	× 0.1440	=	25,920	10,080	− 3,830	32,170
4	36,000	180,000	× 0.1152	=	20,736	15,264	− 5,800	30,200
5	36,000	180,000	× 0.0922	=	16,596	19,404	− 7,374	28,626
6	36,000	180,000	× 0.0737	=	13,266	22,734	− 8,639	27,361
7	36,000	180,000	× 0.0655	=	11,790	24,210	− 9,200	26,800
8	36,000	180,000	× 0.0655	=	11,790	24,210	− 9,200	26,800
9	36,000	180,000	× 0.0656	=	11,808	24,192	− 9,193	26,807
10	36,000	180,000	× 0.0655/2	=	5,895	30,105	− 11,440	24,560
10	30,000					18,210*	− 6,920	23,080

Total $130,196

PW(12%) = −$6,207

IRR = 11.2%

Depreciation recapture = MV − BV = $30,000 − $11,799 = $18,201.

quarter is as long as a firm would wait to claim an ITC, and three months is obviously closer to the beginning of the first year than it is to the end of the year.

A minor complication is introduced in ATCF analyses when the study period is shorter than an asset's MACRS recovery period (e.g., for a 5-year recovery period, the study period is 5 years or less). In such cases, we shall assume that the asset is sold for its MV in the last year of the study period. Due to the half-year convention, only one-half of the normal MACRS depreciation can be claimed in the year of disposal or end of the study period, so there will usually be a difference between an asset's BV and its MV. Resulting income tax adjustments will be made at the time of sale (*note:* see last row in Fig. 7-3), unless the situation clearly specifies that the asset in question is to be relegated to standby service for an indefinite period of time. In such a case, depreciation deductions usually continue through the end of the asset's MACRS recovery period. Our assumption of project termination at the end of the study period makes good economic sense, as illustrated in Example 7-18.

Example 7-18

A highly specialized piece of optical character recognition equipment has a capital investment of $50,000. If this equipment is purchased in 2006, it will be used to produce income (through rental) of $20,000 per year for only 4 years. At the end of year 4, the equipment will be sold for a negligible amount. Estimated annual expenses for upkeep are $3,000 during each of the 4 years. The MACRS recovery period for the equipment is 7 years, and the firm's effective income tax rate is 40%.

   (a)   If the after-tax MARR is 7%, should the equipment be purchased?
   (b)   Rework the problem, assuming that the equipment is placed on standby status such that depreciation is taken over the full MACRS recovery period.

   Solution

EOY	(A) BTCF	(B) Depreciation	(C) = (A) − (B) Taxable income	(D) = −0.4(C) Income taxes	(E) = (A) + (D) ATCF
0	−$50,000				−$50,000
1	17,000	$ 7,145	$ 9,855	−$3,942	13,058
2	17,000	12,245	4,755	− 1,902	15,098
3	17,000	8,745	8,255	− 3,302	13,698
4	17,000	3,123[a]	13,877	− 5,551	11,449
4	0	0	− 18,742[b]	7,497	7,497
					Total = $10,800
					PW(7%) = $ 1,026

[a] Under MACRS, only one half-year of depreciation can be claimed in the year an asset is sold, when disposal occurs before year $N' + 1$.
[b] Remaining BV.

Because the PW > 0, the equipment should be purchased.

EOY	(A) BTCF	(B) Depreciation	(C) Taxable income	(D) Income taxes	(E) ATCF
0	−$50,000	—	—	—	−$50,000
1	17,000	$ 7,145	$ 9,855	−$3,942	13,058
2	17,000	12,245	4,755	− 1,902	15,098
3	17,000	8,745	8,255	− 3,302	13,698
4	17,000	6,245	10,755	− 4,302	12,698
5	0	4,465	− 4,465	1,786	1,786
6	0	4,460	− 4,460	1,784	1,784
7	0	4,465	− 4,465	1,786	1,786
8	0	2,230	− 2,230	892	892
8	0	—	—	—	0
					Total $10,800
					PW(7%) = $ 353

The present worth is $673 higher in part (a), which equals the PW of deferred depreciation credits in part (b). A firm would opt for the situation in part (a) if it had a choice. ■

7.5.2 Computer Spreadsheets for Calculation of ATCFs

This section includes two illustrations of using computer spreadsheets for solving engineering economy problems on an after-tax basis. In the first, a 5-year recovery period is utilized to work Example 7-17 to obtain the printout beneath the cell formulas. As expected, the PW(12%) shown in the printout equals the $2,588 obtained in Table 7-6.

The second illustration reworks Example 7-17 with straight-line depreciation of $30,000 per year for 5 years and with $BV_5 = $30,000$. The cell formulas are again provided, followed by the printout. In the second illustration, we notice that PW(12%) = −$3,133, which is less than the PW with the MACRS depreciation. The difference is due to the fact that straight-line depreciation is "slower" than MACRS, thus shifting income taxes forward in time and lowering the after-tax PW.

7.5.3 Illustration of After-Tax Economic Comparisons of Cost-Only Alternatives

In this section, the after-tax comparison of mutually exclusive alternatives involving costs only is demonstrated and discussed. Example 7-19 presents a break-even type of analysis for purchase versus lease evaluation, and Example 7-20 demonstrates that after-tax analyses may produce different recommendations than do before-tax analyses.

Example 7-19
An engineering consulting firm can purchase a fully configured computer-aided design (CAD) workstation for $20,000. It is estimated that the useful life of the workstation is 7 years, and its MV in 7 years should be $2,000. Operating expenses are estimated to be $40 per 8-hour work day, and maintenance will be performed under contract for $8,000 per year. The MACRS class life is 5 years, and the effective income tax rate is 40%.

As an alternative, sufficient computer time can be leased from a company at an end-of-year cost of $20,000. If the after-tax MARR is 10%, how many work days per year must the workstation be needed in order to justify *leasing* it?

Spreadsheet Illustration #1 (formulas)—Example 7-17

R/C	A	B	C	D	E	F	G
1	Effective Tax Rate		TR				
2	After-tax MARR		DR				
3	MACRS Class Life		LIFE				
4							
5	Cost Basis		CB				
6	Investment Tax Credit		ITC				
7							
8						Cash Flow	
9			MACRS	MACRS	Taxable	for Income	
10	Year	BTCF	Percent	Depr'n	Income	Taxes	ATCF
11	0	(B11)	- -	- -	- -	CB*ITC	B11+F11
12	1	B12	HLOOKUP(LIFE,MACRS,A12+1)	CB*C12	B12-D12	-TR*E12	B12+F12
13	2	B13	HLOOKUP(LIFE,MACRS,A13+1)	CB*C13	B13-D13	-TR*E13	B13+F13
14	3	B14	HLOOKUP(LIFE,MACRS,A14+1)	CB*C14	B14-D14	-TR*E14	B14+F14
15	4	B15	HLOOKUP(LIFE,MACRS,A15+1)	CB*C15	B15-D15	-TR*E15	B15+F15
16	5	B16	HLOOKUP(LIFE,MACRS,A16+1)	CB*C16	B16-D16	-TR*E16	B16+F16
17	6	B17	HLOOKUP(LIFE,MACRS,A17+1)	CB*C17	B17-D17	-TR*E17	B17+F17
18	7	B18	HLOOKUP(LIFE,MACRS,A18+1)	CB*C18	B18-D18	-TR*E18	B18+F18
19	8	B19	HLOOKUP(LIFE,MACRS,A19+1)	CB*C19	B19-D19	-TR*E19	B19+F19
20	9	B20	HLOOKUP(LIFE,MACRS,A20+1)	CB*C20	B20-D20	-TR*E20	B20+F20
21	10	B21	HLOOKUP(LIFE,MACRS,A21+1)	CB*C21	B21-D21	-TR*E21	B21+F21
22	10	B22	- -	- -	B22-(CB-SUM(D12:D21))	-TR*E22	B22+F22
23							
24				PW=	NPV(+DR,G12:G21)+G22/(1+DR)^10+G11		
25				AW=		-PMT(DR,10,+G24)	
26							

Solution

This example involves an after-tax evaluation of purchasing depreciable property versus leasing it. We are to determine how heavily the workstation must be utilized so that the lease option is a good economic choice. A *key* assumption is that the cost of engineering design time (i.e., operator time) is unaffected by whether the workstation is purchased or leased. Variable operations expenses associated with ownership result from the purchase of supplies, utilities, and so on. Hardware and software maintenance cost is contractually fixed at $8,000 per year. It is further assumed that the maximum number of working days per year is 250.

  Lease fees are treated as an expense, and the consulting firm (the lessee) may *not* claim depreciation of the equipment to be an additional expense. (The leasing company presumably

Spreadsheet Illustration #1 (printout) - Example 7-17

R/C	A	B	C	D	E	F	G
1	Effective Tax Rate		0.38				
2	After-tax MARR		0.12				
3	MACRS Class Life		5				
4							
5	Cost Basis		180,000				
6	Investment Tax Credit		0				
7							
8						Cash Flow	
9			MACRS	MACRS	Taxable	for Income	
10	Year	BTCF	Percent	Depr'n	Income	Taxes	ATCF
11	0	($180,000)	- -	- -	- -		($180,000)
12	1	$36,000	0.2	$36,000			$36,000
13	2	$36,000	0.32	$57,600	($21,600)	$8,208	$44,208
14	3	$36,000	0.192	$34,560	$1,440	($547)	$35,453
15	4	$36,000	0.1152	$20,736	$15,264	($5,800)	$30,200
16	5	$36,000	0.1152	$20,736	$15,264	($5,800)	$30,200
17	6	$36,000	0.0576	$10,368	$25,632	($9,740)	$26,260
18	7	$36,000			$36,000	($13,680)	$22,320
19	8	$36,000			$36,000	($13,680)	$22,320
20	9	$36,000			$36,000	($13,680)	$22,320
21	10	$36,000			$36,000	($13,680)	$22,320
22	10	$30,000	- -	- -	$30,000	($11,400)	$18,600
23							
24						PW=	$2,588
25							
26							

has included the cost of depreciation in its fee.) Determination of ATCF for the lease option is relatively straightforward and is not affected by how much the workstation is utilized:

After-tax annual cost of the lease = $20,000(1 − 0.40) = $12,000.

ATCFs for the purchase option involve expenses that are fixed (not a function of equipment utilization) in addition to expenses that vary with equipment usage. If we let X equal the number of working days per year that the equipment is utilized, the variable cost per year of operating the workstation is $40X$. The after-tax analysis of the purchase alternative is shown in Table 7-8.

The after-tax annual cost of purchasing the workstation is

Spreadsheet Illustration #2 (formulas)—Example 7-17 (modified)

R/C	A	B	C	D	E	F
1	Effective Tax Rate		TR			
2	After-tax MARR		DR			
3	ADR Life		LIFE			
4						
5	Cost Basis		CB			
6	Salvage Value		SV			
7	Investment Tax Credit		ITC			
8						
9					Cash Flow	
10			Straight Line	Taxable	for Income	
11	Year	BTCF	Depreciation	Income	Taxes	ATCF
12	0	(B12)	- -	- -	ITC*CB	B12+E12
13	1	B13	IF(A13<=LIFE,(CB-SV)/LIFE,0)	B13-C13	-TR*D13	B13+E13
14	2	B14	IF(A14<=LIFE,(CB-SV)/LIFE,0)	B14-C14	-TR*D14	B14+E14
15	3	B15	IF(A15<=LIFE,(CB-SV)/LIFE,0)	B15-C15	-TR*D15	B15+E15
16	4	B16	IF(A16<=LIFE,(CB-SV)/LIFE,0)	B16-C16	-TR*D16	B16+E16
17	5	B17	IF(A17<=LIFE,(CB-SV)/LIFE,0)	B17-C17	-TR*D17	B17+E17
18	6	B18	IF(A18<=LIFE,(CB-SV)/LIFE,0)	B18-C18	-TR*D18	B18+E18
19	7	B19	IF(A19<=LIFE,(CB-SV)/LIFE,0)	B19-C19	-TR*D19	B19+E19
20	8	B20	IF(A20<=LIFE,(CB-SV)/LIFE,0)	B20-C20	-TR*D20	B20+E20
21	9	B21	IF(A21<=LIFE,(CB-SV)/LIFE,0)	B21-C21	-TR*D21	B21+E21
22	10	B22	IF(A22<=LIFE,(CB-SV)/LIFE,0)	B22-C22	-TR*D22	B22+E22
23	10	SV	- -	B23-(CB-SUM(C13:C22))	-TR*D23	B23+E23
24						
25			PW=	NPV(+DR,F13:F22)+F23/(1+DR)^10+F12		

$$\$20{,}000(A/P, 10\%, 7) + 24X + [\$3{,}200(P/F, 10\%, 1) + \cdots + \$4{,}800(P/F, 10\%, 7)] \cdot$$
$$(A/P, 10\%, 7) - \$1{,}200(A/F, 10\%, 7) = \$24X + \$7{,}511.$$

To solve for X, we equate the after-tax annual costs of both alternatives:

$$\$12{,}000 = \$24X + \$7{,}511.$$

Thus, $X = 187$ days per year. Therefore, if the firm expects to utilize the CAD workstation in its business *more than* 187 days per year, the equipment should be leased. The graphic summary of Example 7-19 shown in Fig. 7-4 provides the rationale for this recommendation. The importance of the workstation's estimated utilization, in work days per year, is now quite apparent.   ■

Spreadsheet Illustration #2 (printout) - Example 7-17 (modified)

R/C	A	B	C	D	E	F
1	Effective Tax Rate		0.38			
2	After-tax MARR		0.12			
3	ADR Life		5			
4						
5	Cost Basis		180,000			
6	Salvage Value		30000			
7	Investment Tax Credit		0			
8					Cash Flow	
9			Straight Line	Taxable	for Income	
10	Year	BTCF	Depreciation	Income	Taxes	ATCF
11	0	($180,000)	- -	- -		($180,000)
12	1	$36,000	$30,000	$6,000	($2,280)	$33,720
13	2	$36,000	$30,000	$6,000	($2,280)	$33,720
14	3	$36,000	$30,000	$6,000	($2,280)	$33,720
15	4	$36,000	$30,000	$6,000	($2,280)	$33,720
16	5	$36,000	$30,000	$6,000	($2,280)	$33,720
17	6	$36,000		$36,000	($13,680)	$22,320
18	7	$36,000		$36,000	($13,680)	$22,320
19	8	$36,000		$36,000	($13,680)	$22,320
20	9	$36,000		$36,000	($13,680)	$22,320
21	10	$36,000		$36,000	($13,680)	$22,320
22	10	$30,000	- -			$30,000
23						
24						
25					PW =	$-3,133

Example 7-20a (Before-Tax Analysis)

A company is going to install a new plastic-molding press. Four different presses are available. The essential differences in initial investment and operating costs are shown in the following chart:

	Press			
	A	B	C	D
Investment (installed):	$6,000	$7,600	$12,400	$13,000
Useful life:	5 yr	5 yr	5 yr	5 yr
Annual operation and maintenance costs:				
Power:	$680	$680	$1,200	$1,260
Labor:	$6,600	$6,000	$4,400	$3,200
Maintenance:	$400	$450	$650	$500
Property taxes and insurance:	$120	$152	$248	$260
Total annual costs:	$7,800	$7,282	$6,298	$5,220

TABLE 7-8   After-Tax Analysis of Purchase Alternative–Example 7-19

End of year, k	(A) BTCF	(B) Depreciation deduction[a]	(C) = (A) – (B) Taxable income	(D) = –t(C) Cash flow for income taxes	(E) = (A) + (D) ATCF
0	–$20,000	—	—	—	–$20,000
1	–40X – $8,000	$4,000	–40X – $12,000	–16X + $4,800	–24X – $3,200
2	–40X – 8,000	6,400	–40X – 14,400	–16X + 5,760	–24X – 2,240
3	–40X – 8,000	3,840	–40X – 11,840	–16X + 4,736	–24X – 3,264
4	–40X – 8,000	2,304	–40X – 10,304	–16X + 4,122	–24X – 3,878
5	–40X – 8,000	2,304	–40X – 10,304	–16X + 4,122	–24X – 3,878
6	–40X – 8,000	1,152	–40X – 9,152	–16X + 3,661	–24X – 4,339
7	–40X – 8,000	0	–40X – 8,000	–16X + 3,200	–24X – 4,800
7	2,000	—	2,000	– 800	1,200

[a] Depreciation deduction$_k$ = $20,000 × (MACRS rate)$_k$. Refer to Table 7-4.

Each press will produce the same number of units. However, because of different degrees of mechanization, each requires different amounts and grades of labor and has different operation and maintenance costs. None is expected to have a market value, and the study period is 5 years. Any capital invested is expected to earn at least 25% per year before taxes. Which press should be chosen?

Solution by the PW Method

When alternatives for which revenues are nonexistent or considered equal are compared with the present worth method, the alternative that has the least expensive total PW is judged to be the most desirable (assuming that one press must be selected). Table 7-9 shows the analysis of the four presses by the PW method. The economic criterion is to choose the alternative that has the minimum PW of costs, which is press A. The order of preference among the alternatives in *decreasing order* is press A, press B, press D, and press C. This rank ordering is identical in a before-tax analysis for all equivalent worth methods, when correctly applied.   ∎

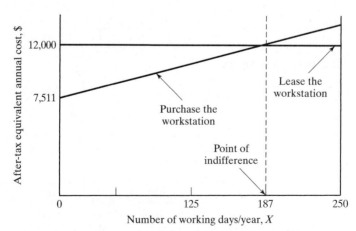

Figure 7-4   Summary of Example 7-19.

TABLE 7-9   Comparison of Four Molding Presses by Using PW Method
(Example 7-20a)

	Press			
	A	B	C	D
Present worth of				
Investment:	$ 6,000	$ 7,600	$12,400	$13,000
Costs:				
(Total annual costs) × (P/A, 25%, 5)	19,594	18,293	15,821	13,867
Total PW (costs)	$25,594	$25,893	$28,221	$26,867

Example 7-20b (After-Tax Analysis)

It is now desired to compare the four presses in Example 7-20a on an after-tax basis by using the present worth (PW) method. Let's assume an effective income tax rate (t) of 40%. Further, we utilize MACRS to compute depreciation with a 3-year property class. None of the presses will have a market value at the end of the 5-year study period. Finally, the after-tax MARR is set equal to $(1 - t)(25\%) = 15\%$ for purposes of discounting after-tax cash flows. The data are given in Table 7-10, which also calculates after-tax cash flows associated with each press.

From Table 7-10, it is seen that the after-tax PW(15%) rank orderings are press D (best), press A, press B, and press C (worst). This is a different ordering of presses than we obtained in the before-tax analysis. (See Table 7-9.)   ∎

Because project desirability in an after-tax analysis can differ from what is obtained in a before-tax analysis, we strongly recommend that after-tax evaluations be performed in engineering economy studies.

7.6 After-Tax Analyses that Include Specific Financing Arrangements

Engineering economy problems encountered thus far have implicitly assumed that the firm's pool of equity and debt capital has been used for investment in engineering and business projects. Engineers and other technical personnel generally do not concern themselves with how funds in a firm's treasury have been sourced, because their mission is to ascertain the best (most profitable) uses of the capital, given that it is available. Consequently, this book typically deals with engineering projects from the standpoint of a firm's *overall* pool of investment capital rather than any specific financing arrangement for the project. As a result, we have focused on *project* cash flows and have excluded financial cash flows specific to the project. Hence, the interest rate (i.e., the MARR) we use for time-value-of-money calculations typically reflects the cost of borrowed money and equity in a particular firm's overall pool of capital (see Equation 7.13.)

By using project cash flows (after taxes) to evaluate investment alternatives, we avoid some serious errors that can arise when explicit treatment of interest expenses is attempted. For instance, if interest is deducted from gross revenues to arrive at operating income, double counting of interest expenses will occur when the MARR includes a component for "return on borrowed funds." *In general, it is advisable to*

TABLE 7-10 After-Tax Cash Flow PW Computations for Comparison of the Four Molding Presses (Example 7-20b)

Press	EOY	(A) BTCF	(B) Depreciation	(C) = (A) − (B) Taxable income	(D) = −0.4(C) Income taxes at 40%	(E) = (A) + (D) ATCF	PW(15%)
A	0	−$6,000	—	—	—	−$6,000	
	1	−7,800	$2,000	−$9,800	$3,920	−3,880	
	2	−7,800	2,667	−10,467	4,187	−3,613	−$19,850
	3	−7,800	889	−8,689	3,476	−4,324	
	4	−7,800	444	−8,244	3,298	−4,502	
	5	−7,800	0	0	3,120	−4,680	
B	0	−$7,600	—	—	—	−$7,600	
	1	−7,282	$2,533	−$9,815	$3,926	−3,356	
	2	−7,282	3,378	−10,660	4,264	−3,018	−$19,919
	3	−7,282	1,126	−8,408	3,363	−3,919	
	4	−7,282	563	−7,845	3,138	−4,144	
	5	−7,282	0	−7,282	2,913	−4,369	
C	0	−$12,400	—	—	—	−$12,400	
	1	−6,298	$4,133	−$10,431	$4,172	−2,126	
	2	−6,298	5,512	−11,810	4,724	−1,574	−$21,269
	3	−6,298	1,836	−8,134	3,254	−3,044	
	4	−6,298	919	−7,217	2,887	−3,411	
	5	−6,298	0	−6,298	2,519	−3,779	
D	0	−$13,000	—	—	—	−$13,000	
	1	−5,220	$4,333	−$9,553	$3,821	−1,399	
	2	−5,220	5,779	−10,999	4,400	−820	−$19,517
	3	−5,220	1,925	−7,145	2,858	−2,362	
	4	−5,220	963	−6,183	2,473	−2,747	
	5	−5,220	0	−5,220	2,088	−3,132	

exclude interest expenses from cash flows and to discount at the firm's after-tax MARR. Nevertheless, there are special situations in which the analysts must consider an investment problem that includes financing specific to the project.

In this section, we shall demonstrate how to analyze a capital investment from the viewpoint of the *equity investor* (i.e., the stockholder). (Later, in Appendix 11-A, we formalize an "adjusted present value" approach to project evaluation.) This viewpoint may be desired when a particular undertaking is quite risky and equity investors insist on an explicit assessment of the project's profitability in terms of *their* capital. To accomplish this, a project's ATCFs must be evaluated from the viewpoint of a firm's stockholders by using an interest rate equal to the *opportunity cost of equity capital* (See Chapter 11.) This can be done simply by adding a column for "Loan and Interest Cash Flow" to Figure 7-3, as is illustrated in Examples 7-21 and 7-22.

Example 7-21

This example is fashioned after Example 7-17. Additional information is given concerning capital borrowed to purchase the machinery, and it is desired to determine the after-tax PW of equity capital. The problem is now restated, with additional information.

Certain new machinery is estimated to cost $180,000 installed. It is expected to reduce net annual operating expenses by $36,000 per year (not including interest) for 10 years and to have a $30,000 MV at the end of the 10th year. Fifty thousand dollars of the investment is to be *borrowed* at 10% interest payable annually, with *all* the principal to be repaid at the end of the 10th year. To simplify the analysis, the machinery is depreciated with the straight-line method over 10 years, using an estimated salvage value of $30,000 in the depreciation calculations.

Determine the after-tax cash flow for this machinery from the stockholders' (equity) viewpoint, and calculate the after-tax profitability of equity capital ($130,000). The opportunity cost of equity capital has been set at 18% per year.

Solution   Table 7-11 shows the recommended tabular format, with the column headings indicating how the calculations are made. Notice that interest on the money borrowed for the project is deductible as a business expense and hence reduces before-tax savings to $31,000. From the results of the right-hand column, the after-tax PW of the equity investment can be calculated:

$$PW(18\%) = -\$130,000 + \$24,920(P/A, 18\%, 10)$$
$$- \$20,000(P/F, 18\%, 10)$$
$$= -\$21,832$$

Because the PW is negative, this investment appears to be a poor one on the basis of ATCFs to owners (stockholders) of the firm.   ■

Example 7-22

A firm is considering purchasing an asset for $10,000, with half of this amount coming from retained earnings (equity) and the other half being borrowed for 3 years at an effective interest rate of 12% per year. Uniform annual payments, consisting of interest and loan principal, will be utilized to repay the $5,000 loan. To simplify the example, straight-line depreciation over a 3-year period will be used, and the asset's MV at this time is expected to be $4,000. Before-tax net revenues minus expenses from the asset are estimated to be $5,000 per year, which *does not* include the cost of borrowed money. Finally, the firm's effective income tax rate is 50%.

TABLE 7-11 After-Tax Equity Cash Flow Analysis for Example 7-21

End of Year, k	(A) BTCF	(B) Depreciation Deduction	(C) Loan Cash Flow (C1) Interest	(C) Loan Cash Flow (C2) Principal	(D) = (A) – (B) + (C1) Taxable Income	(E) = –0.38(D) Cash Flow for Income Taxes	(F) = (A) + (C1) + (C2) + (E) Equity ATCF
0	–$180,000			+$50,000			–$130,000
1–10	+ 36,000[a]	$15,000	–$5,000	0	+$16,000	–$6,080	+ 24,920
10	+ 30,000[b]			–50,000		0[b]	–20,000

[a]Interest expense has not yet been deducted.

[b]The MV of the machinery at the end of year 10 equals its BV. Thus, there is no gain or loss on disposal.

If the opportunity cost of capital to equity investors is 20% per year (after taxes), is this a profitable investment from the stockholder's viewpoint?

Solution

The annual amount of the uniform loan repayment is $5,000(A/P, 12%, 3) = $2,081.75. As described in Chapters 4 and 5, this capital recovery amount consists of repayment of borrowed money (principal) and interest on the unpaid principal at the beginning of each year. Because interest is a deductible business expense but repayment of loan principal is not, a schedule of annual loan principal and interest must be developed as follows (to the nearest dollar):

End of Year, K	Interest	Principal
1	$5,000(0.12) = $600	$2,082 − $600 = $1,482
2	(5,000 − 1,482)(0.12) = 422	2,082 − 422 = 1,660
3	(5,000 − 1,482 − 1,660)(0.12) = 223	2,082 − 223 = 1,859

The equity after-tax cash flow for the asset can now be determined, as indicated in Table 7-12.

The PW of column F at 20% per year is $763. Because the PW is greater than zero, this investment is marginally attractive.   ∎

Example 7-23

In Example 7-22, what is the PW of interest payments in the loan repayment schedule after income taxes ($t = 0.50$) are considered? The after-tax MARR is 20% per year.

Solution

Because interest is deductible from taxable income for legitimate business purposes, its after-tax cost in year k is $(1 - t)(\text{Int}_k)$, and the PW in Example 7-22 is

$$
PW(20\%) = \sum_{k=1}^{3} (1 - t)(\text{Int}_k)(1.20)^{-k}
$$
$$
= 0.5(-\$600)(0.8333) + 0.5(-\$422)(0.6944) + 0.5(-\$223)(0.5787)
$$
$$
= -\$461.03.
$$ ∎

Chapter 11 includes additional approaches for considering the after-tax effects of financial arrangements. Appendix 11-A briefly describes the "Adjusted Present Value" approach, and Appendix 11-B shows how to consider "Bundled Financing."

7.7 Economic Value Added

This section discusses an economic measure for estimating the wealth creation potential of capital investments that is experiencing increased attention and use. The measure, called economic value added (EVA),[6] can be determined from some of the data available in an after-tax analysis of cash flows generated by a capital investment. Through retroactive analysis of a firm's common stock valuation, it has been established that some companies experience a statistically significant relationship between

---

[6] EVA is a registered trademark of Stern Stewart & Company, New York City, NY.

the EVA metric and the historical value of their common stock.[7] For our purposes, EVA can also be used to estimate the profit-earning *potential* of proposed capital investments in engineering projects.

Simply stated, EVA is the difference between the company's adjusted net operating profit after taxes (NOPAT) in a particular year and its after-tax cost of capital (in dollars) during that year. Another way to characterize EVA is "the spread between the return on the capital and the cost of the capital."[8] On a project-by-project basis (i.e., for discrete investments), the EVA metric can be used to gauge the wealth creation opportunity of proposed capital expenditures. We now define annual EVA as

$$
\begin{aligned}
\text{EVA}_k &= (\text{Net Operating Profit After Taxes})_k \\
&\quad - (\text{Cost of Capital Used to Produce Profit})_k \\
&= \text{NOPAT}_k - i \cdot \text{BV}_{k-1},
\end{aligned}
\tag{7-25}
$$

where

$$k = \text{an index for the year in question } (1 \le k \le N),$$
$$i = \text{after-tax MARR based on a firm's cost of capital,}$$
$$\text{BV}_{k-1} = \text{beginning-of-year book value, and}$$
$$N = \text{the study (analysis) period in years.}$$

Figure 7-3 presented in Section 7.5 can be used to relate EVA amounts to after-tax cash flow (ATCF) amounts for a proposed capital investment. The annual EVA amount for year k can be obtained from Figure 7-3 by (1) algebraically *adding* the entry in Column C for year $k(1 \le k \le N)$ to the corresponding entry in Column D to yield net income after taxes (NIAT), which is the same as NOPAT, and then (2) *subtracting* the product of the project's after-tax MARR (based on the cost of capital) and its beginning-of-year book value. This calculation is built into Equation 7-25. Clearly, accurate BTCF estimates (forecasts) in Figure 7-3 are needed for acceptable predictions of annual ATCF and EVA amounts.

Using the notation from Section 7.5, we find that $\text{NOPAT}_k = (1 - t)(R_k - E_k - d_k)$ and $\text{EVA}_k = (1 - t)(R_k - E_k - d_k) - i \cdot \text{BV}_{k-1}$. The ATCF is defined as follows: When $k > 0$, $\text{ATCF}_k = (1 - t)(R_k - E_k - d_k) + d_k$, and $\text{ATCF}_0 = \text{BV}_0$ when $k = 0$. Now we see the relationship between ATCF_k and EVA_k to be $\text{ATCF}_k = \text{EVA}_k + i \cdot \text{BV}_{k-1} + d_k$.

Equation 7-25 and Figure 7-3 are demonstrated in Example 7-24 to determine the ATCF amounts, after-tax AW, and the EVA amounts related to a capital investment.

Example 7-24

Consider the following proposed capital investment in an engineering project and determine its (a) year-by-year ATCF, (b) after-tax AW, and (c) annual equivalent EVA

---

[7] See J. L. Dodd and S. Chen, "EVA: A New Panacea?" *B & E Review*, July–September 1996, pp. 26–28, and W. Freedman "How Do You Add Up?" *Chemical Week*, October 9, 1996, pp. 31–34.

[8] S. Tully, "The Real Key to Creating Wealth," *Fortune*, September 30, 1993, p. 38ff.

Table 7-12 After-Tax Equity Cash Flow Analysis for Example 7-22

End of Year, k	(A) BTC F	(B) Depreciation Deduction	(C) Loan Cash Flow (C1) Interest	(C) Loan Cash Flow (C2) Principal	(D) = (A) − (B) + (C1) Taxable Income	(E) = −0.5(D) Cash Flow for Income Taxes	(F) = (A) + (C1) + (C2) + (E) Equity ATCF
0	−$10,000			+$5,000			−$5,000
1	5,000	$2,000	−$600	− 1,482	$2,400	−$1,200	1,718
2	5,000	2,000	− 422	− 1,660	2,578	− 1,289	1,629
3	5,000	2,000	− 223	− 1,859	2,777	− 1,389	1,529
3	4,000						4,000

Proposed Capital Investment $= \$84,000$
Salvage Value (end of year 4) $= \$0$
Annual Expenses/Year $= \$30,000$
Gross Revenues/Year $= \$70,000$
Depreciation Method $=$ Straight Line
Useful Life $= 4$ years
Effective Income Tax Rate(t) $= 50\%$
After-Tax MARR(i) $= 12\%$ per year

Solution

(a) Year-by-year ATCF amounts are shown, using the computational format of Figure 7-3:

EOY	BTCF	Deprec.	Taxable Income	Income Taxes	ATCF
0	−$84,000	—	—	—	−$84,000
1	70,000 − 30,000	$21,000	$19,000	−$9,500	30,500
2	70,000 − 30,000	21,000	19,000	−9,500	30,500
3	70,000 − 30,000	21,000	19,000	−9,500	30,500
4	70,000 − 30,000	21,000	19,000	−9,500	30,500

(b) The annual equivalent worth of the ATCFs equals $-\$84,000(A/P, 12\%, 4) + \$30,500 = \$2,844$, which we have commonly called the net AW.

(c) From Equation 7-25, the EVA in year k equals $\text{NOPAT}_k - 0.12\text{BV}_{k-1}$. The year-by-year EVA amounts and the annual equivalent worth of EVA ($\$2,844$) are shown in and just after the next table. Hence, the after-tax annual worth and the annual equivalent worth of EVA of the project are *identical*.

EOY_k	NOPAT	$EVA = NOPAT - i \cdot BV_{k-1}$
1	$19,000 − $9,500 = $9,500	$9,500 − 0.12($84,000) = −$580
2	= $9,500	$9,500 − 0.12($63,000) = $1,940
3	= $9,500	$9,500 − 0.12($42,000) = $4,460
4	= $9,500	$9,500 − 0.12($21,000) = $6,980

Annual equivalent EVA $= [-\$580(P/F, 12\%, 1) + \$1,940(P/F, 12\%, 2) + \$4,460(P/F, 12\%, 3) + \$6,980(P/F, 12\%, 4)](A/P, 12\%, 4) = \$2,844$. ∎

The annual equivalent EVA is simply the annual worth, at the after-tax MARR, of a project's after-tax cash flows. This straightforward relationship is also valid when accelerated depreciation methods (such as MACRS) are used in the analysis of a proposed project.

7.8 Summary

This chapter has presented important aspects of TRA 86 and the JGRRA 03 relating to depreciation and income taxes. It is essential to understand these topics so that correct after-tax engineering economy evaluations of proposed projects and ventures may be conducted. Depreciation and income taxes are also integral parts of subsequent chapters in this book.

Many concepts regarding current federal income tax laws have been described—such as taxable income, effective income tax rates, and taxation of ordinary income and depreciation recapture. A general format for pulling together and organizing all these apparently diverse subjects is presented in Fig. 7-3. This format offers the student or practicing engineer a means of collecting on one spreadsheet information required for determining ATCFs for properly evaluating the after-tax financial results of a proposed capital investments.

REFERENCES

Commerce Clearing House, Inc. *Explanation of Tax Reform Act of 1986.* Chicago. (See latest edition.)

Engineering Economist, The. A quarterly journal jointly published by the Engineering Economy Division of the American Society for Engineering Education and the Institute of Industrial Engineers, published by IIE, 25 Technology Park, Atlanta, Georgia 30092.

Kulonda, D.J. "A Note on the Cost of Ownership, Capital Recovery, and Depreciation." *The Engineering Economist,* Vol 48, No 2, 2003, pp 183–189.

Lasser, J. K. *Your Income Tax.* New York: Simon & Schuster. (See latest edition.)

U.S. Department of the Treasury. *Your Federal Income Tax,* IRS Publication 17. Washington, D.C.: U.S. Government Printing Office, revised annually.

U. S. Department of the Treasury. *Tax Guide for Small Business,* IRS Publication 334. Washington, D.C.: U.S. Government Printing Office, revised annually.

U.S. Department of the Treasury. *Depreciation,* IRS Publication 534. Washington, D.C.: U.S. Government Printing Office. (See latest edition.)

PROBLEMS

Note: Terminal book value (BV) and market, or salvage, value (MV) are given in each problem, as appropriate. Unless otherwise stated, use an effective income tax rate of 40%.

7-1. A machine costs $40,000. Its life for depreciation purposes is estimated at 10 years, and its terminal book value is assumed to be $4,000. Determine (1) the depreciation charge for the fifth year and (2) the BV at the end of the fifth year using each of the following methods:

   a. Straight line

   b. Double declining balance

7-2. A new machine has just been purchased by a manufacturer for $25,000. Freight and trucking charges were $500, and the installation cost was $300. The machine has an estimated useful life of 8 years, at which time it is expected that $1,000 dismantling costs will have to be paid in order to sell it for $5,000. Compute (1) the depreciation charge for the first year and (2) the BV at the end of the first year using each of the following methods:

a. Straight line

b. Declining balance

7-3. An asset costs $10,000 and is expected to have $1,000 salvage value at the end of 5 years. Graph its BV as a function of year using each of the following methods:

a. Straight line

b. Double declining balance

7-4. A special-purpose machine is to be depreciated as a linear function of use. It costs $25,000 and is expected to produce 100,000 units and then be sold for $5,000. Up to the end of the third year, it had produced 60,000 units; and during the fourth year, it produced 10,000 units. What are the depreciation charge for the fourth year and the BV at the end of the fourth year?

7-5. Rework Example 7-2 by using the 200% declining balance method. Also assume that it is desired to switch to straight-line depreciation after the fourth year. What is the depreciation charge for all remaining years?

7-6. An asset for drilling that was purchased by a petroleum company in 2006 had a first cost of $60,000 and an estimated salvage value of $12,000. The ADR guideline period for useful life is taken from IRS Publication 534 (Table 7-2), and the MACRS recovery period is 7 years. Compute the depreciation amount in the third year and the BV a at the end of the fifth year of life by each of these methods:

a. the straight-line method over useful (ADR) life;

b. the 200% declining balance method (with $R = 2/N$) over useful (ADR) life; and

c. the MACRS method, using rates from Table 7-4.

7-7. By each of the listed methods, calculate the BV of a highpost binding machine at the end of 4 years if the item originally cost $1,800 and had an estimated salvage value of $400. The ADR guideline period is 8 years, and the MACRS recovery period is 5 years.

a. The machine was purchased in 2005, and MACRS rates in Table 7-4 are applicable.

b. Alternative MACRS is to be utilized over the ADR period of 8 years, with half-year convention.

7-8. An optical scanning machine was purchased in 2005 for $150,000. It is to be used for reproducing blueprints of engineering drawings, and its ADR guideline period is 9 years. The estimated MV of this machine at the end of 9 years is $30,000.

a. What is the MACRS recovery period of the machine?

b. According to your answer to (a), what is the depreciation deduction in 2008?

c. What is the BV at the beginning of 2009?

7-9. An asset was purchased 6 years ago for $6,400. At the same time, its MACRS recovery period and SV were estimated to be 5 years and $1,000, respectively. The ADR guideline period is 9 years. If the asset is sold now for $1,500, what is the difference between its MV of $1,500 and its present BV if depreciation has been by

a. the straight-line method (based on ADR guideline period)?

b. the MACRS method?

7-10. During the first quarter of 2005, a pharmaceutical company purchased a mixing tank that had a retail (fair market) price of $120,000. It replaced an older, smaller mixing tank that had a BV of $15,000 in the second quarter of 2005. Because a special promotion was underway, the old tank was used as a trade-in for the new one, and the cash price (including delivery and installation) was set at $99,500. The ADR guideline period for the new mixing tank is 9.5 years.

   a. Under MACRS, what is the depreciation deduction in 2007?

   b. Under MACRS, what is the BV at the end of 2008?

   c. If 200% declining balance depreciation had been applied to this problem, what would be the cumulative depreciation through the end of 2008?

7-11. Determine the more economical means of acquiring a business machine if you may either (1) purchase the machine for $5,000 with a probable resale value of $2,000 at the end of 5 years, or (2) rent the machine at an annual rate of $900 per year for 5 years with an initial deposit of $500 refundable upon returning the machine in good condition. If you own the machine, you will depreciate it for tax purposes at the annual rate of $600. Of course, all leasing rental charges are deductible from taxable income. As either owner or lessee, you will assume liability for all expenses associated with the operation of the machine. Compare the alternatives using the annual cost method. The after-tax minimum attractive rate of return is 10% per year, and the income tax rate is 40%.

7-12. A corporation in year 2006 expects a gross income of $500,000, total operating (cash) expenditures of $400,000, and capital expenditures of $20,000. In addition, the corporation is able to declare $60,000 depreciation charges for the year. What is the expected taxable income and total federal income taxes owed for the year?

7-13. BIG Corporation is considering making an investment of $100,000 that will increase its annual taxable income from $40,000 to $60,000. What annual federal income taxes will be owed as a result of this investment? What is the average rate to be paid on the taxable income?

7-14. Suppose the investment by BIG Corporation (in Problem 7-13) has expected results as follows:

Annual receipts:	$75,000
Annual disbursements:	$45,000
Life:	8 yr
Terminal book value:	$20,000

   If straight-line depreciation is used and if the after-tax minimum attractive rate of return is 15% per year, determine if the investment is attractive, using the PW method.

7-15. Two alternative machines will produce the same product, but one is capable of higher quality work, which can be expected to return greater revenue. The following are relevant data:

	Machine A	Machine B
First cost:	$20,000	$30,000
Life:	12 yr	8 yr
Terminal BV (and MV):	$4,000	$0
Annual receipts:	$150,000	$188,000
Annual disbursements:	$138,000	$170,000

Determine which is the better alternative, assuming "repeatability" and based on using straight-line depreciation, an income tax rate of 40%, and an after-tax minimum attractive rate of return of 10% per year, using the following methods:

a. Annual worth

b. Present worth

c. Internal rate of return

7-16. Alternative methods I and II are proposed for a plant operation. The following is comparative information:

	Method I	Method II
Initial investment:	$10,000	$40,000
Useful (ADR) life:	5 yr	10 yr
Terminal market value:	$ 1,000	$ 5,000
Annual disbursements:		
Labor:	$12,000	$ 4,000
Power:	$ 250	$ 300
Rent:	$ 1,000	$ 500
Maintenance:	$ 500	$ 200
Property taxes & insurance:	$ 400	$ 2,000
Total annual disbursements:	$14,150	$ 7,000

Determine which method is the better alternative, based on an after-tax annual cost analysis with an effective income tax rate of 40% and an after-tax MARR of 12% per year, assuming the following methods of depreciation:

a. Straight line

b. MACRS

7-17. A manufacturing process can be designed for varying degrees of automation. The following is relevant cost information:

Degree	First cost	Annual labor cost	Annual power and maintenance cost
A	$10,000	$9,000	$ 500
B	14,000	7,500	800
C	20,000	5,000	1,000
D	30,000	3,000	1,500

Determine which degree is best, by after-tax analysis, using an income tax rate of 40%, an after-tax MARR of 15% per year, and straight-line depreciation. Assume each has a life of 5 years and no book or market value. Use each of the following methods:

a. Annual worth

b. Present worth

c. Internal rate of return

7-18. A centerless grinder can be purchased new for $18,000. It will have an 8-year useful life and no MV. Reductions in operating costs (savings) from the machine will be $8,000 for the first 4 years and $3,000 for the last 4 years. Depreciation will be by the MACRS method, using a recovery period of 5 years. A used grinder can be bought for $8,000 and will have no MV in 8 years. It will save a constant $3,000 per year over the 8-year period and will be depreciated at $1,000 per year for 8 years. The effective income tax rate is

40%. Determine the after-tax PW on the *incremental* investment required by the new centerless grinder. Let the MARR be 12% per year after taxes.

7-19. Currently, a firm has annual operating revenues of $190,000, cost of sales of $50,000, and depreciation charges of $40,000 annually. A new project is proposed that will raise annual revenues by $30,000 and increase the cost of sales by $10,000 per year. If this new project necessitates a total capital cost of $50,000, which can be depreciated to zero BV and has no MV at the end of its 6-year life, what is the IRR of the project after federal income taxes are paid? Assume that MACRS depreciation is used with a property class of 5 years.

7-20. A company is considering the purchase of a capital asset for $100,000. Installation charges needed to make the asset serviceable will total $30,000. The asset will be depreciated over 6 years, using the straight-line method and an estimated salvage value for depreciation purposes of $10,000. The asset will be kept in service for 6 years, after which it will be sold for $20,000. During its useful life, it is estimated that the asset will produce annual revenues of $30,000. Operating and maintenance (O&M) expenses are estimated to be $6,000 in the first year. These O&M expenses are projected to increase by $1,000 per year each year thereafter. The after tax MARR is 12% per year and the effective tax rate is 40%.

   a. Use the tabular format given in Figure 7-3 to compute the after-tax cash flows.

   b. Compute the after-tax present worth of the project, and *use a uniform gradient in your formulation.*

7-21. A firm must decide between two systems, A and B, shown in the table. Their effective income tax rate is 40%, and MACRS depreciation is used. If the after-tax desired return on investment is 10% per year, which system should be chosen? Use the AW method and state your assumptions.

	System A	System B
Initial cost:	$100,000	$200,000
Property class:	5 yr	5 yr
Useful ADR life:	7 yr	6 yr
Market value at end of useful life:	$30,000	$50,000
Annual revenues less expenses over useful life:	$20,000	$40,000

7-22. The owners of a small TV repair shop are planning to invest in some new circuit testing equipment. The details of the proposed investment are as follows:

> First cost: $5,000
> Terminal market value: $0
> Extra revenue: $2,000/yr
> Extra expenses: $800/yr
> Expected life: 5 yr (also equal to the MACRS property class)
> Effective income tax rate: 15%

   a. If MACRS depreciation is used, calculate the PW of ATCFs when the MARR (after taxes) is 12% per year. Should the equipment be purchased?

   b. Use the alternative MACRS depreciation method over the ADR recovery period to calculate ATCFs, and recommend whether the new equipment should be purchased if the after-tax MARR is 12% per year. Include the half-year convention when determining depreciation.

7-23. Your firm can purchase a machine for $12,000 to replace a rented machine. The rented machine costs $4,000 per year. The machine that you are considering would have a life of 8 years and a $5,000 terminal BV at the end of its life. By how much could annual operating expenses increase and still provide a return of 10% per year after taxes? The firm is in the 40% income tax bracket, and revenues produced with either machine are identical. Assume that alternate MACRS (straight line with half-year convention) is utilized to recover the investment in the machine and that the ADR period is 8 years.

7-24. Your company has purchased equipment (for $50,000) that will reduce materials and labor costs by $14,000 each year for N years. After N years, there will be no further need for the machine, and since the machine is specially designed, it will have no market value at any time. However, the IRS has ruled that you must depreciate the equipment on a straight-line basis with a tax life of 5 years. If the effective income tax rate is 40%, what is the minimum number of years your firm needs to operate the equipment to earn 10% per year after taxes on its investment?

7-25. Entropy Enterprise, Ltd., is considering a $100,000 heat recovery incinerator (7-year MACRS property class) that is expected to cause a net reduction in out-of-pocket costs of $30,000/year for 6 years (the study period). The incinerator will be depreciated using MACRS recovery allowances from Table 7-4, and no market value is expected. The effective income tax rate is 40%. Determine the PW of the ATCFs when the after-tax MARR is 8% per year.

7-26. Your company has to obtain some new production equipment to be used for the next 6 years, and leasing is being considered. You have been directed to perform an after-tax study of the leasing approach. The pertinent information for the study is as follows:

Lease costs: First year, $80,000; second year, $60,000; third through sixth years, $50,000 per year. Assume that a 6-year contract has been offered by the lessor that fixes these costs over the 6-year period. Other costs (not covered in the contract) are $4,000 per year, and the effective income tax rate is 40%.

a. Develop the annual ATCFs for the leasing alternative.

b. If the MARR after taxes is 8% per year, what is the equivalent annual cost for the leasing alternative?

7-27. A firm can purchase a centrifugal separator for $21,000. Its estimated salvage value for depreciation purposes is $3,000, after a useful life of 6 years, and straight-line depreciation applies to this asset. Operating and maintenance costs for the first year are expected to be $2,000. These costs are projected to increase by $1,000 per year each year thereafter. The income tax rate is 40%, and the MARR is 15% per year after taxes. The separator has *no* market value at the end of six years.

What must the annual benefits be for the purchase of the centrifugal separator to be economical on an after-tax basis? Use a uniform gradient when determining your answer.

7-28. A firm can purchase a centrifugal separator (5-year MACRS property) for $20,000. Its estimated market value is $3,000 after a useful life of 6 years. Operating and maintenance costs for the first year are expected to be $2,000. These costs are projected to increase by $1,000 per year each year thereafter. The income tax rate is 40%, and the MARR is 15% per year after taxes.

What must the annual benefits be for the purchase of the centrifugal separator to be economical on an after-tax basis? Be sure to show your calculations for MACRS depreciation, taxable income, and after-tax cash flows for each year.

7-29. Individual industries will use energy as efficiently as economically possible, and there are several incentives to improve the efficiency of energy consumption. One incentive for the purchase of more energy-efficient equipment is to reduce the time allowed to write off the initial cost. Another "incentive" might be to raise the price of energy in the form of an energy tax.

To illustrate these two incentives, consider the selection of a new motor-driven centrifugal pump for a refinery. The pump is to operate 8,000 hours per year. Pump A costs $1,600, consumes 10 hp, and has an overall efficiency of 65% (it delivers 6.5 hp). The other available alternative, pump B, costs $1,000, consumes 13 hp, and has an overall efficiency of 50% (delivers 6.5 hp). *Note:* 1 hp = 0.746 kW.

Compute the after-tax internal rate of return on extra investment in pump A, assuming an effective income tax rate of 40%, an ADR useful life of 10 years [parts (a) and (c) only], zero market values, and straight-line depreciation for each of these situations:

a. The cost of electricity is $0.04/kWh.

b. A 5-year depreciation write-off period is allowed, the expected life of both pumps is still 10 years, and the cost of electricity is $0.04/kWh.

c. Repeat part (a) when electricity costs $0.07/kWh.

7-30. A firm is considering the introduction of a new product. The marketing department has estimated that the product can be sold over a period of five years at a price of $7.00 per unit. Sales are estimated to be 10,000 units the first year and will increase by 2,000 units each year thereafter. Manufacturing equipment necessary to produce the item will cost $200,000. It is estimated that this equipment can be sold for $50,000 at the end of year five. MACRS depreciation (GDS 5-year recovery period) will be used. The equipment will be financed by borrowing $160,000, and this lost *principal* is to be repaid in five equal end-of-year payments of $32,000 each. Interest at 10% per year will be paid on the load principal outstanding at the beginning of each year. The $40,000 balance will be financed from equity funds. Operating and maintenance costs (not including taxes) will be $50,000 the first year and will decrease by $3,000 each year thereafter. The firm's effective tax rate is 40% and the equity MARR is 15% per year after taxes. Calculate the PW of the after-tax equity cash flow, assuming that the corporation is profitable in its other activities.

7-31. Two corporate vice presidents are trying to decide the less expensive way to produce "widgets." One says that the equipment should be purchased for $10,000 using 100% borrowed funds. These funds and accumulated interest would be repaid *in total* at the end of 5 years and an effective annual interest rate of 10% would be compounded each year. The asset's market and estimated salvage value would be zero, and the capital investment in the equipment would be depreciated with the straight-line method over a 5-year study period. Furthermore, the effective income tax rate is 40% and the after-tax MARR is 15% per year.

The other vice president insists that leasing the machine would be better. Even though operating costs would be about equal, he feels that being able to deduct the entire leasing fee of $3,000 per year for income tax purposes would make this alternative more economical.

If the before-tax annual cash flow for the equipment is estimated to be $4,000, which vice president do you agree with? Show all your calculations.

7-32. Rework Example 7-24 by using the MACRS depreciation method (assume a three-year property class) instead of the straight line depreciation method.

8

Dealing with Price Changes in Capital Investment Analysis

8.1 Introduction

Except for parts of Chapter 3, Regarding Estimating, we have assumed that prices for goods and services in the marketplace are relatively unchanged over substantial periods of time, or that the effect of such changes is constant (or cancels out) among all cash flows for the alternatives under consideration. Unfortunately, these are sometimes not realistic assumptions. *Inflation* is the phenomenon of rising prices bringing about a reduction in the purchasing power of a given unit of money. Similarly, but less commonly experienced, deflation is the phenomenon of decreasing prices. Either can significantly affect the economic comparison of alternatives.

Until the mid-1950s, the U.S. dollar was widely accepted as a fixed measure of the worth of resources. Because the purchasing power of a given sum of money is not constant when inflation is present, individuals and companies alike realize that investment opportunities must be evaluated with money treated as a variable measure of the worth of a resource.

Annual rates of inflation (often referred to as *escalation*) vary widely for different types of goods and services and over different periods of time. For example, the U.S. Government-prepared Consumer Price Index (CPI) rose less than 2% per year during the 1950s, but increased to approximately 7% per year during the 1970s. It is projected that inflation averaging 2% to 4% per year will continue in our economy for at least the near-term future. Although it appears that inflation will extend into the long-term future, it is possible that its opposite, *deflation,* can occur as was true during the depression of the 1930s. Our discussion and examples will focus on inflation.

If all cash flows in an economic comparison of alternatives are inflating at the same rate, inflation can be disregarded in before-tax studies. For cases in which all incomes and all expenses are not inflating at the same rate, inflation gives rise to differences in economic attractiveness among alternatives and should be taken into account. When the effects of inflation are not included in an engineering economy study, an erroneous choice among competing alternatives can result. That is, reversals in preference may occur by assuming that inflation affects all investment opportunities to the same extent. Consequently, the objective of maximizing shareholders' (owners') wealth is inadvertently compromised. To avoid this difficulty, this chapter addresses fundamental concepts and terminology when dealing specifically with inflation in capital investment decisions.[1]

8.1.1 Indexes and Inflation Rates

Numerous indexes have been compiled to represent price changes in many sectors of the U.S. economy. Table 8-1 shows two commonly cited price indexes given for years 1995–2000, together with their year-by-year change rates. The CPI is a composite that measures changes in prices paid for selected goods and services used by ultimate consumers. The Producer Price Index (PPI) is a composite that measures changes in prices paid for selected goods and services used by producers.

TABLE 8-1   Illustrative Consumer Price Index (CPI) and Producer Price Index (PPI) Values and Yearly Change (Inflation or Deflation) Rates

Year	CPI (EOY)	% Change	PPI (EOY)	% Change
1995	153.5	–	129.3	–
1996	158.6	3.32	133.0	2.86
1997	161.3	1.70	131.4	−1.95
1998	163.9	1.61	131.1	−0.23
1999	168.3	2.68	134.9	2.90
2000	174.0	3.39	139.7	3.56

The change rate for any year s, f_s, is easily calculated as

$$f_s = \frac{I_s - I_{s-1}}{I_{s-1}},$$

where I is the index value for the subscripted year.

For example, for the CPI for year 2000,

$$f_s = \frac{174.0 - 168.3}{168.3} = 0.0339 = 3.39\%.$$

---

[1] See also W.G. Sullivan and J.A. Bontadelli, "How an IE Can Account for Inflation in Decision Making," *Industrial Engineering*, Vol. 12, No. 3 (March 1980), pp. 24–33.

If, as is most common, it is desired to find the annual inflation rate f, which is compounded from the end of year k through year n, this can be found by the relationship

$$F = P(F/P, f\%, n - k),$$

or

$$I_n = I_k(F/P, f\%, n - k).$$

For example, for the CPI from the end of year 1995 to 2000,

$$174.0 = 153.5(F/P, f\%, 2000\text{--}1995);$$

or

$$1.34 = (1 + f)^5; \quad f = 2.54\% \text{ per year.}$$

Most price indexes are available as past measures, which can be used for projecting (guesstimating) future inflation for whatever is covered by that index.

8.2 Actual Dollars versus Real Dollars

Inflation describes the situation in which prices of fixed amounts of goods and services are increasing. As prices rise, the value of money, that is, its purchasing power (in real dollars, as defined shortly), decreases correspondingly.

Let us define two distinct kinds of dollars (or other monetary units such as pesos or francs) with which we can work in economic analyses, if done properly:

1. *Actual dollars:* the actual number of dollars as of the point in time they occur and the usual kind of dollar terms in which people think. Sometimes called *then-current dollars,* or *nominal dollars,* or even *inflated dollars,* they will be denoted as "A$" whenever a distinction needs to be made in this book.

2. *Real dollars:* dollars of purchasing power as of some base point in time, regardless of the point in time the actual dollars occur. Sometimes called *constant worth dollars,* or *constant dollars,* or even *uninflated dollars,* they will be denoted as "R$" whenever a distinction needs to be made in this book. If the base point in time k needs to be specified (it is usually the time of the study or the initial investment), it can be shown with a superscript [i.e., $R\$^{(k)}$].

Actual dollars at any time n can be converted into real dollars at time n of purchasing power as of any base time k by

$$R\$_n^{(k)} = A\$_n \left(\frac{1}{1+f}\right)^{n-k} = A\$_n(P/F, f\%, n - k). \tag{8-1}$$

Similarly,

$$A\$_n = R\$_n^{(k)}(1 + f)^{n-k} = R\$_n^{(k)}(F/P, f\%, n - k), \tag{8-2}$$

where f is the average *inflation rate* per period over the $n - k$ periods.

8.3 Real Interest Rate, Combined Interest Rate, and Inflation Rate

We now define several types of rates and show how they are used:

1. *Real interest rate:* Increase in real purchasing power expressed as a percent per period, or the interest rate at which R\$ outflow is equivalent to R\$ inflow. It is sometimes known as *real monetary rate* or *uninflated rate* and denoted as i_r when it needs to be distinguished from i_c (defined shortly).

2. *Combined interest rate:* Increase in dollar amount to cover real interest and inflation expressed as a percent per period; it is the interest rate at which A\$ outflow is equivalent to A\$ inflow. It is sometimes known as *actual rate* or *nominal rate* and is denoted as i_c whenever it needs to be distinguished from i_r.

3. *Inflation rate:* As defined previously, the increase in price of given goods or services as a percent per time period. It is denoted as f. The overall rate for an individual or organization is sometimes called the *general inflation rate*.

Because the real interest rate and the inflation rate have a multiplicative or compounding effect,

$$i_c = (1 + i_r)(1 + f) - 1, \tag{8-3}$$

$$i_c = i_r + f + (i_r \times f). \tag{8-4}$$

Also,

$$i_r = \frac{i_c - f}{1 + f}. \tag{8-5}$$

When f is not large relative to the accuracy desired,

$$i_c \cong i_r + f \tag{8-6}$$

and

$$i_r \cong i_c - f. \tag{8-7}$$

8.4 What Interest Rate to Use in Economy Studies

In general, the interest rate that is appropriate for time-value calculations in engineering economy studies depends on the type of cash flow estimates as follows:

Method	If cash flows are expressed in terms of	Then the interest rate to use is
A	Actual \$, A\$	Combined actual interest rate, i_c
B	Real \$, R\$	Real interest rate, i_r

The preceding is made intuitively consistent if you think in terms of method A as working with inflated (actual) dollars and inflated combined interest and method B as being applicable to uninflated (real) dollars and uninflated (real) interest. Method A is the most natural to use, because we usually think in terms of A$. Since interest paid or earned is based on A$, it is a combined interest rate i_c. However, method B is sometimes easier to use.

8.5 Summary of Formulas for Relating Single Sum A$ and R$ over Time

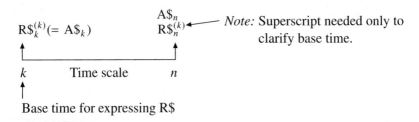

$$R\$_k^{(k)}(= A\$_k)$$

$$A\$_n$$
$$R\$_n^{(k)} \longleftarrow \text{ } \textit{Note:} \text{ Superscript needed only to clarify base time.}$$

k Time scale n

Base time for expressing R$

Method	Type of dollars or conversion	Moving forward in time	Moving backward in time
A	A$ (inflated $)	$A\$_n = A\$_k(F/P, i_c, n-k)$	$A\$_k = A\$_n(P/F, i_c, n-k)$
B	R$ (uninflated $)	$R\$_n^{(k)} = R\$_k^{(k)}(F/P, i_r, n-k)$	$R\$_k^{(k)} = R\$_n^{(k)}(P/F, i_r, n-k)$
		Inflating (at given time)	Deflating (at given time)
	From R$ to A$, or from A$ to R$, at a given time	$A\$_n = R\$_n^{(k)}(F/P, f, n-k)$	$R\$_n^{(k)} = A\$_n(P/F, f, n-k)$
	Using index values	$I_n = I_k(F/P, f, n-k)$	$I_k = I_n(P/F, f, n-k)$

Example 8-1(a)

A certain expense at the end of 2007 is estimated to be $10,000. The end of 2007 is the base point for considering inflation. (Thus, $10,000 = A\$_{07} = R\$_{07}^{(07)}$) Find its equivalent worth at 2012 under the following circumstances (paralleling the preceding formulas) and rates:

$$\text{real rate } i_r = 4\%,$$
$$\text{inflation rate } f = 6\%.$$

Thus,

$$\text{combined rate } i_c = (1 + 0.04)(1 + 0.06) - 1$$
$$= 0.1024 = 10.24\%$$
$$\cong 10\%.$$

Solution In 2012 (denoted 12) dollars,

$$A\$_{12} = A\$_{07}(F/P, i_c\%, 2012\text{--}2007)$$
$$= \$10,000(F/P, 10.24\%, 5) = \$16,282.$$

(*Note:* If 10% approximation were used, the answer would be $16,105.)

In 2007 (denoted 07) dollars,

$$R\$_{12}^{(07)} = R\$_{07}^{(07)}(F/P, i_r\%, 2012\text{–}2007)$$
$$= \$10,000(F/P, 4\%, 5) = \$12,167.$$

In 2012 dollars, beginning with 2012 equivalent in 2007 dollars,

$$A\$_{12} = R\$_{12}^{(07)}(F/P, f\%, 2007\text{–}2002)$$
$$= \$12,167(F/P, 6\%, 5) = \$16,282. \qquad \blacksquare$$

Example 8-1(b)

It is desired to estimate the total cost of a new mainframe computer 5 years from now, given the estimates of cost and inflation rates for each of the following components:

Component	Present cost	Annual inflation rate	Cost 5 years from now
Space	$5,000	8%	
Equipment	$10,000	0%	
Software	–	–	$7,000
Implementation	$8,000	12%	

Solution

$$A\$_s = A\$_0(F/P, f\%, 5), \text{ as applicable.}$$

Component	A$_5
Space $5,000 (F/P, 8%, 5)	$7,350
Equipment $10,000 (F/P, 0%, 5)	$10,000
Software	$7,000
Implementation $8,000 (F/P, 12%, 5)	$14,100
Total	$38,450

$\blacksquare$

8.6 Manipulating Series That Are Uniform in R$

If cash flows expressed in R$ are uniform each year (and thus the A$s increase each year at the average rate of inflation f), they can be conveniently converted to equivalent worth(s) at other point(s) in time by using uniform series formulas at the firm's i_r = real MARR.

Hence, if there is a uniform series in which each end-of-period payment A for n periods is expressed in R$^{(k)}$ (and thus denoted $A^{(k)}$), then

$$P_0^{(k)} = (A^{(k)})(P/A, i_r, n) \qquad (8\text{-}8)$$

and

$$F_n^{(k)} = (A^{(k)})(F/A, i_r, n). \qquad (8\text{-}9)$$

The superscripts make clear the base point in time at which the dollars (like present or future equivalents) are expressed. Normally $k = 0$, but estimates can be converted to any base point in time at the inflation rate f, using Eq. 8-1 or 8-2.

An equivalent formula for finding the present worth of a uniform series that is escalating at the inflation rate f for n years is

$$P_0^{(0)} = \frac{A^{(0)}(F/P, f, 1)[1 - (P/F, i_c\%, n)(F/P, f\%, n)]}{i_c - f}. \tag{8-10}$$

Note that $A^{(0)}(F/P, f, 1) = A^{(1)}$. Thus,

$$P_0^{(0)} = \frac{A^{(1)}[1 - (P/F, i_c\%, n)(F/P, f\%, n)]}{i_c - f}. \tag{8-11}$$

For the special case in which $i_c = f$, so that the real monetary rate $= 0$, Eqs. 8-10 and 8-11 become

$$P_0^{(0)} = A^{(0)}(n) \tag{8-12}$$

and

$$P_0^{(0)} = A^{(1)}(P/F, f, 1)(n). \tag{8-13}$$

Example 8-2

A person who is earning $21,600 salary for (assumed end of) year 1 expects that salary to inflate (escalate) at 10%/year, which is the same as the general rate of inflation. If her real monetary rate i_r is 5%, then $i_c = 0.05 + 0.10 + 0.05(0.10) = 0.155$, or 15.5%. Find the present worth $P_0^{(0)}$, for 3 years of salary.

    (a) Use the approach of Eq. 8-8 to find $P_0^{(1)}$ and then convert to $P_0^{(0)}$.

    (b) Use Eq. 8-11.

Solution

$$\text{(a)} \quad P_0^{(1)} = A^{(1)}(P/A, i_r, 3)$$

$$= \$21,600(P/A, 5\%, 3) = \$58,821;$$

$$P_0^{(0)} = P_0^{(1)}(P/F, f, 1)$$

$$= \$58,821(P/F, 10\%, 1) = \underline{\$53,474}.$$

$$\text{(b)} \quad P_0^{(0)} = \frac{\$21,600[1 - (P/F, i_c, 3)(F/P, f, 3)]}{i_c - f}$$

$$= \frac{\$21,600[1 - (P/F, 15.5\%, 3)(F/P, 10\%, 3)]}{0.155 - 0.10}$$

$$= \$53,496 \text{ (same as for part (a) except for round-off error).} \quad \blacksquare$$

8.7 Manipulating Series That Inflate (Escalate) at Rate Different from General Inflation

When a cash flow series expressed in A$ inflates or escalates at a rate different from general inflation, it will not be a uniform series when expressed in R$. The use of what might be called "differential escalation rates" can be a handy computational convenience in such cases. The next example shows more general ways to convert a series of cash flows subject to whatever rate of inflation (which might vary from year to year) into either R$ or A$, and then to manipulate them (in this case, into present worths) at the general inflation rate.

Example 8-3
Given the same individual salary situation as in Example 8-2, the only difference is that the salary will inflate (escalate) at 8%/year, which differs from the general inflation rate of 10%. To repeat, the end-of-first-year salary $A\$_1 = R\$_1^{(1)} = \$21,600$, $i_r = 5\%$, $f = 10\%$, and $i_c = 15.5\%$.

(a)   Show her salary for three years expressed in R$ and in A$.

(b)   Show the present worth (as of the beginning of the first year) of both ways of expressing this salary.

Solution

In $A\$_n$

Year n	(1) $A\$_n = A\$_1(F/P, 8\%, n-1)$	(2) $(P/F, 15.5\%, n)$	$(3) = (1) \times (2)$ $PW_0^{(0)}$
1	$\$21,600(F/P, 8\%, 0) = \$21,600$	0.8658	$\$18,701
2	$21,600(F/P, 8\%, 1) = 23,328$	0.7495	17,485
3	$21,600(F/P, 8\%, 2) = 25,194$	0.6489	16,348
			$\$52,534

In $R\$_n^{(0)}$

Year n	$R\$_n^{(0)} = A\$_n(P/F, 10\%, n)$	$(P/F, 5\%, n)$	$PW_0^{(0)}$
1	$\$21,600(P/F, 10\%, 1) = \$19,637$	0.9524	$\$18,702
2	$23,328(P/F, 10\%, 2) = 19,278$	0.9070	17,485
3	$25,194(P/F, 10\%, 3) = 18,928$	0.8638	16,350
			$\$52,537

Note, from column (1), that even though her A$ salary is going up (at 8% per year), the R$ (purchasing power) of that salary is going down (approximately 10% − 8% = 2% per year). The present worths of both ways of expressing the salary are the same, except for minor round-off error.   ∎

Example 8-4
Mary Q. Contrary wishes to retire in the year 2030 with personal savings of $500,000 in 2005 spending power. Assume that the expected inflation rate in the economy will average 3.5% during the next 25 years. Mary plans to invest in a 6.0% per year savings account, and her salary is expected to increase by 6.5% per year between 2005 and 2030. Assume that the first deposit occurs at the end of 2006. If Mary puts aside 10% of her salary for retirement purposes, how much will her total salary have to be in 2005 to make her retirement plan a reality?
      This example demonstrates the flexibility of a spreadsheet, even in instances in which all of the calculations are based on a piece of information (the 2005 salary) that we do not yet know. If we deal in actual dollars, the cash flow relationships are straightforward. A spreadsheet

R/C	A	B	C	D	E	F	G
1							
2	2005 Salary	Figure Entry		Annual Salary Increase		F2	
3				Savings (% of Salary)		F3	
4	100000	B4		Savings Interest Rate		F4	
5	10000	B5		2030 Amount (R$)		F5	
6	1000	B6		Average Inflation Rate		F6	
7	100	B7					
8	10	B8					
9	1	B9					
10				Desired 2030 Amount (A$)		+F5*(1+F6)^25	
11	2005 Salary	+B4*A4+B5*A5+B6*A6+B7*A7+B8*A8+B9*A9		Balance Year 2030 (A$)		+E41*(1+$F$4)+D42	
12							
13							
14			Salary	Deposit	Bank Balance		
15					at EOY		
16		Year	(A$)	(A$)	(A$)		
17		2005	+B11				
18		2006	+C17*(1+$F$2)	+C18*($F$3)	+E17*(1+$F$4)+D18		
19		2007	+C18*(1+$F$2)	+C19*($F$3)	+E18*(1+$F$4)+D19		
20		2008	+C19*(1+$F$2)	+C20*($F$3)	+E19*(1+$F$4)+D20		
21		2009	+C20*(1+$F$2)	+C21*($F$3)	+E20*(1+$F$4)+D21		
22		2010	+C21*(1+$F$2)	+C22*($F$3)	+E21*(1+$F$4)+D22		
23		2011	+C22*(1+$F$2)	+C23*($F$3)	+E22*(1+$F$4)+D23		
24		2012	+C23*(1+$F$2)	+C24*($F$3)	+E23*(1+$F$4)+D24		
25		2013	+C24*(1+$F$2)	+C25*($F$3)	+E24*(1+$F$4)+D25		
26		2014	+C25*(1+$F$2)	+C26*($F$3)	+E25*(1+$F$4)+D26		
27		2015	+C26*(1+$F$2)	+C27*($F$3)	+E26*(1+$F$4)+D27		
28		2016	+C27*(1+$F$2)	+C28*($F$3)	+E27*(1+$F$4)+D28		
29		2017	+C28*(1+$F$2)	+C29*($F$3)	+E28*(1+$F$4)+D29		
30		2018	+C29*(1+$F$2)	+C30*($F$3)	+E29*(1+$F$4)+D30		
31		2019	+C30*(1+$F$2)	+C31*($F$3)	+E30*(1+$F$4)+D31		
32		2020	+C31*(1+$F$2)	+C32*($F$3)	+E31*(1+$F$4)+D32		
33		2021	+C32*(1+$F$2)	+C33*($F$3)	+E32*(1+$F$4)+D33		
34		2022	+C33*(1+$F$2)	+C34*($F$3)	+E33*(1+$F$4)+D34		
35		2023	+C34*(1+$F$2)	+C35*($F$3)	+E34*(1+$F$4)+D35		
36		2024	+C35*(1+$F$2)	+C36*($F$3)	+E35*(1+$F$4)+D36		
37		2025	+C36*(1+$F$2)	+C37*($F$3)	+E36*(1+$F$4)+D37		
38		2026	+C37*(1+$F$2)	+C38*($F$3)	+E37*(1+$F$4)+D38		
39		2027	+C38*(1+$F$2)	+C39*($F$3)	+E38*(1+$F$4)+D39		
40		2028	+C39*(1+$F$2)	+C40*($F$3)	+E39*(1+$F$4)+D40		
41		2029	+C40*(1+$F$2)	+C41*($F$3)	+E40*(1+$F$4)+D41		
42		2030	+C41*(1+$F$2)	+C42*($F$3)	+E41*(1+$F$4)+D42		
43							

Figure 8-1   Sample spreadsheet for solving Example 8-4.

model for solving this problem is given in Figure 8-1. The formula in cell F10 converts the desired ending balance into actual dollars. The salary is paid at the end of the year, at which point 10% is placed in a bank account. The interest calculation is based on the cumulative deposits and interest in the account at the beginning of the year, but not on the deposit made at the end of the year. The salary is increased and the cycle repeats.

The spreadsheet model in Figure 8-1 allows us to enter the formulas for the geometric gradient representing the salary increase (column C), the 10% of the salary that goes into the savings account (column D), and the bank balance at the end of the year (column E) without knowing the 2005 salary. Some spreadsheet packages have a "solver" feature that will auto-matically determine the desired 2005 salary. This example illustrates an approach that is not as elegant, but is nonetheless fast and will work for software that does not have a solver feature.

The approach is to revise the 2005 salary systematically and compare the ending bank balance (copied to cell F11 for ease of viewing on the screen) with the desired year 2030 balance. To save keystrokes, the 2005 salary is broken down by powers of 10 into separate cells, in the range B4–B9. The 2005 salary is recombined with the formula in cell B11 and copied to cell C17 for use in the calculations. Starting with the highest power of 10 (cell B4), we bracket the salary that will set cells F10 and F11 equal. The results for this example are shown in Figure 8-2. ■

8.8 Effect of Inflation on Before-Tax and After-Tax Economic Studies

As illustrated previously, if subsequent benefits from an investment bring constant quantities of A$ over time, then inflation will diminish the real value (R$) of the future benefits and, hence, the real rate of return. If, on the other hand, all before-tax costs and benefits are changing at equal rates, then inflation has no net effect on before-tax economic analyses of alternatives. Unfortunately, however, this is not true for after-tax economic analyses.

In general, given two projects with the same before-tax rate of return, it can be shown that inflation results in a smaller after-tax rate of return than that for a project that does not have benefits that increase with inflation. This is because, even though the benefits may increase at the same rate as inflation, the depreciation charges do not increase, which results in larger income tax payments. The net result is that, even though the after-tax cash flow in A$ is increased with inflation, that increase is not large enough to offset both the increased income taxes and inflation. It can be concluded, then, that inflation reduces the real after-tax PW (and IRR) because of the devaluation of fixed depreciation schedules (or other types of unresponsive A$ annuities, such as interest payments). Such unresponsive amounts cause taxable income to increase in an *actual dollar analysis* so that income taxes increase and after-tax cash flows decrease. This observation is confirmed in Example 8-5.

Example 8-5

Your firm has decided it *must acquire* a new piece of machinery that includes the latest safety features required by OSHA. The machinery may either be (1) purchased for cash or (2) leased from another company. Mr. Williams, the president of the firm, has requested that you perform an after-tax analysis of these two means of obtaining the machinery when the following esti-mates and conditions are applicable:

1. The study period is five years and the estimated useful life of the machinery also is 5 years. Straight-line depreciation is elected with an estimated zero book value at the end of the useful life of the purchased machinery. The effective incremental income tax rate (t) is 50%. Also, a 10% investment tax credit can be taken. All recaptured depreciation (if any) is taxed at 50% of the gain.

2. The following interest rates and inflation estimates are used:

    (a)   The real after-tax MARR (i_r) is 10% per year.

R/C	A	B	C	D	E	F	G
1							
2	2005 Salary	Figure Entry		Annual Salary Increase		6.5%	
3				Savings (% of Salary)		10.0%	
4	100000	0		Savings Interest Rate		6.0%	
5	10000	8		2030 Amount (R$)		$500,000	
6	1000	7		Average Inflation Rate		3.5%	
7	100	1					
8	10	7					
9	1	1					
10				Desired 2030 Amount (A$)		$994,894	
11	**2005 Salary**	87171		Balance Year 2030 (A$)		$994,895	
12							
13							
14			Salary	Deposit	Bank Balance		
15					at EOY		
16		Year	(A$)	(A$)	(A$)		
17		2005	$87,171				
18		2006	$92,837	$9,284	$9,284		
19		2007	$98,872	$9,887	$19,728		
20		2008	$105,298	$10,530	$31,441		
21		2009	$112,143	$11,214	$44,542		
22		2010	$119,432	$11,943	$59,158		
23		2011	$127,195	$12,719	$75,427		
24		2012	$135,463	$13,546	$93,499		
25		2013	$144,268	$14,427	$113,535		
26		2014	$153,645	$15,365	$135,712		
27		2015	$163,632	$16,363	$160,218		
28		2016	$174,268	$17,427	$187,258		
29		2017	$185,595	$18,560	$217,053		
30		2018	$197,659	$19,766	$249,842		
31		2019	$210,507	$21,051	$285,883		
32		2020	$224,190	$22,419	$325,455		
33		2021	$238,762	$23,876	$368,859		
34		2022	$254,282	$25,428	$416,418		
35		2023	$270,810	$27,081	$468,484		
36		2024	$288,413	$28,841	$525,435		
37		2025	$307,160	$30,716	$587,677		
38		2026	$327,125	$32,713	$655,650		
39		2027	$348,388	$34,839	$729,828		
40		2028	$371,033	$37,103	$810,721		
41		2029	$395,151	$39,515	$898,879		
42		2030	$420,835	$42,084	$994,895		
43							

Figure 8-2 Spreadsheet results for Example 8-4.

(b) Annual inflation (f) is expected to average 8%.

(c) The combined after-tax MARR (i_c) is 18.8% (i.e., $i_c = 0.10 + 0.08 + 0.10(0.08) = 0.188$ or 18.8%).

3. Annual savings, operating costs, maintenance costs, and the terminal *market value* respond to inflation. In the case of leasing the machinery, the yearly lease payment does *not* grow with inflation. When purchasing the machinery, depreciation write-offs do not respond to inflation.

4. Annual cash flow estimates are as follows:

	Machinery	
	Purchase	Lease
Savings:	$5,000	$5,000
Operating costs:	2,000	$2,000
Maintenance cost:	1,000	(included in lease contract)
Lease fee:	—	$6,000
		(payable at end of year)

All of the preceding annual cash flows have been estimated in real dollars.

5. Investment costs are as follows:

Purchase machine	Lease machine
Initial cost = $20,000	Deposit = $1,500
	(refundable at end of
	5 years with no interest)
Market value = $1,500 at	
end of year 5 (in real	
dollars)	

6. The analysis is to be performed after taxes, and an inflation rate (f) of 8% is estimated to apply to *all* cash flows that respond to inflation.

Based on this information, should your firm purchase or lease the machinery?

Solution   An after-tax cash flow analysis is performed with actual dollar estimates in Table 8-1 for the "purchase machinery" option and in Tables 8-1, 8-2 for the "lease machinery" alternative. Because annual savings, operating costs, and maintenance costs inflate each year at 8%, the real dollar before-tax cash flows are converted to the corresponding actual dollar estimates in column C by using Eq. 8-1. The column C entries are then combined with depreciation write-offs in Table 8-1 and lease payments in Table 8-2 to arrive at the taxable income associated with each alternative. Notice that the firm cannot claim depreciation on the leased machinery.

After-tax cash flow in column G of Table 8-1 is determined in view of an investment tax credit in year 0 and depreciation recapture, which is taxed at 50%, in year 5. The net present worth at $i_c = 18.8\%$ is −$7,601 for the "purchase machinery" alternative. Similarly, the present worth of column G in Table 8-2 is −$4,395 for the leasing alternative. A recommendation should be made to lease the machinery so that present worth of cost is minimized.

It should be noted that a combined interest rate is used to discount the after-tax cash flows in Tables 8-1 and 8-2. In this regard, *the after-tax MARRs of most companies are directly stated as nominal interest rates.* Furthermore, most companies make economic studies in terms of A$ estimates, because decision makers lean toward a measure of financial profitability that includes the effects of inflation.

Referring to Example 8-5, if an after-tax analysis had been performed *ignoring the effects of inflation,* the recommended course of action would have been to purchase the equipment! Thus, assuming a 0% inflation rate in this particular example would have led to an *incorrect* selection between alternatives.   ∎

TABLE 8-1 Example 8-5: Purchase Equipment Alternative—Actual Dollar Analysis

Year	(A) Before-tax cash flow (R$)	(B) Adjustment $(1+f)^{year}$	(C) Before-tax cash flow (A$)	(D) Depreciation (A$)	(E) Taxable income (C) − (D)	(F) Cash flow for income taxes $-t \times (E)$	(G) After-tax cash flow (A$) (C) + (F)
0	−$20,000	1.000	−$20,000			$2,000[a]	−$18,000
1	2,000[b]	1.080	2,160	$4,000	−$1,840	920	3,080
2	2,000	1.166	2,332	4,000	− 1,668	834	3,166
3	2,000	1.260	2,520	4,000	− 1,480	740	3,260
4	2,000	1.360	2,720	4,000	− 1,280	640	3,360
5	2,000	1.469	2,938	4,000	− 1,060	531	3,469
5	1,500	1.469	2,204		2,204	− 1,102	1,102
						PW(18.8%) =	−$7,601

[a] Investment tax credit = 0.10($20,000).
[b] $5,000 (annual savings) − $2,000 − $1,000 = $2,000.

TABLE 8-2 Example 8-5: Lease Equipment Alternative—Actual Dollar Analysis

Year	(A) Before-tax cash flow (R$)	(B) Adjustment $(1+f)^{year}$	(C) Before-tax cash flow (A$)	(D) Lease payments (A$)	(E) Taxable income (C) − (D)	(F) Cash flow for income taxes $-t \times$ (E)	(G) After-tax cash flow (A$) (C) + (F)
0	−$1,500	1.000	−$1,500				−$1,500
1	3,000[a]	1.080	3,240	$6,000	−$2,760	$1,380	− 1,380
2	3,000	1.166	3,498	6,000	− 2,502	1,251	− 1,251
3	3,000	1.260	3,780	6,000	− 2,220	1,110	− 1,110
4	3,000	1.360	4,080	6,000	− 1,920	960	− 960
5	3,000	1.469	4,407	6,000	− 1,593	796.5	− 796.5
5			1,500[b]				1,500
							PW(18.8%) = −$4,395

[a] $5,000 (annual savings) − $2,000 = $3,000.
[b] The deposit on the leased equipment, refunded with no interest.

8.9 Deflation

The previous sections concentrated on inflation because that is the dominant condition experienced in the past and expected in the future. However, we should also recognize deflation. Deflation is the opposite of inflation and can be handled comparably to inflation in economic analyses. That is, estimates can be made in terms of either R$ or A$, and the corresponding interest rate to use should be either the real rate i or the composite rate,

$$i_c = (1 + i_r)(1 - f) - 1 = i_r - f - (i_r \times f), \tag{8-14}$$

where f is the rate of deflation.

8.10 Summary

It is important to be consistent in using the correct interest rate for the type of analysis (actual or real dollars) being done. Two mistakes commonly made are as follows:

Interest rate (MARR)	Type of analysis	
	A$	R$
i_c	(Correct)	Mistake 1 Bias is against capital investment
i_r	Mistake 2 Bias is toward capital investment	(Correct)

In mistake 1, the combined interest rate (i_c), which includes an adjustment for the inflation rate (f), is used in equivalent worth calculations for cash flows estimated in real dollars. Since real dollars have constant purchasing power expressed in terms of the base time period and do not include the effect of general price inflation, we have an inconsistency. There is a tendency to develop future cash flow estimates in terms of dollars with purchasing power at the time of the study and then use the combined interest rate in the analysis. The result of mistake 1 is a bias against capital investment when savings–benefits are considered. The cash flow estimates in real dollars for a project are numerically lower in value than actual dollar estimates with equivalent purchasing power (assuming that $f > 0$). Additionally, the i_c value (which is greater than the i_r value that should be used) further reduces (understates) the equivalent worth of the results of a proposed capital investment.

In mistake 2, the cash flow estimates are in actual dollars, which include the effect of inflation (f), but the real interest rate (i_r) is used for equivalent worth calculations. Since the real interest rate does not include an adjustment for general price inflation, we again have an inconsistency. The effects of this mistake, opposite to those in mistake 1, result in a bias toward capital investment by overstating the equivalent worth of the future benefits.

The economic analysis practitioner typically is faced with a decision as to whether price changes should be explicitly considered in a study. This depends on estimates of prospective price changes and sensitivity of choices and decisions to those changes. When differential price changes are considered, it is common that before-tax and after-tax cash flows be expressed in A$, and then the end results be converted to equivalent R$ if desired.

PROBLEMS

8-1. a. Labor cost is currently $30,000 per year (expressed in year 0 dollars) and is expected to inflate at 5%/yr. Find the PW (at time 0) of the series (assumed to occur at end of each year for 15 years) if the real MARR is 5% and the combined MARR is approximately 15%.

b. The Whackya Bank will pay you 3% for savings account deposits. The average inflation rate is expected to be 5%. What will be your real return rate (use exact formula, not approximation)?

8-2. If the average inflation rate is expected to be 3% per year into the foreseeable future, how many years will it take for the dollar's purchasing power to be one-half of what it is now (that is, the future point in time when it takes two dollars to buy what can be purchased today for one dollar)?

8-3. John and Mary Doe have computed that $21,900 deposited annually into an interest-bearing account with $i_c = 8\%$ will grow to $1 million in 20 years. ($N = 20$ years.) If the average annual inflation rate during this 20-year period is 3%, what is the spending power equivalent, 20 years from now, *in today's dollars* of the $1 million that John and Mary will have accumulated?

8-4. a. It is desired to estimate the 2007 construction cost of a new plant for which 2002 component costs, and applicable rates of inflation, are as follows:

	2002 cost (millions)	Projected annual inflation rate (2002 through 2007) (%)
Labor:	$1	10
Building materials:	5	0
Equipment:	3	15
Total	$9	

b. Using the answer to part (a) and using a (combined) MARR of 20%, find the equivalent worth of the 2007 construction cost as of 2002 (i.e., $R\$_{02}^{(02)}$).

8-5. If you buy a lathe now, it costs $200,000. If you wait 2 years to purchase the lathe, it will cost $260,000. Suppose you decide to purchase the lathe now, reasoning that you can earn $i_c = 20\%$ per year on your $200,000 if you do not purchase the lathe. If the inflation rate (f) in the economy during the next 2 years is expected to average 12% per year, did you make the right decision?

8-6. Your company has just issued bonds with a face value of $10,000. They mature in 10 years and pay annual interest of $1,000. At present they are being sold for $8,770. If the average annual inflation rate over the next 10 years is expected to be 6%, what is the real rate of return per year on this investment?

8-7. The "actual" (then-current) costs for a utility service are expected to be as follows:

End of year	"Actual" costs
2005	$10,000
2006	10,000
2007	12,000

Assume that the average inflation rate is 8%/yr and that the real interest on money is 4%/yr, so the combined interest rate can be approximated as 12%.

a. Find the equivalent worth of the sum of all costs as of the end of 2005.

b. Express each of the three costs in "real dollars" of purchasing power as of the end of 2005.

c. Using the results of (b), find the equivalent worth of the sum of all costs as of the end of 2005. How should this answer compare with the answer to (a)?

8-8. A firm desires to determine the most economic equipment-overhauling schedule alternative to provide for service for the next 9 years of operation. The firm's real minimum attractive rate of return is 5%, and the inflation rate is estimated at 7%. The following are alternatives with all costs expressed in real (constant worth) dollars as of now:

I. Completely overhaul for $12,000 now.

II. A major overhaul for $7,000 now that can be expected to provide 6 years of service, and then a minor overhaul costing $6,000 at the end of 6 years.

III. A minor overhaul costing $5,000 now as well as at the end of 3 years and 6 years from now.

8-9. Ron Future wishes to set aside money for his newborn son's college education. His goal is to have a bank savings account containing an amount equivalent to $50,000 of today's purchasing power at the time of his son's eighteenth birthday. The estimated inflation rate is 6% per year. If the bank pays 3% compounded annually, what lump sum of money should Joe deposit in the bank account on his son's fourth birthday?

8-10. Operation and maintenance costs for alternatives A and B are estimated on different bases as follows:

Year end in which cost incurred	Alternative A in "Actual $"	Alternative B in time 0 "Real $"
1	$110,000	$100,000
2	112,000	100,000
3	114,000	100,000
4	116,000	100,000

If the average inflation rate is 10%/yr and the real interest on money is 5%/yr, show which alternative has the lower equivalent (a) present worth of costs at time 0 and (b) future worth of costs at time 4.

8-11. A large corporation's electricity bill now amounts to $400 million. During the next 10 years, electricity usage is expected to increase by 75%, and the estimated electricity bill 10 years hence has been projected to be $920 million. Assuming electricity usage and rate increases at uniform annual rates over the next 10 years, what is the annual rate of inflation of electricity prices expected by this corporation?

8-12. The AZROC Corporation needs to acquire a small computer system for one of its regional sales offices. The purchase price of the system has been quoted at $50,000, and the system will reduce manual office expenses by $18,000 per year in real dollars. Historically, these manual expenses have inflated at an average rate of 8% per year, and this is expected to continue into the future. A maintenance agreement will also be contracted for, and its cost per year in actual dollars is constant at $3,000.

What is the minimum (integer-valued) life of the system such that the new computer can be economically justified? Assume that the computer's market value is zero at all times. The firm's MARR is 25% and includes an adjustment for anticipated inflation in the economy.

8-13. A high school graduate has decided to invest 5% of her first-year's salary in a mutual fund. This amounts to $1,000 in the first year. She has been told that her savings should keep up with expected salary increases, so she plans to invest an *extra* 8% each year over a 5-year period. Thus, at the end of year 1 she invests $1,000; in year 2, $1,080; and so on through year 5. If the average rate of inflation is expected to be 4% over the next 10 years, and if she expects a 2% *real* return on this investment, what is the worth of the mutual fund at the end of the 5th year?

8-14. The incremental design and installation costs of a total solar system (heating, air conditioning, hot water) in a certain Virginia home were $28,000 in 2006. The annual savings in electricity (in 2006 dollars) have been estimated at $5,000. Assume that the life of the system is 15 years.

a. What is the real internal rate of return on this investment?

b. What average annual inflation rate on electricity would have to be experienced over the system's life to provide a combined rate of return of 25% for this investment?

8-15. A gas-fired heating unit is expected to meet an annual demand for thermal energy of 500 million Btu, and the unit is 80% efficient. Assume that each thousand cubic feet of natural gas, if burned at 100% efficiency, can deliver one million Btu. Suppose further that natural gas is now selling for $2.50 per thousand cubic feet. What is the present worth of fuel cost for this heating unit over a 12-year period if natural gas prices are expected to inflate at an average rate of 10% per year? The firm's combined (nominal) MARR is 15% per year.

8-16. A small heat pump, including the duct system, now costs $2,500 to purchase and install. It has a useful life of 10 years and incurs annual maintenance of $100 per year in real (year 0) dollars over its useful life. A compressor replacement is required at the end of the 8th year at a cost of $500 in real dollars. The yearly cost of electricity for the heat pump is $680 based on prices at the beginning of the investor's time horizon. All costs are expected to escalate at 6%, which is the projected general inflation rate. The firm's MARR, which includes an allowance for general inflation, is 15%. Its real MARR can be approximated as 9% per year. No market value is expected from the heat pump at the end of 15 years.

a. What is the annual equivalent cost, expressed in actual dollars, of owning and operating this heat pump?

b. What is the annual cost in year 0 (real) dollars of owning and operating the heat pump?

8-17. Your company *must* obtain some new production equipment for the next 6 years and is considering leasing. You have been directed to accomplish an actual-dollar after-tax study of the leasing approach. The pertinent information for the study is as follows:

> *Lease costs:* First year, $80,000; second year, $60,000; third through sixth years, $50,000 per year. Assume that a 6-year contract has been offered by the lessor that fixes these costs over the 6-year period.
>
> *Other costs* (not covered in contract): $4,000 in year 0 dollars, and estimated to increase 10% each year.
>
> *Effective income tax rate:* 40%.

a. Develop the actual dollar after-tax cash flow (ATCF) for the leasing alternative.

b. If the real MARR (i_r) after taxes is 5% and the annual inflation rate (f) is 9.524%, what is the actual dollar after-tax equivalent annual cost for the leasing alternative?

8-18. A company has two alternative machines it can purchase to perform a specified task. Machine A costs $150,000 initially while machine B (the deluxe model) costs $200,000. It has been estimated that operating costs will be $1,000 for machine A and $500 for machine B in the first year. Management expects these costs to increase with inflation, which is expected to average 10% per year. The company uses a 10-year study period, and its effective income tax rate is 50%. Both machines qualify as 5-year MACRS property. Which machine should the company choose? Assume neither has a salvage value.

8-19. Because of tighter safety regulations, an improved air filtration system must be installed at a plant that produces a highly corrosive chemical compound. The investment cost of the system is $260,000 in time 0 (2007) dollars. The system has a useful life of 10 years and a recovery period of 5 years. MACRS depreciation is used, with a zero salvage value for tax purposes. However, it is expected that the MV of the system at the end of its 10-year life will be $50,000 in time 0 dollars, and that the replacement investment will be $260,000. Costs of operating and maintaining (O&M) the system, estimated in time 0 dollars, are expected to be $6,000 per year. Annual property tax is 4% of the investment cost and *does not inflate*. Assume that the plant has a remaining life of 20 years, and that O&M costs, replacement costs, and MV inflate at 6% per year.

If the effective income tax rate is 40%, set up a table to determine the ATCF over a 20-year period. The after-tax rate of return desired on investment capital is 12% (including the effect of inflation). What is the PW of cost of this system after income taxes have been taken into account? (Assume that the annual general price inflation rate is 4.5% over the 20-year period.) *Suggestion:* Consider using a computer spreadsheet to solve this problem.

8-20. A certain engine lathe can be purchased for $150,000 and depreciated over 3 years to a zero salvage value with the straight-line method. This machine will produce metal parts that will produce revenues of $80,000 (time 0 dollars) per year. It is a policy of the company that annual revenues will be escalated each year to keep pace with the general inflation rate, which is expected to average 5%/yr ($f = 0.05$). Labor, materials, and utilities totaling $20,000 (time 0 dollars) per year are all expected to increase at

9% per year. The firm's effective income tax rate is 50%, and its after-tax MARR (i_c) is 26%.

a. *Perform an actual-dollar (A$) analysis* and determine the annual ATCFs of the preceding investment opportunity. Use a life of 3 years and work to the nearest dollar.

b. What interest rate would be used for discounting purposes, if needed?

9
Replacement Analysis[1]

9.1 Introduction

A decision situation often encountered in business firms and government organizations, as well as by individuals, is whether an existing asset should be retired from use, continued in service, or replaced with a new asset. As the pressures of worldwide competition continue to increase, requiring higher quality goods and services, shorter response times, and other changes, this type of decision is occurring more frequently. Thus, the *replacement problem*, as it is commonly called, requires careful economic analyses to provide the information needed for sound decisions that improve the operating efficiency and the competitive position of an enterprise.

Engineering economy studies of replacement situations are performed by using the same basic methods as other economic studies involving two or more alternatives. The specific decision situation, however, occurs in different forms. Sometimes it may be whether to retire an asset without replacement (*abandonment*) or whether to retain the asset for backup rather than primary use. Also, the decision may be whether the changed requirements can be met by *augmenting* the capacity or capability of the existing asset(s). Often, however, the decision is whether to replace an existing (old) asset, descriptively called the *defender*, with a new asset. The one or more alternative replacement (new) assets are then called *challengers*.

---

[1] This chapter is adapted from W.G. Sullivan, E.M. Wicks, and J.A. Luxhoj, *Engineering Economy* (12th ed.), Upper Saddle River, NJ, Prentice Hall Inc., 2003. Reprinted by permission of the publisher.

9.2 Reasons for Replacement Analysis

The need to evaluate the replacement, retirement, or augmentation of assets results from changes in the economics of their use in an operating environment. Various reasons can underlie these changes, and unfortunately they are sometimes accompanied by unpleasant financial facts. The four major reasons that summarize most of the factors involved are the following:

1. *Physical Impairment (Deterioration):* These are changes that occur in the physical condition of the asset. Normally, continued use (aging) results in less efficient operation of an asset. Routine maintenance and breakdown repair costs increase, energy use may increase, more operator time may be required, and so forth. Or, some unexpected incident such as an accident occurs that affects the physical condition and the economics of ownership and use of the asset.

2. *Altered Requirements:* Capital assets are used to produce goods and services that satisfy human wants. When the demand for a good or service either increases or decreases, or the design of a good or service changes, the related asset(s) may have the economics of its use affected.

3. *Technology:* The impact of changes in technology varies among different types of assets. For example, the relative efficiency of heavy highway construction equipment is impacted less rapidly by technological changes than that of automated manufacturing equipment. In general, the costs per unit of production, as well as quality and other factors, are favorably impacted by changes in technology, which results in more frequent replacement of existing assets with new and better challengers.

4. *Financing:* Financial factors involve economic opportunity changes external to the physical operation or use of assets and may involve income tax considerations.[2] For example, the rental (lease) of assets may become more attractive than ownership.

Reason 2 (altered requirements) and Reason 3 (technology) are sometimes referred to as different categories of *obsolescence*. Even financial changes (Reason 4) could be considered a form of obsolescence. In any replacement problem, however, factors from more than one of these four major areas may be involved. Regardless of the specific considerations, and even though there is a tendency to regard it with some apprehension, the replacement of assets often represents an economic opportunity for the firm.

For the purposes of our discussion of replacement studies, we next make a distinction among various types of lives for typical assets.

Economic life is the period of time (years) that results in the minimum Equivalent Uniform Annual Cost (EUAC) of owning and operating an asset.[3] If we assume

---

[2] In this chapter, we often refer to Chapter 7 for details on depreciation methods and after-tax analysis.

[3] The *AW* of a primarily cost cash flow pattern is sometimes called the Equivalent Uniform Annual Cost (EUAC). Because this term is commonly used in the definition of the economic life of an asset, we will often use EUAC in this chapter.

good asset management, economic life should coincide with the period of time extending from the date of acquisition to the date of abandonment, demotion in use, or replacement from the primary intended service.

Ownership life is the period between the date of acquisition and the date of disposal by a specific owner. A given asset may have different categories of use by the owner during this period. For example, a car may serve as the primary family car for several years and then serve only for local commuting for several more years.

Physical life is the period between original acquisition and final disposal of an asset over its succession of owners. For example, the car just described may have several owners over its existence.

Useful life is the time period (years) that an asset is kept in productive service (either primary or backup). It is an estimate of how long an asset is expected to be used in a trade or business to produce income.

9.3 Factors That Must Be Considered in Replacement Studies

There are several factors that must be considered in replacement studies. Once a proper perspective has been established regarding these factors, little difficulty should be experienced in making replacement studies. Six factors and related concepts are discussed in this section:

1. Recognition and acceptance of past errors
2. Sunk costs
3. Existing asset value and the *outsider viewpoint*
4. Economic life of the proposed replacement asset (challenger)
5. Remaining (economic) life of the old asset (defender)
6. Income tax considerations

9.3.1 Past Estimation Errors

The economic focus in a replacement study is the future. Any *estimation errors* made in a previous study related to the defender are not relevant (unless there are income tax implications). For example, when an asset's book value (BV) is greater than its current market value (MV), the difference frequently has been designated as an estimation error. Such "errors" also arise when capacity is inadequate, maintenance expenses are higher than anticipated, and so forth.

This implication is unfortunate because in most cases these differences are not the result of errors but of the inability to foresee future conditions better at the time of the original estimates. Acceptance of unfavorable economic realities may be made easier by posing a hypothetical question: "What will be the costs of my competitor, who has no past errors to consider?" In other words, we must decide whether we wish to live in the *past*, with its errors and discrepancies, or to be in a sound competitive position in the *future*. A common reaction is "I can't afford to take the loss in value of the existing asset that will result if the replacement is made." The fact is that the loss already has occurred, whether or not it could be afforded, and it exists whether or not the replacement is made.

9.3.2 The Sunk Cost Trap

Only present and future cash flows should be considered in replacement studies. Any unamortized values (i.e., unallocated value of an asset's capital investment) of an existing asset under consideration for replacement are strictly the result of *past* decisions (i.e., the initial decision to invest in that asset and decisions as to the method and number of years to be used for depreciation purposes). For purposes of this chapter, we define a *sunk cost* to be the difference between an asset's BV and its MV at a particular point in time. Sunk costs have no relevance to the replacement decisions that must be made (*except to the extent that they affect income taxes*). When income tax considerations are involved, we must include the sunk cost in the economic analysis. Clearly, serious errors can be made in practice when sunk costs are incorrectly handled in replacement studies.

9.3.3 Investment Value of Existing Assets and the Outsider Viewpoint

Recognition of the nonrelevance of BVs and sunk costs leads to the proper viewpoint to use in placing value on existing assets for replacement study purposes. In this chapter we use the so-called *"outsider viewpoint"* for approximating the investment amount of an existing asset (defender). In particular, the outsider viewpoint[4] is the perspective that would be taken by an impartial third party to establish the fair MV of a used (secondhand) asset. This viewpoint forces the analyst to focus on present and future cash flows in a replacement study, thus avoiding the temptation to dwell on past (sunk) costs.

The *present realizable* MV is the correct capital investment amount to be assigned to an existing asset in replacement studies.[5] A good way to reason that this is true is to use the *opportunity cost* or *opportunity foregone principle*. That is, if it should be decided to keep the existing asset, we are giving up the opportunity to obtain its net realizable MV at that time. Therefore, this represents the *opportunity cost* of keeping the defender.

There is one addendum to this rationale: If any new investment expenditure (such as for overhaul) is needed to upgrade the exiting asset so that it will be competitive in level of service with the challenger, the extra amount should be added to the

---

[4] The outsider viewpoint is also known as the opportunity cost approach to determining the value of the defender.

[5] In after-tax replacement studies, the before-tax MV is modified by income tax effects related to potential gains (losses) foregone if the defender is kept in service.

present realizable MV to determine the total investment in the existing asset for replacement study purposes.

When using the outsider viewpoint, we find that the total investment in the defender is the opportunity cost of not selling the existing asset for its current MV, *plus* the cost of upgrading it to be competitive with the best available challenger (all feasible challengers are to be considered).

Clearly, the MV of the defender must not also be claimed as a reduction in the challenger's capital investment, because doing so would provide an unfair advantage to the challenger due to double counting the defender's selling price.

Example 9-1

The purchase price of a certain new automobile (challenger) being considered for use in your business is $21,000. Your firm's present automobile (defender) can be sold on the open market for $10,000. The defender was purchased with cash three years ago, and its current book value is $12,000. To make the defender comparable in continued service to the challenger, your firm would need to make some repairs at an estimated cost of $1,500.

Based on this information, what are (a) the total capital investment in the defender using the outsider viewpoint and (b) the unamortized value of the defender?

Solution

(a) The total capital investment in the defender (if kept) is its current market value (an opportunity cost) plus the cost of upgrading the car to make it comparable in service to the challenger. Hence, the total capital investment in the defender is $10,000 + $1,500 = $11,500 (from an outsider's viewpoint). This represents a good starting point for estimating the cost of keeping the defender.

(b) The unamortized value of the defender is the book loss (if any) associated with disposing of it. Given that the defender is sold for $10,000, the unamortized value (loss) is $12,000 − $10,000 = $2,000. This is the difference between the current market value and the current book value of the defender. As discussed in Section 9.3.2, this amount represents a sunk cost and has no relevance to the replacement decision except to the extent that it may impact income taxes. (To be discussed in Section 9.9.)   ■

9.3.4 Economic Life of the Challenger

The economic life of an asset minimizes the equivalent uniform annual cost (EUAC) of owning and operating an asset, and it is often shorter than the useful or physical life. It is essential to know a challenger's economic life in view of the principle that new and existing assets should be compared over their economic (optimum) lives. Economic data regarding challengers are periodically updated (often annually) and replacement studies are then repeated to ensure an ongoing evaluation of improvement opportunities.

9.3.5 Economic Life of the Defender

As we shall see later in this chapter, the economic life of the defender is often one year. Consequently, care must be taken when comparing the defender with a challenger asset because *different lives* are involved in the analysis. We shall see that the defender should be kept longer than its apparent economic life as long as its *marginal cost* is less than the minimum EUAC of the challenger over its economic life. What *assumptions* are involved when two assets having different apparent economic lives are compared, knowing that the defender is a nonrepeating asset? These concepts will be discussed in Section 9.7.

9.3.6 The Importance of Income Tax Consequences

The replacement of assets often results in gains or losses from the sale of *depreciable property*, as discussed in Chapter 7. Consequently, to perform an accurate economic analysis in such cases, the studies must be made on an *after-tax basis*. It is evident that the existence of a taxable gain or loss, in connection with replacement, can have a considerable effect on the results of an engineering study. A prospective gain from the disposal of assets can be reduced by as much as 40% or 50%, depending on the effective income tax rate used in a particular study. Consequently, the decision to dispose of or retain an existing asset can be influenced by income tax considerations.

9.4 Typical Replacement Problems

The typical replacement situations discussed next are used to illustrate several of the factors that must be considered in replacement studies. These analyses use the outsider viewpoint to determine the investment in the defenders.

Example 9-2

A firm owns a pressure vessel that it is contemplating replacing. The old pressure vessel has annual operating and maintenance expenses of $60,000 per year and it can be kept for 5 more years, at which time it will have zero market value. It is believed that $30,000 could be obtained for the old pressure vessel if it is sold now.

A new pressure vessel can be purchased for $120,000. The new pressure vessel will have a market value of $50,000 in 5 years and will have annual operating and maintenance expenses of $30,000 per year. Using a minimum attractive rate of return of 20% per year, determine whether or not the old pressure vessel should be replaced. A study period of 5 years is appropriate.

Solution   The first step in the analysis is to determine the investment value of the defender (old pressure vessel). Using the outsider viewpoint, we find that the investment value of the defender is $30,000, its present MV. We can now compute the PW (or FW or AW) of each alternative and decide whether the old pressure vessel should be kept in service or replaced immediately:

Defender: $PW(20\%) = -\$30,000 - \$60,000(P/A,20\%,5)$
$$= -\$209,436;$$

Challenger: $PW(20\%) = -\$120,000 - \$30,000(P/A,20\%,5) + \$50,000(P/F,20\%,5)$
$$= -\$189,623.$$

The PW of the challenger is greater (less negative) than the PW of the defender. Thus, the old pressure vessel should be replaced immediately. (The EUAC of the defender is $70,035 and that of the challenger is $63,410.) ∎

Example 9-3

The manager of a carpet manufacturing plant became concerned about the operation of a critical pump in one of the processes. After discussing this situation with the supervisor of plant engineering, they decided that a replacement study should be done, and that a 9-year study period would be appropriate for this situation. The company that owns the plant is using a before-tax MARR of 10% per year for its capital investment projects.

The existing pump, *pump A*, including driving motor with integrated controls, cost $17,000 when purchased 5 years ago. An estimated market value of $750 could be obtained for the pump if it is sold now. Some reliability problems have been experienced with pump A, including annual replacement of the impeller and bearings at a cost of $1,750. Annual operating and maintenance (O&M) expenses have been averaging $3,250. Annual insurance and property tax expenses are 2% of the initial capital investment. It appears that the pump will provide adequate service for another nine years if the present maintenance and repair practice is continued. It is estimated that, if this pump is continued in service, its final MV after nine more years will be about $200.

An alternative to keeping the existing pump in service is to sell it immediately and to purchase a replacement pump, *pump B*, for $16,000. An estimated market value at the end of the 9-year study period would be 20% of the initial capital investment. O&M expenses for the new pump are estimated to be $3,000 per year. Annual taxes and insurance would total 2% of the initial capital investment. The data for Example 9-3 are summarized in Table 9-1.

On the basis of these data, should the defender (pump A) be kept [and the challenger (pump B) not purchased] or should the challenger be purchased now (and the defender sold)? Use a before-tax analysis and the outsider viewpoint in the evaluation.

Solution   In an analysis of the defender and challenger, care must be taken to identify correctly the investment amount in the existing pump. On the basis of the outsider viewpoint,

TABLE 9-1 Summary of Information for Example 9-3

MARR (before taxes) = 10% per year		
Existing pump A (defender)		
Capital investment when purchased 5 years ago		$17,000
Annual expenses:		
Replacement of impeller and bearings	$1,750	
Operating and maintenance	3,250	
Taxes and insurance: $17,000 × 0.02	340	
Total annual expenses		$5,340
Present market value		$750
Estimated market value at the end of 9 additional years		$200
Replacement pump B (challenger)		
Capital investment		$16,000
Annual expenses:		
Operating and maintenance	$3,000	
Taxes and insurance: $16,000 × 0.02	320	
Total annual expenses		$3,320
Estimated market value at the end of 9 years: $16,000 × 0.20		$3,200

this would be the current MV of $750; that is, the *opportunity cost* of keeping the defender. Note that the investment amount of pump A ignores the original purchase price of $17,000. By using the principles discussed thus far, a before-tax analysis of EUAC of pump A and pump B can now be made.

The solution of Example 9-3, using EUAC (before taxes) as the decision criterion, follows:

Study period = 9 years	Keep old pump A	Replacement pump B
EUAC(10%):		
Annual expenses	$5,340	$3,320
Capital recovery cost:		
[($750 – $200)(*A/P*, 10%, 9) + $200(0.10)]	115	
[($16,000 – $3,200)(*A/P*, 10%, 9) + $3,200(0.10)]		2,542
Total EUAC(10%)	$5,455	$5,862

Because pump A has the smaller EUAC ($5,455 < $5,862), the replacement pump is apparently not justified and the defender should be kept. We could also make the analysis by using other methods (e.g., *PW*), and the indicated choice would be the same.   ∎

9.5 Determining the Economic Life of a New Asset (Challenger)

Sometimes in practice the useful lives of the defender and the challenger(s) are not known and cannot reasonably be estimated. The period of time an asset is kept in productive service might be extended indefinitely with adequate maintenance and other actions, or it might suddenly be jeopardized by an external factor such as technological change. Under this situation, it is important to know the economic life, minimum EUAC, and total year-by-year (marginal) costs for both the best challenger and the defender, so they can be compared on the basis of an evaluation of their economic lives and the costs most favorable to each.

The economic life of an asset was defined in Section 9.2 as the period of time that results in the minimum EUAC of owning and operating the asset. Also, the economic life is sometimes called the minimum-cost life or optimum replacement interval. For a new asset, its EUAC can be computed if the capital investment, annual expenses, and year-by-year market values are known or can be estimated. The apparent difficulties of estimating these values in practice may discourage performing the economic life and equivalent cost calculations. Similar difficulties, however, are encountered in most engineering economy studies when estimating the *future* economic consequences of alternative courses of action. Therefore, the estimating problems in replacement analysis are not unique and can be overcome in most application studies.

The estimated initial capital investment, as well as the annual expense and market value estimates, may be used to determine the PW through year k of total costs,

PW_k; that is, on a *before-tax* basis,

$$PW_k(i\%) = I - MV_k(P/F, i\%, k) + \sum_{j=1}^{k} E_j(P/F, i\%, j), \qquad (9.1)$$

which is the sum of the initial capital investment (PW of the initial investment amounts, if any occur after time zero), adjusted by the PW of the MV at the end of year k, and of the PW of annual expenses (E_j) through year k. The *total marginal cost* for each year k, TC_k, is calculated, using Equation 9-1, by finding the increase in the PW of total cost from year $k-1$ to year k, and then determining the equivalent worth of this increase at the end of year k. That is, $TC_k = (PW_k - PW_{k-1})(F/P, i\%, k)$. The algebraic simplification of this relationship results in

$$TC_k(i\%) = MV_{k-1} - MV_k + iMV_{k-1} + E_k, \qquad (9.2)$$

which is the sum of the loss in MV during the year of service, the opportunity cost of capital invested in the asset at the beginning of year k, and the annual expenses incurred in year k (E_k). These total marginal (or year-by-year) costs, based on Equation 9-2, are then used to find the EUAC of each year prior to and including year k. The minimum $EUAC_k$ value during the useful life of the asset identifies its economic life, N^*. This procedure is illustrated in Example 9-4.

Example 9-4
A new forklift truck will require an investment of $20,000 and is expected to have year-end market values and annual expenses as shown in columns 2 and 5, respectively, of Table 9-2. If the before-tax MARR is 10% per year, how long should the asset be retained in service?

   Solution   The solution to this problem is obtained by completing Columns 3, 4, 6 (Equation 9-2), and 7 of Table 9-2. In the solution, the customary year-end occurrence of all cash flows is assumed. The loss in market value during year k is simply the difference between the beginning-of-year market value, MV_{k-1}, and the end-of-year market value, MV_k. The opportunity cost of capital in year k is 10% of the capital unrecovered (invested in the asset) at the beginning of each year. The values in column 7 are the equivalent uniform annual costs that would be incurred each year (1 to k) if the asset were retained in service through year k and then replaced (or retired) at the end of the year. The minimum EUAC occurs at the end of year N^*.

   It is apparent from the values shown in column 7 that the new forklift truck will have a minimum EUAC if it is kept in service for only 3 years (i.e., $N^* = 3$). ∎

   The computational approach in the preceding example, as shown in Table 9-2, was to determine the total marginal cost for each year and then to convert these into an EUAC through year k. The before-tax EUAC for any life can also be calculated by using the more familiar capital recovery formulas presented in Chapter 5. For example, for a life of 2 years, the EUAC can be calculated as follows:

$$EUAC_2(10\%) = \$20,000(A/P, 10\%, 2) - \$11,250(A/F, 10\%, 2)$$
$$+ [\$2,000(P/F, 10\%, 1) + \$3,000(P/F, 10\%, 2)](A/P, 10\%, 2)$$
$$= \$8,643.$$

This agrees with the corresponding row in column 7 of Table 9-2.

TABLE 9-2 Determination of the Economic Life N of a New Asset (Challenger A) (Example 9-4)

(1) End of year, k	(2) MV, end of year k	(3) Loss in market value (MV) during year k	(4) Cost of capital = 10% of beginning of year MV	(5) Annual expenses (E_k)	(6) [= (3) + (4) + (5)] Total (marginal) cost for year, (TC_k)	(7) EUAC[a] through year k
0	$20,000	—	—	—	—	—
1	15,000	$5,000	$2,000	$2,000	$9,000	$9,000
2	11,250	3,750	1,500	3,000	8,250	8,643
3	8,500	2,750	1,125	4,620	8,495	**8,598**	$(N^* = 3)$
4	6,500	2,000	850	8,000	10,850	9,084
5	4,750	1,750	650	12,000	14,400	9,954

[a] $EUAC_k = [\sum_{j=1}^{k} TC_j (P/F, 10\%, j)](A/P, 10\%, k)$.

9.6 Determining the Economic Life of a Defender

In replacement analyses, we must also determine the economic life that is most favorable to the defender. This gives us the choice of keeping the defender as long as its minimum EUAC is less than the optimal EUAC of the challenger. When a major outlay for defender alteration or overhaul is needed, the life that will yield the minimum EUAC is likely to be the period that will elapse before the next major alteration or overhaul is needed. Alternatively, *when there is no defender MV now or later (and no outlay for alteration or overhaul) and when the defender's operating expenses are expected to increase annually, the remaining life that will yield the minimum EUAC will be 1 year.*

When MVs are greater than zero and expected to decline from year to year, it is necessary to calculate the apparent remaining economic life, which is done in the same manner as in Example 9-4 for a new asset. Using the outsider viewpoint, the investment value of the defender is considered to be its present realizable MV.

> Regardless of how the remaining economic life for the defender is determined, a decision to keep the defender does not mean that it should be kept only for this period of time. Indeed, the defender should be kept longer than the apparent economic life as long as its *marginal* cost (total cost for an additional year of service) is less than the minimum EUAC for the best challenger.

This important principle of replacement analysis is illustrated in Example 9-5.

Example 9-5
It is desired to determine how much longer a forklift truck should remain in service before it is replaced by the new truck (challenger) for which data were given in Example 9-4 and Table 9-2. The defender in this case is 2 years old, originally cost $13,000, and has a present realizable MV of $5,000. If kept, its market values and annual expenses are expected to be as follows:

End of year, k	MV, end of year k	Annual expenses, E_k
1	$4,000	$5,500
2	3,000	6,600
3	2,000	7,800
4	1,000	8,800

Determine the most economical period to keep the defender *before* replacing it (if at all) with the present challenger of Example 9-4. The cost of capital (MARR) is 10% per year.

Solution   Table 9-3 shows the calculation of total cost for each year (marginal cost) and the EUAC at the end of each year for the defender, based on the format used in Table 9-2. Note that the minimum EUAC of $7,000 corresponds to keeping the defender for one more year. However, the marginal cost of keeping the truck for the second year is $8,000, which is still less than the minimum EUAC for the challenger (i.e., $8,598, from Example 9-4). The *marginal cost* for keeping the defender the third year and beyond is greater than the $8,598 minimum EUAC for the challenger. Based on the available data shown, it would be most

TABLE 9-3 Determination of the Economic Life of an Old Asset (Example 9-5)

(1) End of year, k	(2) MV, end of year k	(3) Loss in market value (MV) during year k	(4) Cost of capital = 10% of beginning of year MV	(5) Annual expenses (E_k)	(6) [= (3) + (4) + (5)] Total (marginal) cost for year, (TC_k)	(7) $EUAC^a$ through year k
0	$5,000	—	—	—	—	—
1	4,000	$1,000	$500	$5,500	$7,000	$7,000
2	3,000	1,000	400	6,600	8,000	7,476
3	2,000	1,000	300	7,800	9,100	7,966
4	1,000	1,000	200	8,800	10,000	8,405

$(N^* = 1)$

$^a \mathrm{EUAC}_k = [\sum_{j=1}^{k} \mathrm{TC}_j \, (P/F, 10\%, j)](A/P, 10\%, k)$.

economical to keep the defender for two more years and then to replace it with the challenger. This situation is portrayed graphically in Figure 9-1.   ■

Example 9-5 assumes that a comparison is being made with the best challenger alternative available. In this situation, if the defender is retained beyond the point at which its marginal costs exceed the minimum EUAC for the challenger, the difference in costs continues to grow and replacement becomes more urgent. This is illustrated to the right of the intersection in Figure 9-1.

Figure 9-2 illustrates the effect of improved new challengers in the future. If an improved challenger X becomes available before replacement with the new asset of Figure 9-1, then a new replacement study should take place to consider the improved challenger. If there is a possibility of a further-improved challenger Y as of, say, 4 years later, it may be better still to postpone replacement until that challenger becomes available. Although retention of the old asset beyond its breakeven point with the best available challenger has a cost that may well grow with time, this cost of waiting can, in some instances, be worthwhile if it permits the purchase of an improved asset having economies that offset the cost of waiting. Of course, a decision to postpone a replacement may also "buy time and information." Because technological change tends to be sudden and dramatic rather than uniform and gradual, new challengers with significantly improved features can arise sporadically and can change replacement plans substantially.

When replacement is not signaled by the economic analysis, more information may become available before the next analysis of the defender. Hence, the next study should include any additional information. *Postponement* generally should mean a

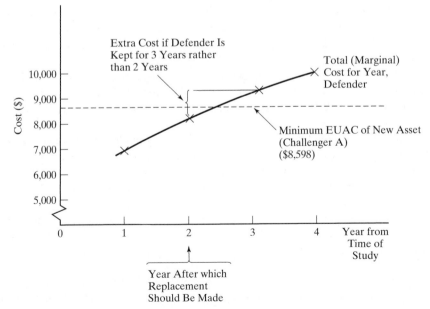

Figure 9-1   Defender versus challenger forklift trucks (based on Examples 9-4 and 9-5).

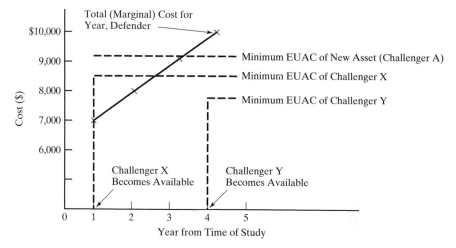

Figure 9-2   Old versus new asset costs with improved challengers becoming available in the future.

postponement of the decision on when to replace, not the decision to postpone replacement until a specified future date.

9.7 Comparisons in Which the Defender's Useful Life Differs from That of the Challenger

In Section 9.4, we discussed a typical replacement situation in which the useful lives of the defender and the challenger were known and were the same, as well as equal to the study period. When this situation occurs, any of the analysis methods, properly applied, may be used.

In the previous two sections (9.5 and 9.6), we discussed the economic lives of a new asset and of a defender and how these results (along with the related cost information) are used in replacement analysis when the useful lives of the assets may or may not be known.

A third situation occurs when the useful lives of the best challenger and the defender are known, or can be estimated, but are not the same. The comparison of the challenger and the defender under these circumstances is the topic of this section.

In Chapter 5, two assumptions used for the economic comparisons of alternatives, including those having different useful lives, were described: (1) *repeatability* and (2) *cotermination*. Under either assumption, the analysis period used is the same for all alternatives in the study. The repeatability assumption, however, involves two main stipulations:

1. The period of needed service for which the alternatives are being compared is either indefinitely long or a length of time equal to a common multiple of the useful lives of the alternatives.

2. What is estimated to happen in the first useful life span will happen in all succeeding life spans, if any, for each alternative.

For replacement analyses, the first of these conditions may be acceptable, but normally the second is not reasonable for the defender. The defender is typically an older and used piece of equipment. An identical replacement, even if it could be found, probably would have an installed cost in excess of the current MV of the defender.

Failure to meet the second stipulation can be circumvented if the period of needed service is assumed to be indefinitely long and *if we recognize that the analysis is really to determine if **now** is the time to replace the defender.* When the defender is replaced, either now or at some future date, it will be by the challenger—the best available replacement.

Example 9-5, involving a before-tax analysis of the defender versus a challenger forklift truck, made implicit use of the *repeatability* assumption. That is, it was assumed that the particular challenger analyzed in Table 9-2 would have a minimum EUAC of $8,598 regardless of when it replaces the defender. Figure 9-3 shows time diagrams of the cost consequences of keeping the defender for 2 more years versus adopting the challenger now, with the challenger costs to be repeated into the indefinite future. Recall that the economic life of the challenger is 3 years. *It can be seen in Figure 9-3 that the only difference between the alternatives is in years 1 and 2.*

The repeatability assumption, applied to replacement problems involving assets with different useful or economic lives, often simplifies the economic comparison of

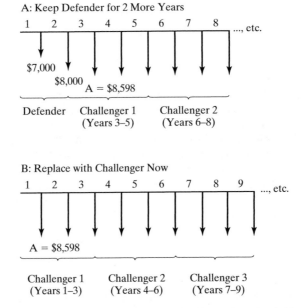

Figure 9-3   Effect of the repeatability assumption applied to alternatives in Example 9-5.

the alternatives. For example, the comparison of the PW values of the alternatives in Figure 9-3 *over an infinite analysis period* (called the capitalized worths) will confirm our previous answer to Example 9-5 that Alternative A (keep defender for 2 more years) is preferred to Alternative B (replace with challenger now). Using a MARR = 10% per year, we have

$$PW_A(10\%) = -\$7,000(P/F,10\%,1) - \$8,000(P/F,10\%,2) - \frac{\$8,598}{0.10}(P/F,10\%,2);$$

$$= -\$84,029$$

$$PW_B(10\%) = -\frac{\$8,598}{0.10} = -\$85,980.$$

The difference $(PW_B - PW_A)$ is –$1,951, which confirms that the additional cost of the challenger over the next 2 years is not justified and it is best to keep the defender for 2 more years before replacing it with the challenger.

Whenever the *repeatability* assumption is not applicable, the *coterminated* assumption may be used; it involves using a finite study period for all alternatives. As described in Section 5.8, use of the *coterminated* assumption requires detailing what and when cash flows are expected to occur for each alternative and then determining which is most economical, using any of the correct economic analysis methods. *When the effects of price changes and taxes are to be considered in replacement studies, it is recommended that the coterminated assumption be used.*

Example 9-6

Suppose that we are faced with the same replacement problem as in Example 9-5, except that the period of needed service is (a) 3 years or (b) 4 years. That is, a finite analysis period under the coterminated assumption is being used. In each case, which alternative should be selected?

Solution

(a) For a planning horizon of 3 years, we might intuitively think that either the defender should be kept for 3 years or it should be replaced immediately by the challenger to serve for the next 3 years. From Table 9-3, the EUAC for the defender for 3 years is $7,966, and from Table 9-2, the EUAC for the challenger for 3 years is $8,598. Thus, following this reasoning, the defender would be kept for 3 years. However, that is not quite right. Focusing on the "Total (Marginal) Cost for Year" columns, we can see that the defender has the lowest cost in the first 2 years, but in the third year its cost is $9,100; the EUAC of 1 year of service for the challenger is only $9,000. Hence, it would be more economical to replace the defender after the second year. This conclusion can be confirmed by enumerating all replacement possibilities and their respective costs and then computing the EUAC for each, as will be done for the 4-year planning horizon in Part (b).

(b) For a planning horizon of 4 years, the alternatives and their respective costs for each year and the EUAC of each are given in Table 9-4. Thus, the most economical alternative is to keep the defender for 2 years and then replace it with the challenger, to be kept for the next 2 years. The decision to keep the defender for 2

TABLE 9-4 Determination of When to Replace the Defender for a Planning Horizon of 4 Years Example 9-6, Part (b)

Keep defender for	Keep challenger for	Total (marginal) costs for each year				EUAC at 10% for 4 years	
		1	2	3	4		
0 years	4 years	$9,000[a]	$8,250[a]	$8,495[a]	$10,850[a]	$9,084	
1	3	7,000	9,000	8,250	8,495	8,140	
2	2	7,000	8,000	9,000	8,250	**8,005**	← Least cost alternative
3	1	7,000	8,000	9,100	9,000	8,190	
4	0	7,000[b]	8,000[b]	9,100[b]	10,000[b]	8,405	

[a]Column 6 of Table 9-2.
[b]Column 6 of Table 9-3.

years happens to be the same as when the repeatability assumption was used, which, of course, would not be true in general.   ■

When a replacement analysis involves a defender that cannot be continued in service because of changes in technology, service requirements, and so on, a choice among two or more new challengers must be made. Under this situation, the repeatability assumption may be a convenient economic modeling approach for comparing the alternatives and making a present decision. Note that when the defender is not a feasible alternative, the replacement problem is no different than any other analysis involving mutually exclusive alternatives.

9.8 Retirement without Replacement (Abandonment)

Consider a project for which the period of required service is finite and that has *positive* net cash flows following an initial capital investment. Market values, or abandonment values, are estimated for the end of each remaining year in the project's life. In view of an opportunity cost (MARR) of $i\%$ per year, should the project be undertaken? Given that we have decided to implement the project, what is the best year to abandon the project? In other words, what is the "economic life" of the project?

For this type of problem, the following assumptions are applicable:

1. Once a capital investment has been made, the firm desires to postpone the decision to abandon a project as long as its present equivalent value (PW) is not decreasing.

2. The existing project will be terminated at the best abandonment time and will not be replaced by the firm.

Solving the abandonment problem is similar to determining the economic life of an asset. In abandonment problems, however, annual benefits (cash inflows) are present; but in economic life analysis, costs (cash outflows) are dominant. In both cases, the objective is to increase the overall wealth of the firm by finding the life that maximizes profits or, equivalently, minimizes costs.

Example 9-7

A $50,000 baling machine for recycled paper is being considered by the XYZ Company. Revenues less expenses for each year and end-of-year abandonment (market) values for the machine have been estimated for the project. The firm's MARR is 12% per year. When is the best time to abandon the project if the firm has already decided to acquire the baling machine and use it for no longer than 7 years?

	End of year						
	1	2	3	4	5	6	7
Annual revenues less expenses	$10,000	$15,000	$18,000	$13,000	$ 9,000	$ 6,000	$ 5,000
Abandonment value of machine[a]	40,000	32,000	25,000	21,000	18,000	17,000	9,000

[a]Estimated market value.

Solution   The PWs that result from deciding now to keep the machine exactly 1, 2, 3, 4, 5, 6, and 7 years are shown:

Keep for 1 year:

$$PW(12\%) = -\$50,000 + (\$10,000 + \$40,000)(P/F,12\%,1)$$
$$= -\$5,355.$$

Keep for 2 years:

$$PW(12\%) = -\$50,000 + \$10,000(P/F,12\%,1)$$
$$+ (\$15,000 + \$32,000)(P/F,12\%,2)$$
$$= -\$3,603.$$

In the same manner, the PW for years 3 through 7 can be computed. The results are as follows:

$$Keep\ for\ 3\ years:\ \ PW(12\%) = \$1,494;$$
$$Keep\ for\ 4\ years:\ \ PW(12\%) = \$5,306;$$
$$Keep\ for\ 5\ years:\ \ PW(12\%) = \$7,281;$$
$$Keep\ for\ 6\ years:\ \ PW(12\%) = \$8,719;$$
$$Keep\ for\ 7\ years:\ \ PW(12\%) = \$7,613.$$

As you can see, PW is maximized (\$8,719) by retaining the machine for 6 years. Thus, the best abandonment time would be in 6 years.   ∎

In some cases, management may decide that, although an existing asset is to be retired from its current use, it will not be replaced or removed from all service. Although the existing asset may not be able to compete economically at the moment, it may be desirable and even economical to retain the asset as a standby unit or for some different use. The cost to retain the defender under such conditions may be quite low, because of its relatively low realizable MV and perhaps low annual expenses. Often, income tax considerations (to be discussed in the next section) also bear on the true cost of retaining the defender.

9.9 After-Tax Replacement Studies

As discussed in Chapter 7, income taxes associated with a proposed project may represent a major cash flow for a firm. Therefore, income taxes should be considered along with all other relevant cash flows when assessing the economic profitability of a project. This fact also holds true for replacement decisions. The replacement of an asset often results in gains or losses from the sale of the existing asset (defender). The income tax consequence resulting from the gain (loss) associated with the sale (or retention) of the defender has the potential to impact the decision to keep the defender or to sell it and purchase the challenger. The remainder of this section is devoted to demonstrating the procedure for performing replacement analyses on an after-tax basis. Note that after-tax replacement analyses require knowledge of the depreciation schedule already in use for the defender as well as the appropriate depreciation schedule to be used for the challenger.

9.9.1 After-Tax Economic Life

In earlier sections, the economic life of a new asset (Example 9-4) and the economic life of an existing asset (Example 9-5) were determined on a before-tax basis. However, an *after-tax* analysis can also be used to determine the economic life of an asset by extending Equation 9-1 to account for income tax effects. This extension is shown in Equation 9-3:

$$PW_k(i\%) = I + \sum_{j=1}^{k} [(1-t)E_j - td_j](P/F, i\%, j) - [(1-t)MV_k + t(BV_k)](P/F, i\%, k).$$

$$(9\text{-}3)$$

This computation finds the PW of the after-tax cash flows (expressed as costs) through year k, PW_k, by (1) adding the initial capital investment (PW of investment amounts, if any occur after time zero) and the sum of the after-tax PW of annual expenses through year k, including adjustments for annual depreciation amounts (d_j), and then (2) adjusting this total after-tax PW of costs by the after-tax consequences of gain or loss on disposal of the asset at the end of year k. Similar to the previous before-tax analysis using Equation 9-1, Equation 9-3 is used to determine the after-tax total marginal cost for each year k, TC_k. That is, $TC_k = (PW_k - PW_{k-1})(F/P, i\%, k)$. The algebraic simplification of this relationship results in Equation 9-4:

$$TC_k(i\%) = (1-t)(MV_{k-1} - MV_k + iMV_{k-1} + E_k) + i(t)(BV_{k-1}). \qquad (9\text{-}4)$$

This is $(1-t)$ times Equation 9-2, plus interest on the tax adjustment from the book value of the asset at the beginning of year k. A tabular format incorporating Equation 9-4 is used in the solution of the next example to find the economic life of a new asset on an after-tax basis (N^*_{AT}). This same procedure can also be used to find the after-tax economic life of an existing asset.

Example 9-8

Find the economic life on an after-tax basis for the new forklift truck (challenger) described in Example 9-4. Assume that the new forklift is depreciated as a MACRS 3-year property class asset, the effective income tax rate is 40%, and the after-tax MARR is 6% per year.

Solution   The calculations using Equation 9-4 are shown in Table 9-5. The expected year-by-year market values and annual expenses are repeated from Example 9-4 in columns 2 and 5, respectively. In column 6, the *sum* of the loss in MV during year k, cost of capital based on the MV at the beginning-of-year (BOY) k, and annual expenses in year k are multiplied by $(1-t)$ to determine an *approximate* after-tax total marginal cost in year k.

The BV amounts at the end of each year, based on the new forklift trucks being a MACRS (GDS) 3-year property class asset, are shown in column 7. These amounts are then used in column 8 to determine an annual tax adjustment (last term in Equation 9-4), based on the BOY book values (BV_{k-1}). This annual tax adjustment is algebraically added to the entry in column 6 to obtain an *adjusted* after-tax total marginal cost in year k, TC_k. The total marginal cost amounts are used in column 10 to calculate, successively, the equivalent uniform annual cost $EUAC_k$ of retirement of the asset at the end of year k. In this case, the after-tax economic life (N^*_{AT}) is 3 years, the same result obtained on a before-tax basis in Example 9-4.∎

TABLE 9-5 Determination of the After-Tax Economic Life for the new Asset Described in Example 9-4

(1) End of year, k	(2) MV, end of year k	(3) Loss in MV during year k	(4) Cost of capital = 6% of BOY MV in col. 2	(5) Annual expenses	(6) Approximate after-tax total (marginal) cost for year k $(1-t)\cdot$ (col. 3 + 4 + 5)
0	$20,000	0	0	0	0
1	15,000	$5,000	$1,200	$2,000	$4,920
2	11,250	3,750	900	3,000	4,590
3	8,500	2,750	675	4,620	4,827
4	6,500	2,000	510	8,000	6,306
5	4,750	1,750	390	12,000	8,484

(1) End of year, k	(7) MACRS BV at end of year k	(8) Interest on tax adjustment = 6% · t · BOY BV in col. 7	(9) Adjusted after-tax total (marginal) cost (TC_k) (col. 6 + col. 8)	(10) EUAC^a (after-tax) through year k
0	$20,000	0	0	0
1	13,334	$480	$5,400	$5,400
2	4,444	320	4,910	5,162
3	1,482	107	4,934	5,090
4	0	36	6,342	5,377
5	0	0	8,484	5,928

$N^*_{AT} = 3$

^{a}EUAC$_k = [\sum_{j=1}^{k} (\text{Col. 9})_j \cdot (P/F, 6\%, j)](A/P, 6\%, k)$.

It is not uncommon for the before-tax and after-tax economic lives of an asset to be the same (as occurred in Examples 9-4 and 9-8).

9.9.2 After-Tax Investment Value of the Defender

The outsider viewpoint has been used in this chapter to establish a before-tax investment value of an existing asset. Using this viewpoint, the present realizable MV of the defender is the appropriate before-tax investment value. This value (although not an actual cash flow) represents the opportunity cost of keeping the defender. In determining the after-tax investment value, we must also include the opportunity cost of gains (losses) not realized if the defender is kept.

Consider, for example, a printing machine that was purchased 3 years ago for $30,000. It has a present market value of $5,000 and a current book value of $8,640. If the printing machine were sold now, the company would experience a loss on disposal of $5,000 − $8,640 = −$3,640. Assuming a 40% effective income tax rate, this loss would translate into a $(−0.40)(−$3,640) = $1,456$ tax savings. Thus, if it is decided to keep the printing machine, not only would the company be giving up the opportunity to obtain the $5,000 market value, it would also be giving up the opportunity to obtain the $1,456 tax savings that would result from selling the printing machine at a price less than its current BV. Hence, the total after-tax investment value of the existing printing machine is $5,000 + $1,456 = $6,456.

The computation of the after-tax investment value of an existing asset is quite straightforward. Using the general format for computing after-tax cash flows (ATCF) presented previously in Figure 7-3, we would have the entries shown in the table that follows if the defender were sold now (year 0). Note that MV_0 and BV_0 represent the MV and BV, respectively, of the defender at the time of the analysis.

End of year, k	BTCF	Depreciation	Taxable income	Cash flow for income taxes	ATCF (if defender is sold)
0	MV_0	None	$MV_0 - BV_0$	$-t(MV_0 - BV_0)$	$MV_0 - t(MV_0 - BV_0)$

Now, if it was decided to keep the asset, the preceding entries become the opportunity costs associated with keeping the defender. The appropriate year 0 entries for analyzing the after-tax consequences of keeping the defender are shown in Figure 9-4. Note that the entries in Figure 9-4 are simply the same values previously shown, only reversed in sign to account for the change in perspective (keep versus sell).

	(A)	(B)	(C)	(D) = $-t$(C)	(E) = (A) + (D)
End of year, k	BTCF	Depreciation	Taxable income	Cash flow for income taxes	ATCF (if defender is kept)
0	$-MV_0$	None	$-(MV_0 - BV_0)$	$-t[-(MV_0 - BV_0)]$ $= t(MV_0 - BV_0)$	$-MV_0 + t(MV_0 - BV_0)$

Figure 9-4   General procedure for computing the after-tax investment value of a defender.

Example 9-9

An existing asset being considered for replacement has a current market value of $12,000 and a current book value of $18,000. Determine the after-tax investment value of the existing asset (if kept), using the outsider viewpoint and an effective income tax rate of 34%.

    Solution   Given that $MV_0 = \$12,000$, $BV_0 = \$18,000$, and $t = 0.34$, we can easily compute the ATCF associated with keeping the existing asset by using the format of Figure 9-4.

End of year, k	BTCF	Depreciation	Taxable income	Cash flow for income taxes	ATCF
0	−$12,000	—	−($12,000 − $18,000) = $6,000	(−0.34)($6,000) = −$2,040	−$12,000 − $2,040 = −$14,040

The appropriate after-tax investment value for the existing asset is $14,040. Note that this is higher than the before-tax investment value of $12,000 due to the tax credit given up by *not* selling the existing machine at a loss.   ∎

Example 9-10

An engineering consulting firm is considering the replacement of its computer-aided design (CAD) workstation. The workstation was purchased 4 years ago for $20,000. Depreciation deductions have followed the MACRS (GDS) 5-year property class schedule. The workstation can be sold now for $4,000. Assuming the effective income tax rate is 40%, compute the after-tax investment value of the CAD workstation if it is kept.

    Solution   To compute the ATCF associated with keeping the defender, we must first compute the current BV, BV_0. The workstation has been depreciated for 4 years under the MACRS system with a 5-year property class. Thus,

$$BV_0 = \$20,000(1 - 0.2 - 0.32 - 0.192 - 0.1152) = \$3,456.^{[6]}$$

    By using the format presented in Figure 9-4, the ATCF associated with keeping the defender can be computed as follows:

End of year, k	BTCF	Depreciation	Taxable income	Cash flow for income taxes	ATCF
0	−$4,000	—	−($4,000 − $3,456) = −$544	(−0.4)(−$544) = $218	−$4,000 + $218 = −$3,782

The after-tax investment value of keeping the existing CAD workstation is $3,782. Note that, in the case in which MV_0 is higher than BV_0, the after-tax investment value is lower than the before-tax investment value. This is because the gain on disposal (and resulting tax liability) would not occur at this time if the defender were retained.   ∎

---

    [6] Current tax law dictates that gains and losses be taxed as ordinary income. As a result, it is not necessary to explicitly account for the MACRS half-year convention when computing the "if sold" BV (the increase in taxable income due to a higher BV is offset by the half-year of depreciation that could be claimed if the defender is kept). This allows us to simplify the procedure for computing the after-tax investment value of the defender.

9.9.3 Illustrative After-Tax Replacement Analyses

The examples that follow represent typical after-tax replacement analyses. They illustrate the appropriate method for including the effect of income taxes, as well as several of the factors that must be considered in general replacement studies.

Example 9-11 (Restatement of Example 9-3 with tax information)

The manager of a carpet manufacturing plant became concerned about the operation of a critical pump in one of the processes. After discussing this situation with the supervisor of plant engineering, he decided that a replacement study should be done and that a 9-year study period would be appropriate for this situation. The company that owns the plant is using an after-tax MARR of 6% per year for its capital investment projects. The effective income tax rate is 40%.

The existing pump, *pump A*, including driving motor with integrated controls, cost $17,000 when purchased 5 years ago. The accounting records show the depreciation schedule to be following that of an asset with a MACRS (ADS) recovery period of 9 years. Some reliability problems have been experienced with pump A, including annual replacement of the impeller and bearings at a cost of $1,750. Annual operating and maintenance (O&M) expenses have been averaging $3,250. Annual insurance and property tax expenses are 2% of the initial capital investment. It appears that the pump will provide adequate service for another 9 years if the present maintenance and repair practice is continued. An estimated market value of $750 could be obtained for the pump if it is sold now. It is estimated that if this pump is continued in service, its final MV after 9 more years will be about $200.

An alternative to keeping the existing pump in service is to sell it immediately and to purchase a replacement pump, *pump B*, for $16,000. A 9-year class life (MACRS 5-year property class) would be applicable to the new pump under the GDS. An estimated market value at the end of the 9-year study period would be 20% of the initial capital investment. O&M expenses for the new pump are estimated to be $3,000 per year. Annual taxes and insurance would total 2% of the initial capital investment. The data for Example 9-11 are summarized in Table 9-6.

Based on these data, should the defender (pump A) be kept [and the challenger (pump B) not purchased] or should the challenger be purchased now (and the defender sold)? Use an after-tax analysis and the outsider viewpoint in the evaluation.

TABLE 9-6   Summary of Information for Example 9-11

MARR (after taxes) = 6% per year	
Effective income tax rate = 40%	
Existing pump A (defender)	
MACRS (ADS) recovery period	9 years
Capital investment when purchased 5 years ago	$17,000
Total annual expenses	$5,340
Present market value	$750
Estimated market value at the end of 9 additional years	$200
Replacement pump B (challenger)	
MACRS (GDS) property class	5 years
Capital investment	$16,000
Total annual expenses	$3,320
Estimated market value at the end of 9 years	$3,200

TABLE 9-7 ATCF Computations for the Defender (Existing Pump A) in Example 9-11

	(A)	(B)	(C) = (A) – (B)	(D) = –0.4(C)	(E) = (A) + (D)
End of year, k	BTCF	MACRS (ADS) depreciation	Taxable income	Income taxes at 40%	ATCF
0	–$750		$ 7,750	–$3,100	–$3,850
1–4	–5,340	$1,889	–7,229	2,892	–2,448
5	–5,340	944	–6,284	2,514	–2,826
6–9	–5,340	0	–5,340	2,136	–3,204
9	200		200[a]	–80	120

[a]Gain on disposal (taxable at the 40% rate).

Solution   The after-tax computations for keeping the defender (pump A) and not purchasing the challenger (pump B) are shown in Table 9-7. *Year 0* of the *analysis period* is at the *end of the current (fifth) year of service* of the defender. The year 0 entries of Table 9-7 are computed by using the general format presented in Figure 9-4 and are further explained here:

1. BTCF (–$750)—The same amount is used in the before-tax analysis of Example 9-3. This amount is based on the outsider viewpoint and is the opportunity cost of keeping the defender instead of replacing it (and selling it for the estimated present MV of $750).

2. Taxable income ($7,750)—This amount is the result of an increase in taxable income of $7,750 due to the tax consequences of keeping the defender instead of selling it. Specifically, *if we sold the defender now*, the loss on disposal would be as follows:

$$\text{Gain or loss on disposal (if sold now)} = MV_0 - BV_0;$$
$$BV_0 = \$17,000[1 - 0.0556 - 4(0.1111)] = \$8,500;$$
$$\text{Loss on disposal (if sold now)} = \$750 - \$8,500 = -\$7,750.$$

But, *since we are keeping the defender (pump A) in this alternative*, we have the reverse effect on taxable income, an increase of $7,750 due to an opportunity foregone.

3. Cash flow for income taxes (–$3,100)—The increase in taxable income due to the tax consequences of keeping the defender results in an increased tax liability (or tax savings foregone) of –0.4($7,750) = –$3,100.

4. ATCF (–$3,850)—The total after-tax investment value of the defender is the result of two factors: the present MV ($750) and the tax credit ($3,100) foregone by keeping the existing pump A. Therefore, the ATCF representing the investment in the defender (based on the outsider viewpoint) is –$750 – $3,100 = –$3,850.

The remainder of the ATCF computations over the 9-year analysis period for the alternative of keeping the defender are shown in Table 9-7. The after-tax computations for the alternative of purchasing the challenger (pump B) are shown in Table 9-8.

TABLE 9-8   ATCF Computations for the Challenger (Replacement Pump B) in Example 9-11

End of year, k	(A) BTCF	(B) MACRS (GDS) depreciation	(C) = (A) – (B) Taxable income	(D) = –0.4(C) Income taxes at 40%	(E) = (A) + (D) ATCF
0	–$16,000				–$16,000
1	– 3,320	$3,200	–$6,520	$ 2,608	– 712
2	– 3,320	5,120	– 8,440	3,376	56
3	– 3,320	3,072	– 6,392	2,557	– 763
4	– 3,320	1,843	– 5,163	2,065	– 1,255
5	– 3,320	1,843	– 5,163	2,065	– 1,255
6	– 3,320	922	– 4,242	1,697	– 1,623
7–9	– 3,320	0	– 3,320	1,328	– 1,992
9	3,200		3,200[a]	– 1,280	1,920

[a]Gain on disposal (taxable at the 40% rate).

The next step in an after-tax replacement study involves equivalence calculations by using an after-tax MARR. The following is the after-tax EUAC analysis for Example 9-11:

$$EUAC(6\%) \text{ of pump A} = \$3,850(A/P,6\%,9)$$
$$\text{(defender)} \quad + \$2,448(P/A,6\%,4)(A/P,6\%,9)$$
$$+ [\$2,826(F/P,6\%,4) + \$3,204(F/A,6\%,4)]$$
$$- \$120](A/F,6\%,9)$$
$$= \$3,332.$$

$$EUAC(6\%) \text{ of pump B} = \$16,000(A/P,6\%,9)$$
$$\text{(challenger)} \quad + [\$712(P/F,6\%,1) - \$56(P/F,6\%,2)$$
$$+ \$763(P/F,6\%,3)$$
$$+ \cdots + \$1,992(P/F,6\%,9)](A/P,6\%,9)$$
$$- \$1,920(A/F,6\%,9)$$
$$= \$3,375.$$

Because the EUACs of both pumps are very close, other considerations, such as the improved reliability of the new pump, could detract from the slight economic preference for pump A. The after-tax annual costs of both alternatives are considerably less than their before-tax annual costs.

The after-tax analysis does not *reverse* the results of the before-tax analysis for this problem. (See Example 9-3.) However, due to income tax considerations, identical before-tax and after-tax recommendations should not necessarily be expected. ∎

The next example involves the determination of the economic life of a defender on an after-tax basis and the use of after-tax marginal costs to determine the most economical time to replace the defender.

Example 9-12

The Carolina Metal Stamping Company is considering the replacement of a spray system. The new improved system will cost $60,000 installed and will have an estimated economic life of 12 years. The market value of the new system at the end of 12 years is expected to be $6,000. Further, it is estimated that annual operating and maintenance expenses will average $32,000 per year for the new system and that straight-line depreciation (with a $6,000 terminal market value) will be used.

The present system has a remaining useful life of 3 years. It has a current book value of $12,000 and a present realizable market value of $8,000. The estimated operating expenses, market values, and book values of the present system for the next 3 years are as follows:

Year	Market value at end of year	Book value at end of year	Operating expenses during year
1	$6,000	$9,000	$40,000
2	5,000	6,000	50,000
3	4,000	3,000	60,000

A spray system will be needed for as long as the company remains in business (which it hopes is a long, long time). Perform an after-tax analysis to determine the most economical period to keep the defender *before* replacing it with the new system. The after-tax MARR is 15% per year and the effective income tax rate is 50%.

Solution   This analysis begins with the determination of the after-tax economic life of the present system (the economic life of the challenger was given to be 12 years). Table 9-9 shows the calculations of the year-by-year after-tax marginal costs (Equation 9-4) of the defender and the corresponding EUAC. It can be seen in column 10 that the economic life of the defender is 1 year.

Table 9-10 contains the ATCF calculations for the challenger. The ATCFs are used to compute the after-tax EUAC of the challenger as follows:

$$EUAC = \$60,000(A/P,15\%,12) + \$13,750 - \$6,000(A/F,15\%,12) = \$24,613.$$

Comparing just the EUACs of the defender with the EUAC of the challenger, you may be tempted to conclude that the old system should be kept at least one more year, perhaps even

TABLE 9-9   Determination of the After-Tax Economic Life for the Defender Described in Example 9-12

(1)	(2)	(3)	(4) Cost of capital = 15% of	(5)	(6) Approximate after-tax total (marginal) cost
End of year, k	MV, end of year k	Loss in MV during year k	BOY MV in Col. 2	Annual expenses	for year k $(1-t) \cdot (\text{Col. } 3+4+5)$
0	$8,000	0	0	0	0
1	6,000	$2,000	$1,200	$40,000	$21,600
2	5,000	1,000	900	50,000	25,950
3	4,000	1,000	750	60,000	30,875

TABLE 9-9 (*cont'd*)

	(7)	(8)	(9)	(10)	
End of year, k	BV at end of year k	Interest on tax adjustment = $15\% \cdot t \cdot$ BOY BV in col. 7	Adjusted after-tax total (marginal) cost (TC_k) (col. 6 + col. 8)	$EUAC^a$ (after-tax) through year k	
0	$12,000	0	0	0	
1	9,000	$900	$22,500	**$22,500**	$N*_{AT} = 1$
2	6,000	675	26,625	24,418	
3	3,000	450	31,325	26,408	

$^a EUAC_k = [\sum_{j=1}^{k} (\text{Col. 9})_j \cdot (P/F, 15\%, j)](A/P, 15\%, k)$

TABLE 9-10   ATCF Computations for the Challenger in Example 9-12

End of year, k	(A) BTCF	(B) Straight-line depreciation	(C) = (A) − (B) Taxable income	(D) = −0.4(C) Income taxes at 40%	(E) = (A) + (D) ATCF
0	−$60,000				−$60,000
1–12	−32,000	$4,500^a	−$36,500	$18,250	−13,750
12	6,000		0^b	0	6,000

^{a}Straight-line depreciation amount = ($60,000 − $6,000)/12 = $4,500.
$^b BV_{12} = $60,000 − (12)($4,500) = $6,000; MV_{12} − BV_{12} = 0.$

two more years. However, in this situation you should examine marginal costs. The valid economic criterion when operating expenses are increasing over time is to keep the old system as long as the marginal cost of an additional year of service is less than the equivalent uniform annual cost of the new system. The marginal cost of keeping the old system for the first year is $22,500. This $22,500 is less than the $24,613 EUAC of the new system, thus justifying keeping the old system for the first year. The marginal cost of keeping the old system for the second year is $26,625. The $26,625 is greater than the $24,613 average annual cost of the new system, thus indicating that the old system should not be kept the second year, but rather that it be replaced at the end of the first year.

9.10 A Comprehensive Example

Sometimes, in engineering practice, a replacement analysis involves an existing asset that cannot meet future service requirements without *augmentation* of its capabilities. When this is the case, the defender with increased capability should be competitive with the best available challenger. The analysis of this situation is included in the following comprehensive example:

Example 9-13

The emergency electrical supply system of a hospital owned by a medical service corporation is presently supported by an 80-kW diesel-powered electrical generator that was put into service 5 years ago [capital investment = $210,000; MACRS (GDS) 7-year property class]. An engineering firm is designing modifications to the electrical and mechanical systems of the hospital as part of an expansion project. The redesigned emergency electrical supply system will require 120-kW of generating capacity to serve the increased demand. Two preliminary designs for the system are being considered. The first involves the augmentation of the existing 80-kW generator with a new 40-kW diesel-powered unit (GDS 7-year property class). This alternative represents the augmented defender. The second design includes replacement of the existing generator with the best available alternative, a new turbine-powered unit with 120-kW of generating capacity (the challenger). Both alternatives will provide the same level of service to the operation of the emergency electrical supply system.

The challenger, if selected, will be leased by the hospital for a 10-year period. At that time, the lease contract would be renegotiated either for the original piece of equipment or for a replacement generator with the same capacity. The following additional estimates have been generated for use in the replacement analysis:

	Alternative		
	Defender		Challenger
	80-kW	40-kW	
Capital investment	$90,000[a]	$140,000	$10,000[b]
Annual lease amount	0	0	$39,200
Operating hours per year	260	260	260
Annual expenses (year zero $):			
Operating and maintenance (O&M) expense per hour	$80	$35	$85
Other expenses	$3,200	$1,000	$2,400
Useful life	10 years	15 years	15 years

[a]Opportunity cost based on present market value of the defender (outsider viewpoint).
[b]Deposit required by the terms of the contract to lease the challenger. It is refundable at the end of the study period.

The annual lease amount for the challenger will not change over the 10-year contract period. The operating and maintenance (O&M) expense per hour of operation and the other annual expense amounts for both alternatives are estimated in year zero dollars, and are expected to escalate at the rate of 4% per year.

The present estimated market value of the 80-kW generator is $90,000 and its estimated market value at the end of an additional 10 years, in year zero dollars, is $30,000. The estimated market value of the new 40-kW generator, 10 years from now, in year zero dollars, is $38,000. Both future market values are estimated to escalate at the rate of 2% per year.

The corporation's after-tax, market-based MARR (i_c) is 12% per year, and its effective income tax rate is 40%. A 10-year planning horizon (study period) is considered appropriate for this decision situation (note that, with income tax considerations and price changes in the analysis, a study period based on the coterminated assumption is being used).

Based on an after-tax, actual-dollar analysis, which alternative (augmentation of the defender or lease of the challenger) should be selected as part of the design of the modified emergency electrical power system?

Solution   The after-tax analysis of the first alternative (defender), keeping the existing 80-kW generator and augmenting its capacity with a new 40-kW generator is shown in Table 9-11. The initial $230,000 before-tax capital investment amount is the sum of (1) the present $90,000 market value of the existing 80-kW generator, which is an opportunity cost based on an outsider viewpoint, and (2) the $140,000 capital investment for the new 40-kW generator. The −$43,149 of taxable income at time zero is due to the gain on disposal, *which is not incurred* when the 80-kW generator is kept instead of sold.

The after-tax PW of keeping the defender and augmenting its capacity is

$$PW_D(12\%) = -\$212,740 - \$5,783(P/F,12\%,1) - \cdots$$
$$+ (\$49,735 - \$30,286)(P/F,12\%,10)$$
$$= -\$282,468.$$

Under the contract terms for leasing the challenger, there is an initial $10,000 deposit, which is fully refundable at the end of the 10-year period. There are no tax consequences associated with the deposit transaction. The annual before-tax cash flow (BTCF) for the challenger is the sum of (1) the annual lease amount, which stays constant over the 10-year contract period, and (2) the annual O&M and other expenses, which escalate at the rate of 4% per year.

TABLE 9-11   Defender Augmented with a New 40-kW Generator (Example 9-13)

End of year, k	BTCF	Depreciation 80-kW	Depreciation 40-kW	Taxable income	Cash flow for income taxes	ATCF
0	−$230,000			−$43,149[c]	$ 17,260	−$212,740
1	−35,464	$18,732	$20,006	−74,202	29,681	−5,783
2	−36,883[a]	18,753	34,286	−89,922	35,969	−914
3	−38,358	9,366	24,486	−72,210	28,884	−9,474
4	−39,892		17,486	−57,378	22,951	−16,941
5	−41,488		12,502	−53,990	21,596	−19,892
6	−43,147		12,488	−55,635	22,254	−20,893
7	−44,873		12,502	−57,375	22,950	−21,923
8	−46,668		6,244	−52,912	21,165	−25,503
9	−48,535			−48,535	19,414	−29,121
10	−50,476			−50,476	20,190	−30,286
10	82,892[b]			82,892	−33,157	49,735

[a] −[260($80 + $35) + ($3,200 + $1,000)](1.04)^2 = −$36,883.
[b] MV_{10} = ($30,000 + $38,000)(1.02)^{10} = $82,892.
[c] If the defender were sold now, gain on disposal = $90,000 − $46,851 = $43,149; where BV_0 = $46,851.

For example, the BTCF for the challenger in year 1 is $-\$39,200 - [\$85(260) + \$2,400](1.04)$ $= -\$64,680$. These annual BTCF amounts for years 1 through 10 are fully deductible from taxable income by the corporation, and they are also the taxable income amounts for the alternative. (The corporation cannot claim any depreciation on the challenger, because it does not own the equipment.) Hence, the after-tax PW of selecting the challenger, assuming it is leased under these contract terms, is

$$\begin{aligned} PW_C(12\%) = &-\$10,000 + \$10,000(P/F,12\%,10) \\ &-(1-0.4)(\$39,200)(P/A,12\%,10) \\ &-(1-0.4)[\$85(260) + \$2,400](P/A,i_V = 7.69\%,10) \\ = &-\$239,705, \end{aligned}$$

where $i_{Vr} = (0.12 - 0.04)/(1.04) = 0.0769$, and $(P/A, 7.69\%, 10) = 6.8049$.

Based on an after-tax analysis, the challenger is economically preferable for use in the emergency electrical supply system, because its PW has the less negative value.   ■

9.11 Spreadsheet Applications

A vital ingredient of many replacement studies is an asset's economic life. The next example provides a general spreadsheet model that can be used to determine the economic life of an asset, given the initial capital investment, year-by-year market values, and the annual operating expenses. This spreadsheet can also be used to determine the best time to abandon a project.

Example 9-14
The estimated year-by-year market values and operating expenses for a replacement piece of equipment are shown in Figure 9-5 (column B and column E, respectively). The market values are used to compute year-by-year loss in value (column C) and the cost of capital (column D). The resulting capital recovery amount is combined with the expenses for the year (shown as a net cash flow in column E) to determine the total marginal cost for the year (column F). Column G calculates the equivalent annual worth of the cash flows in column F successively through each year. Column K contains an IF( ) function that places the label "Economic Life" next to the maximum equivalent annual worth (which corresponds to the minimum equivalent uniform annual cost) found in Column G. The formulas for the highlighted cells in Figure 9-5 are given in the following table:

Cell	Contents
C10	= B9 – B10
D10	= B9 * $B$1
E10	User input of net cash flow for year
F10	= E10 – (C10 + D10)
G10	= –PMT($B$1,A10,NPV($B$1,F$8:F10))
H10	= IF(G10 = MAX(G8:G12), "Economic Life, "

MARR	15%						
End of year	MV at end of year	Loss in MV during year K	Cost of capital	Net cash flow for year (R - E)	Total (marginal) cash flow for year (R - E - CR)	Equivalent annual worth through year	
0	$ 15,000						
1	$ 12,000	$ 3,000	$ 2,250	$ (1,000)	$ (6,250)	($6,250)	Economic life
2	$ 10,000	$ 2,000	$ 1,800	$ (1,100)	$ (4,900)	($5,622)	
3	$ 7,000	3,000	1,500	(1,300)	(5,800)	(5,673)	
4	$ 3,000	$ 4,000	$ 1,050	$ (2,000)	$ (7,050)	($5,949)	
5	$ 500	2,500	450	(2,500)	(5,450)	($5,875)	

Figure 9-5 Spreadsheet for determining economic life in Example 9-14.

9.12 Summary

To summarize, there are several important points to keep in mind when conducting a replacement or retirement study:

1. The MV of the defender must *not* also be deducted from the purchase price of the challenger *when using the outsider viewpoint* to analyze a replacement problem. This error double-counts the defender's MV and biases the comparison toward the challenger.

2. A sunk cost (i.e., $MV - BV < 0$) associated with keeping the defender must *not* be added to the purchase price of the best available challenger. This error results in an incorrect penalty that biases the analysis in favor of retaining the defender.

3. In Section 9.6, we observed that the economic life of the defender is often 1 year, which is generally true if annual expenses are high relative to the defender's investment cost when using the outsider viewpoint. Hence, the marginal cost of the defender should be compared with the EUAC at the *economic life* of the challenger to answer the fundamental question "Should the defender be kept for one or more years or disposed of now?" The typical pattern of the EUAC for a defender and a challenger is illustrated in Figure 9-6.

4. The income tax effects of replacement decisions should not be ignored. The foregone income tax savings associated with keeping the defender may swing the economic preference away from the defender, thus making the challenger the better choice.

5. The *best available* challenger(s) must be determined. Failure to do so represents unacceptable engineering practice.

6. Any excess capacity, reliability, flexibility, safety, and so on of the challenger may have value to the owner and should be claimed as a dollar benefit if a dollar estimate can be placed on it. Otherwise, this value would be treated as a *nonmonetary* benefit.

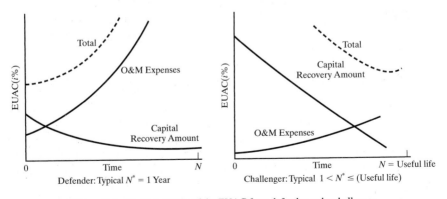

Figure 9-6 Typical pattern of the EUAC for a defender and a challenger.

REFERENCES

Bean, J. C., J. R. Lohmann, and R. L. Smith. "A Dynamic Infinite Horizon Replacement Economy Decision Model." *The Engineering Economist*, vol. 30, no. 2, 1985, pp. 99–120.

Bernhard, R. H. "Improving the Economic Logic Underlying Replacement Age Decisions for Municipal Garbage Trucks: Case Study." *The Engineering Economist*, vol. 35, no. 2, Winter 1990, pp. 129–147.

Hartman, J. C. "A General Procedure for Incorporating Asset Utilization Decisions into Replacement Analysis." *The Engineering Economist*, vol. 44, no. 3, 1999, pp. 217–238.

Kulonda, D. J. "Replacement Analysis with Unequal Lives—The Study Period Method." *The Engineering Economist*, vol. 23, no. 3, Spring 1978, pp. 171–179.

Lake, D. H. and A. P. Muhlemann. "An Equipment Replacement Problem." *Journal of the Operational Research Society*, vol. 30, no. 5, 1979, pp. 405–411.

Leung, L. C. and J. M. A. Tanchoco. "Multiple Machine Replacement within an Integrated Systems Framework." *The Engineering Economist*, vol. 32, no. 2, 1987, pp. 89–114.

Matsuo, H. "A Modified Approach to the Replacement of an Existing Asset." *The Engineering Economist*, vol. 33, no. 2, Winter 1988, pp. 109–120.

Park, C. S. and G. P. Sharp-Bette. *Advanced Engineering Economics*. New York: John Wiley & Sons, 1990.

PROBLEMS

9-1. An industrial lift truck has been in service for several years and management is contemplating replacing it. A planning horizon of 5 years is to be used in the replacement study. The old lift truck (defender) has a current market value of $1,500. If the defender is retained, it is anticipated to have annual operating and maintenance costs of $7,300. It will have a zero market value at the end of 5 additional years of service. The new lift truck (challenger) will cost $10,000 and will have operating and maintenance costs of $5,100. It will have a market value of $2,500 at the end of the planning horizon. Determine the preferred alternative by using a present worth comparison and a minimum attractive rate of return (before taxes) of 20% per year.

9-2. Suppose that you have an old car, that is a real gas-guzzler. It is 10 years old and could be sold to a local dealer for $400 cash. Assume its MV 2 years from now is zero. The annual maintenance expenses will average $800 into the foreseeable future, and the car averages only 10 miles per gallon. Gasoline costs $1.50 per gallon and you average 15,000 miles per year. You now have an opportunity to replace the old car with a better one that costs $8,000. If you buy it, you will pay cash. Because of a 2-year warranty, the maintenance expenses are expected to be negligible. This car averages 30 miles per gallon. Use the IRR method to determine which alternative you should select. Utilize a 2-year analysis period and assume that the new car can be sold for $5,000 at the end of year 2. Let your MARR be 15% per year. State any other assumptions you make.

9-3. The Ajax Corporation has an overhead crane that has an estimated remaining life of 10 years. The crane can be sold now for $8,000. If the crane is kept in service, it must be overhauled immediately at a cost of $4,000. Operating and maintenance costs will be $3,000 per year after the crane is overhauled. The overhauled crane will have zero market value at the end of the 10-year study period. A new crane will cost $18,000, will last

for 10 years, and will have a $4,000 market value at the time. Operating and mainte-
nance costs are $1,000 per year for the new crane. The company uses a before-tax inter-
est rate of 10% per year in evaluating investment alternatives. Should the company
replace the old crane?

9-4.

a. Find the economic life of an asset having the following projected cash flows (the
MARR is 0% per year):

$$\text{Capital investment} = \$5,000$$
$$\text{MV} = \$0 \text{ (at all times)}$$
$$\text{Annual expenses} = \$3,000(\text{EOY 1}),$$
$$\$4,000(\text{EOY 2}),$$
$$\$5,000(\text{EOY 3}),$$
$$\text{and } \$6,000(\text{EOY 4}).$$

b. Find the economic life of another asset having these cash flow estimates (the
MARR is 12% per year):

$$\text{Capital investment} = \$10,000$$
$$\text{MV} = \$10,000 \text{ (at all times)}$$
$$\text{Annual expenses} = \$3,000 \text{ (EOY 1)},$$
$$\$4,000 \text{ (EOY 2)},$$
$$\$5,000 \text{ (EOY 3)},$$
$$\text{and } \$6,000 \text{ (EOY 4)}.$$

c. Repeat (b), except that MV = $0 at all times.

9-5. Robert Roe has just purchased a 4-year-old used car, paying $3,000 for it. A friend has
suggested that he determine in advance how long he should keep the car so as to ensure
the greatest overall economy. Robert has decided that, because of style changes, he
would not want to keep the car longer than 4 years, and he has estimated the annual
expenses and market values for years 1 through 4 as follows:

	Year 1	Year 2	Year 3	Year 4
Annual expenses	$ 950	$1,050	$1,100	$1,550
Market value at end of year	2,250	1,800	1,450	1,160

If Robert's capital is worth 12% per year, at the end of which year should he dispose of
the car?

9-6. A present asset (defender) has a current market value of $87,000 ($\text{MV}_0$). Based on the
used equipment market, the estimated market values at the end of the next 3 years are
$\text{MV}_1 = \$76,000$, $\text{MV}_2 = \$60,000$, $\text{MV}_3 = \$40,000$. The annual expenses are $18,000 in
present (year zero) dollars, and these expenses are estimated to increase at 4.1% per
year. The before-tax MARR is 10% per year. The best challenger available has an economic life of 6 years, and its EUAC over this period is $44,210. Based on this infor-
mation and a before-tax analysis, when should you plan to replace the defender with the
challenger?

9-7. The replacement of a planing machine is being considered by the Reardorn Furniture Company. (There is an indefinite future need for this type of machine.) The best challenger will cost $30,000 installed and will have an estimated economic life of 12 years and a $2,000 MV at that time. It is estimated that annual expenses will average $16,000 per year. The defender has a present BV of $6,000 and a present MV of $4,000. Data for the defender for the next 3 years are as follows:

Year	MV at end of year	BV at end of year	Expenses during the year
1	$3,000	$4,500	$20,000
2	2,500	3,000	25,000
3	2,000	1,500	30,000

a. Using a before-tax interest rate of 15% per year, make a comparison to determine whether it is economical to make the replacement now.

b. If the annual expenses for the present machine had been estimated to be $15,000, $18,000, and $23,000 in years 1, 2, and 3, respectively, what replacement strategy should be recommended?

9-8. A construction firm currently owns a heavy-duty tractor that has a present market value (MV) of $80,000. Estimates of the tractor's operating and maintenance (O&M) expenses and MV at the end of each of the remaining 6 years of useful life are as follows:

	End of year k					
	1	2	3	4	5	6
O&M expenses	$20,000	$25,000	$38,000	$45,000	$47,000	$50,000
Market value	70,000	60,000	50,000	40,000	30,000	20,000

The firm is considering a new heavy-duty tractor to replace the one presently owned. The new tractor's purchase price is $220,000 and its estimated O&M and MV for each of the next 6 years of the study period are these:

	End of year k					
	1	2	3	4	5	6
O&M expenses	$10,000	$12,000	$16,000	$17,000	$20,000	$25,000
Market value	180,000	150,000	120,000	100,000	90,000	75,000

If the MARR = 0% per year, should the new tractor be purchased? If so, when?

9-9. An existing robot is used in a commercial material laboratory to handle ceramic samples in the high-temperature environment that is part of several test procedures. Due to changing customer needs, the robot will not meet future service requirements unless it is upgraded at a cost of $2,000. Because of this situation, a new advanced technology robot has been selected for potential replacement of the existing robot. The following estimates have been developed from information provided by some current users of the new robot and data obtained from the manufacturer:

Defender		Challenger	
Current market value	$38,200	Purchase price	$51,000
Upgrade cost (year 0)	2,000	Installation cost	5,500
Annual expenses	$1,400 in year 1, and increasing at the rate of 8%/yr thereafter.	Annual expenses	$1,000 in year one, and increasing by $150/yr thereafter.
Useful life (years)	6	Useful life (years)	10
Market value at end of useful life	−$1,500	Market value at end of useful life	$7,000

The firm's before-tax MARR is 25% per year. Based on this information, should the existing robot be replaced? Assume that a robot will be needed for an indefinite period of time.

9-10. A diesel engine (defender) was installed 10 years ago at a cost of $50,000. It has a present realizable MV of $14,000. If kept, it can be expected to last for 5 more years, have annual expenses of $14,000, and have a market value of $8,000 at the end of the 5 years. This engine can be replaced with an improved version that costs $65,000 and has an expected life of 20 years. The challenger will have estimated annual expenses of $9,000 and a final market value of $13,000. It is thought that an engine will be needed indefinitely and that the results of the economy study would not be affected by the consideration of income taxes. Using a before-tax MARR of 15% per year, perform an analysis to determine whether to keep or replace the old engine.

9-11. A steel pedestrian overpass must be either reinforced or replaced. Reinforcement would cost $22,000 and would make the overpass adequate for an additional 5 years of service. If the overpass is torn down now, the scrap value of the steel would exceed the removal cost by $14,000. If it is reinforced, it is estimated that its net salvage (market) value would be $16,000 at the time it is retired from service. A new prestressed concrete overpass would cost $140,000 and would meet the foreseeable requirements of the next 40 years. Such a design would have no net scrap or market value. It is estimated that the annual expenses of the reinforced overpass would exceed those of the concrete overpass by $3,200. Assume that money costs the state 10% per year and that the state pays no taxes. What would you recommend?

9-12. A small high-speed commercial centrifuge has the following net cash flows and abandonment values over its useful life:

	End of year				
	1	2	3	4	5
Annual revenues less expenses	$2,000	$2,000	$2,000	$2,000	$2,000
Abandonment value of machine[a]	$6,200	$5,200	$4,000	$2,200	0

[a]Estimated market value.

The firm's MARR is 10% per year. Determine the optimal time for the centrifuge to be abandoned if it is acquired for $7,500 and not to be used for more than 5 years.

9-13. Consider a piece of equipment that initially cost $8,000 and has these estimated annual expenses and market values:

End of year, k	Annual expenses	MV at end of year
1	$3,000	$4,700
2	3,000	3,200
3	3,500	2,200
4	4,000	1,450
5	4,500	950
6	5,250	600
7	6,250	300
8	7,750	0

If the after-tax MARR is 7% per year, determine the after-tax economic life of this equipment. MACRS (GDS) depreciation is being used (5-year property class). The effective income tax rate is 40%.

9-14. A current asset (defender) is being evaluated for potential replacement. It was purchased 4 years ago at a cost of $62,000. It has been depreciated as a MACRS (GDS) 5-year property class asset. The present market value of the defender is $12,000. Its remaining useful life is estimated to be 4 years, but it will require additional repair work now (a one-time $4,000 expense) to provide continuing service equivalent to the challenger. The current effective income tax rate is 39% and the after-tax MARR = 15% per year. Based on an outsider viewpoint, what is the after-tax initial investment in the defender if it is kept (*not* replaced now)?

9-15. The present worth of the after-tax cash flows through year k, PW_k, for a defender (3-year remaining useful life) and a challenger (5-year useful life) are as follows:

	PW of ATCF through year k, PW_k	
Year	Defender	Challenger
1	−$14,020	−$18,630
2	−28,100	−34,575
3	−43,075	−48,130
4		−65,320
5		−77,910

Assume the after-tax MARR is 12% per year. Based on this information, answer the following:

a. What is the economic life and the related minimum equivalent uniform annual cost, EUAC, when $k = N^*_{AT}$ for both the defender and the challenger?

b. When should the challenger (based on the present analysis) replace the defender? Why?

c. What assumption(s) have you made in answering Part (b)?

9-16. Four years ago the Attaboy Lawn Mower Company purchased a piece of equipment for its assembly line. Because of increasing maintenance costs for this equipment, a new piece of machinery is being considered. Information about the defender (present equipment) and the challenger are as follows:

Defender	Challenger
Original cost = $9,000	Purchase cost = $13,000
Maintenance = $300 in first year of use 4 years ago, increasing by 10% per year thereafter	Maintenance = $100 in year 1, increasing by 10% per year thereafter
MACRS (ADS) with a 9-year recovery period	MACRS (GDS) 5-year property class
MV = 0 five years from now	MV = $3,000 at the end of year 5

Assume that a $3,200 MV is available now for the defender. Perform an after-tax analysis using an after-tax MARR of 10% per year and a 5-year analysis period to determine which alternative to select. The effective income tax rate is 40%.

9-17. It is being decided whether or not to replace an existing piece of equipment with a newer, more productive one that costs $80,000 and has an estimated MV of $20,000 at the end of its useful life of 6 years. Installation charges for the new equipment will amount to $3,000; this is not added to the capital investment, but will be an expensed item during the first year of operation. MACRS (GDS) depreciation (5-year property class) will be used. The new equipment will reduce direct costs (labor, maintenance, rework, etc.) by $10,000 in the first year, and this amount is expected to increase by $500 each year thereafter during its 6-year life. It is also known that the BV of the old machine is $10,000 (no additional depreciation will be taken) but that its present fair MV is $14,000. The MV of the old machine will be zero in 6 years. The effective income tax rate is 40%.

a. Determine the prospective after-tax *incremental* cash flow associated with the new equipment if it is believed that the existing machine could perform satisfactorily for 6 more years.

b. Assume the after-tax MARR is 12% per year. Based on the ERR method, should you replace the defender with the challenger? Assume ε = MARR.

9-18. Ten years ago a corporation built a facility at a cost of $400,000 in an area that has since developed into a major retail location. At the time the facility was constructed, it was estimated to have a depreciable life of 20 years with no market value at the end of that time. Straight-line depreciation has been used. The corporation now finds it would be more convenient to be in a less congested area and can sell the old facility for $250,000. A new facility in the desired location would cost $500,000 and have a MACRS (GDS) property class of 10 years. There would be annual savings of $4,000 per year in expenses. Taxes and insurance on the old facility have been 5% of the initial capital investment per year, while for the new facility they are estimated to be only 3% of the initial capital investment per year. The study period is 10 years and the estimated MV of the new facility at the end of 10 years is $200,000. The corporation has a 40% income tax rate and capital is worth 12% per year after taxes. What would you recommend on the basis of an after-tax IRR analysis?

9-19. Use the PW method to select the better of the alternatives shown in the table. Assume that the defender was installed 5 years ago and that its MACRS (GDS) property class is 7 years. The after-tax MARR is 10% per year, and the effective income tax rate is 40%.

Annual expenses	Defender: alternative A	Challenger: alternative B
Labor	$300,000	$250,000
Material	250,000	100,000
Insurance and property taxes	4% of initial capital investment	None
Maintenance	$ 8,000	None
Leasing cost	None	$100,000

Definition of alternatives:

a. Retain an already owned machine (defender) in service for 8 more years.

b. Sell the defender and lease a new one (challenger) for 8 years.

Alternative A (additional information):

$$\text{Cost of defender 5 years ago} = \$500,000$$
$$\text{BV now} = \$111,550$$
$$\text{Estimated market value 8 years from now} = \$50,000$$
$$\text{Present MV} = \$150,000$$

9-20. Suppose that it is desired to make an after-tax analysis for the situation posed in Problem 9-10. The defender is being depreciated by the straight-line method over 15 years; an estimated market value of $8,000 is being used for depreciation purposes. Assume that, if the replacement is made, the challenger will be depreciated as a MACRS (GDS) 5-year property class asset. Also, assume that the effective income tax rate is 40%. Use the AW method to determine if the replacement is justified by earning an after-tax MARR of 10% per year or more.

9-21. You have a machine that was purchased 4 years ago and depreciated under MACRS (ADS) using a 5-year recovery period. The original cost was $150,000 and the machine can last for 10 years or more in its current application. A new machine is now available at a cost of only $100,000. It can be depreciated with the MACRS (GDS) method (5-year property class). The annual expenses of the challenger are only $5,000 while those of the defender are $20,000. The challenger has a useful life greater than 10 years. You find that $40,000 is the best price you can get if you sell the present machine now. Your best projection for the future is that you will need the service provided by one of the two machines for the next 5 years. The MV of the defender is estimated at $2,000 in 5 years, but that of the challenger is estimated at $5,000 in 5 years. If the after-tax MARR is 10% per year, should you sell the defender and purchase the challenger? You do not need both. Assume that the company is in the 40% income tax bracket.

9-22. Five years ago, an airline installed a baggage conveyor system in a terminal, knowing that within a few years it would have to be moved. The original cost of the installation was $120,000 and, through accelerated depreciation methods, the company has been able to depreciate the entire cost. It now finds that it will cost $40,000 to move and upgrade the conveyor. This capitalized cost would be recovered over the next 6 years using the MACRS (ADS) with half-year convention and a 5-year recovery period. The airline believes this is a good estimate of the remaining useful life of the system, if moved. As an alternative, the airline finds that it can purchase a somewhat more efficient conveyor system for an installed cost of $120,000. The new system would result in an estimated reduction in annual expenses of $6,000 in year zero dollars. Annual expenses are expected to escalate by 6% per year. The new system is in the 5-year MACRS (GDS) property class, and its estimated market value 6 years from now, in year zero dollars, is 50% of its installed cost. This MV is estimated to escalate 3% per year. A small airline company, which will occupy the present space, has offered to buy the old conveyor for $90,000.

Annual property taxes and insurance on the present equipment have been $1,500, but it is estimated that they would increase to $1,800 if the equipment were moved and upgraded. For the new system, it is estimated that these costs would be about $2,750 per year. All other expenses would be about equal for the two alternatives. The company is in the 40% income tax bracket. It wishes to obtain at least 10% per year, after taxes, on any invested capital. What would you recommend?

9-23. A manufacturing company has some existing semiautomatic production equipment that it is considering replacing. This equipment has a present MV of $57,000 and a BV of $27,000. It has 5 more years of depreciation available under MACRS (ADS) of $6,000 per year for 4 years and $3,000 in year 5. (The original recovery period was 9 years.) The estimated MV of the equipment 5 years from now (in year zero dollars) is $18,500. The MV escalation rate on this type of equipment has been averaging 3.2% per year. The total annual expenses are averaging $27,000 per year.

New automated replacement equipment would be leased. Estimated annual expenses for the new equipment are $12,200 per year. The *annual* leasing costs would be $24,300. The MARR (after taxes) is 9% per year, $t = 40\%$, and the analysis period is 5 years. (Remember: The owner claims depreciation and the leasing cost is an operating expense.)

On the basis of an after-tax, actual-dollar analysis, should the new equipment be leased? Base your answer on the IRR of the incremental cash flow.

9-24. A company is considering replacing a turret lathe (defender) with a single-spindle screw machine (challenger). The turret lathe was purchased 4 years ago at a cost of $80,000, and depreciation has been based on MACRS (GDS) 5-year property class calculations. It can be sold now for $15,000 and, if retained, would operate satisfactorily for 4 more years and have zero market value. The screw machine is estimated to have a useful life of 10 years. MACRS (GDS) depreciation would be used (5-year property class). It would require only 50% attendance of an operator who earns $12.00 per hour. The machines would have equal capacities and would be operated eight hours per day, 250 days per year. Maintenance on the turret lathe has been $3,000 per year; for the screw machine, it is estimated to be $1,500 per year. Taxes and insurance on each machine would be 2% annually of the initial capital investment. If capital is worth 10% per year to the company after taxes, and the company has a 40% income tax rate, what is the maximum price it can afford to pay for the screw machine? Assume a 4-year analysis period and market value of $0.7679 investment cost for the challenger at the end of 4 years.

9-25. *Brain Teaser*

There are two customers requiring three-phase electrical service, one existing at location A and a new customer at location B. The load at location A is known to be 110 KVA, and at location B, it is contracted to be 280 KVA. Both loads are expected to remain constant indefinitely into the future. Already in service at A are three 100-KVA transformers that were installed some years ago when the load was much greater. Thus, the alternatives are as follows:

Alternative A: Install three 100-KVA transformers (new) at B now and replace those at A with three 37.5-KVA transformers only when the existing ones must be retired.

Alternative B: Remove the three 100-KVA transformers now at A and relocate them at B. Then install three 37.5-KVA transformers (new) at A.

Data for both alternatives are provided in the table that follows. The existing transformers have 10 years of life remaining. Suppose that the before-tax MARR = 8% per year. Recommend which action to follow after calculating an appropriate criterion for comparing these alternatives. List all assumptions necessary and ignore income taxes.

	Existing and new transformers	
	Three 37.5-KVA	Three 100-KVA
Capital Investment:		
Equipment	$900	$2,100
Installation	$340	$ 475
Property tax	2% of capital investment	2% of capital investment
Removal cost	$100	$ 110
Market value	$100	$ 110
Useful life (years)	30	30

10

New Product and Expansion Analyses

10.1 Introduction

Developing and financing a plan for continued growth is a key concern in any enterprise. For a company to remain financially viable and healthy as discussed in Chapter 1, it must continuously pursue opportunities that add economic value to replace those segments of the business which are no longer viable in an evolving marketplace. Economic value is added whenever we complete investment projects whose risk is acceptable and whose present worth is at least zero when evaluated at the interest rate that reflects the firm's cost of capital (or MARR). This rate may differ according to the organization, project, and timing.

New product investments, and the organizational capacity to develop them and distinguish them from competitor's offerings, are the lifeblood of the enterprise. Because of this, current financial health is a necessary, but not sufficient, condition for long-term success. A balanced scorecard approach, as described in Chapter 1, which includes a strategic map and new product development direction, is a much more comprehensive measure of success.

As new product markets grow in size, decisions on manufacturing and delivery capacity must be made. Both new product and capacity expansion decisions are complex issues requiring consideration of many strategic issues and always involving a rigorous assessment of the cost and benefits of each alternative. Both typically require investment in new facilities and equipment in order to realize a new or enlarged income stream and corresponding cash flows. Since both also involve the use of forecasts and difficult estimates, rather than certain data, the advanced techniques of Part Three of the text are frequently applied to analyses of these kinds of opportunities.

10.2 Investment Opportunities and System Analysis

The paradigm for analyses of these opportunities continues to follow the nine-step outline developed in Chapter 5. Following the steps in that paradigm, our objective is to assess an investment opportunity, and the alternatives are the different options available to realize a payoff from that opportunity. For the examples that follow, we will use the present worth approach as our measure of profitability. Much of the business planning literature breaks the cash flow into two categories: the initial investment (an outflow) and the net cash receipts, expressed as the present worth of the stream of inflows. The difference between the two is called the Net Present Worth (NPW), or Net Present Value (NPV).

This chapter will use spreadsheets in the examples developed, as they are convenient for handling irregular cash flows in an after-tax analysis. Variations are simply inserted into the spreadsheet table at the appropriate line and column. Care is required to ensure that the financial functions are properly interpreting the cash flows. For example, the Microsoft Excel function NPV calculates the present worth of the inflows at EOY 1 and must be adjusted and offset by the initial investment to compute the net present worth as of time zero.

The possibilities for variations in cash flow adjustments are unlimited, and hence it is impractical to make an exhaustive list of all possibilities. It is far more important to realize that the real challenge lies not with the computation, but rather with the recognition of relevant cash flows that must be considered in the analysis.

In Chapters 7 and 9, we developed the after-tax cash flow (ATCF) as the appropriate measure of the inflows resulting from the investment in a replacement machine. In business expansion and new product analyses, we must enlarge this concept to include additional cash flows, such as those necessary for infusion of added working capital and for sustaining capital expenditures (Capex) to support large scale investments. The after-tax cash flow from operations, and the changes in working capital, Δ WC, if any, sustaining capital expenditures, if any, and the after-tax salvage value, if any, are all relevant cash flows that must be included. The sum of all relevant cash flows, normally adjusted for income tax effects, is called the after-tax cash flow (ATCF), which is sometimes equal to free cash flow, FCF.

This chapter examines a series of increasingly complex examples, ranging from simple service provision, through manufacturing expansion, up to acquisition of a new company. Emphasis is placed upon the use of accounting concepts, and more realistic issues in the acquisition and interpretation of both the estimates used in, and the results obtained from, the analysis.

10.3 A Service Opportunity

Example 10-1. Starr Real Estate

Starr Real Estate Company is considering the purchase of an apartment building that has recently become available. Because of impending environmental restraints, the building must be demolished 5 years from now. The local government will pay for the demolition and will pay the owner $500,000 for the property, including the land valued at $100,000 at that time. The current owner is asking $500,000 for the building and land, which contains 20 identical

apartments that rent for $7,200 each annually. Because of the impending demolition, mainte-
nance costs are minimal at $4,000 per year. Management duties can be absorbed by the current
staff person, who earns $30,000 annually and currently manages a similar building. Starr
believes that a 15% after-tax return is appropriate for this kind of investment, and they pay
income taxes at an effective rate of 30%. Should Starr buy the building?

Solution   First, it is obvious that the study period should be set at 5 years, as all the
relevant cash flows will occur within that time frame. Cash flows for the revenues can be forecast on the basis of the number of units and the projected occupancy rate, which is assumed
to be 80%. Costs for maintenance have been estimated, perhaps accurately, on the basis of
Starr's experience in the business. Management costs are excluded on the basis that there is
no incremental cash flow, as the new management tasks are being picked up by the staff person. This choice may be controversial with those who would argue that, with two buildings
under her wing, half of the staff person's $30,000 salary should be "charged" to the new build-
ing. However, unless there is some constructive opportunity for alternative use of that time, it
is correct to value it at zero for this analysis. IRS regulations suggest straight-line depreciation
over a 40-year period for buildings, and land is not depreciated, so the annual depreciation is
$400,000 / 40 = $10,000 per year. When these are input to a spreadsheet, the result is as shown
in Table 10-1.

The value in the bottom row, $12,119, is the net present worth of the after-tax cash
flows, using the NPV function with an annual interest rate of 15% less the required investment
of $500,000, providing a small positive NPV. However, this conclusion rests upon belief in the
80% occupancy rate; a doubtful number in a very uncertain area. It is premature to conclude the
analysis and recommend purchase of the building.

As one approach, Starr could sharpen its revenue forecast by estimating a more realistic
occupancy rate on the basis of its experience and repeating the analysis. This is relatively pain-
less in a spreadsheet.

If Starr has no firm estimate, it might establish the "break-even" occupancy rate, either
by trial and error, using the spreadsheet, or by solving the expression

$$500,000 = \{[20 \times 7,200 \times R - 4,000 - 10,000] \, (1 - .3) + 10,000\} [P/A, 15\%, 5]$$
$$+ 485,000 \, [P/F, 15\%, 5]$$

for the occupancy rate R, which just equates the value of the property to its asking price. The
break-even rate is found to be 76.5%. In other words, the project must support an occupancy
rate at this level in order to generate the required 15% after-tax MARR. That may be a rather
high occupancy requirement, casting doubt on the desirability of the investment. More elaborate ways to conduct such sensitivity tests are described in Part Three. The main point here is
simply that the initial conclusion resulting from an investment analysis is usually not a final
answer, but merely a sound starting point for further probing.   ■

10.4 The Effects of Inflation

Up to this point, we have dealt mostly with dollars as if they were an absolute measure of economic gain. But we intuitively understand that the dollar, or any currency,
is a rubber yardstick that changes in value with inflation (or deflation) in the overall
economy.

TABLE 10-1 · Starr Real Estate (Example 10-1)

Interest Rate   15.00%
Purchase Price $ 500,000
Income Tax Rate   30.00%
Sale Price $500,000

Year	A Rent per unit	B Number of units	C % Occupancy	$D = A \times B \times C$ Gross rent	E Maintenance	$F = D - E$ Before-tax cash flow
0						-$500,000
1	$7,200	20	80%	$115,200	$4,000	$111,200
2	$7,200	20	80%	$115,200	$4,000	$111,200
3	$7,200	20	80%	$115,200	$4,000	$111,200
4	$7,200	20	80%	$115,200	$4,000	$111,200
5	$7,200	20	80%	$115,200	$4,000	$111,200
5						$500,000

Year	G Depreciation	$H = D - E - G$ Taxable income	$I = H \times$ Tax rate Income tax	$J = H - I$ Net operating profit after tax	$K = F - I$ After tax cash flow
0					-$500,000
1	$10,000	$101,200	$30,360	$70,840	$80,840
2	$10,000	$101,200	$30,360	$70,840	$80,840
3	$10,000	$101,200	$30,360	$70,840	$80,840
4	$10,000	$101,200	$30,360	$70,840	$80,840
5	$10,000	$101,200	$30,360	$70,840	$80,840
5*		$50,000*	$15,000		$485,000

NPW = $12,119

*Market Value – Book Value = $ 50,000, a taxable gain.

Overall observed interest rates increase when inflation is expected, as lenders try to correct for the fact that their loans will be repaid in cheaper dollars. Borrowers, cognizant of the fact that their nominal income will increase with inflation, are willing to pay the higher interest rates. They know that they will repay the loan with cheaper dollars. In a free market, this interaction between lenders and borrowers will result in an interest rate that includes an inflationary premium which best reflects the net beliefs of informed buyers and sellers. Thus, marketplace interest rates always include inflationary expectations. They are referred to as nominal (or actual or combined) rates to distinguish them from "real" interest rates.

The issue here is how to include the effects of inflation in investment analyses, which involve a forecast of future cash flows. As discussed in Chapter 8, there are two choices. One is to forecast the cash flows at their expected amounts for each point in time and then discount them, using actual (nominal or combined) interest rates. A second approach is to adjust the interest rate to a real, noninflated value, and then forecast the cash flows at their real, noninflated values. Specifics are detailed in Chapter 8, but the important principle is that the interest rate choice and the cash flow forecast choice must be consistent. To illustrate, we return to the Starr Real Estate example and use the first approach.

Example 10-2. Starr Real Estate Revisited

Suppose the Starr Real Estate opportunity is recast. The current economic climate includes inflation at an estimated 3% annually, and this is reflected in the 15% after-tax MARR that Starr has applied to the project. The city's price for the building remains fixed at $500,000 in the 5-year horizon. However, rents may be raised to the going market rate, and maintenance expenses can be assumed to increase with economic inflationary pressures. What is the NPW for this project?

   Solution   Since the interest rate chosen is a nominal rate, the cash flow forecast must be done in nominal (actual) dollars. If we assume that rents are increased at exactly 3% per year effective in year 2 and adjust maintenance expenses at the inflation rate, then the revised spreadsheet shown in Table 10-2 properly reflects the cash flow situation. Note that depreciation is fixed in actual dollars and does not respond to inflation. As stated in the scenario, the building repurchase by the city is fixed at the $500,000 level. Recognizing the effect of inflation in this way raises the net present worth of the after-tax cash flow to $25,986, as shown in the bottom row of Table 10-2.   ■

10.5 Manufacturing Expansion

In Examples 10-1 and 10-2, our models ignored some realities, such as late receipt of rent payments, supplies inventories, and deferred payments that accountants would place under the label "working capital." These are probably small impact items, so their omission is not a major flaw in the analysis. However, in a manufacturing scenario, in which inventories of raw materials, in-process goods, and finished goods are significant, and customers purchase on trade credit, resulting in high accounts receivable, the impact of investments in working capital to support the business is very relevant. In particular, a capacity expansion to increase sales

TABLE 10-2 Starr Real Estate Revisited (Example 10-2)

Interest Rate	15.00%
Purchase Price	$500,000
Tax Rate	30.00%
Sale Price	$500,000
Inflation Rate	3.00%
	effective in year 2

	A	B	C	D = A × B × C	E	F = D − E
Year	Rent per unit	Number of units	% Occupancy	Gross rent	Maintenance	Before-tax cash flow
0						−$500,000
1	7200	20	80%	$115,200	$4,000	$111,200
2	7416	20	80%	$118,656	$4,120	$114,536
3	7638	20	80%	$122,216	$4,244	$117,972
4	7868	20	80%	$125,882	$4,371	$121,511
5	8104	20	80%	$129,659	$4,502	$125,157
5						$500,000

	G	H = F − G	I = H × Tax rate	J = H − I	K = F − I
Year	Depreciation	Taxable income	Income tax	Net operating profit after tax	After-tax cash flow
0					−$500,000
1	$10,000	$101,200	$30,360	$70,840	$80,840
2	$10,000	$104,536	$31,361	$73,175	$83,175
3	$10,000	$107,972	$32,392	$75,580	$85,580
4	$10,000	$111,511	$33,453	$78,058	$88,058
5	$10,000	$115,157	$34,547	$80,610	$90,610
5*		$50,000*	$15,000		$485,000
					NPW = $25,986

*Market Value − Book Value = $50,000, a taxable gain.

321

must invariably be supported by increased working capital to support the business. The money tied up in working capital must be considered as part of the cash flow associated with the project.

As the business ramps up, inventories accumulate to support operations and new sales are made on open account. These are partially offset by credit purchases of raw materials and supplies on open account. As these spontaneous investments are made, funds are in effect loaned to the project. As they are recovered (for example, as receivables are collected), they are replaced by new extensions of credit. They remain on loan to the project, varying daily with business activity, but hovering about an average related to the firm's financial practices. Ultimately, they are recovered when operations cease and remaining inventories are sold, outstanding receivables are collected, and vendor invoices are paid. These are illustrated in Example 10-3.

Example 10-3. Manufacturing Expansion

HD Motorsports is on a roll. Currently, they fully utilize five manufacturing cells to produce midsize motorbikes at the rate of 2,000 bikes per cell per year. Lately, they are facing an increasing order backlog despite working 6 days per week. Their policy is to avoid overtime for normal operations, and use it only as a buffer for unplanned emergencies; therefore, a capacity increase must be considered. They are confident that they can meet the new demand by adding a sixth cell with capacity of 2,000 bikes per year. There is space available, and plant support resources for a new cell would cost about $200,000 annually. The equipment required would cost $5 million. The cell would have a useful life of 10 years and would be depreciated on a straight-line basis to zero. The cost and selling price for a typical bike are as follows:

Selling Price	$5,000
Standard Cost	
Material	2,000
Labor	1,000
Variable OH	600
Fixed OH	800
Total	$4,400

Should HD expand to add the sixth cell? The company requires an annual 20% after-tax return on projects with this kind of risk. The effective income tax rate is 40%.

Solution Evaluating the expansion project simply requires determination of the present value of the cash flows resulting from the new sales, and comparison to the required investment. Since the equipment life is 10 years, that is a reasonable time horizon for the study. If, for the moment, working capital requirements are ignored, annual after-tax cash flows are determined directly from Net Operating Profit After Taxes (NOPAT) from the alternate approach tabulated next. This approach of adding depreciation to NOPAT is a commonly used shortcut for the determination of after-tax cash flow when NOPAT has already been determined.

	Annual Amount
Sales (2,000 for one cell at $5,000 each)	$10,000,000
Expenses	
Material	4,000,000
Labor	2,000,000
Variable Overhead	1,200,000
Plant Support	200,000
Depreciation	500,000
Total Expenses	7,900,000
Taxable Income	2,100,000
Less: Income Taxes	840,000
Net Operating Profit After Tax (NOPAT)	1,260,000
Add: Depreciation	500,000
After Tax Cash Flow	$1,760,000

On the basis of this cash flow, the PW of the increased capacity is $1,760,000 \times (P/A,20\%,10)$, or $7,379,000, a handsome excess over the investment of $5,000,000, resulting in an NPV of $2,379,000. The same result is obtained by using the spreadsheet shown in Table 10-3 through Column J. (Columns K and L will be explained shortly.) But this ignores the reality that working capital is required to support the sales increase. To resolve it, a way to forecast working capital requirements is needed. Since adding a sixth cell augments capacity by 20%, we might simply project 20% increases in the balance sheet values for inventories, receivables, and payables.

An alternative approach might be to estimate the needs on the basis of policy variables. For example, suppose that inventory turns were a healthy 20 times per year due to cellular manufacturing, receivables were extended to 90 days to support the dealer network, and payables were kept to 30 days.

Using the cost data provided and sales of 10,000 bikes annually, sales are $50 million and cost of goods sold (COGS) is $44 million annually. Twenty inventory turns implies that

$$20 = \text{COGS/Inventory} = \$44 \text{ million}/20 = \$44 \text{ million/Inventory}.$$

$$\text{Therefore, Inventory} = \$2.2 \text{ million}.$$

Ninety days of receivables implies that

$$\text{Accounts receivable} = \text{Annual sales} \times 90/360 = \$50 \text{ million} \times 90/360$$
$$= \$12.5 \text{ million}.$$

TABLE 10-3 Analysis of Expansion Opportunity for HD Motorsports (Example 10-3)

All dollars in thousands

| | | Income tax rate | 40.00% | | | | | | | | | |
| | | MARR | 20.00% | | | | | | | | | |

Year	A Sales	B Material	C Labor	D Variable OH	E Plant costs	F Depreciation	G = A − B − C − D − E − F Taxable income	H = .4G Taxes	I = G − H NOPAT	J = I + F After-tax cash flow	K Change in working capital	L = J + K Net After-tax cash flow
0										−5,000		−5,000
1	$10,000	$4,000	$2,000	$1,200	$200	$500	$2,100	$840	$1,260	$1,760	−$2,506	−$746
2	$10,000	$4,000	$2,000	$1,200	$200	$500	$2,100	$840	$1,260	$1,760		$1,760
3	$10,000	$4,000	$2,000	$1,200	$200	$500	$2,100	$840	$1,260	$1,760		$1,760
4	$10,000	$4,000	$2,000	$1,200	$200	$500	$2,100	$840	$1,260	$1,760		$1,760
5	$10,000	$4,000	$2,000	$1,200	$200	$500	$2,100	$840	$1,260	$1,760		$1,760
6	$10,000	$4,000	$2,000	$1,200	$200	$500	$2,100	$840	$1,260	$1,760		$1,760
7	$10,000	$4,000	$2,000	$1,200	$200	$500	$2,100	$840	$1,260	$1,760		$1,760
8	$10,000	$4,000	$2,000	$1,200	$200	$500	$2,100	$840	$1,260	$1,760		$1,760
9	$10,000	$4,000	$2,000	$1,200	$200	$500	$2,100	$840	$1,260	$1,760		$1,760
10	$10,000	$4,000	$2,000	$1,200	$200	$500	$2,100	$840	$1,260	$1,760	$2,506	$4,266
									NPW =	$2,379	NPW =	$695

If raw materials and all variable overhead items are purchased, then 30 days of payables implies that

$$\text{Accounts Payable} = \text{Annual Purchases} \times 30/360 = \$26 \text{ million} \times 30/360$$
$$= \$2.17 \text{ million.}$$

The theoretical net working capital to support the existing five cells is

Inventory	2.20 million
Accounts Receivable	+ $12.50 million
Accounts Payable	–2.17 million
Total = Net Working Capital	$12.53 million

The forecast amount needed to support the sixth cell and the added sales would be approximately one-fifth of $12.53 million, or $2.506 million. If this amount were projected to accumulate in the current accounts during the 1st year of the expansion project and be recovered fully in the 10th year, the resulting investment spreadsheet would require the addition of columns K and L as shown in Table 10-3.

By including the requirement for working capital to support the expansion, the NPV drops to $695,000—still positive, but substantially less than $2,379,000. Note that, in this table, the analysis proceeds directly to the computation of after-tax cash flows from NOPAT and depreciation, rather than subtracting income taxes from before-tax cash flow.

Observe that the working capital is assumed to build during the first year of operation and, following our usual computation convention, is treated as an end-of-year cash flow. A problem at the end of this chapter asks you to determine the NPV if the initial influx of working capital is required immediately.   ■

10.6 New Product Development

Like product expansion, new product development involves investment of working capital to support manufacturing operations. To the extent that it involves entry into new product markets, product development may generate further downstream opportunities that create added opportunities for growth. It may also require new organizational capabilities, which, if unrecognized, can destroy value by placing the firm in a precarious situation. Here, there are more unknowns and more strategic implications affecting the survival of the firm. More than ever, the economic analyses must be correct, as the stakes are often high. However, the data are more uncertain. Unlike in the case of capacity expansion, we have no sales experience to extrapolate; our understanding of the manufacturing processes or the market channels involved may be limited, and receptiveness of customers and our employees to the changes involved may be lukewarm. For these reasons, the capital investment analysis is but one aspect of an overall strategic decision. There are several implications:

- The investment analysis is not an event, but rather a series of analyses, each progressively more comprehensive as new knowledge is gained as a result of on-going refinement of the project plan.

- The methods for handling uncertainty discussed in Part Three take on a larger role in the analysis.
- The impact of intangibles is more carefully weighed. Subjective factors, resulting from executive opinion, may overrule a favorable cash flow appraisal. To maintain perspective, the real options approach, discussed in Chapter 17, and multiattribute analysis approaches, described in Chapter 19, escalate in applicability and importance.

Cast in this light, capital investment analysis is one input to an overall strategic assessment of a proposed major investment. It does not necessarily provide the go–no-go decision that marks the end of the analysis process, but rather it provides the investment basics as the framework for building a total impact assessment of a proposed major investment. Our discussion of these considerations will evolve in subsequent chapters. For now, let's illustrate these ideas through an example analyzed in the conventional benefit–cost framework.

Example 10-4. A New Product

Ajax Pool Products is a specialty producer for home swimming pools and spas. One major product is chlorine tablets for home pool dispensers. They are made in a simple, but messy, process of blending dry powders, pelletizing them in hopper-fed presses, and packaging the resulting pellets in various sizes of drums and pails for sale in independent pool care stores.

In an effort to stimulate sales, which have been relatively flat at 500 tons per year, Ajax is considering the introduction of a premium-grade tablet with added sodium bromide as an algae preventer. It is felt that some customers will gladly pay for the convenience that the combination tablet offers. The initial estimate is that sales would start at 100,000 pounds annually, and ramp up to 400,000 pounds over a 4-year period. Half of these sales would cannibalize existing standard product sales, and half would represent new market penetration.

The powder blender is somewhat dated, although serviceable for at least 6 to 8 years, but will be totally depreciated to a zero salvage value in 6 years at the current rate of $30,000 annually. It has a capacity of 800 tons per year and, ultimately, the new product would use a quarter of that capacity in producing 400,000 pounds annually.

The process bottleneck is pelletizing, and a new press and packaging conveyor would be needed to support any expansion. A rough estimate of $100,000 for this equipment was provided by an equipment vendor. Although it is useable for 10 years, it has very little salvage value after a few years of use because of the corrosive nature of the chemicals that are processed. Representative price and cost data for the standard product on a per-pound basis are as follows:

	Per pound
Selling Price	$1.00
Material	0.40
Labor	0.03
Variable Overhead	0.02
Fixed Overhead	0.12
Total Standard Cost	0.57

The new product is anticipated to sell at a 20% price premium, and represents a lucrative opportunity, as the net added cost for the sodium bromide amounts to a nickel per pound. Net working capital to support the current 500 tons-per-year output is $300,000 and can be expected to increase in proportion to any increase in output. The after-tax MARR is 15% per year.

Solution   This proposal appears to be in its formative stages and requires a preliminary assessment of its merits before developing into a full formal investment analysis. We will attempt to forecast the incremental sales and cash flows that would result from the new product, and compare them in present worth terms with the required $100,000 investment. The results, assuming a 6-year life, are shown in Table 10-4. First, note that the table develops the incremental sales dollars and associated costs resulting from the introduction of the new product. These are developed in the upper half of the table as the sales increases of the new product at $1.20 per pound offset by the cannibalized sales of the existing product at $1.00 per pound. The second half of the spreadsheet develops the net after-tax cash flow. One point worth discussing is the deliberate omission of a depreciation charge for the blender, despite the commitment of 25% of its capacity to the new product. This recognizes the fact that no new cash flows are associated with the use of the blender. Any increase in maintenance of the equipment can reasonably be expected to be included in the variable overhead. The fixed overhead allocation to the existing product is irrelevant, as all fixed costs should be covered by existing product and, with the exception of depreciation of the new press and conveyor, are not expected to change as a result of the new product decision. Increases in working capital reduce cash flow at the rate of $15,000 for every 25-ton increase in sales until fully recovered at the end of the 6-year horizon. As shown in the bottom row, the present worth of these cash flows is $153,800, after deducting the $100,000 investment required to generate them. This preliminary analysis suggests that the new product, Algae-Pruf, is a viable opportunity. But before diving in headfirst (pun intended, sorry), it makes good sense to explore the intangibles and risk factors involved. One systematic way to do this is to review the changes and risks to consider, and the assumptions to test, at each stage in the value-adding process. Table 10-5 illustrates this process. Similar tables can be developed for stakeholders in the company (Table 10-6) and for other entities in the company's environment (Table 10-7).   ■

Stakeholders are those who have an interest in the company, but who are not identified in the value stream. Examples of such stakeholders and the impact upon them appear in Table 10-6.

In addition to stakeholders, there usually are environmental entities that may be affected by a project or can affect a company's actions. Table 10-7 identifies representative environmental entities for Ajax.

The combined result of this assessment of current business factors is that there is much yet to be done before final approval to proceed with Algae-Pruf. There remains one other fit assessment that must be considered (i.e., the fit between the proposed investment and the strategic plan for Ajax). If Ajax has developed a strategy and has documented its impact in a balanced scorecard format, the fit can readily be assessed by examining the project's expected impact on each of the metrics in the scorecard. If not, then the assessment is more subjective, but the comparisons focus on the fit between the proposed project and the strategic direction of the company. For example, fit with the product market strategy, fit with the customer value proposition, fit with

TABLE 10-4 Ajax Pool Products New Product (Example 10-4)

All dollar amounts in thousands

	A	B	C	D	E	F = B − D	G	H	I	J = G + H + I
Year	New products sales (1,000 lbs)	Sales $	Lost sales (1,000 lbs)	Lost sales $	Net tonnage increases	Net sales $	Material costs	Labor cost	Variable overhead	Total variable costs
1	100	$120	50	$50	25	$70	$25	$3	$2	$30
2	200	$240	100	$100	50	$140	$50	$6	$4	$60
3	300	$360	150	$150	75	$210	$75	$9	$6	$90
4	400	$480	200	$200	100	$280	$100	$12	$8	$120
5	400	$480	200	$200	100	$280	$100	$12	$8	$120
6	400	$480	200	$200	100	$280	$100	$12	$8	$120

	K = F − J	L	M	N = K − L − M	O = N × 0.4	P = N − O	Q = K − O	R	S = R + Q
Year	Before-tax cash Flow	Depreciation blender	Depreciation press & conveyor	Taxable income	Income tax @ 40%	NOPAT	After-tax cash flow	(Increase) in working capital	Net after-tax cash flow
0	−$100						−$100		−$100
1	$40	$0	$10	$30	$12	$18	$28	−$15	$13
2	$80	$0	$10	$70	$28	$42	$52	−$15	$37
3	$120	$0	$10	$110	$44	$66	$76	−$15	$61
4	$160	$0	$10	$150	$60	$90	$100	−$15	$85
5	$160	$0	$10	$150	$60	$90	$100		$100
6	$160	$0	$10	$150	$60	$90	$100	$60	$160
6*				−40*	−16*		$16		$16
								NPW =	$153.80

*Book (Capital) Loss of $100 − 6($10) = $40 creates a tax savings of $16.

TABLE 10-5   Value Chain Factors at Ajax

Value chain factor	Changes/risks to consider	Assumptions to test
Customers	New value proposition	Customers will pay 20% premium
	New price	No problem with educating dealers
Market Channel	Same	No issues with added item
Distribution	Same	No issues with added item
Production	New item	No significant costs associated
	Must segregate	
	New Q/A process	
	Employee training	
Suppliers	New ingredient to purchase and store	No issues with assured supply

TABLE 10-6   Stakeholders Impacted at Ajax

Stakeholder	Changes/risks to consider	Assumptions to test
Stockholders, Owners	Increased size of company	Ability of management to cope
Creditors, Debt holders	Increased debt	Robustness of ability to repay
Workforce	Ability/willingness to expand	Willingness to work overtime if needed to support demand surge
	Safety issues with new chemical	
Union	Support of new effort	
Competitors	Will they copy?	4-year window for growth

TABLE 10-7   Environmental Factors

Environmental factor	Changes/risks to consider	Assumptions to test
Local regulations (e.g. OHSA, local fire codes, zoning)	Impact of new product	No impact assumption
Social acceptability	Expanding a messy workplace	Community will not react adversely
Environmental regulations	Expanding a messy workplace	No pending new laws that might curtail operations
Product technology	Growth of Ozonators as substitutes for Chlorantors	6 year window is viable
Process technology	Basic, no change	Compatibility of added chemical with equipment

core competencies in engineering, operations and other key areas, fit with the intended future direction, etc. This is important in the sense that it can be very sensible to accept a marginal project that provides future business opportunities which are especially lucrative or provide a competitive advantage. For now, we will be content to recognize the reality that the assessment of quantifiable costs and benefits is not an end point, but rather a starting point in capital investment decisions. Next, we will consider another investment with strong strategic implications: the merger-acquisition decision.

10.7 Valuing a Business Acquisition Prospect

Valuing a business acquisition prospect is conceptually no more difficult than valuing a new machine or a new product opportunity. All that is needed is a proper valuation of the cash flows over the life of the project at the appropriate interest rate. In Chapter 11, we will examine firm-wide considerations that affect the determination of the MARR.

In analyzing acquisitions, two issues arise in development of the cash flows. The first has to do with the project horizon. The acquisition prospect can be expected to have a long, perhaps indefinite, life. This requires that we project not only cash flow from operations and working capital infusions, but also future capital expenditures to sustain and grow the business. All three are part of the net cash flow, NCF. It may be possible to project these over a 3 year to 5 year horizon, but the future may be very blurry beyond that time frame. One solution is to project a growth rate as a percentage of existing business, and estimate continuing capital expenditures (Capex) and working capital changes (ΔWC) as a percentage of sales levels. This assumes some degree of stability in these ratios and a belief that the future will mimic the past. Obviously, such projections may be altered by technology changes or by falling costs, etc.

A common practice is to estimate cash flows over a reasonable horizon and then value the remaining indefinite cash flow as a perpetuity that is realized at the end of the horizon. This terminal value and the annual after-tax cash flows in the horizon are then simply discounted to the present to obtain a value for comparison.

Therein lies the second problem. Frequently, the cash flows in the planning horizon are small if the acquisition target is a new company just building capability, or if it is a distressed company facing profitability problems. In those cases, the bulk of the net present worth is driven by the terminal value and the assumptions behind its determination. This stands in stark contrast with the analysis for a new machine, wherein the terminal value is a small percentage of the initial outlay and a minor part of the value of the project. Clearly, the critical tasks here are arriving at a sound means for establishing and testing a terminal value. Often, this is done by specialists with industry expertise and/or a good facility for comparative financial projections. Such clairvoyance is well beyond the scope of this book; however, it is useful to examine an example that uses simple projections commonly observed in practice.

Example 10-5. Save-A-Buck Pool Supplies

Save-A-Buck operates a chain of pool supply stores in the Southeast. They franchise local operators who then purchase their wares from the company. They have learned about the plans Ajax has developed for Algae-Pruf, and believe that it has substantial market appeal. Further, because of competitive pressure from discount stores, they literally make no profit on the tablets they sell in their stores, but must supply them as a convenience to their franchisees. They see an opportunity here. Were they to buy the entire tablet business from Ajax, they could earn some margin on the existing product, capture the benefits of Algae-Pruf for their chain, and increase the sales of the Ajax product as they expand their chain. They have contacted Ajax for a copy of the information contained in the capital project analysis. Since the company presidents, Smith and Jones, are old friends and Jones, the president of Ajax, is nearing retirement, they agree on an open and friendly discussion. Jones would like to have $3 million for his company. Smith needs an assessment of the value of Ajax to begin the negotiation. He believes that his after-tax MARR is 15% per year.

   Solution   Since Save-A-Buck receives no profit from its existing chlorine tablet business, all the cash flow that would emanate from the Ajax acquisition would contribute value to Save-A-Buck stores. To make that assessment, Smith modifies the spreadsheet to incorporate all tablet sales and associated cash flows. The spreadsheet is shown in Table 10-8.

   Again, this analysis is most easily described in segments. The upper half shows all the sales of tablets, including the existing product line, but now growing at 5% annually. These sales are augmented by the new Algae-Pruf tablets with the same cannibalization impact as assumed by Ajax, resulting in the Net Sales dollars shown.

   The remaining columns develop the operating costs and resultant net operating profit after taxes (NOPAT). Annual depreciation has been included for all company assets to be acquired, including the Blender, at $30,000, other plant and manufacturing equipment at $80,000, and $10,000 for the new press and conveyor for a total of $120,000.[1]

   The final section of Table 10-8 develops the net cash flow from the sum of its components. After tax cash flow is NOPAT plus the $120,000 depreciation just described. Working capital is, as before, based upon the funds needed to support the increased production, and reduces cash flow. The newly proposed palletizing press is no longer assumed to be retired at the end of a 6-year period. Finally, the company routinely budgets and spends $100,000 for capital equipment (Capex) to meet environmental regulations, replace existing equipment, etc. Summing these inflows and outflows yields a net after-tax or free cash flow with a present value of $1,223,000, quite distant from the $3,000,000 asking price. But this amount represents only the results from 6 years of operation. The analysis normally should account for the value of ongoing operations after the planning horizon. One way might be to estimate the market value of the company at the end of the horizon. This might be accomplished by determining a terminal value at the end of year 6 that reflects the perpetual cash flows beginning in year 7 and continuing indefinitely. The after-tax net cash flows in any year are

$$\text{ATNCF} = \text{NOPAT} + d + \Delta\text{WC} + \text{Capex},$$

or, equivalently,

$$\text{ATNCF} = \text{BTCF} - \text{Income Taxes} + \Delta\text{WC} + \text{Capex}, \qquad (10\text{-}1)$$

---

[1] For computational convenience, these depreciation amounts are shown as unchanged from those incurred by Ajax Manufacturing. In reality, they are likely to change under the new ownership, if the acquisition is executed.

TABLE 10-8 Save-A-Buck Pool Supplies (Example 10-5)

						All dollar amounts in thousands				
	A	B	C	D	E	F	G = C-E	H	I	J	K = H + I + J
Year	Sales existing products (1,000's)	New products sales (1,000 lbs)	Sales $	Lost sales (1,000 lbs)	Lost sales $	Net tonnage increases	Net sales $	Material costs	Labor cost	Variable overhead	Total variable costs
1	1,000	100	$1,120	50	$50	25	$1,070	$425	$63	$42	$530
2	1,050	200	$1,290	100	$100	50	$1,190	$470	$69	$46	$585
3	1,103	300	$1,463	150	$150	51	$1,313	$516	$75	$50	$641
4	1,158	400	$1,638	200	$200	53	$1,438	$563	$81	$54	$699
5	1,216	400	$1,696	200	$200	29	$1,496	$586	$85	$57	$728
6	1,276	400	$1,756	200	$200	30	$1,556	$611	$89	$59	$758

TABLE 10-8 (Continued)...

All dollar amounts in thousands

Year	L = G − K Before-tax cash flow	M Depreciation blender	N Other depreciation	O Depreciation press & conv	P = L − M − N − O Taxable income	Q = P × 0.4 Income Tax @ 40%	R = P − Q NOPAT	S = R + M + N + O After-tax cash flow	T working capital	U Capex	V = S + T + U Net After-tax cash flow
1	$540	$30	$80	$10	$420	$168	$252	$372	−$15	−$100	$257
2	$605	$30	$80	$10	$485	$194	$291	$411	−$30	−$100	$281
3	$671	$30	$80	$10	$551	$221	$331	$451	−$31	−$100	$320
4	$739	$30	$80	$10	$619	$248	$371	$491	−$32	−$100	$359
5	$768	$30	$80	$10	$648	$259	$389	$509	−$17	−$100	$391
6	$798	$30	$80	$10	$678	$271	$407	$527	−$18	−$100	$409
										NPV =	$1,223

where the annual components are as follows:

BTCF is the before-tax cash flow,

NOPAT is the net operating profit after tax,

d is the depreciation,

Δ WC is the change in working capital, normally negative for an increase, and

Capex is the added capital expenditures, normally negative for an increase.

The annual growth rate (inflation or otherwise) for each of these components of ATNCF might be estimated as an annual percentage. Alternatively, each component might be estimated for year 7 and the growth rate calculated as shown in Table 10-9. Each can then be projected indefinitely using the perpetuity relationship

$$P_N = A/(i-g) \quad \text{for} \quad i > g, \tag{10-2}$$

where

P_N is the present value of a perpetuity at the beginning of year $N + 1$ or, equivalently, the end of year N,

A is an annual cash flow starting in year $N + 1$, assuming year-end payments,

i is the nominal cost of capital per year, as a percentage of the invested amount, and

g is the growth rate of A per year, as a percentage of the annual amount.

The terminal value, previously calculated as P_N, should then be discounted to the present as $P_0 = P_N (P/F, i, N)$ and added to the present value of the projections for the initial N years (here, 6) to determine the NPW of the acquisition target.

Applying these to four relevant components of ATCF results in Table 10-9.

Table 10-9   Analysis of Terminal Value of the Ajax Pool Company Acquisition

Component	Year 6 value	Year 7 value	g	$(i{-}g)$	P_6	P_0
NOPAT	$407.00	$427.35	0.05	0.10[1]	$4,274	$1,848
d	$120.00	$120.00	0	0.15	$800	346
Δ WC	–$18.00	–$18.90	0.05	0.10	–$189	–$82
Capex	–$100.00	–$100.00	0	0.15	–$667	–$288
Total	$409.00	$428.45			$4,218	$1,823

[1]For example, $g = (427.35 - 407)/407 = 0.05$; $\therefore i - g = 0.15 - 0.05 = 0.10$.

As shown in Table 10-9, the present worth of the terminal value is $1,823,000. This terminal value assumes that the organization generates returns forever, based upon the influx of new capital expenditures. The present worth of the after-tax cash flow over the 6-year planning horizon developed in Table 10-8 is $1,223,000 for a combined total of $3,046,000. This generates an NPW of $46,000 relative to Jones' asking price of $3,000,000. This appears to be a marginally viable proposition; however, that can be only a tentative conclusion, as sensitivity testing, risk analysis, and analysis of business fit or strategic fit have not yet been considered. The problem exercises leave that important task to the readers.   ■

10.8 Summary

At this point, after-tax present worth computations should be a routine matter. However, the determination of the appropriate cash flows to include will always require careful thought. The analysis is seldom a single event, but rather an evolution of thought as our understanding of a proposed investment becomes clearer as more details emerge throughout the project life cycle.

This chapter has enlarged the investment framework from the consideration of investment in individual machines or projects to one that considers prospects which affect the entire business. The notion of cash flow after taxes has expanded to include the impact of working capital changes, as well as undefined future capital expenditures to sustain an ongoing business venture. This enlargement progressed to the acquisition of businesses, wherein terminal values after an abbreviated planning horizon may have a major impact on the valuation results.

Equally important, the chapter developed a framework for consideration of intangibles and identification of risks in investment analysis. These are pursued through a qualitative analysis of fit with value chain factors, stakeholders, and environmental entities in the business environment and fit with the strategic direction of the company.

Finally, these enlargements allow us to expand the systems analysis framework outlined in Chapter 1 and Chapter 5. A further amplification, now displayed as the Capital Investment Analysis Worksheet in Figure 10-1, shows that process with the updates needed to incorporate the new concepts added in this chapter and a preview of further steps to be presented in the remainder of the text.

SYSTEM ANALYSIS STEPS:

1 SPECIFY PROJECT OBJECTIVE

What must be accomplished? By when?

2 DEFINE ALTERNATIVES AND EDIT

Independent projects can be compared to the do-nothing alternative. Mutually exclusive alternatives may require special handling for IRR and B/C approaches.

Figure 10-1   Capital investment analysis worksheet.

3 DETERMINE CASH FLOWS ASSOCIATED WITH EACH ALTERNATIVE, INCLUDING TAXES AND INCREMENTAL SUPPORTING CASH FLOWS

Focus on cash flows, not on accounting data masquerading as cash flows. It is preferable to include related taxes, as they are cash flows. Exclusion of income taxes through use of a before-tax interest rate is often unacceptable. Working capital cash flows are significant for expansion projects, new products, and new businesses. Valuations of business units require consideration of ongoing capital expenditures as a normal part of their cash flow.

4 SPECIFY ANALYSIS APPROACH AND CASH FLOW SIGN CONVENTION

Usually done with cash inflows as positive and outflows as negative; however, in annual cost analyses for machine replacement or for lease vs. buy comparisons that do not involve traceable revenues, it is often more natural and convenient to use outflows as positive to avoid handling negative revenue numbers.

5 DETERMINE TIME HORIZON FOR ANALYSIS AND IDENTIFY ANY CASH FLOWS OUTSIDE HORIZON

This becomes particularly important in equipment replacement analyses for which unequal lives are involved. One way to avoid the repeating cycle assumptions in the common life multiple approach, which is implicit in equivalent worth comparisons, is to obtain a reasonable salvage or horizon value for one or more of the alternatives at the end of the "study period." In acquisition analyses, the terminal value often dominates the analysis, and hence is often the subject of intense study.

6 SPECIFY INTEREST RATE FOR DISCOUNTING

Normally, in an investment situation, the correct interest rate is the investor's minimum attractive rate of return, or MARR. Some books refer to this as the 'desired ROI." Others may call it the "hurdle rate," since it specifies the threshold that a proposed project should attain. Nominal cash flows are discounted by using nominal rates. Real cash flows are discounted by using real interest rates. Classically, we work under the assumption of separating investment decisions (which investments/projects to choose) from financing decisions (how to obtain the funds to undertake investments/projects). Chapter 11 develops the notion of a "cost of capital" as a floor on the MARR and discusses the implications of adjusting the MARR for business risk.

7 CONDUCT ECONOMIC COMPARISON BY USING A CALCULATOR, TABLED FACTORS, OR SPREADSHEET

Non-DCF Methods	DCF Methods
Accounting rate of return	Present worth
Payback	Annual worth
	Internal rate of return (IRR)
	Benefit–Cost ratios (B/C)

Non-DCF methods are not recommended, but are commonly used for rough screening and riskiness indicators. The present worth is always the conceptually correct method and is the most natural approach on a spreadsheet. IRR and B/C methods require special handling for mutually exclusive alternatives. It is most important to watch sign conventions built into calculators and spreadsheets.

8 SELECT MOST FAVORABLE ALTERNATIVE (TENTATIVELY)

9 ASSESS RISKS: OPERATIONAL AND FINANCIAL

For example, consider the following questions: Does the project make operational sense? Have we done this before? Do we understand the technology involved? Are the cost estimates realistic? Are all costs considered? Are the market projections realistic? Have adequate market tests been performed?

10 PERFORM SENSITIVITY ANALYSES

Which variables must be known with precision to identify a clear-cut choice? Spreadsheets offer many "what if" possibilities. Chapter 13 shows numerous approaches to sensitivity testing.

Figure 10-1   (Continued).

11 DETERMINE NEED TO EXPLICITLY MODEL RISK CONSIDERATIONS

These approaches are developed in Chapters 12–17:
Worst-case Analysis
Unrecovered Investment
Decision Trees
Simulation Models
Options Treatment

12 COMPARE RISKS/RETURNS

13 ASSESS FIT WITH OVERALL STRATEGY: BUSINESS, PRODUCT MARKET, OPERATIONS, AND FINANCIAL

14 MAKE FINAL RECOMMENDATION(S)

Figure 10-1   (Continued).

PROBLEMS

The problems are deliberately broadly stated and may require some judgment and assumptions. Although a variety of solution methods may be applied, the most natural one is the net present worth method, using a spreadsheet. Unless otherwise instructed, this approach is recommended for consistency of classroom communication. Similarly, it is suggested that you assume end-of-year payments, straight-line depreciation, and a 40% effective income tax rate, unless there are other implications in the problem statement.

10-1. In Starr Real Estate (Example 10-1), what is the NPW for an occupancy rate of (a) 90%? (b) 60%?

10-2. In Starr Real Estate Revisited (Example 10–2), suppose that the rental market is tight and rents are likely to increase by 5% annually, rather than the overall inflation rate of 3% presumed in the example. What is the NPW in this situation?

10-3. Mary Ortiz has recently developed a highly successful business distributing high-fashion ladies' handbags. She believes that she can double her $5 million annual sales by developing a South American market. Since most of her production is outsourced, her investment in physical plant and equipment is minimal at $200,000. However, her current working capital requirements are $1,400,000, broken down as follows:

Accounts Receivable	$ 1,000,000
Inventory	800,000
Accounts Payable	400,000
Net Working Capital	$ 1,400,000

How much must she be prepared to invest before taxes to accomplish the expansion? It's likely that physical facilities and payables would remain proportional to business levels. The distance and delays associated with payment of international accounts could lengthen the payment cycle by 50 to 100%. Transportation delays for international business could require added inventory of 20–30% over normal levels for this segment of the business. Mary estimates the marketing and development costs at $1,000,000.

10-4. Suppose in Problem 10-3 that facilities cost and business development costs are not capitalized and are expensed in the year that they are incurred. Suppose also that the South American market builds within the first year and dies just as rapidly after the 4th full year of business. Mary Ortiz expects to recover 60% of her investment in net working capital at that time. What profit margin must she have to realize a 20% return on her investment in the business after taxes?

10-5. The valuation for Save-A-Buck of Ajax Pool Supplies in Example 10-5 shows that an NPW of $46,000 is attainable. Review the situation as described and determine the changes and risks to consider and assumptions to test. Place yourself in the role of a consultant to Save-A-Buck in assessing this opportunity. Use the framework of Tables 10-5, 10-6, and 10-7 to develop your analysis.

10-6. (Adapted, Horngren) Linkhammer Company has been offered a new special-purpose riveting machine that would automate a manual process for $165,000. This machine is expected to have a useful life of 8 years, with a final salvage value of $45,000. Before-tax savings are expected to be $50,000 annually. Linkhammer has an effective tax rate of 40% and requires a 15% after-tax return on this kind of project. One complication is that additional working capital is needed to keep the machine running efficiently and without exorbitant downtime. This includes spare rivet heads, belts, cleaners, lubricants, and spare parts. These items are replaced as used and require a continuous investment of $8,000 in an inventory of such items. At the conclusion of the machine's useful life, the remaining inventory can be liquidated at 80% of full value. Should the machine be purchased? Consider both the present worth of the cash flows and intangible considerations in making your recommendation.

10-7. (Adapted, Horngren) Waterman Corporation, an apparel producer, has a plant that will become idle at the end of 2006 because of a loss of business to foreign sourcing. There are three options regarding the future of the plant at that point in time:

Option 1. The plant, now fully depreciated, can be sold for $9 million at the time of closing.

Option 2. The plant can be leased to a supplier, Huntsville Industries, for 4 years at a rent of $2.4 million per year, payable monthly. Further, the supplier would discount its monthly bills with a net impact of $40,000 per month. Huntsville would be responsible for maintenance, taxes, and insurance during the term of the lease. At that time, the building could be sold for an estimated $2.0 million.

Option 3. Waterman could secure a contract to manufacture souvenir jackets. Keeping the plant open would incur "fixed" overhead costs of $200,000 annually that would not be required in either of the other two options. Jackets would sell for $42, and variable costs of production would amount to $33 per jacket. Refurbishing the plant and its equipment would require $1.5 million up front, but would boost the value of the plant in four years to $3.0 million. Expected annual sales and production of jackets are as follows:

Year	Sales
2007	200,000
2008	300,000
2009	400,000
2010	100,000

Waterman uses an after-tax MARR of 12.57% per year (1% per month) and pays taxes at an effective rate of 40%. Which option would you recommend? Consider both the present worth of the cash flows and intangible considerations in making your recommendation.

10-8. Wingo Corporation has just hit with its latest fad product and anticipates a sharp rise in sales over its short product life. To capitalize on this new opportunity, they are anticipating the purchase of an additional molding machine for $200,000. New molds and installation will add another $20,000. The opportunity will require additional current assets of $50,000 and added current liabilities of $30,000. Both the fad and the machine have a life of about 4 years. Wingo expects that the machine will have no salvage value in excess of its removal and disposition costs and that all working capital will be recovered.

Additional Sales in the four years are expected as follows:

Year	Sales
1	$200,000
2	$250,000
3	$125,000
4	$75,000

Operating expenses are entirely variable and are expected to be consistent at 20% of sales. Wingo's effective tax rate is 40%. Should they make the investment to capture the market potential if their after-tax MARR is 15%?

10-9. Suppose, in Problem 10-8, that the IRS classifies the molding machine as a 3-year MACRS asset. (a) What is your evaluation? (b) How would it affect your recommendation? (c) Do you need the answer to (a) in order to answer (b)? (d) Answer (b) anyway.

10-10. Returning to the scenario of 10-8, suppose that the impact of the expansion on working capital had not been considered. (a) How would that have affected your evaluation? (b) How would that have affected your recommendation? (c) Do you need the answer to (a) in order to answer (b)?

10-11. Having successfully completed your graduate studies and parlayed your life's dream for developing an affordable hearing aid based on nanotechnology into a sale of your company that has made you independently wealthy, you are bored to tears. An investment banker approaches you with a patented design for a low-cost defibrillator that would save many lives if deployed in offices and stores globally. Intrigued by the prospect, you obtain the following information from the bank's accountant regarding current operations:

Total Revenue	$2,000,000
Cost of Goods Sold	1,000,000
SGA Expense	200,000
EBITDA	800,000
Depreciation and Amortization	200,000
Current Assets	400,000
Current Liabilities	160,000
Long-Term Debt	500,000
Book Value of Assets	1,800,000

On the basis of your experience, you believe that the year-to-year revenue growth will be 10%, 15%, 20%, 25%, and 30%, respectively, over each of the next 5 years, before stabilizing at a 15% growth rate. You expect that the Cost of Goods Sold and SGA, as a percentage of revenue will remain at current levels. You estimate depreciation and amortization, capital expenditures, and net working capital to grow at 10% of revenues annually. Your tax bracket is 40% and you want a 20% after-tax return on a project with this degree of risk. (a) What is the maximum amount you are willing to pay for these assets? (b) How does this compare with the book value of the assets? (c) How does this compare with the value of the equity?

10-12. In HD Motorsports (Example 10-3), working capital to support the capacity expansion was shown as occurring at the end of year 1. Suppose it was needed at the beginning of year 1 (time 0). (a) What would be the impact of that requirement on the expansion project NPV? (b) Show that, under the assumption in (a), the impact on project NPV of working capital that is fully recovered at the end of the project is simply the present value of the interest on funds tied up in working capital. (c) Returning to the original assumption of showing working capital at the end of the 1st year, suppose that only 80% of the working capital is recovered in the terminal year. What is the impact on NPV? (d) How might you decide on which assumptions make the most sense?

APPENDIX 10-A

Sample Forms for Summary Presentations

Exhibits 10-A-1, -2, and -3 provide excellent examples of well-designed forms for summarizing the results of thorough "Special Expenditure Request" analyses, often for new products or expansions. They were developed and provided courtesy of the Eastman Chemical Company.[2] The numbers shown are for an actual project, but the project description and assumptions, etc., are not shown for confidential business reasons.

Exhibit 10-A-1 is the main summary page, which sometimes would include a place for various persons to "sign off," if desired. Note how many different measures of economic desirability are summarized in the upper middle part of the form. Also, a spreadsheet on the lower left displays many parameters over time. Exhibit 10-A-2, labeled "Performance Commitment," also contains a time-tracking spreadsheet with parameters, which would be useful primarily for operating managers.

Exhibit 10-A-3 shows very effective uses of graphs for time-tracking, as well as for probabilistic and sensitivity studies. Such studies are described further in Chapters 12 and 13. The forms used by Eastman make good use of color printouts, which facilitate even easier reading and understanding than are possible here.

[2] Special thanks are due to Officers of the Eastman Chemical Company and to Mr. William Fortenberry, Jr., Director of Corporate Investment Analysis.

Eastman

SER/CP#: 9999

Business Organization: *Eastman Corporate*

Draft

Project Name: **Example**
Print Date 1/12/04 7:53 PM
Project Stage: **Interim Project Definition**

Project Description:
Describe project, what it is, what is involved, etc.

Capital:		
This Project Request:	$	94.0
Other Related Projects:	$	8.0
Allocated Support Capital:	$	18.4
Total:	$	120.4
Net Working Capital:	$	63.4

Eastman Corporate Information   Business: SBU:

Pres./GM/Div. Supt.:	Employees:	
Financial Manager:	LTM Revenue:	
SBU Representative:	Three year CAGR:	
Tech./Mfg. Rep.:	LTM EBIT:	
Manufacturing Div.:	Three year CAGR:	

Business Rationale: Explain business reason for project and how value is created through this project.

Project Parameters:

Hurdle Rate:		12.5%
IRR:		20%
Expected NPV:	$	72.6
CROC:		26%
Profitability Index:		0.60
Cash Payback Year:		6 Yrs
NPV Payback Year:		9 Yrs
Capital Intensity:		3.5
Revenue at Maturity:	$	422.2
EBITDA at Maturity:	$	100.5
LT Avg EBITDA%:		20%

Project Profile:

Yes	Is this an approved Corporate Need?
Yes	Is this part of an approved Bus. Org. strategy?
Yes	Does this provide a new technology platform?
No	Does this provide a new service platform?
No	Is this an entree into new markets?
Yes	Is this an entree into new geographic regions?
	Other Info:

Critical Assumptions: Initials

Assumption A

Assumption B

Assumption C

Capital Spending per Projections

Capacity

Other

Distribution:

BO GM/Div. Supt.:	Bus. / Mfg. Champion:
Dir. BG Finance.:	Comptroller:
	Project Manager:

Performance Commitment Review Date: **Q4/2005**   Evaluation Date: **11/31/2003**
Prepared by: **Bill Fortenberry**   Tel.: **2512**   Revision Date: **None**

Fixed Asset Location(s): Location of the affected fixed assets?

Summary	2004	2005	2006	2007	2008	2009
SER Capital:	$ 62.0	$ 32.0	$ -			
Support Capital:	$ -	$ 16.0	$ 10.4			
Working Capital:	$ -	$ 12.0	$ 17.9	$ 19.4	$ 14.2	$ (0.1)
Totals:	$ 62.0	$ 60.0	$ 28.3	$ 19.4	$ 14.2	$ (0.1)
Revenue:	$ -	$ 80.0	$ 199.1	$ 328.1	$ 422.5	$ 422.0
Gross Profit:	$ -	$ 4.8	$ 28.7	$ 70.6	$ 101.9	$ 99.0
Gross Margin:	na	6.0%	14.4%	21.5%	24.1%	23.5%
R&D%:	na	3.0%	3.0%	3.0%	3.0%	3.0%
SGA%:	na	6.0%	6.0%	6.0%	6.0%	6.0%
EBIT:	$ (9.0)	$ 8.3	$ 38.0	$ 72.9	$ 91.1	$ 74.9
EBITDA:	$ -	$ 4.8	$ 28.7	$ 70.6	$ 101.9	$ 99.0
EBITDA%:	na	6.0%	14.4%	21.5%	24.1%	23.5%
Cash Flow:	$ (67.2)	$ (62.1)	$ (14.8)	$ 21.3	$ 46.9	$ 59.2

Exhibit 10-A-1

Eastman — Performance Commitment

SER # 9999

Print Date: 11/31/2003

Date for post-completion assessment: Q4/2005

Project Name: Example							
Bus. / Mfg. Champion:	Hurdle Rate: 12.5%	IRR: 20%	NPV: $72.6				Eastman Corporate
		P. Index: 0.60	CROC: 26%				

Key Assumptions Crucial to Achieving Commitments

	Initials	Signature
Pricing		
Market/Volume Growth		
Competitive Advantages		
Capital Spending per Projections		
Capacity		
Other		

Key Commitments

		2003	2004	2005	2006	2007	2008	Total
					Forecast			
SER Capital Cost		$ 62.0	$ 32.0	$ -	$ -	$ -	$ -	$ 94.0
Minimum Annual Capacity	Thousands of KGs	0.0	150.0	550.0	550.0	550.0	550.0	550.0
Sales Volume	Thousands of KGs	0.0	100.0	250.0	400.0	500.0	500.0	
Average Price	US$ per KG	$ 0.85	$ 0.80	$ 0.80	$ 0.82	$ 0.84	$ 0.84	
Cash Conversion Cost	US$ per KG	na	$ 0.28	$ 0.21	$ 0.17	$ 0.17	$ 0.17	
Raw Material Content Cost	US$ per KG	$ 4.89	$ 0.40	$ 0.40	$ 0.40	$ 0.40	$ 0.40	
Gross Margin	Percent	na	19%	28%	31%	31%	27%	
Other key factor:								
Other Commitments:	Quality (Fitness for use in markets served)							
	Safety (All corporate standards met or exceeded)							
	Other:							

	Initials	Signature

Project Description:

Describe project, what it is, what is involved, etc.

Financial Manager Bill Fortenberry

BO VP/GM, Division Superintendent

Exhibit 10-A-2

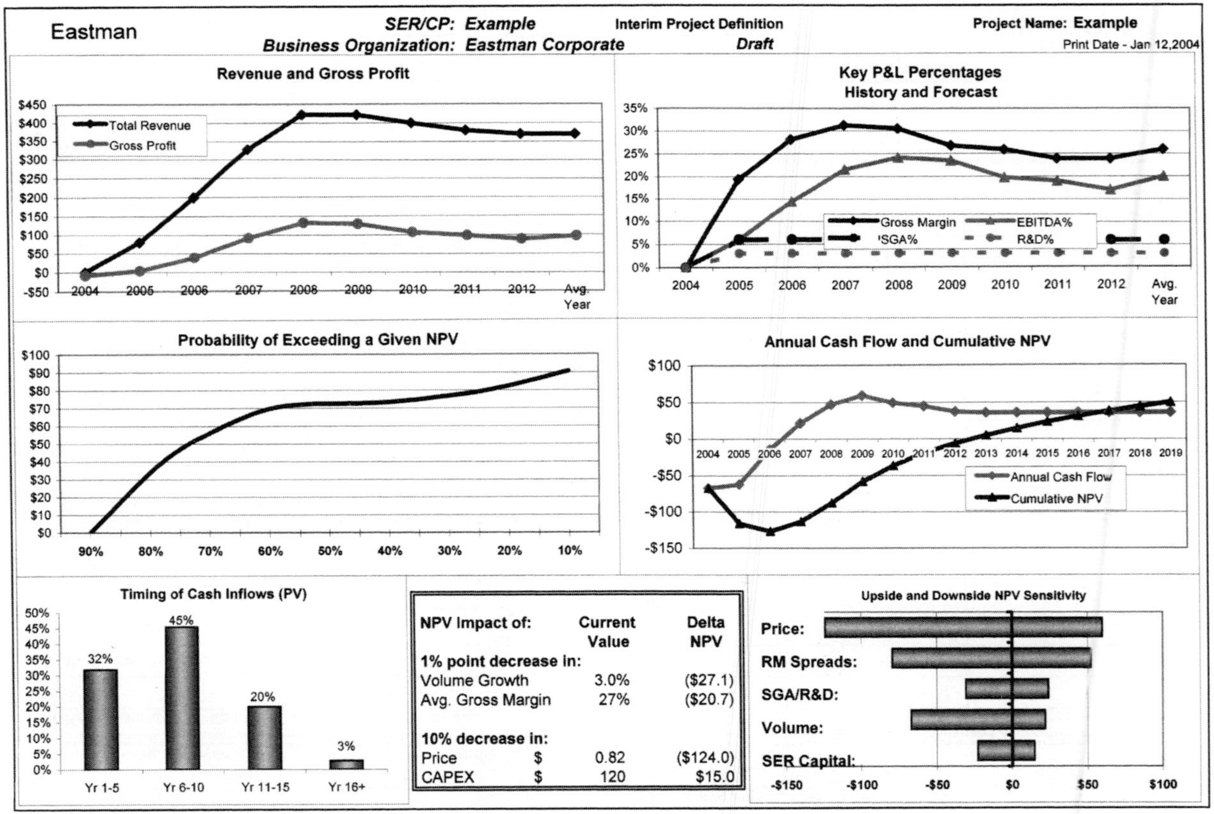

Exhibit 10-A-3

11
Capital Planning and Budgeting

11.1 Introduction

Previous chapters in Part Two have discussed a systems analysis framework for the evaluation of capital investment projects. The culmination of that effort is summarized in the process outlined in the conclusion to Chapter 10. In this chapter, we return to the "total company" level of Part One to study capital investment decision processes employed at that level. These processes include obtaining funding for capital investment projects, allocating those funds to competing investment proposals, and managing the capital investment evaluation process.

An entrepreneurial firm must obtain capital funds from investors and lenders (*capital financing*) and then invest these funds in equipment, tools, and other resources (*capital allocation*) to produce goods and services for sale. The revenues from the engineering and other capital projects involved must earn an adequate return on the funds invested in terms of profit (additional wealth) if the firm is to achieve economic growth and be competitive in the future. Thus, the decision by a company to implement an engineering project involves the expenditure of current capital funds to obtain future economic benefits, or to meet safety, regulatory, or other operating requirements. This implementation decision is normally made, in a well-managed company, as part of a capital budgeting process. The capital financing and allocation functions are primary components of this process.

11.1.1 Capital Financing and Allocation

The capital financing and allocation functions are closely linked, as illustrated in Figure 11-1, and they are simultaneously managed as part of the capital budgeting process. The amounts of new funds needed from investors and lenders, as well as funds

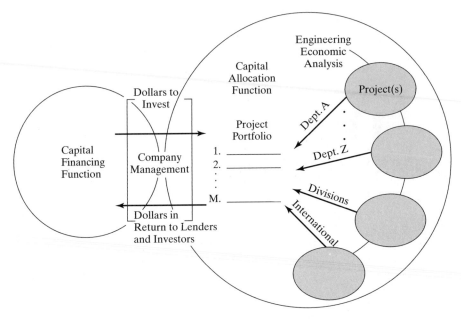

Figure 11-1 Overview of capital financing and allocation activities in a typical organization.

available from internal sources to support new capital projects, are determined in the capital financing function. Also, the *sources* of any new externally acquired funds—issuing additional stock, selling bonds, obtaining loans, and so on—are determined. These amounts, in total, as well as the ratio of debt to equity capital, must be commensurate with the financial status of the firm and balanced with the current and future capital investment requirements.

The selection of the engineering projects for implementation occurs in the capital allocation function. The total capital investment in new projects is constrained by the amount decided upon for this purpose during the capital financing considerations. The capital allocation activities begin in the various organizations in the company—departments (say, engineering), operating divisions, research and development, and so on. During each capital budgeting cycle, these organizations plan, evaluate, and recommend projects for funding and implementation. Economic and other justification information is required with each project recommendation. Engineering economy studies are accomplished as part of this process to develop much of the information required.

As shown in Figure 11-1, the available capital is allocated among the projects selected on a companywide basis in the capital allocation function. Management, through its integrated activities in both functions, is responsible for ensuring that a reasonable return (in dollars) is earned on these investments so providers of debt and equity capital will be motivated to furnish more capital when the need arises. Thus, it should be apparent why the informed practice of engineering economy is an essential element in the foundation of an organization's competitive culture.

In sum, the capital financing and allocation functions in Figure 11-1 are closely linked decision processes regarding *how much* and *where* financial resources will be obtained and expended on future engineering and other capital projects to achieve economic growth and to improve the competitiveness of a firm.

11.1.2 Sources of Capital Funds

Outside sources of capital include, for example, equity partners, stockholders, bondholders, banks, and venture capitalists, all of whom expect a return on their investment. If a company seeks to obtain new outside capital for investment in the business, it must attract new investors. It does so by paying a rate of return that is attractive in a competitive market in which many differing securities are offered. The variety of securities available and their special provisions are limited only by human creativity. A complete exploration is beyond the scope of this text, but an understanding of the cost of capital is a key concept. Hence, the theory of capital cost is developed for the most important equities and debts or liabilities that form the capitalization, or total capital base, for a company. Here, we focus on those financing concepts that guide the capital planning phase, as compared to the specific details needed in the execution phase (i.e., the actual acquisition of funds). Many refinements to the models we develop are appropriate when considering an action such as selling equity shares (stocks) or issuing new bonds. These kinds of actions and more complex models lie in the realm of finance and are not explicitly detailed here. The fundamental models presented in this chapter convey the understanding needed by engineers and nonfinancial managers. Those readers interested in the specifics of the capital acquisition process and more complex models should look to the references provided at the end of the chapter for details and procedures involved in the capital acquisition process.

Some perspective on the costs of various sources of capital can be gained by examining the returns provided to investors for various kinds of securities. To provide some stable benchmarks, it is helpful to review historical values of rates of returns on various securities. A widely circulated reference is the yearbook, published by Ibbotson Associates,[1] which tracks the long-term performance of five different portfolios of securities:

1. A portfolio of treasury bills
2. A portfolio of long-term U.S. government bonds
3. A portfolio of long-term corporate bonds
4. The S&P 500 index of common stocks
5. A portfolio of the common stocks of small firms

Information from 1926 to 2002 is compiled in Table 11-1. Returns fluctuate significantly over time, but the long-term perspective used by Ibbotson provides some valuable insights. For example, it is obvious that the rates earned by investors increase as we move down the rows in Table 11-1 toward increasingly risky securities. The standard deviation of annual returns is our proxy measure of *risk*. Since the averages

---

[1] *Stocks, Bonds, Bills and Inflation Yearbook*, R.G. Ibbotson Associates, Chicago, IL, 2003.

TABLE 11-1   Rates of Return from 1926-2002 on Five Types of Securities

Type of security	Average annual return, R_m	Standard deviation of returns
Treasury bills	3.8%	4.4%
Long-Term U.S. Bonds	5.8	9.4
Corporate Bonds	6.2	8.7
S & P 500 Stocks	12.2	20.5
Small-Firm Stocks	16.9	33.2

Source: R.G. Ibbotson Associates, Chicago, IL, 2003.

are the annual returns that investors expect, they are indicative of the rates that a company must offer in order to attract investors. During this same period, the average annual rate of inflation has been 3.1%. Investors, of course, are very aware of inflation, and the rates in the table are therefore nominal annual returns, already adjusted for the inflation expected by the marketplace.

11.2 Debt Capital

One source of capital is borrowed money. The money may be loaned by banks as a line of credit, or it may be obtained from the sale of bonds or debentures. The return to the buyer or bondholder is the interest on the bond and the eventual return of principal. In order to market the bonds successfully, the return must be attractive to buyers. The attractiveness will depend upon the interest rate offered and the perceived riskiness of the offering. In the United States, two major investment companies, (1) Moody's and (2) Standard and Poor's Investment Services, publish ratings of various bond issues on the basis of their analyses of each company's financial health. The higher the rating, the lower the interest rate required to attract investors. The four highest rating categories are associated with investment grade bonds; the lowest, with "junk" bonds. The junk bonds have a significant risk of default and therefore offer higher interest rates to attract investors. Many investors pool that risk by developing investment portfolios that include such risky bonds. At the other end of the spectrum, treasury bills are backed by the U.S. government and are regarded as the safest investment, because they have never defaulted. As might be guessed, the interest rate paid by the government for "T-bills" is very low. The current treasury bill rate is often used as a proxy for the risk-free rate of return, R_F.

11.2.1 Tax Deductibility of Debt

Regardless of the rate that companies pay for the money they borrow, the interest that they pay is tax deductible. This tax deductibility results in a tax savings which generally should be considered in engineering economic analysis of capital investments. There are basically three ways to accomplish this.

One way would be first to compute the interest during each year and deduct it from the project cash flows before computing income taxes and next to consider interest and principal reductions to the cash outflows in lieu of the lump sum investment amount. This adds a computational chore, as the interest amount changes from period to period with level debt service or uniform annual amounts. More importantly, it is generally not easy to assign debt repayment to a specific project, as debt is determined at the corporate, not the project, level. In fact, it is undesirable to attach a specific financing package to an individual project, because costs of capital are tied to the overall financial structure of the firm as a whole and are seldom attached to individual projects. This approach is usually not recommended or used because of companywide issues. An exception occurs when a large capital asset is uniquely considered with especially favorable financial terms. An example is given in Appendix 11-B, entitled "Bundled Financing."

A second way to include the tax deductibility of interest would be to calculate the present worth of the interest tax shield (savings) and adjust the present worth of the project accordingly. The interest tax shield in any year is the product of the interest amount and the effective income tax rate t. The present worth of the tax shield can then be added to the PW as a "side benefit" resulting from the use of debt financing. This approach is gaining increasing popularity as a means of analyzing large and complex projects. It is known as "Adjusted Present Worth" or "Adjusted Present Value," and it is discussed in Appendix 11-A.

The third, and most commonly used, approach is to modify the cost of debt to account for its tax deductibility in the interest rate (MARR) used. To make it clear that the interest tax shield is being picked up in the discount rate, rather than in the cash flows, financial analysts sometimes modify the descriptors for the elements of after-tax cash flow. Specifically, in year k, let

ATCF_k = after-tax cash flow (excluding interest on debt),

NOPAT_k = the net operating profit after taxes, and

EBIT_k = the earnings before interest and taxes.

Then in year k, we have

$$\text{ATCF}_k = \text{NOPAT}_k + d_k = (1-t)\,\text{EBIT}_k + d_k. \tag{11-1}$$

When calculating the weighted average cost of capital (WACC), as in Section 11.4, the cost of debt capital (i_b) is offset by the tax deduction it provides, and the *after-tax* cost of a company's debt, as a percent of the amount borrowed, is $(1-t)\,i_b$, where

i_b is the cost of debt as an annual rate and

t is the effective income tax rate.

11.3 Equity Capital

Equity sources of funding include not only stockholder's capital, but also earnings retained by the company for reinvestment in the business and the cash flow resulting from depreciation charges against income. Just as with bonds, the percentage cost of equity funds e_a can be determined by market forces whenever new equity is issued. However, this is an infrequent event. Furthermore, the cost of internal equity funds should be based upon opportunity costs associated with the best use of those funds within the firm. Here, on the basis of historical performance as proxy for anticipated costs, we must resort to some means of infering that cost. For example, one way might be to consider the trend in past values of return on equity, ROE. If these have been satisfactory, that might be an indication of the value of e_a. Another approach is to look at the return demanded by the shareholders; however, this fluctuates widely and creates an estimating challenge. Table 11-1 shows representative long-term values for e_a, but these are not especially useful as indicators at a given point in time. We need a way to obtain a fairly current value and yet to remove the impact of daily and other fluctuations. When such equity is issued as common stock, a theoretical construct, the capital asset pricing model (CAPM), offers a fruitful approach.

11.3.1 The Capital Asset Pricing Model (CAPM)

The CAPM appeared in the 1960s as a classic piece of economic research, developed independently by three eminent economists. Their work is based upon the seminal portfolio theory developed by Harry Markowitz.[2] It forms the basis for much of contemporary thought about risk and return. The economic logic underpinning this work is elegant. Dedicated readers can consult the original works cited in the footnote or a good finance text. Our immediate needs are well served with an intuitive explanation of their results as shown in Figure 11-2.

Markowicz showed that the best combinations of risk and return on investment can be achieved by investing in a portfolio that includes a mixture of stocks. Further, the most efficient portfolios lie on a curve that specifies the maximum return for a given level of risk. This concave-shaped curve is called the "Markowicz efficient market portfolio" and is shown in Figure 11-2. Return is measured as expected return and risk is measured as the standard deviation of returns. An investor could then diversify his or her stock investments to achieve the maximum return, commensurate with the level of risk he or she is willing to tolerate. Or, conversely, the investor could minimize the risk associated with a required return by choosing a portfolio on the Markowicz curve as shown in Figure 11-2.

By using some rather idealistic assumptions and sound economic reasoning, the CAPM asserts that the best combinations of return and risk must lie along a straight line called the security market line, SML. That line is established by the risk-free rate R_F and the point of tangency with the Markowicz efficient market portfolio. That is, rather than choosing an equity portfolio on the basis of risk and return preference, investors can always achieve the best combination of risk and return by investing in the efficient market portfolio and then borrowing or lending at the risk-free rate to

---

[2] This work first appeared in the article, "Portfolio Selection," *Journal of Finance*, Vol. 7, pp. 77–91, March, 1952. The articles that developed the CAPM include W.F. Sharpe, "Capital Asset Prices: A Theory of Market Equilibrium under Conditions of Risk," *Journal of Finance*, Vol. 19, pp. 425–442, September 1964; J.Lintner, "The Valuation of Risk Assets and the Selection of Risky Investments in Stock Portfolios and Capital Budgets," *Review of Economics and Statistics*, Vol. 47, pp. 13–37, February 1965.

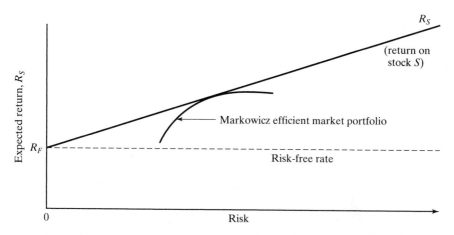

Figure 11-2   Markowitz efficient portfolio and the security market line.

achieve their personal preference for combined risk and return values. Because of this ability, any security must lie on the security market line. As shown in Figure 11-3, the SML can be used to develop a standardized estimate of the risk and hence, the return associated with any security.

By using this model, risk is calibrated to the risk associated with a market port-folio (i.e., a portfolio consisting of all the stocks in the market). The return on the mar-ket portfolio R_M is keyed to a level of risk that we will call β_M and assign a value of 1.0. This is one point on the SML, which identifies the market return R_M and the level of market risk β_M. The risk-free rate R_F and a beta value of zero indicate that the risk-free rate does not fluctuate with the market.

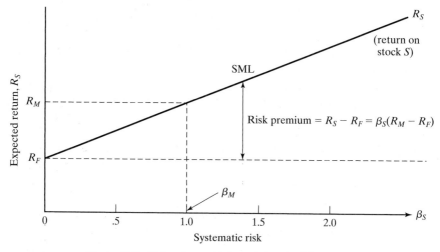

Figure 11-3   Risk and return in the capital asset pricing model.

The CAPM asserts that the return R_S on any stock S depends upon its risk relative to the market. That risk, β_S, is the stock's contribution to the riskiness of the market portfolio. CAPM suggests that the risk premium of any stock is proportional to its beta as follows:

$$R_S - R_F = \beta_S(R_M - R_F). \tag{11-2}$$

Equation 11-2 states that the quantity $(R_S - R_F)$ is the current *risk premium* associated with a stock S relative to a risk-free investment. The quantity $(R_M - R_F)$ is the *market premium* for the average stock market risk. Over many years, the market premium has steadily hovered at 8.4% above the risk-free rate, as suggested in Table 11-1 and confirmed in studies showing an annual range between 8% and 10%. In this chapter, the long-term average market premium of 8.4% is used.

β_S is the beta value of stock S. It is a widely published and periodically updated statistic for any stock. It is measured by regressing changes in the price of the stock against changes in the market index. The slope of that regression line is labeled as beta and is a widely used measure of a stock's riskiness. Many investment advisory services compute and publish estimates of beta.

11.3.2 Estimating the Cost of Equity

The bottom line of all this is that we can estimate the cost of equity e_a as the return on equity R_S required in the market by solving Equation 11-2.

Example 11-1
Microsoft has a beta value of 1.68. They have no long-term debt. What is their cost of equity based upon the CAPM?

Solution   Using CAPM, a long-term market premium $(R_M - R_F)$ of 8.4% and an estimated risk-free rate (R_F) of 2% in 2002, we see that

$$e_a = R_S = R_F + \beta_S\ (R_M - R_F) = 0.02 + 1.68(0.084) = 0.161 = 16.1\%.$$

In 2002, Microsoft reported a return on equity (ROE) of 15.7%. This is consistent with the preceding estimate.   ∎

11.4 The Weighted Average Cost of Capital (WACC)

The WACC is the product of the fraction of total capital from each source and the cost of capital from that source, summed over all sources. To illustrate the concept, we will simply look at debt and equity sources and their respective costs. This usually provides an adequate figure for planning. Again, a more precise estimate can be developed in specific situations. For our purposes, Equation 7-13 is used to compute a firm's WACC, and it is repeated here for the reader's convenience.

Let

$$\lambda = \text{the fraction of the total capital}$$
$$\text{obtained from debt,}$$
$$(1-\lambda) = \text{the fraction of the total capital}$$
$$\text{obtained from equity,}$$
$$t = \text{effective income tax rate as a decimal,}$$
$$i_b = \text{the cost of debt financing, as measured}$$
$$\text{from appropriate bond rates, and}$$
$$e_a = \text{the cost of equity financing, as measured}$$
$$\text{from historical performance of the CAPM.}$$

Then,

$$\text{WACC} = \lambda(1-t)i_b + (1-\lambda)e_a. \tag{11-3}$$

Equation 11-3 says that the average cost of funds is a weighted average of the costs of each of its capital sources. Further, if we assess or value investment projects by discounting their returns at the WACC, then any project with a PW ≥ 0 provides value in excess of the cost of the capital required to accomplish it.

Example 11-2

What is the weighted average cost of capital for Microsoft? What is the WACC for Duke Energy?

Solution for Microsoft: Since Microsoft has no debt, it is 100% equity financed. Therefore,

$$\text{WACC} = e_a = R_S = 16.1\%, \text{ as calculated in Example 11-1.}$$

Solution for Duke Energy   Since the Enron debacle, energy companies have been viewed cautiously by investors. Research sources show that Duke Energy's stock prices have been more stable than most, with a beta of 0.32, but their post-Enron bond rating of BBB has become much lower than the pre-Enron value. Their effective income tax rate is 0.35. As of 2003, Duke's balance sheet shows $20 billion in debt and roughly $15 billion in equity. But there are 900 million shares outstanding and the price per share is approximately $20, resulting in a market value of the equity of $18 billion. Long-term bonds rated BBB currently earn 6% per year. For Duke energy,

$$\lambda = \$20 \text{ billion} \div (\$20 \text{ billion} + \$18 \text{ billion}) = 0.526,$$
$$i_b = 0.06,$$
$$e_a = R_S = R_F + \beta_S (R_M - R_F) = 0.02 + 0.32 \ (0.084) = 0.047,$$
$$t = 0.35, \text{ and}$$
$$\text{WACC} = 0.526 \ (1 - 0.35)(0.06) + 0.474 \ (0.047) = 0.043 = 4.3\%.$$

Notice that the market value of the equity ($18 billion) rather than the book value ($15 billion) is used to compute λ, the debt–equity ratio. This reflects the relevance of current market information vis-a-vis the historical book value of equity. Market values of debt do not fluctuate as dramatically as equity values.   ■

11.4.1 The Separation Principle

A related point is that, for every firm, there is an optimal mix of debt and equity financing that minimizes the WACC to that firm. The task of the company treasurer is to identify that mix and maintain it as a permanent part of the capital structure of the company. One difficulty is that it is not easy to determine or maintain the optimal mix. Because any new bond or stock issue incurs underwriting and marketing costs, there is a fixed charge of "going to the well." New capital is procured in "chunks," rather than in small amounts that would permit close adherence to the optimal ratio. It does not make business sense for the hurdle rate (MARR) for investments to be low when the cost of capital is low because this year's projects are to be funded with debt. A more consistent long-term view is obtained by evaluating investment projects at the WACC or the MARR, regardless of how they are being financed. This concept is referred to as the *separation principle*. This, in effect, separates the investment decision from the financing decision and requires projects to have a PW ≥ 0, at the WACC or MARR; or an IRR $\geq$ WACC or MARR, thus creating value for the investors.

11.4.2 WACC and Risk

The economic theory behind CAPM was hardly needed to persuade experienced managers that some capital projects are riskier than others, and that prudence required a higher return (or very conservative estimates of the cash flows) for riskier projects. In reality, every project should be evaluated on the basis of the risk it adds to the company as well as its return. If the risks are roughly "average" and if there are no significant capital limitations, then WACC is an appropriate hurdle rate (i.e., the MARR). If the project is more risky than the current business, then an upward adjustment in the MARR might be appropriate. For instance, in the examples of Chapter 10, HD Motorsports might use its WACC as the MARR, because its capacity expansion has more or less the same risks as its overall business. Ajax Chemicals may use a higher MARR in evaluating its new product venture. As intuitively appealing as this may be, the problem becomes one of operationalizing the rate adjustment. An alternative is to use more formal techniques of risk assessment. Part Three of this book is dedicated to that important task.

11.4.3 Relation of WACC to the MARR

Thus far, we have been comparing investment alternatives, using the MARR as the hurdle rate. The WACC merely establishes a floor on the MARR. In many circumstances, a higher MARR than the WACC may be chosen, because of a shortage of investment funds as well as differences in risk.

The theory developed so far has implicitly assumed that the firm can obtain as much capital as it needs at the WACC, and thus should go forward with any normal risk project that has a positive PW when evaluated at the WACC. In the real world, however, opportunities often occur in lumps rather than in a continuous stream. Management may wish to conserve capital in anticipation of a large future opportunity, or it may wish to encourage investment in a new division while discouraging investments in mature product lines. For that reason, the MARR may be deliberately raised above the WACC in some years or for some divisions. There may be instances in which unusually risk-free cash flows could be evaluated at a MARR lower than the WACC.

11.4.4 Opportunity Costs and Risk Categories in Determining MARRs

A commonly overlooked viewpoint on the determination of the MARR is the *opportunity cost* viewpoint; it comes as a direct result of the capital rationing necessary when there is a limitation of funds relative to prospective proposals to use the funds. This limitation may be either internally or externally imposed. Its parameter is often expressed as a fixed sum of capital; but when the prospective returns from investment proposals, together with the fixed sum of capital available to invest, are known, then the parameter can be expressed as a minimum acceptable rate of return, or cut-off rate.

Ideally, the cost of capital by the opportunity cost principle can be determined by ranking prospective projects of similar risk according to a ladder of profitability and then establishing a cut-off point such that the capital is used on the better projects. The return earned by the last project before the cut-off point is the minimum attractive rate of return by the opportunity cost principle.

To illustrate the preceding, Fig. 11-4 ranks projects according to prospective internal rate of return and the cumulative investment required. For purposes of illustration, the amount of capital shown available is $4 million. By connecting up (to the next whole project within the $4 million) and across, one can read the minimum rate of return under the given conditions, which turns out to be 25%.

It is not uncommon for firms to set two or more MARR levels, according to risk categories. For example, one major industrial firm defines risk categories for income-producing projects and "normal" MARR standards for each of these catagories as follows:

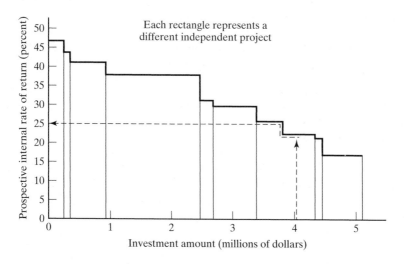

Figure 11-4   Schedule of prospective returns and investment amounts.

1. *High Risk* (MARR = 40%)
   New products
   New business
   Acquisitions
   Joint ventures

2. *Moderate Risk* (MARR = 25%)
   Capacity increase to meet forecasted sales

3. *Low Risk* (MARR = 15%)
   Cost improvements
   Make versus buy
   Capacity increase to meet existing orders

To illustrate how the preceding set of MARR standards could be determined, the firm could rank prospective projects in each risk category according to prospective rates of return and investment amounts. After tentatively deciding how much investment capital should be allocated to each risk category, the firm could then determine the MARR for each risk category, as illustrated in Figure 11-4 for a single category. Of course, the firm might reasonably shift its initial allocation of funds according to the opportunities available in each risk category, thereby affecting the MARR for each category allocated more or fewer investment funds.

In principle, it would be desirable for a firm to invest additional capital as long as the return from that capital were greater than the cost of obtaining that capital. In such a case, the opportunity cost would equal the marginal cost (in interest and/or stockholder returns) of the last capital used. In practice, however, the amount of capital actually invested is more limited due to risk and conservative money policies; thus, the opportunity cost is usually higher than the marginal cost of the capital.

It is often reasonable to argue that, since one cannot truly know the opportunity cost for capital in any given period (such as a budget year), it is useful to proceed as if the MARR for the upcoming period were the same as in the previous period. In addition to the normal difficulty of projecting the profitability of future projects and the availability of capital, there also may be pressures to manipulate the standards of acceptability to permit the approval of favored, even if economically undesirable, projects or classes of projects.

11.5 Project Selection

To the extent that project proposals can be justified through profitability measures, the most common basis of selection is to choose those proposals that offer the highest prospective profitability, subject to allowances for intangibles or nonmonetary considerations, risk considerations, and limitations on the availability of capital. If the minimum acceptable rate of return has been determined correctly, one can choose proposals according to a rate of return method, annual worth method, or present worth method.

For certain types of project proposals, monetary justification is not feasible—or at least any monetary return is of minor importance compared with intangible or nonmonetary considerations. These types of projects should require careful judgment and analysis, including how they fit in with long-range strategy and plans. Factor-weighting methods, such as those discussed in Chapter 19, are particularly suited to projects for which monetary justification is not feasible.

The capital budgeting concepts discussed in this chapter are based on the presumption that the projects under consideration are *not* mutually exclusive (i.e., the adoption of one does not preclude the adoption of others, except with regard to the availability of funds). Whenever projects are mutually exclusive, the alternative should be chosen on the basis of justification through the incremental return on any incremental investment(s), as well as on proper consideration of nonmonetary factors.

11.5.1 Classifying Investment Proposals

For purposes of study of investment proposals, there should be some system or systems of classification into logical, meaningful categories. Investment proposals have so many facets of objective, form, and competitive design that no one classification plan is adequate for all purposes. Several classification plans are possible according to the following criteria:

1. the kinds and amounts of scarce resources used, such as equity capital, borrowed capital, available plant space, the time required of key personnel, etc.;

2. whether the investment is tactical or strategic—a *tactical investment* does not constitute a major departure from what the firm has been doing in the past and generally involves a relatively small amount of funds; *strategic investment* decisions, on the other hand, may result in a major departure from what a firm has done in the past and may involve large sums of money;

3. the business activity involved, such as marketing, production, product line, warehousing, etc.;

4. priority, such as absolutely essential, necessary, economically desirable, or general improvement;

5. type of benefits expected to be received, such as increased profitability, reduced risk, community relations, employee benefits, etc.;

6. whether the investment involves facility replacement, facility expansion, or product improvement;

7. the way benefits from the proposed project are affected by other proposed projects; this is generally a most important classification consideration, for there quite often exist interrelationships or dependencies among pairs or groups of investment projects.

Of course, all of the preceding classification systems probably are not needed or desirable. As an example, one major corporation uses the following four major categories for higher management screening:

1. expanded facilities;

2. research and development;

3. improved facilities—for process improvement, cost savings, or quality improvement; and

4. necessity—for service facilities, emergency replacements, or for the removal or avoidance of a hazard or nuisance.

11.5.2 Degrees of Dependency among Projects

Several main categories of dependency among projects are briefly defined in Table 11-2. Actually, the possible degrees of dependency among projects can be expressed as a continuum from "prerequisite" to "mutually exclusive," with the degrees "complement," "independent," and "substitute" between these extremes, as shown in Figure 11-5.

In developing a project proposal to be submitted for review and approval, the sponsor should include whatever complementary projects seem desirable as part of a

TABLE 11-2   Degrees of Dependence between Pairs of Projects

"If the results of the first project would _____ by acceptance of second project . . .	. . . then the second project is said to be _____ the first project."	Example
be technically possible or would result in benefits only	a prerequisite of	Car stereo purchase feasible only with purchase of car
have increased benefits	a complement of	Additional hauling trucks more beneficial if automatic loader purchased
not be affected	independent of	A new engine lathe and a fence around the warehouse
have decreased benefits	a substitute for	A screw machine that would do part of the work of a new lathe
be impossible or would result in no benefits	mutually exclusive with	A brick building or a wooden building for a given need

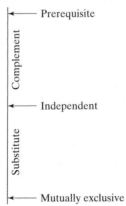

Figure 11-5   Continuum of degrees of dependence between pairs of projects.

single package. Also, if a proposed project will be a partial substitute for any projects to which the firm is already committed or that are under consideration, this fact should be noted in the proposal.

For cases in which choices involved in planning a proposed project are considered sufficiently important that the final decision should be made by higher levels of management, the project proposal should be submitted in the form of a set of mutually exclusive alternatives. For example, if it is to be decided whether to move a plant to a new location, and several alternative sites are possible, then separate proposals should be made for each site so as to facilitate the choice of which site, if any, should be chosen.

Whenever capital budgeting decisions involve several groups of mutually exclusive projects and independent projects to be considered within capital availability constraints, the mathematical programming models presented later in this chapter can be useful for selecting the optimal combination of projects.

11.5.3 Organization for Capital Planning and Budgeting

In most large organizations, project selections are accomplished by sequential review through various levels of the organization. The levels required for approval should depend on the nature and importance of the individual project, as well as on the particular organizational makeup of the firm. In general, a mix of central control and coordination, together with authority to make project commitments delegated to operating divisions, is considered desirable. Three typical basic plans for delegating investment decisions are as follows:

1. Whenever proposals are clearly "good" in terms of economic desirability according to operating division analysis, the division is given the power to commit as long as appropriate controls can be maintained over the total amount invested by each division and as long as the division analyses are considered reliable.

2. Whenever projects represent the execution of policies already established by headquarters, such as routine replacements, the division is given the power to commit within the limits of appropriate controls.

3. Whenever a project requires a total commitment of more than a certain amount, this request is sent to higher levels within the organization. This is often coupled with a budget limitation regarding the maximum total investment that a division may undertake in a budget period.

To illustrate the concept of larger investments requiring higher administrative approval, the limitations for a medium-sized firm might be as follows:

If the total investment is . . .		then approval is required through
more than	but less than	
$ 50	$ 5,000	Plant manager
5,000	50,000	Division vice president
50,000	125,000	President
125,000	—	Board of directors

11.5.4 Communication

The importance of effective communication of capital investment proposals is often overlooked. No matter how great are the merits of a proposed project, if those merits are not communicated to the decision maker(s) in understandable terms, with emphasis on the proper matters, that proposal may well be rejected in favor of less desirable, though better communicated, proposals. Klausner[3] provides good insight into this problem, with emphasis on the differing perspectives of engineers responsible for technical design and proposal preparation, and management decision makers responsible for monitoring the firm's capital resources. The proposal preparer should be as aware as possible of the decision maker's perspective and related informational needs. For example, in addition to basic information such as investment requirements, measures of merit, and other expected benefits, the decision maker may well want clearly presented answers to such questions as the following:

1. What bases and assumptions were used for estimates?
2. What level of confidence does the proposer have regarding these estimates?
3. How would the investment outcome be affected by variations in these estimated values?

If project proposals are to be transmitted from one organizational unit to another for review and approval, there must be effective means for communication. The means can vary from standard forms to personal appearances. In communicating proposals to higher levels, it is desirable to use a format that is as standardized as possible to help assure uniformity and completeness of evaluation efforts. In general, the technical and marketing aspects of each proposal should be completely described in a format that is most appropriate to each individual case. However, the financial implications of all proposals should be summarized in a standardized manner so that they may be uniformly evaluated.

11.6 Postmortem Review

The provision of a system for periodic postmortem reviews (postaudits) of the performance of consequential projects previously authorized is an important aspect of a capital budgeting system. That is, the earnings or costs actually realized on each such project undertaken should be compared with the corresponding quantities estimated at the time the project investment was committed.

This kind of feedback review serves at least three main purposes, as follows:

1. It determines if planned objectives have been obtained.
2. It determines if corrective action is required.
3. It improves estimating and future planning.

---

[3] R. F. Klausner, "Communicating Investment Proposals to Corporate Decision Makers," *The Engineering Economist*, Vol.17, No.1 (Fall 1971): 45–55.

Postmortem reviews should tend to reduce biases in favor of what individual divisions or units preparing project proposals see as their own interests. When divisions of a firm have to compete with each other for available capital funds, there is a tendency for them to evaluate their proposals optimistically. Estimating responsibilities can be expected to be taken more seriously when the estimators know that the results of their estimates will be checked. However, this checking function should not be overexercised, for there is a human tendency to become overly conservative in estimating when one fears severe accountability for unfavorable results.

It should be noted that a postmortem audit is inherently incomplete. That is, if only one of several alternative projects is selected, it can never be known exactly what would have happened if one of the other alternatives had been chosen. "What might have been if . . ." is at best conjecture, and all postmortem audits should be made with this reservation in proper perspective.

11.7 Budgeting of Capital Investments and Management Perspective

The approved capital budget is limited typically to a 1- or 2-year period or less, but this should be supplemented by a long-range capital plan, with provision for continual review and change as new developments occur. The long-range plan (or plans) can be for a duration of from 2 years to 20 years, depending on the nature of the business and the desire of management to force preplanning.

Even when the technological and market factors in the business are so changeable that plans are no more than guesses to be continually revised, it is valuable to plan and budget as far ahead as possible. Planning should encourage the search for investment opportunities, provide a basis for adjusting other aspects of management of the firm as needed, and sharpen management's forecasting abilities. Long-range budget plans also provide a better basis for establishing minimum rate of return figures that properly take into account future investment opportunities.

An aspect of capital budgeting that is difficult and often important is deciding how much to invest now as opposed to later. If returns are expected to increase for future projects, it may be profitable to withhold funds from investment for some time. The loss of immediate return, of course, must be balanced against the anticipation of higher future returns.

In a similar vein, it may be advantageous to supplement funds available for present projects whenever returns for future projects are expected to become less than those for present projects. Funds for present investment can be supplemented by the reduction of liquid assets, the sale of other assets, and the use of borrowed funds.

The procedures and practices discussed in this book are intended to aid management in making sound investment decisions. Management's ability to sense the opportunities for growth and development and to time their investments to achieve optimum advantage is a primary ingredient of success for an organization.

11.8 Leasing Decisions

Leasing of assets is a business arrangement that makes assets available without incurring initial capital investment costs of purchase. By the term *lease,* we normally are referring to the financial type of lease; that is, a lease in which the firm has a legal obligation to continue making payments for a well-defined period of time for which it has use, but not ownership, of the asset(s) leased. Financial leases usually have fixed durations of at least 1 year. Many financial leases are very similar to debt and should be treated in essentially the same manner as debt. A significant proportion of assets in some firms is acquired by leasing, thus making the firm's capital available for other uses. Buildings, railroad cars, airplanes, production equipment, and business machines are examples of the wide array of facilities that may be leased.[4]

Lease specifications are generally detailed in a formal written contract. The contract may contain such specifications as the amount and timing of rental payments; cancellability and sublease provisions, if any; subsequent purchase provisions; and lease renewability provisions.

Although the financial lease provides an important alternative source of capital, it carries with it certain subtle, but important, disadvantages. Its impact is similar to that of added debt capital. The acquisition of financial leases or debt capital will reduce the firm's ability to attract further debt capital and will increase the variability (leverage) in prospective earnings on equity (owner) capital. Higher leverage results in more fixed charges for debt interest and repayment, and thus will make good conditions even better and poor conditions even worse for the equity owners.

Confused reasoning may be introduced into economic studies in which calculations appropriate for justifying long-run economy of proposed investments in physical assets are combined with calculations related to the financing of those assets. It is recommended that analyses (and decisions) such as these be made *separately,* not mixed, whenever possible.

Figure 11-6 depicts the types of analyses that should be made for lease-related decisions and also shows what conclusion (final choice) should be made for various combinations of conditions (analysis outcomes). The "buy" versus "status quo" (do-nothing) decision is to determine long-run economy or feasibility and is sometimes referred to as an *equipment* decision. In general practice, only if this is investigated and "buy" is found to be preferable, should we be concerned with the "buy" versus "lease" question, which is considered a financing decision. However, we cannot always separate the *equipment* and *financing* decisions. For example, if the *equipment* decision results in "status quo" being preferable and yet it is possible that the equipment might be favorable if leased, then we should compare "status quo" with "lease." This, by definition, is a *mixed* decision involving both *equipment* and *financing* considerations.

---

[4] So-called leases with "an option to buy" are treated as conditional sales contracts and are not considered to be true financial leases.

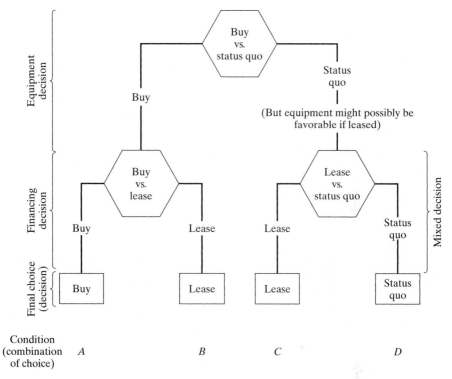

Figure 11-6   Separability of "buy" versus "lease" versus "status quo" decisions, and the choices that should result.

The main point we should retain from the preceding is that we should not merely compare "buy" versus "lease" alternatives; we should also compare choices, if possible, against the "status quo" alternative to determine if the asset is justified under any financing plan.

A major factor in evaluating the economics of leasing versus buying is the tax deductions (reductions in taxable income) allowable. In the case of leasing, the tax payer is allowed to deduct the full cost of normal financial lease payments. In the case of buying (ownership), only depreciation charges and interest payments, if any, are deductible. Of course, other disbursements for operating the property are tax deductible under either financing plan.

Example 11-3

Suppose a firm is considering purchasing equipment for $200,000 that would last for 5 years and have zero market value. Alternatively, the same equipment could be leased for $52,000 at the *beginning* of each of those 5 years. The effective income tax rate for the firm is 55% and straight-line depreciation is used. The net before-tax cash benefits from the equipment are $56,000 at the end of each year for 5 years. If the after-tax MARR for the firm is 10%, use the present worth method to show whether the firm should buy, lease, or maintain the status quo.

Solution

Alternative	Yr	(A) Before-tax cash flow	(B) Depreciation	(C) = (A) – (B) Taxable income	(D) = –0.55(C) Income taxes	(E) = (A) + (D) After-tax cash flow
Buy	0	–$200,000				–$200,000
	1–5		–$40,000	–$40,000	+$22,000	+ 22,000
Lease	0	– 52,000		– 52,000	+ 28,600	– 23,400
	1–4	– 52,000		– 52,000	+ 28,600	– 23,400
Status quo	{1–5	– 56,000		– 56,000	+ 30,800	– 25,200

Thus, the present worths (after taxes) for the three alternatives are as follows:

$$\text{Buy:} \qquad -\$200,000 + \$22,000(P/A, 10\%, 5) = -\$116,600;$$
$$\text{Lease:} \qquad -\$23,400 - \$23,400(P/A, 10\%, 4) = -\$97,600;$$
$$\text{Status quo:} \quad -\$25,200(P/A, 10\%, 5) \qquad\qquad = -\$95,500.$$

If the sequence of analysis follows that in Fig. 11–6, we would first compare "buy" (PW = –$116,600) against "status quo" (PW = –$95,500) and find the status quo to be better. Then we would compare "lease" (PW = –$97,600) against "status quo" (PW = –$95,500) and find the "status quo" to be better. Thus, "status quo" is the better final choice. It should be noted that, had we merely compared "buy" versus "lease," "lease" would have been the choice. But the final decision should not have been made without comparison against "status quo."   ∎

Another way the problem could have been solved would have been to include the +$56,000 before-tax cash benefits with both the "buy" and the "lease" alternatives. Then the two alternatives being considered would have been "buy rather than status quo" (PW = –$116,600 + $95,500 = –$21,100) and "lease rather than status quo" (PW = –$97,600 + $95,500 = –$2,100). Of these alternatives, "lease rather than status quo" is the better, but "lease" is still not justified because of the negative PW, indicating that the costs of "lease" are greater than the benefits.

11.9 Capital Allocation

We have been discussing capital financing topics in Sections 11.2 through 11.8 that deal with (1) how a company obtains capital (and from what sources) and (2) how much capital the company has available, and at what cost, to maintain a successful business enterprise in the years ahead.

An outstanding phenomenon of present-day industrialized civilizations is the extent to which engineers and managers, by using capital (money and property), are able to create wealth through activities that transform various types of resources into goods and services. Historically, the largest industrialized nations in the world consume a significant portion of their gross national product each year by investing in wealth-creating assets such as equipment and machinery (so-called intermediate goods of production).

The remainder of this section examines the capital-expenditure decision-making process, also referred to as *capital allocation*. This process involves the planning, evaluation, and management of capital projects. In fact, much of this book has dealt with concepts and techniques required to make correct capital-expenditure decisions involving engineering projects. Now, our task is to place them in the broader context of upper management's responsibility for proper planning, measurement, and control of the firm's overall portfolio of capital investments.

11.9.1 Allocating Capital among Independent Projects

Companies are constantly presented with independent opportunities in which they can invest capital across the organization. These opportunities usually represent a collection of the best projects for improving operations in all areas of the company (e.g., manufacturing, research and development, etc.). In most cases, the amount of available capital is limited, and additional capital can be obtained only at increasing incremental cost. Thus, companies have a problem of budgeting, or allocating, available capital to numerous possible uses.

One popular approach to capital allocation among projects uses the net PW criterion, as discussed in Chapter 5. If project risks are about equal, the procedure is to compute the PW for each investment opportunity and then to determine the combination of projects that maximizes PW, subject to various constraints on the availability of capital. The next example provides a general review of this procedure.

Example 11-4

Consider these four independent projects and determine the best allocation of capital among them if no more than $300,000 is available to invest:

Independent project	Initial capital outlay	PW
A	$100,000	$25,000
B	125,000	30,000
C	150,000	35,000
D	75,000	40,000

TABLE 11-3   Project Combinations for Example 11-4

Combination	Total PW ($\times 10^3$)	Total capital outlay ($\times 10^3$)
AB	$55	$225
AC	60	250
AD	65	175
BC	65	275
BD	70	200
CD	75	225
ABC	90	375
ACD	100	325
BCD	105	350
ABD	95	300* **Best**
ABCD	130	450

Solution   All possible combinations of these independent projects taken two, three, and four at a time are shown in Table 11-3, together with the total PW and initial capital outlay of each. After eliminating those combinations that violate the $300,000 funds constraint, the proper selection of projects would be ABD, and the maximum PW is $95,000. The process of enumerating combinations of projects having nearly identical risks is best accomplished with a computer when large numbers of projects are being evaluated.   ■

Methods for determining which possible projects should be allocated available funds seem to require the exercise of judgment in most realistic situations. Example 11-5 illustrates such a problem and possible methods of solution.

Example 11-5
Assume that a firm has five investment opportunities (projects) available, which require the indicated amounts of capital and which have economic lives and prospective after-tax IRRs as shown in Table 11-4. Further, assume that the five ventures are independent of each other, investment in one does not prevent investment in any other, and none is dependent upon the undertaking of another.

Now, suppose that the company has unlimited funds available, or at least sufficient funds to finance all these projects, and that capital funds cost the company 6% per year after taxes. For these conditions, the company probably would decide to undertake all projects that offered a return of at least 6% per year, and thus projects A, B, C, and D would be acceptable. However, such a conclusion would assume that the risks associated with each project are reasonable in light of the prospective IRR or are no greater than those encountered in the normal projects of the company.

Unfortunately, in most cases, the amount of capital is limited, either by absolute amount or by increasing cost. If the total of capital funds available is $60,000, the decision becomes more difficult. Here, it would be helpful to list the projects in order of *decreasing* profitability, as we have done in Table 11-5 (omitting the undesirable project E). Here, it is clear that a complication exists. We naturally would wish to undertake those ventures that have the greatest profit potential. However, if projects B and C are undertaken, there will not be sufficient capital for financing project A, which offers the next greatest rate of return. Projects B, C, and D could be undertaken and would provide an approximate annual return of $4,600 (= $15,000 × 10% + $20,000 × 8% + $25,000 × 6%). If project A were undertaken, together with either B or C, the

TABLE 11-4 Prospective Projects for a Firm[a]

Project	Capital investment	Life (years)	Rate of return (% per year)
A	$40,000	5	7
B	15,000	5	10
C	20,000	10	8
D	25,000	15	6
E	10,000	4	5

[a]Here, we assume that the indicated rates of return for these projects can be repeated indefinitely by subsequent "replacements."

TABLE 11-5   Prospective Projects of Table 11-4 Ordered by IRR

Project	Capital investment	Life (years)	Rate of return (%)
B	$15,000	5	10
C	20,000	10	8
A	40,000	5	7
D	25,000	15	6

TABLE 11-6   Prospective Projects of Table 11-5 Ordered
According to Overall Desirability

Project	Capital investment	Life (years)	Rate of return (%)	Risk rating
C	$20,000	10	8	Lower
A	40,000	5	7	Average
B	15,000	5	10	Higher
D	25,000	15	6	Average

total annual return would not exceed $4,600.[5] A further complicating factor is the fact that project D involves a longer life than the others. It is thus apparent that we might *not* always decide to adopt the alternative that offers the greatest profit potential.

The problem of allocating limited capital becomes even more complex when the risks associated with the various available projects are not the same. Assume that the risks associated with project B are determined to be higher than the average risk associated with projects undertaken by the firm, and that those associated with project C are lower than average. The company thus might rank the projects according to their overall desirability, as in Table 11-6. Under these conditions, the company might decide to finance projects C and A, thus avoiding one project with a higher than average risk and another having the lowest prospective return and longest life of the group.   ∎

11.9.2 Linear Programming Formulations of Capital Allocation Problems

For large numbers of independent or interrelated investments, the "brute force" enumeration and evaluation of all combinations of projects, as illustrated in Example 11-5, is impractical. This section describes a mathematical procedure for efficiently determining the optimal *portfolio* of projects in industrial capital allocation problems (Figure 11-1). Only the formulations of these problems will be presented in this section; their solution is beyond the scope of the book.

Suppose that the goal of a firm is to maximize its net PW by adopting a capital budget that includes a large number of mutually exclusive combinations of projects. When the number of possible combinations becomes fairly large, manual methods for determining

---

[5] This return amount is given, assuming that the leftover capital could earn no more than 6% per year.

the optimal investment plan tend to become complicated and time consuming, and it is worthwhile to consider linear programming as a solution procedure. The remainder of this section describes how simple capital allocation problems can be formulated as linear programming problems. Linear programming is a mathematical procedure for maximizing (or minimizing) a linear objective function, subject to one or more linear constraint equations. Hopefully, the reader will obtain some feeling for how more involved problems might also be modeled.

Linear programming is a useful technique for solving certain types of multiperiod *capital allocation problems* when a firm is not able to implement all projects that may increase its PW. For example, constraints often exist on how much investment capital can be committed during each fiscal year, and interdependencies among projects may affect the extent to which projects can be successfully carried out during the planning period.

The *objective function* of the capital allocation problem can be written as

$$\text{Maximize net PW} = \sum_{j=1}^{m} B_j^* X_j,$$

where

B_j^* = net PW of investment opportunity (project) j during the
       planning period being considered;

X_j = fraction of project j that is implemented during the planning period
  (*Note* : In most problems of interest, X_j will be either 0 or 1; the X_j
       values are the decision variables);

m = number of mutually exclusive combinations of projects under
       consideration.

In computing the net PW of each mutually exclusive combination of projects, a MARR must be specified.

The following notation is used in writing the constraints for a linear programming model:

c_{kj} = cash outlay (e.g., initial capital investment or annual operating
       budget) required for project j in time period k

C_k = maximum cash outlay that is permissible in time period k

Typically, two types of constraints are present in capital budgeting problems:

1. *Limitations on cash outlays for period k of the planning horizon,*

$$\sum_{j=1}^{m} c_{kj} X_j \le C_k.$$

2. *Interrelationships among the projects.* The following are examples:

(a) If projects p, q, and r are mutually exclusive, then

$$X_p + X_q + X_r \leq 1.$$

(b) If project r can be undertaken only if project s has already been selected, then

$$X_r \leq X_s \quad \text{or} \quad X_r - X_s \leq 0.$$

(c) If projects u and v are mutually exclusive and project r is dependent (contingent) on the acceptance of u or v, then

$$X_u + X_v \leq 1$$

and

$$X_r \leq X_u + X_v.$$

To illustrate the formulation of linear programming models for capital allocation problems, Example 11-6 and Example 11-7 are presented.

Example 11-6

Five engineering projects are being considered for the upcoming capital budget period. The interrelationships among the projects and the estimated net cash flows of the projects are summarized in the following table:

Project	Cash flow ($000s) for end of year, k					PW at MARR = 10% per year
	0	1	2	3	4	
B1	−50	20	20	20	20	13.4
B2	−30	12	12	12	12	8.0
C1	−14	4	4	4	4	−1.3
C2	−15	5	5	5	5	0.9
D	−10	6	6	6	6	9.0

Projects B1 and B2 are mutually exclusive. Projects C1 and C2 are mutually exclusive *and* dependent on the acceptance of B2. Finally, project D is dependent on the acceptance of C1.

Using the PW method and MARR = 10% per year, determine which combination (portfolio) of projects is best if the availability of capital is limited to $48,000.

Solution   The objective function and constraints for this problem are written as follows:
Maximize

$$\text{Net PW} = 13.4X_{B1} + 8.0X_{B2} - 1.3X_{C1} + 0.9X_{C2} + 9.0X_D,$$

subject to

$$50X_{B1} + 30X_{B2} + 14X_{C1} + 15X_{C2} + 10X_D \le 48;$$

(constraint on investment funds)

$$X_{B1} + X_{B2} \le 1;$$

(B1 and B2 are mutually exclusive)

$$X_{C1} + X_{C2} \le X_{B2};$$

(C1 or C2 is contingent on B2)

$$X_D \le X_{C1};$$

(D is contingent on C1)

$$X_j = 0 \text{ or } 1 \text{ for } j = \text{B1, B2, C1, C2, D.}$$

(No fractional projects are allowed.)

A problem such as this could be solved readily by using the simplex method of linear programming if the last constraint ($X_j = 0$ or 1) were not present. With that constraint included, the problem is classified as a linear *integer* programming problem. (Many computer programs are available for solving large linear integer programming problems.) ∎

Example 11-7
Consider a three-period capital allocation problem having the net cash flow estimates and PW values shown in Table 11-7. The MARR is 12% per year and the ceiling on investment funds available is $1,200,000. In addition, there is a constraint on operating funds for support of the combination of projects selected—$400,000 in year 1. From these constraints on funds outlays and the interrelationships among projects indicated in Table 11-7, we shall formulate this situation in terms of a linear integer programming problem.

   Solution   First, the net PW of each investment opportunity at 12% per year is calculated (Table 11-7). The objective function then becomes

$$\text{Maximize PW} = 135.3X_{A1} + 146.0X_{A2} + 119.3X_{A3} + 164.1X_{B1}$$
$$+ 151.9X_{B2} + 8.7X_{C1} - 13.1X_{C2} + 2.3X_{C3}.$$

The budget constraints are the following:

   Investment funds constraint is

$$225X_{A1} + 290X_{A2} + 370X_{A3} + 600X_{B1} + 1,200X_{B2}$$
$$+ 160X_{C1} + 200X_{C2} + 225X_{C3} \le 1,200.$$

Table 11-7 Project Interrelationships and PW Values (Example 11-7)

Project		0	1	2	3	Net PW ($000s) at 12% per year[b]
			Net cash flow ($000s). end of year[a]			
mutually exclusive	A1		150	150	150	
		−225	(60)	(70)	(70)	+135.3
	A2		200	180	160	
		−290	(180)	(80)	(80)	+146.0
	A3		210	200	200	
		−370	(290)	(170)	(170)	+119.3
independent	B1		100	400	500	
		−600	(100)	(200)	(300)	+164.1
	B2		500	600	600	
		−1,200	(250)	(400)	(400)	+151.9
mutually exclusive and dependent on acceptance of A1 or A2	C1		70	70	70	
		−160	(80)	(50)	(50)	+8.1
	C2		90	80	60	
		−200	(65)	(65)	(65)	−13.1
	C3		90	95	100	
		−225	(100)	(60)	(70)	+2.3

[a]Estimates in parentheses are annual operating expenses (which have already been subtracted in the determination of net cash flows).

[b]For example, net PW for A1 = −$225,000 + $150,000$(P/A, 12\%, 3)$ = +$135,300.

First year's operating cost constraint is

$$60X_{A1} + 180X_{A2} + 290X_{A3} + 100X_{B1} + 250X_{B2}$$
$$+ 80X_{C1} + 65X_{C2} + 100X_{C3} \leq 400.$$

Interrelationships among the investment opportunities give rise to these constraints on the following problem:

$$X_{A1} + X_{A2} + X_{A3} \leq 1 \qquad \text{A1, A2, A3 are mutually exclusive}$$

$$\left. \begin{array}{c} X_{B1} \leq 1 \\ X_{B2} \leq 1 \end{array} \right\} \qquad \text{B1, B2 are independent.}$$

accounts for dependence of

$$X_{C1} + X_{C2} + X_{C3} \leq X_{A1} + X_{A2} \qquad \text{C1, C2, C3 (which are mutually exclusive) on A1 } or \text{ A2}$$

Finally, if all decision variables are required to be either 0 (not in the optimal solution) or 1 (included in the optimal solution), the last constraint on the problem would be written

$$X_j = 0, 1 \quad \text{for} \quad j = \text{A1, A2, A3, B1, B2, C1, C2, C3.}$$

As can be seen in Example 11-7, a fairly simple problem such as this would require a large amount of time to solve by listing and evaluating all mutually exclusive combinations. Consequently, it is recommended that a suitable computer program be used to obtain solutions for all but the most simple capital allocation problems.   ■

11.10 Summary

This chapter has provided an overview of the capital financing and capital allocation functions, as well as the total capital budgeting process. Our discussion of capital financing has dealt with where companies get money to continue to grow and prosper and the costs to obtain this capital. Also included was a discussion of the weighted average cost of capital. In this regard, differences between debt capital and owner's (equity) capital were made clear. Leasing, as a source of capital, was also described, and a lease-versus-purchase example was analyzed.

Our treatment of capital allocation among independent investment opportunities has been built on two important observations. First, the primary concern in capital expenditure activity is to ensure the survival of the company by implementing ideas to maximize future shareholder wealth, which is equivalent to maximization of shareholder PW. Second, engineering economic analysis plays a vital role in deciding which projects are recommended for funding approval and included in a company's overall capital investment portfolio.

REFERENCES

Baumol, W. J., and R. E. Quandt. "Investment and Discount Rates Under Capital Rationing—A Programming Approach." *Economic Journal*, vol. 75, no. 298, June 1965, pp. 317–329.

Bussey, L. E., and T. G. Eschenbach. *The Economic Analysis of Industrial Projects*, 2nd ed. (Englewood Cliffs, NJ: Prentice Hall, 1992).

GURNANI, C. "Capital Budgeting: Theory and Practice." *The Engineering Economist*, vol. 3, no. 1 (Fall 1984), pp. 19–46.

Luehrman, Timothy R. "What's It Worth? A General Manager's Guide to Valuation." *Harvard Business Review*, May–June, 1997.

Luehrman, Timothy R. "Using APV: A Better Tool for Valuing Operations. "*Harvard Business Review*, May–June, 1997.

Myers, Stewart C. "Interactions of Corporate Financing and Investment Decisions—Implications for Capital Budgeting." *Journal of Finance*, vol. 29 (March 1974) pp. 1–25.

Park, C. S., and G. P. Sharpe-Bette. *Advanced Engineering Economics* (New York: John Wiley & Sons, Inc., 1990).

PROBLEMS

11-1. Describe how an organization's capital financing and allocation activities affect the practice of engineering economy.

11-2. Why do most engineering economic analyses normally assume that a company's pool of capital is being used to fund a capital project instead of a specific source of capital (e.g., equity versus borrowed funds)?

11-3. List five possible sources of funds to a corporation for funding capital projects and continuing operations.

11-4. Briefly describe the five basic steps associated with a company's capital budgeting process.

11-5. Under what circumstances is it reasonable to have more than one minimum acceptable rate of return for a given firm?

11-6. Select the classification systems for investment proposals that you think would be most useful for the typical small (say, fewer than 200 employees) enterprise. Do the same for the typical large (say, more than 2,000 employees) enterprise. Explain the reasonableness of any differences in your selections for the two size groups.

11-7. What is the purpose of a postmortem review? Can it be a means of correcting unwise commitments made as a result of past project analyses?

11-8. Explain the various degrees of dependency among two or more projects. If one or more projects seems desirable and is complementary to a given project, should the complementary project(s) be included in a proposal package or kept separate for review by management?

11-9. A 4-year-old truck has a present net realizable value of $6,000 and is now expected to have a market value of $1,800 after its remaining 3-year life. Its operating disbursements are expected to be $720 per year.

An equivalent truck can be leased for $0.40 per mile plus $30 a day for each day the truck is kept. The expected annual utilization is 3,000 miles and 30 days. If the before-tax MARR is 15%, find which alternative is better by comparing before-tax equivalent annual costs by using

a. only the preceding information;

b. further information that the annual cost of having to operate without a truck is $2,000.

11-10. Work Problem 11-9 by comparing after-tax equivalent present worths if the effective income tax rate is 40%, the present book value is $5,000, and the depreciation charge is $1,000 per year if the firm continues to own the truck. Any gains or losses on disposal of the old truck affect taxes at the full 40% rate, and the after-tax MARR is 5%.

11-11. A lathe costs $56,000 and is expected to result in net cash inflows of $20,000 at the end of each year for 3 years and then have a market value of $10,000 at the end of the 3rd year. The equipment could be leased for $22,000 a year, with the first payment due immediately.

a. If the organization does not pay income taxes and its MARR is 10%, show whether the organization should lease or purchase the equipment.

b. If the lathe is thought to be worth only, say, $18,000 per year to the organization, what is the better economic decision?

11-12. The Shakey Company can finance the purchase of a new building costing $2 million with a bond issue, for which it would pay $100,000 interest per year, and then repay the $2 million at the end of the life of the building. Instead of buying in this manner, the company can lease the building by paying $125,000 per year, the first payment being due 1 year from now. The building would be fully depreciated for tax purposes over an expected life of 20 years. The income tax rate is 40% for all expenses and capital gains or losses, and the firm's after-tax MARR is 5%. Use annual worth analysis based on equity (nonborrowed) capital to determine whether the firm should borrow and buy or lease if, at the end of 20 years, the building has the following market values for the owner: (a) nothing, (b) $500,000. Straight-line depreciation will be used, but is allowable only if the company purchases the building.

11-13. The Capitalpoor Company is considering purchasing a business machine for $100,000. An alternative is to rent it for $35,000 at the beginning of each year. The rental would include all repairs and service. If the machine is purchased, a comparable repair and service contract can be obtained for $1,000 per year.

The salesperson of the business machine firm has indicated that the expected useful service life of this machine is 5 years, with zero market value, but the company is not sure how long the machine will actually be needed. If the machine is rented, the company can cancel the lease at the end of any year. Assuming an income tax rate of 25%, a straight-line depreciation charge of $20,000 for each year the machine is kept, and an after-tax MARR of 10%, prepare an appropriate analysis to help the firm decide whether it is more desirable to purchase or rent.

11-14. An existing piece of equipment has been performing poorly and needs replacing. More modern equipment can be *purchased* or it can be *leased*. If purchased, the equipment will cost $20,000 and will have a depreciable life of 5 years with no market value. For simplicity, assume straight-line depreciation is used by the firm. Because of improved operating characteristics of the equipment, raw materials savings of $5,000 per year are expected to result relative to continued use of the present equipment. However, annual labor expenses for the new equipment will most likely increase by $2,000 and annual maintenance will go up by $1,000. To lease the new equipment requires a refundable deposit of $2,000, and the end-of-year leasing fee is $6,000. Annual materials savings and extra labor expenses will be the same when purchasing or leasing the equipment, but the lessee company will provide maintenance for its equipment as part of the leasing fee. The after-tax MARR is 15% per year, and the effective income-tax rate is 50%. If purchased, it is believed that the equipment can be sold at the end of 5 years for $1,500 even though $0 was used in calculating straight-line depreciation. Determine whether the company should buy or lease the new equipment, assuming that it has been decided to replace the present equipment.

11-15. A firm is considering the development of several new products. The products under consideration are listed in the next table. Products in each group are mutually exclusive. At most, one product from each group will be selected. The firm has an MARR of 10% per year and a budget limitation on development costs of $2,100,000. The life of all products is assumed to be 10 years, with no salvage value. Formulate this capital allocation problem as a linear integer programming model.

Group	Product	Development Cost	Annual net cash income
A	A1	$500,000	$90,000
	A2	650,000	110,000
	A3	700,000	115,000
B	B1	600,000	105,000
	B2	675,000	112,000
C	C1	800,000	150,000
	C2	1,000,000	175,000

11-16. Four proposals are under consideration by your company. Proposals A and C are mutually exclusive; proposals B and D are mutually exclusive and cannot be implemented unless proposal A *or* C has been selected. No more than $140,000 can be spent at time zero. The before-tax MARR is 15% per year. The estimated cash flows are shown in the accompanying table. Form all mutually exclusive combinations in view of the specified contingencies, and formulate this problem as an integer linear programming model.

End of year	Proposal			
	A	B	C	D
0	-$100,000	-$20,000	-$120,000	-$30,000
1	40,000	6,000	25,000	6,000
2	40,000	10,000	50,000	10,000
3	60,000	10,000	85,000	19,000

11-17. Three alternatives are being considered for an engineering project. Their cash-flow estimates are shown in the accompanying table. A and B are mutually exclusive, and C is an optional add-on feature to alternative A. Investment funds are limited to $5,000,000. Another constraint on this project is the engineering personnel needed to design and implement the solution. No more than 10,000 person-hours of engineering time can be committed to the project. Set up a linear integer programming formulation of this resource allocation problem.

	Alternative		
	A	B	C
Initial investment (10^6)	4.0	4.5	1.0
Personnel requirement (hours)	7,000	9,000	3,000
After-tax annual savings, years 1 through 4 (10^6)	1.3	2.2	0.9
PW at 10% per year (10^6)	0.12	2.47	1.85

11-18. Four proposals are under consideration by your company. Proposals A and C are mutually exclusive; proposals B and D are mutually exclusive and cannot be implemented unless proposal A *or* C has been selected. No more than $140,000 can be spent at time zero. The before-tax MARR is 15% per year. Formulate this situation in terms of a linear integer programming problem. Relevant data are provided in the table below.

End of Year	Estimated cash flow for proposals			
	A	B	C	D
0	−50,000	−20,000	−120,000	−30,000
1	0	10,000	55,000	15,000
2	0	10,000	55,000	15,000
3	83,000	10,000	55,000	15,000
PW (15%)	4,574	2,832	5,577	4,248
IRR	18.4%	23.4%	17.8%	23.4%

11-19. Refer to the associated graph. Identify when the WACC approach to project acceptability agrees with the CAPM approach. When do recommendation of the two approaches differ? Explain why.

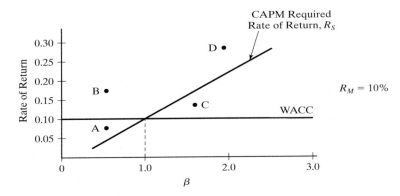

11-20. A firm is considering a capital investment. The risk premium is 0.04, and it is considered to be constant through time. Riskless investments may now be purchased to yield 0.06 (6%). If the project's beta (β) is 1.5, what is the expected return for this investment?

11-21. (Refer to Appendix 11-B) Martin Machine Tool is trying to convince Adams Corporation, a hundred million dollar company, to purchase a new numerically controlled inspection system for $150,000. Adams estimates the after-tax cash savings, including the depreciation tax shield, at $40,000 annually. However, he is concerned that technology will change and demand for this product line will decline rapidly after 5 years, and the machine will have no salvage value at that time. Adams' WACC is 15% per year (after tax) and their pretax cost of debt is 8%.

NPV(15%) = −150,000 + $40,000 [*P/A*, 15%,5] = −$15,914, and Adams' engineers are ready to drop the idea. However, Martin offers to finance the entire venture with a 5% loan, payable in five annual installments. The installments are calculated at $150,000 [*A/P*,5%,5] = $34,646. Will this change Adams' view toward purchasing the new inspection system?

APPENDIX 11-A

Adjusted Present Worth (or Value)

In previous chapters, we have learned how to use discounted cash flow techniques to evaluate the profitability of proposed capital investments. Our conclusion was that a firm should accept any project that has a PW, AW, or FW greater than or equal to zero (funds permitting, of course). Moreover, we learned that a project with an IRR or ERR greater than or equal to the MARR should be accepted. In this appendix, we extend our previous analysis to illustrate how to take into account a company's capital structure when evaluating capital investment projects on an after-tax basis. The technique we adopt for this purpose is called "adjusted present worth," and it provides the same recommendation regarding project acceptability as our earlier methods (e.g., PW) did.

The adjusted present worth or value (APW or APV) of a project is equal to the PW of the project, assuming equity financing plus the PW of its interest tax shield (i.e., savings in taxes due to deductibility of interest) from additional debt financing created by implementing the project. The "unlevered" PW is computed at the firm's unlevered (debt free) cost of equity, as might be determined by the CAPM method discussed previously in this chapter. The APW technique was alluded to in Section 7.6 and is demonstrated in this appendix with an example.

Example 11-8

Harbison Power Products has designed a new high-efficiency snowblower. It will cost $10 million to start the business, which would generate an after-tax net cash flow of $1.8 million annually for the anticipated 10-year product life. The unlevered cost of equity is 12% per year, and bond financing, if needed, would cost 8% per year. Harbison currently uses a 50–50 mix of debt and equity in its financial structure. With this leverage, the cost of equity is 16% per year. Harbison pays taxes at an effective income tax rate of 40%. Following its financial debt policy, Harbison would use $5.0 million in retained earnings and $5.0 million in debt to finance this project. It is desired to determine the after-tax profitability of the project using (a) APW analysis, and (b) conventional PW analysis.

Solution (a) APW Analysis The PW of this project at the unlevered cost of equity (12% per year) is

$$PW(12\%) = -\$10 + \$1.8\ (P/A, 12\%, 10) = \$0.170 \text{ million.}$$

This signals an acceptable project. Because there is the debt financing of $5.0 million involved, we need to compute the PW of its interest tax shield and add it to $0.170 million to obtain the APW of this project. To determine these financial side effects, we first calculate the annual payment on the borrowed capital to be

$$\text{Payment} = \$5\ (A/P, 8\%, 10) = \$0.745 \text{ million.}$$

Calculations to determine the PW of the loan's tax shield are shown in Table 11-A-1. Notice that interest and principal (balance reduction) are broken out in Columns D and E. The interest tax shield is shown in Column F, and its PW at 8% per year equals $0.722 million. Therefore, the APW of this project is $0.170 + $0.722 = $0.892 million.

Solution (b) Conventional PW Analysis   The after-tax WACC (using Equation 7-13) and net present worth, using the conventional separation principle, is

$$\text{WACC} = \lambda(1-t)i_b + (1-\lambda)e_a = (0.5)(1-0.4)(0.08) + (0.5)(0.16) = 0.104 = 10.4\%.$$

And,

$$\text{NPW } (10.4\%) = -\$10.0 + \$1.8(P/A,10.4\%,10) = -\$10.0 + \$10.872 = \$0.872 \text{ million},$$

making the project reasonably attractive.

Thus, the acceptability of the Harbison snowblower project is confirmed by both methods of analysis. If we assume that debt is rebalanced every year to maintain the 50–50 debt–equity ratio, the APW method and the conventional after-tax PW method will provide the same numerical answer.

TABLE 11-A-1   Interest Tax Shield Calculations ($ in millions)

A	B	C	D = 0.08 B	E = C – D	F = 0.40D	G = C – F
Year	Beginning of year balance	Payment	Interest at 8%	Balance reduction	Interest tax shield	After-tax payment
1	$5,000	$0.745	0.400	$0.345	$0.160	$0.585
2	$4,655	$0.745	0.372	$0.373	0.149	0.596
3	$4,282	$0.745	0.343	$0.403	0.137	0.608
4	$3,880	$0.745	0.310	$0.435	0.124	0.621
5	$3,445	$0.745	0.276	$0.470	0.110	0.635
6	$2,975	$0.745	0.238	$0.507	0.095	0.650
7	$2,468	$0.745	0.197	$0.548	0.079	0.666
8	$1,920	$0.745	0.154	$0.592	0.061	0.684
9	$1,329	$0.745	0.106	$0.639	0.043	0.703
10	$0.690	$0.745	0.055	$0.690	0.022	0.723
				Total = $5.000		
				PW (8%) =	$0.722	4.728

APPENDIX 11-B

Bundled Financing

Although separation of investment from financing decisions is the recommended approach for most projects, sometimes sellers bundle attractive financing terms with their sales package. This forces the analysis to include financing considerations along with the investment. Our approach here is to determine the present worth of the financial subsidy and treat it as an adjustment to the investment outlay in the conventional after-tax PW analysis. The mechanics are straightforward in concept, but somewhat cumbersome in execution. They are best illustrated through an example.

Example 11-9

Returning to the same Harbison example given in Appendix 11-A, suppose that an equipment vendor offers special financing at 6% per year over the 10-year project life for the $5.0 million that Harbison expects to finance. This is a 2-percentage point reduction compared with the 8% per year rate that Harbison pays through its normal credit channels. How would this impact the project decision?

Solution   With a 6% rate, the debt payment is $5.0 [$A/P$,6%,10] = $0.679 million. To determine the financial advantage, we need to project the financial cash flows at 6%. Calculations are shown in Table 11-B-1. Columns D and E show the annual split of debt payments into interest and balance reduction. The total of Column E is $5.0 million, indicating that the stream repays the debt. Column F is the interest tax shield obtained by multiplying the interest payment by the effective tax rate. Note in particular, the last column, G, shows the after-tax financing stream resulting from the subsidized, 6% loan. This column is simply the difference between the annual loan payment amount and the annual interest tax shield (i.e., tax savings due to interest deductibility), which changes as the unpaid balance is reduced. If the subsidized loan were accepted, this stream would replace the up-front payment of $5.0 million obtained from borrowed funds (at 8% under usual circumstances) and would save money. The value of that savings is simply the present worth of the 10-year financing cash flow stream at the after-tax cost of borrowed capital. The financing cash flow stream, using the normal after-tax cost of debt of $(1-t)$ 8% = 4.8% results in a present worth of $4.703 million, a saving of $0.297 million when compared to the $5.0 million loan proceeds. This saving would be deducted from the total initial investment to reduce it (in round numbers) from $10.0 million to approximately $9.7 million, thus increasing the conventional after-tax net present value of the project at the 10.4% WACC from $0.872 million [solution (b) in Appendix 11-A] to $1.272 million.

TABLE 11-B-1   Interest Tax Shield and Financing Cash Flow
for a 6% Loan ($ in millions)

A	B	C	D = .06 B	E = C − D	F = .40 D	G = C − F
Year	Beginning of year balance	Payment	Interest $ at 6%	Balance reduction	Interest tax shield	After-tax payment
1	$5.000	$0.679	0.300	$0.379	0.120	$0.559
2	$4.621	$0.679	0.277	$0.402	0.111	$0.568
3	$4.219	$0.679	0.253	$0.426	0.101	$0.578
4	$3.792	$0.679	0.228	$0.452	0.091	$0.588
5	$3.341	$0.679	0.200	$0.479	0.080	$0.599
6	$2.862	$0.679	0.172	$0.508	0.069	$0.611
7	$2.354	$0.679	0.141	$0.538	0.056	$0.623
8	$1.816	$0.679	0.109	$0.570	0.044	$0.636
9	$1.245	$0.679	0.075	$0.605	0.030	$0.649
10	$0.641	$0.679	0.038	$0.641	0.015	$0.664
			Total	$5.000		
			PW at 4.8% = $4.703			

PART THREE

Capital Investment Analyses in an Uncertain World: Formal Assessments

12
Introduction to Risk and Uncertainty

All of the economic study methods and illustrations in Parts One and Two were for conditions of "assumed certainty" [i.e., all elements (parameters) considered were estimated or specified by a single figure]. Generally, such elements as life, salvage value, and periodic incomes and costs are random variables rather than known constants. Hence, in many economic studies, it is necessary or desirable to extend the results of assumed certainty analyses by directly considering the risk and uncertainty involved, due to variability in the outcome of elements. Part Three of this book treats the subject of capital investment evaluation under risk and uncertainty conditions.

12.1 The Difference between Risk and Uncertainty

The classical distinction between risk and uncertainty is that an element or analysis involves *risk* if the probabilities of the alternative possible outcomes are known, while it is characterized by *uncertainty* if the frequency distribution of the possible outcomes is not known. The distinction between conditions of assumed certainty, risk, and uncertainty for a given element such as project life is portrayed graphically in Fig. 12-1.

Another less restrictive distinction between risk and uncertainty is that risk is the dispersion of the probability distribution of the element being estimated or calculated outcome(s) being considered, while uncertainty is the degree of lack of confidence that the estimated probability distribution is correct. The word *risk* can be used to apply to the outcome of any element or measure of merit. Colloquially, "risk" is often used merely to denote variability of outcome, and often the only variability of concern is variability in an unfavorable direction.

Figure 12-1   Illustrations of assumed certainty, risk, and uncertainty as applied to life of a project.

There are several combinations of risk and assumed certainty that can specify a given element estimate over time. For example, Fig. 12-2 represents an assumed-certain outcome amount (e.g., cash flow) at an assumed-certain point in time, while Fig. 12-3 shows a risky amount at an assumed-certain point in time. Figure 12-4 represents an assumed-certain amount at risky (discrete) points in time, while Fig. 12-5 represents random amounts at each risky (discrete) point in time.

12.2 Causes of Risk and Uncertainty

Risk and uncertainty in project investment decisions are attributable to many possible sources. A brief description of some main causes is presented next.

1. *Insufficient number of similar investments.* In general, a firm will have only a few investments of a particular type. This means that there will be insufficient opportunity for the results of a particular investment type to "average out" (i.e., for the effect of unfavorable outcomes to be virtually cancelled by favorable outcomes). This type of risk is dominant when the magnitude of

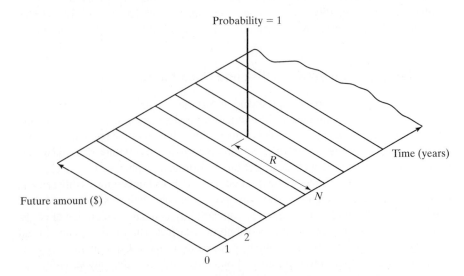

Figure 12-2   Assumed-certain outcome at assumed-certain point in time.

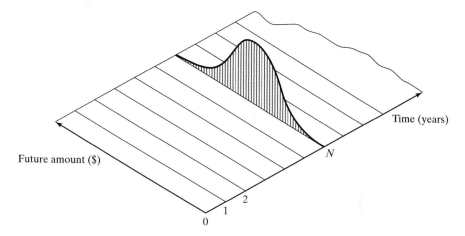

Figure 12-3   Risky amount at assumed-certain point in time.

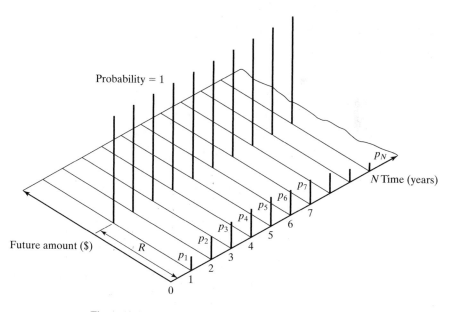

Figure 12-4   Assumed-certain outcome at risky points in time.

the individual investment commitment is large compared with the financial resources of the firm.

2. *Bias in the data and its assessment.* It is common that individuals making or reviewing economic analyses have biases of optimism or pessimism or are unconsciously influenced by factors that should not be a part of an objective study. A pattern of consistent undue optimism or pessimism on the part of an analyst should be recognized through analysis review procedures.

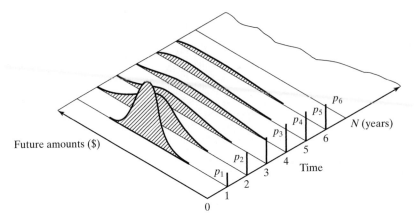

Figure 12-5   Risky amounts at risk points in time.

3. *Changing external economic environment, invalidating past experience.*
   Whenever estimates are made of future conditions, the usual bases are past
   results for similar quantities, whenever available. While the past information is often valuable, there is risk in using it directly without adjustment for
   expected future conditions.

4. *Misinterpretation of data.* Misinterpretation may occur if the underlying
   factors behind elements to be estimated are so complex that the relationship of one or more factors to the desired elements is misunderstood.

5. *Errors of analysis.* Errors can occur either in analysis of the technical operating characteristics of a project or in the analysis of the financial implications
   of a project.

6. *Managerial talent availability and emphasis.* The performance of an industrial
   investment project or set of projects usually depends in substantial part on the
   availability and application of managerial talent once the project has been
   undertaken. In general, management talent is a very limited resource within a
   firm; hence, it follows that the results of some projects are going to suffer compared to the results of other projects. Thus, there is risk due to lack of availability or neglect in needed managerial talent applied to investment projects.

7. *Salvageability of investment.* Of prime consideration in judging risk is the
   relative recoverability of investment commitments if a project, for performance considerations or otherwise, is to be liquidated. For example, an
   investment in special-purpose equipment that has no value to other firms
   entails more risk than an investment in general-purpose equipment that
   would have a high percentage salvage value if sold because of poor operating results. A descriptive synonym is "bailoutability."

8. *Obsolescence.* Rapid technological change and progress are characteristic
   of our economy. Not only do products become superseded, thus rendering
   those products' productive facilities less needed or useless, but also changes
   in process technology can render existing facilities obsolete.

12.3 Weakness in Probabilistic Treatment of Project Analyses Involving Risk

While the use of probabilities is freely made in analyses of projects involving risk, it should be pointed out that these probabilities are not generally objectively verifiable, and hence are generally *subjective* (sometimes called *personal*) probabilities. A further weakening fact is that the evidence supporting any given probability in an analysis may differ markedly in both quality and quantity from that for any other probability.

When probabilities are used, the risk and uncertainty concerning outcomes in question are not eliminated, but rather the uncertainty then becomes uncertainty connected with the probabilities on which the analysis is based. Nevertheless, it is often worthwhile to express degree of confidence in estimates through the use of probability distributions, rather than through subjective verbal expressions.

12.4 Ways to Change or Influence Degree of Uncertainty

It is usually possible for the firm to take actions that will decrease the degree of uncertainty to which it is subject as a result of investment project selection. Several notable ways are as follows:

1. by increasing information obtained before decision, such as through additional market research or investigation of technical performance characteristics,

2. by increasing size of operations so as to have enough different investment projects to increase expectation that results will "average out," and

3. by diversifying products, particularly by choosing product lines for which sales are affected differently by changes in business activity (i.e., when sales of some products decrease, then sales of other products can be expected to increase).

12.5 Return, Risk, and Choice

It is generally accepted that the riskier a project, the higher is the apparent return it must promise to warrant its acceptance. It would be desirable to determine differential risk allowances that would reduce all projects to a common basis. This cannot be done precisely, however, for the statement of differential risk allowances is very much a matter of subjective judgment.

Before a firm can make investment decisions to include allowances for risk, the firm's policy toward risk should be determined. The amount of risk a firm is prepared to undertake to secure a given actual or apparent monetary return is a general question of values. There is no rational or logical criterion by which the choice can be made. Rather, this is largely a function of the preferences of the decision makers of the firm and the amount of risk to which the firm is already exposed.

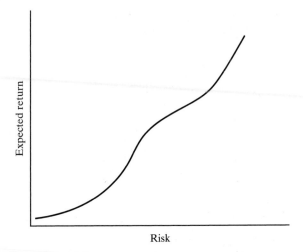

Figure 12-6   Relationship between return and risk.

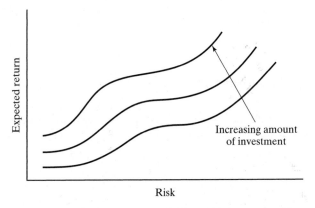

Figure 12-7   Relationship between return and risk, considering amount of investment.

In general, the relationship between expected return and risk (degree of variability of the return) can be represented as in Fig. 12-6. A refinement of the concept of Fig. 12-6 can be made if risk is subcategorized into the quantitative components of variability of returns and amount of investment. This is shown in Fig. 12-7.

12.6 Decision Guides on When and How Much to Consider Risk and Uncertainty

The question of the extent to which risk and uncertainty should be directly considered in economic analyses is of great concern and cannot be answered categorically. The concern stems from the fact that the risk and uncertainty of the future pervade most capital projects for which analyses are made. The impossibility of categorical answers stems from the fact that there is an infinite variety of sources of risk and relative degrees of risk for various projects and firms.

It is sometimes felt that the risk and uncertainty inherent in most investment decisions make it not worthwhile to engage in any complex or "sophisticated" methods of analysis. While this may be true in particular situations, the position is strongly suggested that it is, in general, very worthwhile to supplement judgment quantitatively and explicitly in the analysis of risk and uncertainty for investment projects. Analysis is needed to ensure that the implementation of judgment is not accompanied by errors in quantification and omissions in factors considered. Indeed, with the increasing complexity and economic size of individual projects compared with the worth of the firm as a whole, quantitative consideration of risk and uncertainty in capital project analyses becomes increasingly important. This section provides some qualitative decision guides regarding this question.

A conceptual answer to the question of when and how much risk and uncertainty should be considered can be reduced to simple economics—that is, put more study effort into the analysis as long as the savings from further study are greater than the cost of further study (i.e., as long as marginal savings are greater than marginal cost). Since the amount of marginal savings (and possibly marginal cost also) for a given amount of added study is a variable, it is necessary to modify the rationale. A reasonable modification is suggested then to be the consideration of expected values. Thus, the rationale can be restated as follows: Put more study effort into the analysis as long as the expected savings from further study are greater than the expected cost of that further study.

The great problem in applying this rationale in practice is that it is quite difficult to estimate the expected savings from further study. In economic analyses of mutually exclusive projects (i.e., when at most one project can be chosen), savings from further study occur if the further study correctly causes a reversal or change in decision as to the project accepted. In economic analyses of nonmutually exclusive projects (i.e., when the choice of one project does not affect the desirability of choice of any other project), savings from further study occur if the further study correctly causes the decision maker to drop one or more projects previously accepted or correctly to add one or more projects not previously accepted. By "correctly" is meant "with favorable consequences." Other savings can be created by the added study. For example, the added study may provide information that will prove useful in future operating decisions or investment analyses.

The savings from further study can be determined conceptually as the discounted present worth of the new project(s) accepted after the further study minus the present worth of the project(s) accepted before the further study. However, the practical problem of determination of the expected savings from further study, as based on the amount of savings and the likelihood or probability of those savings, is generally quite difficult. It should be noted that the expected savings from added study may well not be a continuous function of the amount of the added study, but rather it is likely to change in discrete steps.

The expected cost of added study is more readily determinable than the expected savings from that study; nevertheless, it is not always apparent. Two common viewpoints on this cost are that it is equal to the direct cost of the resources devoted to the added study or that it is essentially zero on the grounds that the resources are available and paid for regardless of whether or not they are used on that added study. The most defensible cost of added study is based on the opportunity cost principle; that is, the cost of the added study should be determined by the value to the company of those study resources if put to best productive use on work other than that added study. While this opportunity cost is often hard to evaluate, it seems reasonable that in a well-managed company the cost will be at least as great as the direct cost of those resources.

Figure 12-8 shows a flow diagram that depicts a general recommended sequence of steps in making economic analyses and shows qualitative test points regarding the extent of the analysis. This sequence would be applicable to analyses of groups of either mutually exclusive or nonmutually exclusive projects. Note that the recommended sequence displayed in the figure shows four different points at which the decision could be made concerning which project(s) to accept. Also, there are four stages at which provision is made for dropping from further consideration projects that analysis indicates are clearly not contenders worthy of further study.

The meaning of the test points included is worthy of discussion. The test points are depicted as diamond shapes and are numbered in parentheses. Test point 1 considers the magnitude of the fixed monetary commitments involved in the decision for purposes of deciding whether further study is justified. The relevant amount of money to consider is the total present worth of the nonsalvageable investment costs as well as other fixed costs the company would incur if it should accept that project. If the magnitude of the fixed commitments for each of the projects being considered is low compared to the cost of further study, then it may be decided that further study is not justified and that the choice(s) should be made. The break-even point concerning the size of fixed commitments to use as this criterion is rather subjective. Determined intuitively, it appears that this point would be related to the company's financial health, the size of the projects usually considered, and the availability of resources for further analysis.

Test point 2 in Fig. 12-8 considers how close is the choice between projects. In this case, "close" can be defined as the nearness of the measure of merit for the most preferred alternative to the next most preferred alternative. If the assumed-certainty analysis results up to that point show that the decision is not at all close [i.e., the choice(s) is (are) apparent], then further study is hardly justified and the choice(s) should be made.

Test point 3 is concerned with the decision of whether the results of an initial analysis considering variation of elements (which would be essentially a risk analysis such as described in the next chapters) is based on sufficient study considering the economic importance of the decision and the closeness of the analysis results for the projects considered. If the decision is important enough in terms of worth of the fixed commitments for the projects considered and the analysis results are somewhat close, then further study should be performed before the choice(s) is (are) made. The further study would take the form of closer estimations of elements and sensitivity analyses.

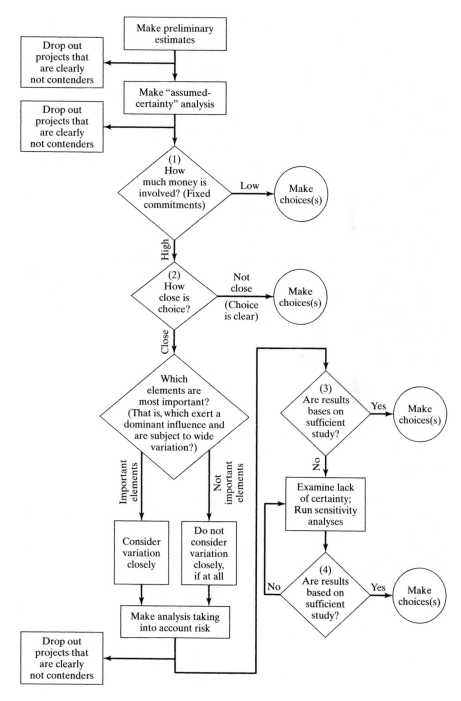

Figure 12-8   Recommended sequence of steps for economic analyses.

Test point 4 in Fig. 12-8 is repetitious of test point 3. It shows that closer estimations and sensitivity analyses would be continued until it is decided that the results of the analyses are based on sufficient study and the choice(s) can be made.

The sequence suggested in Fig. 12-8 is subject to shortcuts in cases where warranted. For example, if a given analysis involves projects that are of extreme importance to the future of the company, it may be decided to perform directly a risk analysis that considers variation of multiple elements, without bothering to perform an initial "assumed certainty" study.

The sequence of steps shown in Fig. 12-8 provides a conceptual basis for determining the extent to which economic analyses should be performed. The decisions in the sequence are rather intangible; nevertheless, the sequence of steps represents a formalized structure for thinking, which the analyst or decision makers can follow in determining the extent to which analyses should be carried out in particular situations. It should be recognized that all of these steps of analyses involving a particular set of alternatives are not necessarily performed at a single level in the organization. In fact, the more money involved, the higher is the level in the organization to which the analysis is referred, in general, before a decision is made.

In summary, the justifiable extent of an economic analysis depends on the economic importance of the study. When the risks or potentials are substantial, systematic procedures are needed to reduce the inherent uncertainties and to give them appropriate weight in arriving at decisions.

12.7 General Model for Risk and Uncertainty Problems

In order to provide a framework for the subsequent discussion, it is useful to employ a general model of decision problems in which there are various possible outcomes (called *states of nature*) in combination with several alternative actions. As depicted in Table 12-1, m mutually exclusive investment alternatives $\{A_i\}$ and k mutually exclusive and collectively exhaustive states of nature $\{S_j\}$ have been identified. The combination of action A_i and state of nature S_j yields a net result R_{ij}. The outcome R_{ij} is normally expressed in equivalent returns or costs, but it can be in any measure. (In fact, it may well be that the outcome is multidimensional; however, it is generally assumed that, in such cases, a one-dimensional utility measure can be obtained.) This tabular model is often descriptively called a *payoff table*.

When the decision problem is considered a decision under risk, then for $j = 1, \ldots,$ k, estimates are provided of $P(S_j)$, which is the probability of state of nature S_j occurring. In the absence of such probabilities, the decision is considered to be a decision under uncertainty.

The general model is very useful in the case of a single decision. However, when sequential decisions are made, with each subsequent decision being influenced by a partial realization of the future state, then an alternative model is generally used. In Chapter 16, a decision tree model is used to represent decision problems under risk that involve sequential decisions.

TABLE 12-1   General Model for Risk and Uncertainty Problems

Alternative	State of nature (probability of state)			
	S_1 $P(S_1)$	$S_2 \ldots$ $P(S_2) \ldots$	$S_j \ldots$ $P(S_j) \ldots$	S_k $P(S_k)$
A_1	R_{11}	$R_{12} \ldots$	$R_{1j} \ldots$	R_{1k}
A_2	R_{21}	$R_{22} \ldots$	$R_{2j} \ldots$	R_{2k}
.	.	.	.	.
.	.	.	.	.
.	.	.	.	.
A_i	R_{i1}	$R_{i2} \ldots$	$R_{ij} \ldots$	R_{ik}
.	.	.	.	.
.	.	.	.	.
.	.	.	.	.
A_m	R_{m1}	$R_{m2} \ldots$	$R_{mj} \ldots$	R_{mk}

12.8 Estimating in Terms of Probability Distributions

It is reasonable to estimate many element outcomes in terms of subjective, usually continuous, probability distributions. This can be most useful, either for purposes of calculating measures of merit that directly take these distributions into account or for purposes of merely judging the degree and effect of probable outcome variation. When it is desired to estimate the subjective probability distribution of an element and that element is not thought to fit one of the computationally convenient distributions, one good way is to estimate in terms of a cumulative probability distribution function and then convert the results to other forms if needed. For example, suppose you desire to estimate the life of a project (such as the length of time before your car will have a major breakdown). After considering the experience records for similar cars and making your subjective adjustments for future conditions, you may decide that, say, there is practically nil probability that the life will be equal to or less than 2 years; 0.10 probability that the life will be equal to or less than 3 years; and you keep making similar estimates until the life is reached at which you feel there is 100% chance that the life will not be exceeded. A complete set of estimates for this example is given in Table 12-2 and then graphed, assuming a continuous distribution, in Fig. 12-9. Figure 12-10 shows the same estimates converted into the more commonly portrayed probability density form. The reader may recall that the probability density function (height of the curve) for a continuous distribution equals the slope of the cumulative probability distribution function over the entire range of the element estimated.

Another way to estimate in terms of a cumulative probability distribution, and which is probably even easier to do than that just depicted, is to estimate median, quartile, and extreme values. We next present typical suggested questions and examples of answers for estimating, say, the labor cost for a new product.

TABLE 12-2 Examples of Estimates
Expressed in Cumulative Probability Form

Life (yr)	Probability that life will be equal to or less than life given
2	0.0
3	0.1
4	0.3
5	0.7
6	0.9
7	1.0

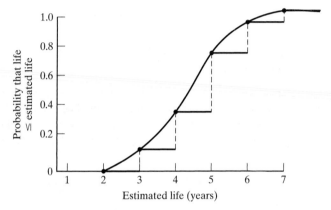

Figure 12-9 Cumulative probability distribution approximation of data in Table 12-2.

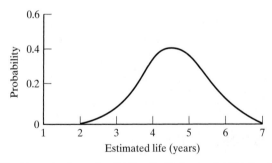

Figure 12-10 Alternative form of probability distribution (probability density) for data given in Table 12-2.

For Median (50% Cumulative Probability):
At what value is the labor cost as likely to be above as below that value? (*Ex. Ans.:* $18)

For Upper Quartile (75% Cumulative Probability):
Given that the labor cost is above the median, $18, at what value is the labor cost as likely to be above as below that value? (*Ex. Ans.:* $28)

For Lower Quartile (25% Cumulative Probability):
Given that the labor cost is below the median, $18, at what value is the labor cost as likely to be above as below that value? (*Ex. Ans.*: $12)

For Upper Extreme (99% Cumulative Probability):
What value would the labor cost exceed only 1% of the time? (*Ex. Ans.*: $45)

For Lower Extreme (1% Cumulative Probability):
What value would the labor cost be lower than only 1% of the time? (*Ex. Ans.*: $8)

After having obtained the preceding answers, you can either plot directly on a cumulative probability distribution or first check the answers for consistency and make adjustments by asking questions such as the following:

1. Is the labor cost more likely to be within the two quartiles (i.e., $12 to $28) or outside the two quartiles (i.e., < $12 or > $28)?

2. Is the labor cost more likely to be less than the lower quartile estimate, $12, or greater than the upper quartile estimate, $28?

If the answers to the preceding questions are not ". . . equally likely . . . ," then adjustments in one or more of the original estimates should be made until you are satisfied that the estimates represent the best judgments that can be made within the time and talent resources available. Let us suppose that, after such adjustments, the final estimates are as follows:

		Cum. Probability (Percentile)
Median (M)	$19	50
Upper quartile (UQ)	30	75
Lower quartile (LQ)	12	25
Upper extreme (UE)	48	99
Lower extreme (LE)	8	1

Figure 12-11 shows these results on a cumulative probability graph.

The next two sections give simplified approximation procedures for estimating parameters of elements thought to be distributed according to the Beta distribution and to the normal distribution.

12.8.1 Beta II Distribution

The Beta II distribution is of interest because it can describe a wide range of left-skew and right-skew conditions of differing variances.[1]

The Beta estimation procedure presented next is based on a system developed for the PERT network planning and scheduling technique. It involves first making an "optimistic" estimate, a "pessimistic" estimate, and a "most likely" estimate for the element. These estimates are to correspond to the lower (or upper) bound, upper

---

[1] Whereas the Beta I (normally called just Beta) distribution applies to variables ranging between 0 and 1, the Beta II distribution applies to variables ranging over any set of outcomes.

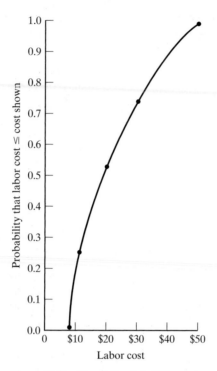

Figure 12-11   Cumulative probability graph.

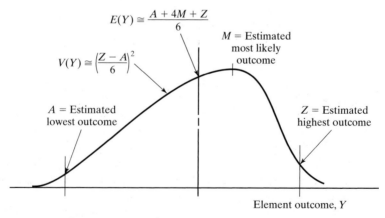

$$E(Y) \cong \frac{A + 4M + Z}{6}$$

$$V(Y) \cong \left(\frac{Z - A}{6}\right)^2$$

M = Estimated most likely outcome

A = Estimated lowest outcome

Z = Estimated highest outcome

Element outcome, Y

Figure 12-12   Demonstration of estimates with Beta distribution.

(or lower) bound, and mode, respectively, of the assumed Beta II distribution describing the element. Figure 12-12 shows an assumed Beta distribution for a typical element, together with the meaning of the preceding types of estimates. In this case, the distribution happens to be left skewed.

Once the three estimates of element outcome have been made, the approximate mean and variance of the Beta distribution for the element may be calculated as

$$E[Y] \approx \frac{A + 4M + Z}{6} \qquad (12\text{-}1)$$

and

$$V[Y] \approx \left(\frac{Z - A}{6}\right)^2, \qquad (12\text{-}2)$$

where

$E[Y]$ = estimated expected outcome,
$V[Y]$ = estimated variance of outcome,
A = estimated lowest outcome,
M = estimated most likely outcome (mode), and
Z = estimated highest outcome.

It is worthy of note that the difference between the approximate expected values as calculated by Eq. 12-1 and the exact formula is relatively small for a wide range of Beta II distribution conditions. On the other hand, the difference between the approximate variance as calculated by Eq. 12-2 and the exact formula can be quite high, and the difference usually is in the direction of underestimation of the exact value.

If several elements, as estimated by the preceding procedure, are assumed to be independent and are added together, the distribution of the total outcome so obtained is approximately normal according to the central limit theorem. The mean of this total outcome distribution can be calculated by adding the means of the individual elements. Further, the variance of this total outcome distribution can be calculated by adding the variances of the distributions of the individual elements.

12.8.2 Normal Distribution

Quite often, the best subjective estimate of the shape of the distribution of an element that can be made in practice is that the distribution is normal. It can be observed from tables of area under the normal distribution that the middle 50% of a normal distribution (i.e., between the upper quartile and the lower quartile) is within ± 0.675 standard deviations of the mean of that distribution, as shown in Fig. 12-13. Thus, for a normally distributed element, if you are willing to estimate the smallest range r (= UQ − LQ) within which that variable is expected to occur with 50% probability, then the standard deviation σ for that variable can be calculated by the relation $0.675\sigma = r/2$. In practice, it is generally sufficiently close to approximate the 0.675 with the fraction $\frac{2}{3}$.

This same idea for estimating the variance for normally distributed variables could be applied by using any other number of standard deviations and the associated probabilities. The values suggested here, however, are probably most useful because of the relative ease of visualizing the minimum range that would include 50% probability of occurrence.

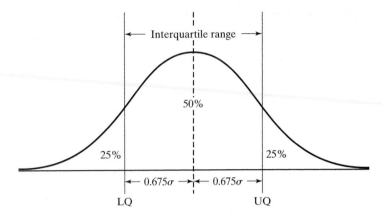

Figure 12-13   Estimating relationships for normal distribution.

As an example, suppose that the investment for a project is estimated to be normally distributed, and it is thought that there is a 50% chance it will be between $9,000
and $12,000. The standard deviation for this distribution is calculated as $\frac{2}{3}\sigma =$
($12,000 − $9,000)/2, or $\sigma = \$2,250$. The mean for the distribution is, of course,
($12,000 + $9,000)/2 = $10,500.

12.9 Human Problems in Reflecting Degree of Uncertainty in Estimates

An interesting, but troublesome, phenomenon that should be taken into account when
estimating variability is a human tendency to reflect greater certainty (i.e., less variability or spread) in estimates than is justified. This has been shown in numerous
empirical studies, perhaps the most notable of which used thousands of Harvard MBA
students as subjects over many years. Each student was asked to estimate the values of
numerous physical phenomena, which they presumably did not know for certain, in
terms of upper quartile and lower quartile values. When the exact values (outcomes) of
the phenomena were later made known to them, the students then determined if the
exact value fell within their estimated interquartile range (= UQ − LQ, as in Fig. 12-13)
without necessarily assuming a normal distribution. On the average, only about 35% of
the exact values were found to fall within their respective interquartile ranges. You
would expect that, if the estimates had fairly reflected the true degree of uncertainty,
the average should have been closer to 50%! The conclusion is that estimators in general tend to think they know more than they really do. The moral is that variability estimates generally need to be spread out sufficiently to counteract human tendencies to
compress the range of uncertainty.

Problem 12-18 is a very abbreviated version of some of the questions asked of
the Harvard MBA students. It might interest you to see how close the interquartile
range for your estimate comes to including the correct answer for the respective questions. (This problem also makes an instructive in-class experience for students.)

12.10 Developing a Risk Adjusted Interest Rate

In Chapter 11, we developed the cost of equity financing on the basis of a measure of risk that we called β. That measure reflects the market's perception of the riskiness of the company's stock (equity) relative to the riskiness of the equity market as a whole. If a particular stock had a $\beta > 1.0$, its risk, as measured by variability of returns, is greater than the risk of a portfolio consisting of all the stocks in the market. Conversely, a $\beta < 1.0$ indicates that a stock is less risky than the market. The net result was that a company's cost of equity, and hence its cost of capital, depends upon its financial and business risk for the company as a whole. This means that the usual business risks are accounted for if we evaluate projects by discounting at the WACC. But this practice can lead to erroneous results for a specific project if the project itself is more or less risky than the business itself. Safe, conservative projects might be discarded by discounting their returns at too high a rate. Speculative projects may be foolishly accepted by discounting their risky returns at too low a rate. In other words, the appropriate β for a specific project could be vastly different than the company's value of β. If a value for each project β were available, then the task would be straightforward. The problem is that there is no reliable way to estimate a project's β. Some companies overcome this difficulty by developing rules of thumb for adjusting the MARR to account for risk involved in the project. For example, one industrial company defines risk categories for income-producing projects and "normal" MARR standards as shown earlier in Section 11.4.4.

While such a tiered structure makes more sense than a single MARR based on the company WACC, there are some conceptual difficulties. Using a risk-adjusted rate implies that we understand precisely the relationship between risk and return. This is seldom the case. It is more natural to correct for risk by reflecting upon and enumerating potential variability in the cash flows themselves, rather than by adjusting the interest rate by a "fudge factor." The purpose of this and subsequent chapters is to develop the techniques that provide that capability. Those may seem tedious, but they are well worth the effort on large, capital-intensive projects.

Rules of thumb may overcompensate for risk on long-range projects. If project abandonment is possible when a new line of business does not "pan out," then risk can be curtailed. In this case, the real options approach, as described in Chapter 17, may be a better approach to the analysis.

12.11 Responses to Risk and Uncertainty

Although it is generally agreed that risks and uncertainties exist in the conduct of economic analyses, there does not exist complete agreement on how much information is to be incorporated in the analysis. Some prefer to rely heavily on the recommendations obtained from analytical models; others choose to rely solely on intuition and judgment.

There is great appeal in being able to specify quantitatively, completely, and accurately an "optimum" course of action in the face of risk and uncertainty. However, what is optimum to one individual might be very unsatisfactory to another.

Hence, two approaches have emerged in performing economic analyses that explicitly incorporate risk and uncertainty conditions in the analysis.

One approach is to develop a *descriptive* model that *describes* the economic performance of an individual investment alternative. As an illustration, you might descriptively model the present worths of each of several alternatives. No recommendation would be forthcoming *from the model*. Rather, the decision maker would be furnished descriptive information concerning each alternative; the final choice among the alternatives would require a separate action.

The second approach is to develop a *prescriptive* or *normative* model that includes an objective function to be maximized or minimized. The output from the model *prescribes* the course of action to be taken. In previous analyses under assumed certainty, the objective was implicitly stated as the maximization of, say, present worth.

Chapters 13 and 14 are concerned with *descriptive* modeling of cash flow profiles in the face of risk and uncertainty. Chapters 15 and 16 present *prescriptive* modeling approaches that can be used to evaluate economic investment alternatives under risk and uncertainty conditions. It will be found that the analyses in Chapters 13 and 14 can be useful not only when the formal analysis is to terminate with descriptive modeling, but also in providing the required inputs for the prescriptive models.

PROBLEMS

12-1. What distinction between risk and uncertainty do you think is most useful? Why?

12-2. Which of the eight causes of risk and uncertainty listed in this chapter may be said to be generally within the control (power to affect) of the economic analyst or the people from whom he or she obtains estimates?

12-3. Which of the three listed ways for a firm to change or influence degree of uncertainty is generally within the control of the economic analyst?

12-4. Under what circumstances might it be reasonable to start the analysis at the point immediately following "test point 2" in the sequence of steps shown in Fig. 12-8?

12-5. In Fig. 12-8, are there any points at which the step "Drop out projects that are clearly not contenders" might reasonably be added or deleted to make it a more reasonable representation of ideal general practice? Explain your reasoning.

12-6. What are the pros and cons of descriptive modeling versus prescriptive modeling? Which do you prefer? Why?

12-7. Classify each of the following models as descriptive or prescriptive and justify your classifications:

  a. Linear programming model

  b. Replacement model

  c. Present worth model

  d. Dynamic programming model

  e. Inventory (EOQ) model

  f. Queueing model

    g. Simulation model

    h. Quality control model

    i. Reliability model

    j. Warehouse location model

    k. Rate of return model

    l. Forecasting model

12-8. a. If you were faced with the decision problem shown in the table under risk, which action would you choose? Why?

	S_1 (0.1)	S_2 (0.7)	S_3 (0.2)
A_1	−$40	$30	$20
A_2	− 20	20	40
A_3	0	20	10

    b. Answer part (a) assuming all payoffs are in thousands of dollars.

    c. Answer part (a) assuming the decision problem is for a large, wealthy corporation and all payoffs are in tens of thousands of dollars. Further assume that it is unacceptable in that corporate environment to recommend a project with a potential payoff loss greater than, say, $250,000.

12-9. a. If you were faced with the personal decision problem shown in the table under risk, which would you choose? Why?

	S_1 (0.4)	S_2 (0.6)
A_1	$110	−$20
A_2	− 40	85
A_3	0	50

    b. Answer part (a) assuming all the payoffs are in hundreds of thousands of dollars.

12-10. If the decision problems in Problems 12-8 and 12-9 had been decisions under uncertainty, would your preferences change? Why or why not?

12-11. Estimate the expected total remaining life of a given car, either one you happen to own or some other car with which you are familiar, in terms of the following:

    a. Single best estimate

    b. Subjective continuous probability distribution (cumulative and probability density function)

    c. Beta II distribution (calculate the approximate mean and variance)

    d. Normal distribution (calculate the approximate mean and variance)

12-12. Rework Problem 12-11, except change the variable to be estimated to the number of basketball (or football) games to be won next year by the team of your choice. Of course, since the variable is discrete, the Beta II and normal distributions are only approximations.

12-13. Repair costs have been recorded as follows:

$2,250	$4,000	$ 750	$1,250
1,000	500	1,000	750
500	250	250	2,750
750	1,500	1,750	250
500	1,750	1,500	1,000

On the basis of the data in the table,

a. Find the median value.

b. Find the upper and lower quartile values.

c. Estimate the probability that a given repair will cost $1,000 or more.

d. Show the data in graph form with "Cost to repair in $" as the horizontal axis and "Estimated probability that repair cost will not exceed $ _____ " as the vertical axis.

12-14. A heat exchanger is being installed as part of a plant modernization program. It costs $10,000, including installation, and is expected to reduce the overall plant operating cost by $2,500 per year. Estimates of the useful life of the heat exchanger range from an optimistic 12 years to a pessimistic 4 years. The most likely value is 5 years.

a. Using the Beta II approximation formula, determine the before-tax rate of return at the expected life.

b. At what life would the investment just be recovered (i.e., the payback period)?

12-15. Estimated maintenance expenses are very uncertain, but it is thought that there is a 50% chance they will be less than $15,000 and more than $9,000. Assuming the estimate can best be described by a normal distribution, what is

a. the estimated mean and standard deviation?

b. the probability the expense will be more than $12,000?

c. the probability the expense will be more than $16,000? *Hint:* Use Appendix E and the relation

$$S = \frac{\text{Upper limit cost} - \text{Expected (mean) cost}}{\text{Standard deviation}}.$$

12-16. Using the Beta II distribution approximation formula and the fact that, for independent elements (variables) being added, the expected total = the sum of the expected elements and the variance of the total = the sum of the variances of the independent elements, find the expected total cost and variance of the total cost for the following (all numbers are in thousands of dollars):

Cost element	Optimistic cost	Most likely cost	Pessimistic cost
1. Direct labor	$79	$95	$95
2. Direct material	60	66	67
3. Indirect expenses	93	93	96

12-17. Estimate the age of your instructor (or someone else whose exact age is unknown to you) in terms of the following:

    a. Most likely value.

    b. Interquartile range (UQ and LQ). Assuming a normal distribution for your estimate of the age, determine the mean and standard deviation. Should your mean correspond to your answer to (a)? What does a relatively low versus high value indicate?

12-18. For each of the following physical phenomena, treat it as a variable (to you) and estimate its value in terms of an interquartile range (UQ and LQ):

    a. Area of the United States (in millions of square miles).

    b. Weight of the human heart (in ounces, not including blood).

    c. Maximum speed of Boeing 747 "jumbo jet" (in mi/hr).

    d. Elevation of Mt. McKinley in Alaska (in ft above sea level).

    e. Number of muscles in the human body (not including the brain!).

    f. Proportion of the world's surface covered by sea (in %).

Allow yourself less than half a minute to estimate each, thus eliminating the possibility of "looking up" the value (outcome). Your UQ and LQ should reflect a range within which you think there is a 50% probability that the value (outcome) will be contained. For your interest in determining your (and perhaps a group's) variable estimating tendencies, the correct answers (values) for parts (a) through (f) are given at the end of the problems in Chapter 13.

13
Sensitivity Analysis

13.1 Introduction

There are a number of procedures for *describing* analytically the effects of risk and uncertainty on capital projects. Such procedures are generally categorized as *sensitivity* or *risk analyses*. In this chapter, sensitivity analysis procedures are described.

Sensitivity analyses are performed when conditions of uncertainty exist for one or more parameters. The objectives of a sensitivity analysis are to provide the decision maker with information concerning (1) the behavior of the measure of economic effectiveness due to errors in estimating various values of the parameters and (2) the potential for reversals in the preferences for economic investment alternatives. The term "sensitivity analysis" is derived from the desire to measure the sensitivity of a decision to changes in the values of one or more parameters.

13.2 One-at-a-Time Procedure and Break-Even Analysis

One-at-a-time procedures consider the sensitivity of the measure of economic effectiveness (e.g., AW) caused by changes in a single parameter. A popular form of this type of sensitivity analysis is called *break-even analysis,* which is useful in situations in which there is uncertainty regarding a single element (or parameter) in an engineering economy study. The break-even point for an element in the analysis of a single project is defined as that value of the element at which the project is marginally acceptable (barely justified). Example 13-1 illustrates break-even analysis and the more general one-at-a-time procedure of sensitivity analysis.

Example 13-1

An investment project is being considered by a firm. The following most likely (or expected) estimates have been provided:

Parameter	Estimate
Investment:	$10,000
Project life:	5 yr
Salvage value:	$2,000
Annual receipts:	$5,000
Annual disbursements:	$2,200
MARR = 8%	
AW(8%) = $636	

(a) Suppose that, for the single project for which most likely estimates are shown, the parameter for which there is particular uncertainty is the life of the project; hence, it is desired to find the minimum project life at which the project will barely be justified (i.e., the break-even life).

(b) A sensitivity analysis is to be performed for those parameters whose estimated values are most uncertain. In this case, the effect of up to $\pm 100\%$ changes in project life, MARR, and annual disbursements on the AW are to be considered one at a time.

Solution

(a) A break-even analysis on project life results in this equation:

$$\$5,000 - \$2,200 - (\$10,000 - \$2,000)(A/P, 8\%, N) - \$2,000(8\%) = 0;$$

$$(A/P, 8\%, N) = \frac{-\$2,640}{-\$8,000} = 0.330.$$

By interpolating in the interest tables, we find that $N = 3.7$ years. Thus, a project life of more than 3.7 years results in a positive AW and an acceptable project.

(b) The approach to be used in the one-at-a-time procedure depends on whether or not the investment will be repeated within the planning horizon. If the investment will be repeated a sufficient number of times that either an indefinite planning horizon or a least-common-multiple-of-lives planning horizon is appropriate, then it is easier to perform an AW analysis than, say, a PW analysis. However, if the investment will not be repeated, then a present worth analysis may be easier to perform. Both approaches will be used to illustrate the differences in the results.

In the annual worth approach,

$$AW = -\$10,000(A/P, i\%, N) + \$5,000 - D + \$2,000(A/F, i\%, N),$$

where D denotes the annual disbursement. The results of a one-at-a-time sensitivity analysis are depicted graphically in Fig. 13-1. The abscissa expresses the changes in i, N, and D as a percentage of the estimated values.

In the present worth approach,

$$PW = -\$10,000 + \$5,000(P/A, i\%, N) - D(P/A, i\%, N)$$
$$+ \$2,000(P/F, i\%, N).$$

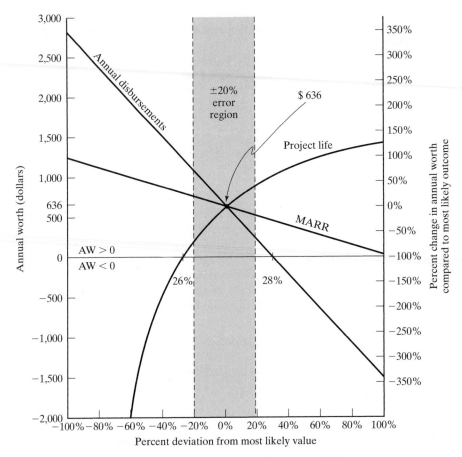

Figure 13-1   One-at-a-time sensitivity analysis of AW.

The results of a one-at-a-time sensitivity analysis are depicted in Fig. 13-2. It appears, from the ± 20% error regions, that both the AW and the PW are equally sensitive to changes in the project life and annual receipts. Both AW and PW are relatively insensitive to changes in the MARR. Furthermore, the project will be profitable (AW > 0 and PW > 0) so long as either the project life is at least 74% of the estimated value, the MARR is no more than twice the estimated value, or the annual disbursements do not increase by more than approximately 28%. The primary difference in the AW and PW analyses is the effect of project life. With AW, the effect of changes in project life decreases at a faster rate (i.e., levels off sooner) than with PW. In the ± 20% error region, however, the sensitivity of project life is relatively the same for both measures of economic profitability. ■

Computer spreadsheets are extremely useful when conducting sensitivity studies. As an example, Figure 13-3 shows sample spreadsheet formulas for one-at-a-time sensitivity analyses (in PWs) for Example 13-1. Figure 13-4 then shows calculated results for a large number of percentage deviations in life, MARR, and annual disbursements

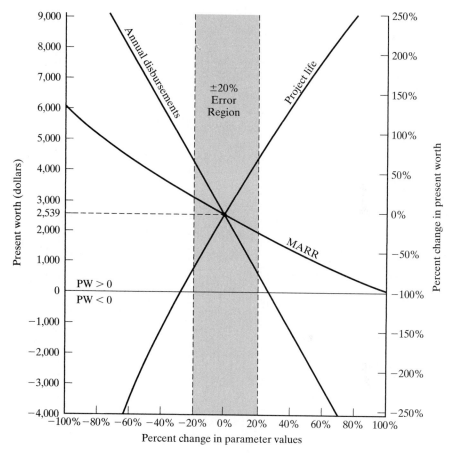

Figure 13-2   One-at-a-time sensitivity analysis of PW.

(which could have been used for plotting Figure 13-2). Notice that the PW results shown in Figure 13-2 are somewhat similar to the AW results shown in Figure 13-1. However, Figure 13-2 better illustrates the nonlinearity of PW to changes in the MARR and useful life.

Example 13-2

Four alternative incinerators are to be compared on the basis of the following estimates:

	Incinerator			
	A	B	C	D
First cost:	$3,000	$3,800	$4,500	$5,000
Life:	10 yr	10 yr	10 yr	10 yr
Salvage value:	$0	$0	$0	$0
Annual operating disbursements:	$1,800	$1,770	$1,470	$1,320
MARR = 10%				

R/C	A	B	C	D
1	Nominal Values			
2				
3	Investment:		INV	
4	Salvage Value:		SV	
5	Annual Receipts:		AR	
6	Annual Disbursements:		AD	
7	Project Life (yrs):		N	
8	MARR:		I	
9				
10				Annual
11	% Change	Life	MARR	Disbursements
12				
13	-100%	-INV+SV/(1+i)^(N*(1+A13))-PV(i,N*(1+A13),AR-AD)	-INV+SV/(1+i*(1+A13))^N-PV(i*(1+A13),N,AR-AD)	-INV+SV/(1+i)^N-PV(i,N,AR-AD*(1+A13))
14	-90%	-INV+SV/(1+i)^(N*(1+A14))-PV(i,N*(1+A14),AR-AD)	-INV+SV/(1+i*(1+A14))^N-PV(i*(1+A14),N,AR-AD)	-INV+SV/(1+i)^N-PV(1,N,AR-AD*(1+A14))
15	-80%	-INV+SV/(1+i)^(N*(1+A15))-PV(i,N*(1+A15),AR-AD)	-INV+SV/(1+i*(1+A15))^N-PV(i*(1+A15),N,AR-AD)	-INV+SV/(1+i)^N-PV(1,N,AR-AD*(1+A15))
16	-70%	-INV+SV/(1+i)^(N*(1+A16))-PV(i,N*(1+A16),AR-AD)	-INV+SV/(1+i*(1+A16))^N-PV(i*(1+A16),N,AR-AD)	-INV+SV/(1+i)^N-PV(1,N,AR-AD*(1+A16))
17	-60%	-INV+SV/(1+i)^(N*(1+A17))-PV(i,N*(1+A17),AR-AD)	-INV+SV/(1+i*(1+A17))^N-PV(i*(1+A17),N,AR-AD)	-INV+SV/(1+i)^N-PV(1,N,AR-AD*(1+A17))
18	-50%	-INV+SV/(1+i)^(N*(1+A18))-PV(i,N*(1+A18),AR-AD)	-INV+SV/(1+i*(1+A18))^N-PV(i*(1+A18),N,AR-AD)	-INV+SV/(1+i)^N-PV(1,N,AR-AD*(1+A18))
19	-40%	-INV+SV/(1+i)^(N*(1+A19))-PV(i,N*(1+A19),AR-AD)	-INV+SV/(1+i*(1+A19))^N-PV(i*(1+A19),N,AR-AD)	-INV+SV/(1+i)^N-PV(1,N,AR-AD*(1+A19))
20	-30%	-INV+SV/(1+i)^(N*(1+A20))-PV(i,N*(1+A20),AR-AD)	-INV+SV/(1+i*(1+A20))^N-PV(i*(1+A20),N,AR-AD)	-INV+SV/(1+i)^N-PV(1,N,AR-AD*(1+A20))
21	-20%	-INV+SV/(1+i)^(N*(1+A21))-PV(i,N*(1+A21),AR-AD)	-INV+SV/(1+i*(1+A21))^N-PV(i*(1+A21),N,AR-AD)	-INV+SV/(1+i)^N-PV(1,N,AR-AD*(1+A21))
22	-10%	-INV+SV/(1+i)^(N*(1+A22))-PV(i,N*(1+A22),AR-AD)	-INV+SV/(1+i*(1+A22))^N-PV(i*(1+A22),N,AR-AD)	-INV+SV/(1+i)^N-PV(1,N,AR-AD*(1+A22))
23	0%	-INV+SV/(1+i)^(N*(1+A23))-PV(i,N*(1+A23),AR-AD)	-INV+SV/(1+i*(1+A23))^N-PV(i*(1+A23),N,AR-AD)	-INV+SV/(1+i)^N-PV(1,N,AR-AD*(1+A23))
24	10%	-INV+SV/(1+i)^(N*(1+A24))-PV(i,N*(1+A24),AR-AD)	-INV+SV/(1+i*(1+A24))^N-PV(i*(1+A24),N,AR-AD)	-INV+SV/(1+i)^N-PV(1,N,AR-AD*(1+A24))
25	20%	-INV+SV/(1+i)^(N*(1+A25))-PV(i,N*(1+A25),AR-AD)	-INV+SV/(1+i*(1+A25))^N-PV(i*(1+A25),N,AR-AD)	-INV+SV/(1+i)^N-PV(1,N,AR-AD*(1+A25))
26	30%	-INV+SV/(1+i)^(N*(1+A26))-PV(i,N*(1+A26),AR-AD)	-INV+SV/(1+i*(1+A26))^N-PV(i*(1+A26),N,AR-AD)	-INV+SV/(1+i)^N-PV(1,N,AR-AD*(1+A26))
27	40%	-INV+SV/(1+i)^(N*(1+A27))-PV(i,N*(1+A27),AR-AD)	-INV+SV/(1+i*(1+A27))^N-PV(i*(1+A27),N,AR-AD)	-INV+SV/(1+i)^N-PV(1,N,AR-AD*(1+A27))
28	50%	-INV+SV/(1+i)^(N*(1+A28))-PV(i,N*(1+A28),AR-AD)	-INV+SV/(1+i*(1+A28))^N-PV(i*(1+A28),N,AR-AD)	-INV+SV/(1+i)^N-PV(1,N,AR-AD*(1+A28))
29	60%	-INV+SV/(1+i)^(N*(1+A29))-PV(i,N*(1+A29),AR-AD)	-INV+SV/(1+i*(1+A29))^N-PV(i*(1+A29),N,AR-AD)	-INV+SV/(1+i)^N-PV(1,N,AR-AD*(1+A29))
30	70%	-INV+SV/(1+i)^(N*(1+A30))-PV(i,N*(1+A30),AR-AD)	-INV+SV/(1+i*(1+A30))^N-PV(i*(1+A30),N,AR-AD)	-INV+SV/(1+i)^N-PV(1,N,AR-AD*(1+A30))
31	80%	-INV+SV/(1+i)^(N*(1+A31))-PV(i,N*(1+A31),AR-AD)	-INV+SV/(1+i*(1+A31))^N-PV(i*(1+A31),N,AR-AD)	-INV+SV/(1+i)^N-PV(1,N,AR-AD*(1+A31))
32	90%	-INV+SV/(1+i)^(N*(1+A32))-PV(i,N*(1+A32),AR-AD)	-INV+SV/(1+i*(1+A32))^N-PV(i*(1+A32),N,AR-AD)	-INV+SV/(1+i)^N-PV(1,N,AR-AD*(1+A32))
33	100%	-INV+SV/(1+i)^(N*(1+A33))-PV(i,N*(1+A33),AR-AD)	-INV+SV/(1+i*(1+A33))^N-PV(i*(1+A33),N,AR-AD)	-INV+SV/(1+i)^N-PV(1,N,AR-AD*(1+A33))

Figure 13-3 Sample spread sheet formulas for Example 13-1 (one-at-a-time sensitivity analysis of PW).

R/C	A	B	C	D
1	Nominal Values			
2				
3	Investment:		$10,000	
4	Salvage Value:		$2,000	
5	Annual Receipts:		$5,000	
6	Annual Disbursements:		$2,200	
7	Project Life (yrs):		5	
8	MARR:		8%	
9				
10				Annual
11	% Change	Life	MARR	Disbursements
12				
13	-100%	($8,000)	$6,000	$11,325
14	-90%	($6,754)	$5,592	$10,446
15	-80%	($5,556)	$5,200	$9,568
16	-70%	($4,402)	$4,822	$8,690
17	-60%	($3,292)	$4,459	$7,811
18	-50%	($2,224)	$4,109	$6,933
19	-40%	($1,196)	$3,772	$6,054
20	-30%	($208)	$3,447	$5,176
21	-20%	$744	$3,134	$4,298
22	-10%	$1,660	$2,832	$3,419
23	0%	$2,541	$2,541	$2,541
24	10%	$3,389	$2,260	$1,662
25	20%	$4,204	$1,988	$784
26	30%	$4,989	$1,726	($94)
27	40%	$5,745	$1,473	($973)
28	50%	$6,472	$1,228	($1,851)
29	60%	$7,171	$992	($2,730)
30	70%	$7,844	$763	($3,608)
31	80%	$8,492	$542	($4,486)
32	90%	$9,115	$328	($5,365)
33	100%	$9,715	$120	($6,243)

Figure 13-4   Calculated results for Example 13-1 (one-at-a-time sensitivity analysis of PW).

It is desired to determine the sensitivity of the preferred alternative to either estimation errors or changes in the annual operating disbursements.

Solution   The equivalent uniform annual cost is given in Fig. 13-5 for various percent changes in the annual operating disbursements. The choice will be between alternatives A and D, because one or the other will have the minimum annual cost over the entire range. Setting $AC(A) = AC(D)$ and solving for the percent change in annual disbursements gives

$$\$3,000(A/P, 10\%, 10) + \$1,800(1 + x) = \$5,000(A/P, 10\%, 10)$$
$$+ \$1,320(1 + x),$$
$$\$2,000(A/P, 10\%, 10) = \$480(1 + x),$$
$$\$325.50 = \$480(1 + x),$$
$$x = -0.322 \text{ or } -32.2\%.$$

Thus, so long as the annual disbursements are at least 67.8% of the estimated value, then alternative D will be preferred. On the basis of the analysis, it might be concluded that the decision

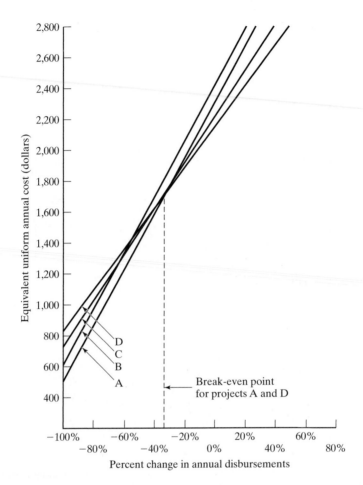

Figure 13-5   Sensitivity analysis of changes in operating disbursements for Example 11-2.

is relatively insensitive to changes in annual disbursements. Furthermore, this example illustrates that the break-even point for an element in the comparison of two or more alternatives is
the value of the element at which the alternatives are equally desirable.   ∎

A related sensitivity test that is often quite valuable is to determine the relative (or
absolute) change in one or more elements (parameters) which will just reverse the decision. Applied to the investment project example (Example 13-1), this means the relative
change that will decrease the net AW by $636 so that it reaches $0. Table 13-1 shows
this and emphasizes that decision reversal is most sensitive to relative changes in annual
savings and least sensitive to relative changes in the salvage value. To further illustrate
the computations, let y = estimated outcome for the investment at which AW(8%) = $0:

$$-y(A/P, 8\%,5) + \$2,000(A/F,8\%, 5) + \$5,000 - 2,200 = 0.$$

TABLE 13-1   Example of Sensitivity to Decision Reversal

	Most likely estimate	To reverse decision (decrease AW to $0)		
		Estimate outcome	Change amount	Change amount as % of most likely
Investment:	$10,000	−$12,540	+$2,540	+ 25%
Life:	5 yr	3.7 yr	−1.3 yr	− 26%
Salvage value:	$2,000	−$1,740	−$3,740	−187%
Annual savings:	$5,000	$4,363	− $637	− 13%
Annual disbursements:	$2,200	$2,837	+ $637	+ 29%
Minimum attractive rate of return:	8%	16.2%	+8.2%	+103%

From this equation, we find $y \simeq \$12,540$, which is 25% (rounded) more than the $10,000 most likely estimate.

For examples of tabular and graphical displays of sensitivity analyses applied to comparison of alternatives, the interested reader is referred to Appendix 13-A.

13.3 Multiparameter Procedures

The preceding discussion concentrated on the sensitivity of the measure of merit to changes in a single parameter one at a time. Such an approach allows the relative sensitivity of parameters to be identified. However, it overlooks the possibility of interaction among parameters. Additionally, since estimation errors will generally occur in more than one parameter, it is important to examine the sensitivity of the measure of effectiveness to multiple parameters.

Two approaches will be described in the consideration of multiparameter sensitivity analysis. First, the approach used in the one-at-a-time analysis will be extended to multiple parameters; second, an optimistic–pessimistic approach will be described.

13.3.1 Sensitivity Surface Approach

In the one-at-a-time analysis depicted in Figs. 13-1 and 13-2, sensitivity curves were obtained. If combinations of more than one parameter are analyzed, a sensitivity surface is required.

Example 13-3
Consider the data used to develop Figs. 13-1 and 13-2. Suppose that the most critical parameters are believed to be the initial investment and the annual receipts. Perform a multiparameter sensitivity analysis involving the two parameters.

Solution   Let x denote the percent change in the initial investment, and let y denote the percent change in the annual receipts. The annual worth can be given as

$$\text{AW} = -\$10,000(1 + x)(A/P, 8\%, 5) + \$5,000(1 + y) - \$2,200$$
$$+ \$2,000(A/F, 8\%, 5)$$
$$= \$636.32 - \$2,504.60x + \$5,000y.$$

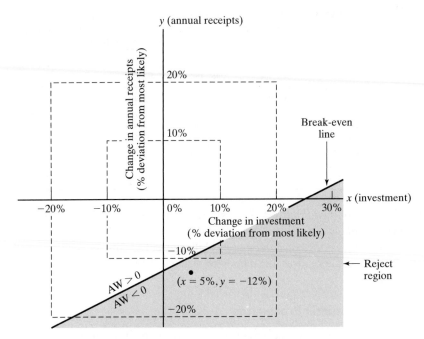

Figure 13-6   Two-parameter sensitivity analysis.

The investment will be profitable so long as AW ≥ 0 or

$$y \geq -0.127264 + 0.50092x.$$

Plotting the inequality relationship yields the two regions depicted in Fig. 13-6. The favorable region (AW > 0) is dominant. If errors in estimating the values of the investment and annual receipts were statistically independent and uniformly distributed over the interval ±10%, then the probability of AW < 0 would be given by the ratio of the shaded area below the break-even line, which is contained within the ±10% error region to the total area in the error region. However, it is not generally the case that estimation errors are either statistically independent or uniformly distributed. Furthermore, the incorporation of probabilistic considerations in sensitivity analyses lies in the domain of risk analysis, to be considered in Chapter 14. If it is anticipated that ±20% estimation errors will be made, then there is greater concern for the profitability of the investment. For example, the combination of a 5% increase in the initial investment and a 12% reduction in annual receipts would result in a negative annual worth. ∎

Example 13-4
Suppose that, in the previous example, the life of the investment were also a critical parameter. Perform a multiparameter sensitivity analysis for the three parameters.

    Solution   It is difficult to develop a three-dimensional representation of the sensitivity surface. But it is possible to gain an insight into the sensitivity of annual worth to errors in

estimating the three parameters by plotting a family of break-even lines for each possible value of project life. Letting AW(N) denote the annual worth as a function of N, the following
results are obtained:

$$AW(N) = -\$10,000(1 + x)(A/P, 8\%, N) + \$5,000(1 + y) - \$2,200$$
$$+ 2,000(A/F, 8\%, N) \geq 0;$$

$$AW(2) = -\$1,846.62 - \$5,607.70x + \$5,000y \geq 0;$$
$$y \geq 0.369324 + 1.12154x;$$

$$AW(3) = -\$464.24 - \$3,880.30x + \$5,000y \geq 0;$$
$$y \geq 0.092848 + 0.77606x;$$

$$AW(4) = \$224.64 - \$3,019.20x + \$5,000y \geq 0;$$
$$y \geq -0.044928 + 0.60384x;$$

$$AW(5) = \$636.32 - \$2,504.60x + \$5,000y \geq 0;$$
$$y \geq -0.127264 + 0.50092x;$$

$$AW(6) = \$909.44 - \$2,163.20x + \$5,000y \geq 0;$$
$$y \geq -0.181888 + 0.43264x;$$

$$AW(7) = \$1,121.15 - \$1,920.70x + \$5,000y \geq 0;$$
$$y \geq -0.22423 + 0.38414x.$$

The family of break-even lines is given in Fig. 13-7. For $N = 4$, 5, 6 (a $\pm$ 20% range), the profitability of the investment continues to appear promising. However, for $N = 4$, there is very little
opportunity for error in estimating the investment required and the annual receipts. For example, if $N = 4$, then with a 10% increase in the investment there must be at least a 1.55% increase
in annual receipts for the annual worth to be positive.   ■

13.3.2 Optimistic–Pessimistic Approach

The optimistic–pessimistic approach, as its name implies, involves changing
estimates of one or more elements (parameters) in a favorable outcome (optimistic)
direction and in an unfavorable outcome (pessimistic) direction to determine the
effect of these various changes on the economic study result.

In using this method, it is desirable for the estimator to adopt a guideline philosophy of "how optimistic" and "how pessimistic" in making the estimates. One convenient way to do this is to adopt a probabilistic statement such as "An optimistic
estimate will mean a value of the element that we would expect to be bettered or
exceeded in outcome no more than, say, 5% of the time, while a pessimistic estimate is
a value of the element that we would expect to be worse than the final outcome no more
than, say, 5% of the time." If, for example, the 5% were changed to 25%, the optimistic
and pessimistic estimates would correspond to quartile estimates described in the latter
part of Chapter 12. Generally, though, 1% to 10% "tail" or "outside of" probabilities
are used. Figure 13-8 illustrates this.

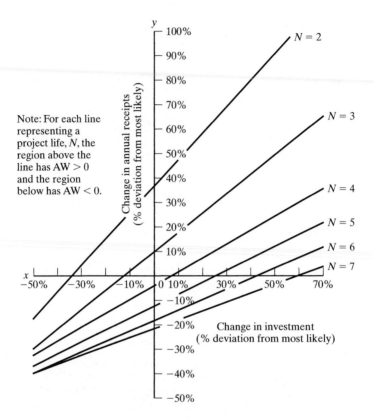

Figure 13-7   Three-parameter sensitivity analysis.

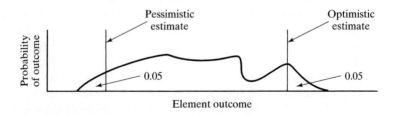

Figure 13-8   Illustration of estimates for optimistic–pessimistic approach.

Example 13-5

For Example 13-1, suppose the optimistic, pessimistic, and most likely estimates are as follows:

	Estimation condition		
	Optimistic	Most likely	Pessimistic
Investment:	$10,000	$10,000	$10,000
Life:	7 yr	5 yr	4 yr
Salvage value:	$2,000	$2,000	$2,000
Annual receipts:	$6,000	$5,000	$4,500
Annual disbursements:	$2,200	$2,200	$2,400
Interest rate:	8%	8%	8%

Solution   Using the annual worth method, we have

$$AW(\text{optimistic}) = -\$10,000(A/P, 8\%, 7) + \$2,000(A/F, 8\%, 7) + \$3,800$$
$$= \$2,103;$$
$$AW(\text{most likely}) = -\$10,000(A/P, 8\%, 5) + \$2,000(A/F, 8\%, 5) + \$2,800$$
$$= \$636;$$
$$AW(\text{pessimistic}) = -\$10,000(A/P, 8\%, 4) + \$2,000(A/F, 8\%, 4) + \$2,100$$
$$= -\$475.$$

These sample calculations are based on the assumption that the elements will all equal the pessimistic estimates, all equal the most likely estimates, or all equal the optimistic estimates. Of course, we can investigate the effect on calculated results when various elements equal the optimistic, most likely, or pessimistic estimates in various combinations. When this is done, it is usually helpful to summarize the results in tabular form.   ■

Example 13-6

For Example 13-5, note that estimates for the three conditions differed only for the project life, annual receipts, and annual disbursements. Table 13-2 shows the calculated results for all combinations of estimating conditions—optimistic (*O*), most likely (*M*), and pessimistic (*P*)—for only the three elements (parameters). Displays such as Table 13-2 can be made more

TABLE 13-2   Calculated Results for All Combinations
of Estimating Conditions in Example 13-6

		Net annual worth								
		Annual disb.—*O*			Annual disb.—*M*			Annual disb.—*P*		
Life		*O*	*M*	*P*	*O*	*M*	*P*	*O*	*M*	*P*
Annual	O	$2,103*	$1,636*	$1,235	$2,103*	$1,636*	$1,235	$1,903*	$1,436	$1,035
receipts	M	1,103	636	235	1,103	636	235	903	436	35
	P	603	136	− 275	603	136	− 275	403	− 64	− 475

Key: *O* is optimistic outcome.   *Indicates net AW > $1,500
    M is most likely outcome.
    P is pessimistic outcome.

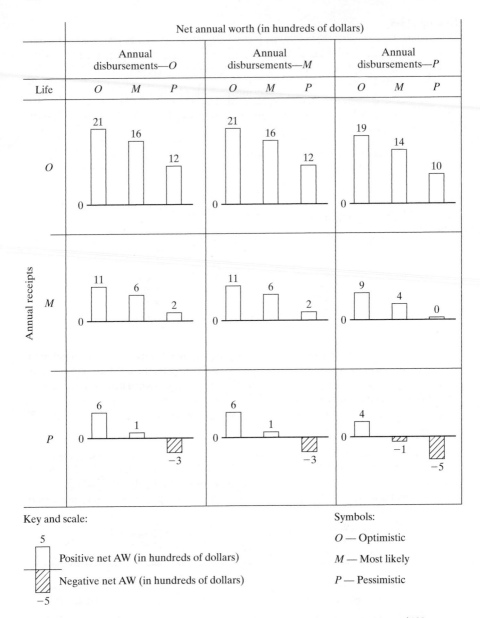

Figure 13-9   Example 13-6 results shown in histogram form with rounding to nearest $100.

informative by the use of graphical symbols. For example, all outcomes greater than $1,500 are marked for emphasis in Table 13-2. Figure 13-9 shows results in histogram bar form with each outcome rounded to the nearest $100. Such rounding can aid decision makers, especially when showing results in a table. Other devices such as color coding, shading, etc., can be very useful for communicating in terms of tables and graphs.   ∎

13.4 Summary

Several popular descriptive techniques for exploring the sensitivity of estimates in capital investment evaluations were presented in this chapter. Specifically, single-parameter and multiparameter sensitivity analyses were illustrated. In addition, break-even analysis was covered as a subset of one-at-a-time sensitivity analysis. Illustration of tabular and graphical approaches to sensitivity analysis for comparison of alternatives has been provided as an appendix to this chapter.

PROBLEMS

13-1. A certain potential investment project is critical to a firm. The following are "best" or "most likely" estimates:

Investment:	$100,000
Life:	10 yr
Salvage value:	$20,000
Net annual cash flow:	$30,000
Minimum required rate of return:	10%

It is desired to show the sensitivity of a measure of merit (net annual worth) to variation, over a range of ±50% of the expected values, in the following elements: (a) life, (b) net annual cash flow, (c) interest rate. Graph the results. To which element is the decision most sensitive?

13-2. Suppose that, for a certain potential investment project, the optimistic–pessimistic estimates are as follows:

	Optimistic	Most likely	Pessimistic
Investment:	$90,000	$100,000	$120,000
Life:	12 yr	10 yr	6 yr
Salvage value:	$30,000	$20,000	0
Net annual cash flow:	$35,000	$30,000	$20,000
Minimum required rate of return:	10%	10%	10%

a. What is the net annual worth for each of the three estimation conditions?

b. It is thought that the most critical elements are life and net annual cash flow. Develop a table showing the net annual worth for all combinations of estimates for those two elements, assuming all other elements to be at their "most likely" values. Also, show results with histogram bars for entries in your table.

13-3. Two pumps, A and B, are being considered for a given drainage need. Both pumps operate at a rated output of 8 kW (10.7 hp), but differ in initial cost and electrical efficiency. Electricity costs 5¢ per kWh, and the minimum before-tax rate of return is 15%. The following data are relevant:

	Pump A	Pump B
Cost installed:	$3,500	$4,500
Expected life to termination at zero market value:	10 yr	15 yr
Maintenance cost per 1,000 hr of operation:	$50	$30
Efficiency:	60%	80%

a. The critical variable hardest to estimate is the number of hours of operation per year. Determine the break-even point for this variable.

b. If the operating time is greater than the break-even point in part (a), which pump is better?

c. Plot the total annual costs of each pump as a function of hours of operation.

13-4. A new all-season hotel will require land costing $300,000 and a structure costing $500,000. In addition, fixtures will cost $150,000, and working capital of 30 days' gross income at 100% capacity will be required. While the land is nondepreciable, the investment in fixtures should be recovered in 8 years, and the investment in the structure should be recovered in 25 years.

When the hotel is operating at 100% capacity, the gross income will be $1,400 per day. Fixed operating expenses, exclusive of depreciation and interest, will amount to $120,000 per year. Operating expenses, which vary linearly in proportion to the level of operation, are $80,000 per year at 100% capacity.

a. At what percentage of capacity must the hotel operate to earn a before-tax minimum attractive rate of return of 20%?

b. Plot the revenue and total cost (including cost of capital) as a function of percentage of capacity. (*Note:* The point at which revenue and the total of all costs, including capital, are equal is commonly called the *tipping point.*)

13-5. It is desired to determine the most economical thickness of insulation for a large cold-storage room. Insulation is expected to cost $150 per 1,000 sq ft of wall area per in. of thickness installed and to require annual property taxes and insurance of 5% of first cost. It is expected to have $0 net salvage value after a 20-year life. The following are estimates of the heat loss per 1,000 sq ft of wall area for several thicknesses:

Insulation, in.	Heat loss, Btu per hr
3	4,400
4	3,400
5	2,800
6	2,400
7	2,000
8	1,800

The cost of heat removal is estimated at $0.02 per 1,000 Btu per hr. The minimum required yield on investment is 20% per year. Assuming continuous operation throughout the year, analyze the sensitivity of the optimal thickness to errors in estimating the cost of heat removal. Use the AW technique. (A computer spreadsheet should be considered here.)

13-6. An industrial machine costing $10,000 will produce net cash savings of $4,000 per year. The machine has a 5-year economic life but must be returned to the factory for major repairs after 3 years of operation. These repairs cost $5,000. The company's cost of capital is approximately 10%. What rate of return will be earned on purchase of this machine? Analyze the sensitivity of the internal rate of return to ±$2,000 changes in the repair cost.

13-7. The following alternatives are available to fill a given need that is expected to exist indefinitely:

	Plan A	Plan B	Plan C
First cost:	$2,000	$6,000	$12,000
Life cycle:	6 yr	3 yr	4 yr
Annual disbursements:	$3,500	$1,000	$400

Each is expected to have $0 salvage value at the end of each life cycle.

a. Analyze the sensitivity of the preferred plan due to ±30% errors in estimating the annual disbursements. Use a MARR of 10% per year.

b. Analyze the sensitivity of the preferred plan due to ±50% errors in estimating the MARR (i.e., the MARR will vary from 5% to 15%).

13-8. An improved facility costing $50,000 has been proposed. Construction time will be 2 years, with expenditures of $20,000 the first year and $30,000 the second year. Cash flows are as follows:

Year	Savings
−1	−$20,000
0	−30,000
1	10,000
2	14,000
3	18,000
4	22,000
5	26,000

The facility will not be required after 5 years and will have a market value of $5,000. Analyze the sensitivity of annual worth due to errors in estimating both the savings in the first year and the magnitude of the gradient amount. Use a table to show results of ±50% changes in both variables. The MARR is 10% per year.

13-9. It is desired to determine the optimal height for a proposed building that is expected to last 40 years and then be demolished at zero salvage. The following are pertinent data:

	Number of floors			
	2	3	4	5
Building first cost:	$200,000	$250,000	$320,000	$400,000
Annual revenue:	40,000	60,000	85,000	100,000
Annual cash disbursements:	15,000	25,000	25,000	45,000

In addition to the building first cost, the land requires an investment of $50,000 and is expected to retain that value throughout the life period. Analyze the sensitivity of the decision due to changes in estimates of the MARR at values of 10%, 15%, and 20%. Use the PW method and ignore income taxes.

13-10. There are five alternative machines to do a given job. Each is expected to have a market value of 100% of the investment amount at the end of its life of 4 years. The firm's minimum attractive rate of return is 12%. Analyze the sensitivity of the best choice due to the salvage value ranging from 0% to 100% of the initial investment.

			Machine		
	A	B	C	D	E
Investment:	$10,000	$14,000	$21,000	$27,000	$34,000
Net cash flow per yr:	1,100	1,800	2,800	3,400	4,450

13-11. Best estimates of the parameters for an investment are given in the table. It is expected that the investment will be repeated indefinitely.

Initial investment:	$15,000
Net annual receipt:	$2,500
Project life:	10 yr
Salvage value:	$0
MARR = 15%	

a. Perform a one-at-a-time sensitivity analysis to help determine the most critical parameter(s). One way to do this is to calculate "sensitivity to decision reversal" as in Table 13-1.

b. Perform a multiparameter sensitivity analysis by using the sensitivity surface approach. Base the analysis on the initial investment and the net annual receipts.

13-12. An office building is considering converting from a coal burning furnace to one that burns either fuel oil or natural gas. The cost of converting to fuel oil is estimated to be $80,000 initially; annual operating costs are estimated to be $4,000 less than that experienced using the coal furnace. Approximately 140,000 Btu are produced per gallon of fuel oil; fuel oil is anticipated to cost $1.10 per gallon.

    The cost of converting to natural gas is estimated to be $60,000 initially; additionally, annual operating and maintenance costs are estimated to be $6,000 less than that for the coal-burning furnace. Approximately 1,000 Btu are produced per cubic foot of natural gas; it is estimated that natural gas will cost $0.02 per cu ft.

    A planning horizon of 20 years is to be used. Zero salvage values and a 10% MARR are appropriate. Perform a sensitivity analysis for the annual Btu requirement for the heating system. [*Hint:* First, calculate the break-even number of Btu (in thousands). Then, determine AWs if the Btu requirement is ±30% of the break-even amount.]

13-13. A company is considering investing $10,000 in a heat exchanger. The heat exchanger will last 5 years, at which time it will be sold for $2,000. The maintenance cost at the end of the first year is estimated to be $1,000 and then will increase by $500 per year over its life. As an alternative, the company may lease the equipment for $3,500 per year, including maintenance. The MARR (before tax) is 5% per year.

    Perform a multiparameter sensitivity analysis for this problem, where x = change in initial investment of the heat exchanger and y = change in annual leasing expense. Draw a diagram to illustrate your answer.

13-14. Black Diamond Coal Company is considering an investment in a piece of property (land including mineral rights). The property will cost $24 million. It is estimated that 40 million tons of run-of-mine coal are recoverable from the property. The run-of-mine coal will be cleaned by a neighboring coal cleaning plant. The cleaning recovery is 70%, and the cleaning costs will be $5 per ton of *cleaned* coal.

  The project requires a $60 million investment in mine development and equipment and will deplete the coal reserves at a rate of 2 million tons of cleaned coal per year. The investment in mine development and equipment will be capitalized and recovered by the straight-line depreciation method over a 14-year period. The depletion allowance is 10% of net before-tax income, and depletion serves to reduce taxable income (just as depreciation does). The selling price of cleaned coal is $38 per ton. Working capital is estimated at $5 million for the project. Working capital and one-half the purchase price of the property are salvageable at the end of the project's life. The operating costs are estimated at $18.50 per ton of cleaned coal. The income tax is 50% of taxable income (mining companies are subject to more taxes than are manufacturing companies). Assume all investments occur at time zero and all cash flows are end-of-year cash flows. The MARR for Black Diamond Coal Company is 15% per year.

  a. Determine the annual worth of the project.

  b. How sensitive is the annual worth to plus and minus 10% changes in the average operating cost?

13-15. A single-stage centrifugal blower is to be purchased for a manufacturing plant. Suppliers have been consulted and the choice has been narrowed down to two new units of modern design, both made by the same company and both having the same rated capacity and pressure. Both are driven at 3,600 rpm by identical 200-hp electric motors.

  One blower has a guaranteed efficiency of 72% at full load and is offered installed for $5,000. A second blower is more expensive because of aerodynamic refinement, which gives it a guaranteed efficiency of 75% at full load.

  Except for those differences in efficiency and price, the units are equally desirable in all respects such as durability, maintenance, ease of operation, and quietness. In both cases, plots of efficiency versus amount of air handled are flat in the vicinity of full rated load. The application is such that, whenever the blower is running, it will be at full load.

  Ignore income tax differences that may arise between the two blowers. Assume both blowers have negligible salvage values. The firm's MARR is 10% per year.

  a. Develop a formula for calculating how much the manufacturer could afford to pay for the more efficient unit. (*Hint:* You need to specify important variables and use them in your formula.)

  b. Sketch a graph, for your response to (a), in which the affordable amount is a function of the efficiency of the second (more expensive) blower.

13-16. (*Note:* This problem is based on Appendix 13-A.) The following alternatives will fill a given need ("repeatability" can be assumed and MARR = 10%):

	Plan	
	A	B
Investment:	$2,000	$6,000
Life:	6 yr	3 yr
Annual disbursements:	$3,500	$1,000
Salvage value:	$0	$4,000

a. Calculate the "assumed-certain" equivalent annual cost for each.

b. Set up a "Sensitivity to Decision Reversal" table (recognizing that the decision is reversed when AW(A) ≥ AW(B)), and show calculations for
(i) life for project A only;
(ii) annual disbursements for project A only.

c. Set up a sensitivity surface equation and graph for showing the following proportion changes:

$$x = \text{proportion change in project A investment only;}$$

$$y = \text{proportion change in project A annual disbursements only.}$$

Show the area (domain) in which project A is preferred to project B.

13-17. Consider these two alternatives for solid-waste removal:

a.

Alternative A: Build a Solid-Waste Processing Facility

Capital cost:	$108 million in 1996 (commercial operation starts in 1996)
Expected life of facility:	20 years
Annual operating costs:	$3.46 million (expressed in 1996 dollars)
Estimated salvage value:	40% of initial capital cost at all times

Alternative B: Contract with Vendors for Solid-Waste Disposal After Intermediate Recovery

Capital cost:	$17 million in 1996 (this is for *intermediate* recovery from the solid waste stream)
Expected contract period:	20 years
Annual operating costs:	$2.10 million (in 1996 dollars)
Repairs to intermediate recovery system every 5 years:	$3.0 million (in 1996 dollars)
Annual fee to vendors:	$10.3 million (in 1996 dollars)
Estimated salvage value at all times:	$0

Related Data

MACRS class life:	15 years (see Table 7-4)
Study period:	20 years
Effective income tax rate:	40%
Company MARR (after tax):	10.0%
Inflation rate:	0% (ignore inflation)

a. How much more expensive (in terms of capital cost only) could alternative B be in order to break even with alternative A?

b. How sensitive is the after-tax PW of alternative B to cotermination of both alternatives at the end of *year 10?*

c. Is the initial decision to adopt alternative B in (a) reversed if our company's annual operating costs for alternative B only ($2.10 million per year) unexpectedly double? Explain why (or why not).

[Answers to Problem 12-18: (a) 3.62 million square miles; (b) 10.6 ounces; (c) 640 miles per hour; (d) 20,320 ft; (e) 639 muscles; and (f) 70.2%.]

Tabular and Graphical Illustrations of Sensitivity Analysis for the Comparison of Alternatives*

In this appendix, we illustrate sensitivity analysis with a series of examples involving simple graphical and tabular techniques applied to a two-alternative case. Most of these examples could be extended to the comparison of three or more alternatives; and, of course, measures of merit other than net AW could be used.

We might conclude that use of all the types of tabular and graphical techniques to be illustrated would be "overkill" for comparison of a given set of alternatives. However, it is very important to use visual displays to communicate effectively the results of an engineering economy study. Consequently, we can select from among the techniques shown (or other variations we might create) according to which are judged to be most useful/practicable for decision-making purposes.

Assumed-Certain Example

Table 13-A-1 is a typical assumed-certain or "expected" (or most likely) set of estimates for examples of alternatives A and B. The net annual worth (net AW) measure of merit for each alternative is shown for a before-tax minimum attractive rate of return (MARR) of 20%. As an example of the calculations, the net AW for alternative B at $i = 20\%$ can be determined as

$$-[(\$200{,}000 - \$40{,}000)(A/P, 20\%, 15) + \$40{,}000(20\%)] + \$72{,}000 - \$20{,}000$$
$$= \$+9{,}775.$$

The bracketed term is, of course, the capital recovery cost for alternative B. The net AW for alternative B is not only positive, thus indicating an acceptable project, but it is $3,550 greater than the +$6,225 net AW for alternative A. On the basis of economic considerations alone, alternative B would be the recommended choice assuming certainty.

TABLE 13-A-1 Examples of "Expected" Estimates with Calculated Net Annual Worths

Variable	Alternative A	Alternative B
Investment:	+$150,000	$200,000
Life:	10 yr	15 yr
Salvage value:	$0	$40,000
Annual receipts or savings:	$87,000	$72,000
Annual disbursements:	$45,000	$20,000
Net AW at $i = 20\%$:	+$6,225	+$9,775

Note: Alt. A or Alt. B must be chosen.

*Adapted from J. R. Canada, and W. G. Sullivan, *Economic and Multiattribute Evaluation of Advanced Manufacturing Systems* (Englewood Cliffs, NJ: Prentice Hall, 1989). Reprinted by permission of the publisher.

It should be noted that the analyses of mutually exclusive alternatives having different economic lives as shown herein all use the common "repeatability" assumptions.

Exploration of Sensitivity to Interest Rates

Any parameter (variable) could be varied by ±20%, for example, to establish a rank ordering of impact (significance on the chosen measure of merit). Here, we suppose that the interest rate (MARR) is a major contributor to the uncertainty of project economic desirability. Figure 13-A-1 shows plots of the net AWs for both alternatives as a function of the interest rate varying from 15% to 30%. It is instructive to note that the two curves intersect at an interest rate, $i = 26\%$. When $i < 26\%$, alternative B has a higher net AW and is preferred, but alternative A is preferred for $i > 26\%$. For all examples presented in the remainder of this appendix, the MARR is held constant at 20%.

Optimistic–Pessimistic Estimates

The optimistic and pessimistic estimates shown in Table 13-A-2 happen to be ±10% of expected estimates in Table 13-A-1 for alternative A and ±20% of expected estimates for alternative B, merely for convenience in illustration. It is reasonable

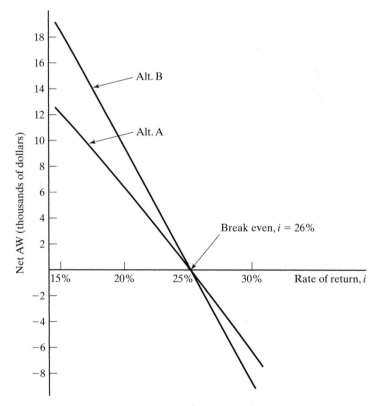

Figure 13-A-1   New AWs as function of minimum attractive rate of return.

TABLE 13-A-2 Example Optimistic and Pessimistic Estimates[a] with Calculated Net AWs

	Optimistic estimates		Pessimistic estimates	
	Alternative A	Alternative B	Alternative A	Alternative B
Investment:	$135,000	$160,000	$165,000	$240,000
Life:	11 yr	18 yr	9 yr	12 yr
Salvage value:	$0	$40,000	$0	$40,000
Annual receipts or savings:	$95,700	$86,400	$78,300	$57,600
Annual disbursements:	$40,500	$16,000	$49,500	$24,000
Net AW at $i = 20\%$:	+$24,002	+$37,464	−$12,137	−$19,460
AW(B) − AW(A)	+$13,462		−$7,323	

[a] For this example, the optimistic and pessimistic estimates happen to be ±10% of expected estimates for alternative A and ±20% of expected estimates for alternative B. (*Exception:* Salvage value does not vary.)

that, for given "tail probabilities," some of the estimates will vary widely and some little or not at all in a favorable or unfavorable direction.

The results of Table 13-A-2 are useful if we think that outcomes for the two alternatives will be perfectly correlated. That is, if both sets of estimates take on their "optimistic" values, B is better than A by a net AW of $13,462. On the other hand, if both sets assume their "pessimistic" values, then alternative A is better than alternative B by a net AW of $7,323.

Without perfectly correlated outcomes, there are $3^2 = 9$ combinations of estimates to consider. Each combination requires a separate set of calculations to compute the measure of merit. Probability statements regarding the final measure of merit cannot be formulated unless much more information is given.

Table 13-A-3 allows you to explore the effects of different combinations of optimistic and pessimistic outcomes on the difference between the two alternatives. The diagonal that is highlighted in Table 13-A-3 reflects the results obtained from Tables 13-A-1 and 13-A-2. As an example, suppose that you were concerned with the

TABLE 13-A-3 Difference in Net Annual Worth of Alternative B Compared with Alternative A [AW(B) − AW(A)] for All Combinations of Optimistic, Expected, and Pessimistic Estimate Sets[a]

	Alternative A		
Alternative B	Optimistic (AW = +$24,002)	Expected (AW = +$6,225)	Pessimistic (AW = −$12,137)
Optimistic (AW = +$37,464)	+$13,462	+$31,239	+$49,601
Expected (AW = +$9,775)	−14,227	+ 3,550	−21,912
Pessimistic (AW = −$19,460)	−43,462	−25,685	− 7,323

[a] Positive-valued entries favor alternative B; negative-valued entries favor alternative A.

difference between the alternatives if alternative A results in its "optimistic" outcome and alternative B results in its "expected" outcome. For this case, alternative A would be better than alternative B by a net AW of $14,227. As another example, you might be interested in the *extremes* that alternative B could be better than alternative A by a net AW of $49,601, or the reverse could be true by a net AW of $43,462.

Sensitivity Tables

Table 13-A-4 shows "sensitivity to decision reversal" caused by changes in expected estimates for alternative A, then alternative B, and then the difference between the alternatives. For each variable, the amount and the percent change that would just make the AWs for the two alternatives equal are shown. This shows, for example, that relatively low percentage changes in receipts, disbursements, or investment required for either alternative would reverse the decision. Thus, you could say that the decision is relatively sensitive to changes in those three variables. Furthermore, you might reasonably conclude that the decision is only moderately sensitive to changes in the lives and quite insensitive to changes in the salvage values.

Sensitivity Graphs

Figure 13-A-2 is a useful type of graph for exploring sensitivity to changes in two variables at a time. In this case, the net AW for both alternatives is shown as a function of the net annual cash flows (i.e., annual receipts minus annual disbursements) varying over a range on the x axis and for various assumed lives. All other variables are held constant at their "expected" values. Thus, you can choose particular combinations of outcomes for the variables plotted in Fig. 13-A-2 to observe how the net AWs compare for alternatives A and B. For instance, a 25% increase in net annual cash flows and a life of 15 years for alternative A will cause roughly the same net AW as a 25% increase in net annual cash flows and a life of 10 years for alternative B.

TABLE 13-A-4   Sensitivity to Decision Reversal (Based on Expected Estimates)

a. Alternative A

| | | To reverse decision (Net AW same as for alternative B) | | |
	Expected estimate	Estimated outcome	Change amount	Change as % of expected
Investment:	$150,000	$135,115	−$14,885	−10.0
Life:	10 yr	14.7 yr	+4.7 yr	+47.0
Salvage value:	$0	$92,208	+$92,208	—
Annual receipts				
or savings:	$87,000	$90,550	+$3,550	+ 4.1
Annual				
disbursements:	$45,000	$41,450	−$3,550	− 7.9

b. Alternative B

| | | To reverse decision (Net AW same as for alternative A) | | |
	Expected estimate	Estimated outcome	Change amount	Change as % of expected
Investment:	$200,000	$216,601	+$16,601	+ 8.3
Life:	15 yr	10.3 yr	−4.7 yr	− 31.3
Salvage value:	$40,000	−$215,468	−$255,468	−639
Annual receipts				
or savings:	$72,000	$68,450	−$3,550	− 4.9
Annual				
disbursements:	$20,000	$23,550	+$3,550	+ 17.8

c. Difference (Alt. B − Alt. A), or (Alt. A → Alt. B)

| | | To reverse decision (Equal net AWs) | | |
	Expected estimate	Estimated outcome	Change amount	Change as % of expected
Investment:	$50,000	Meaningful only if lives same		
Life:	5 yr	Meaningful only if lives same		
Salvage value:	$40,000	Meaningful only if lives same		
Annual receipts				
or savings:	−$15,000	−$18,550	$−3,550	23.7
Annual				
disbursements:	−$25,000	−$28,550	$−3,550	14.2
MARR ($i\%$)	Both alts. 20%	25.9%	+5.9%	29.5

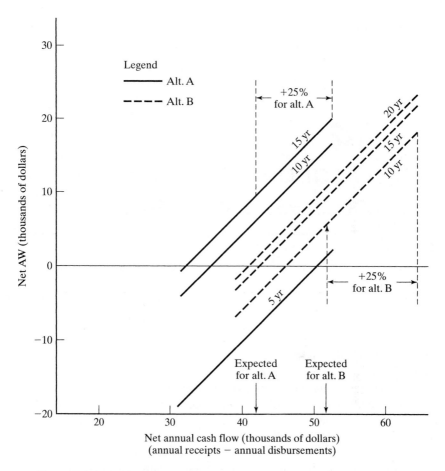

Figure 13-A-2   Sensitivity of net annual worths to net annual cash flows varying ±25% of expected and project lives varying ± 5 years from expected for alternatives A and B.

14
Analytical and Simulation Approaches to Risk Analysis

14.1 Introduction

As was noted in Chapter 13, *sensitivity analyses* are performed under for an investment in order to *describe* the effects on the measure of merit due to either estimation errors or changes in the values of one or more parameters.

The term "risk analysis" has different interpretations among various government agencies and units, as well as nongovernment organizations. However, there is a growing acceptance that risk analysis involves the development of probability(ies) associated with the measures of merit. The risk associated with an investment alternative is generally either given as the probability of an unfavorable value for the measure of merit or is determined by some measure(s) of the variability of the measure of merit. This chapter concentrates on very basic analytical methods for describing risk, and on Monte Carlo methods for simulating risky economic situations.

14.2 Analytical Methods of Risk Analysis

Typical parameters for which conditions of risk can reasonably be expected to exist include the initial investment, yearly operating and maintenance expenses, annual revenues, the life of an investment, the planning horizon, and the minimum attractive rate of return. The parameters can be statistically independent, correlated with time and/or correlated with each other. We will not address partial correlations, which are typically difficult to quantify.

14.2.1 Expected Values and Variances

Three very common statistical measures for probabilistic estimates for calculations of measures of merit are expected values, variances, and standard deviations. It is assumed that most readers are familiar with basic calculations involving these measures, but if needed, the reader may refer to general formulas and sample calculations in Sections 15.6 and 15.7 for expected values and variances, respectively.

14.2.2 Random Cash Flows

In order to determine analytically the probability distribution for the measure of merit, a number of simplifying assumptions is normally made. The simplest situation is one involving a known number of random and statistically independent cash flows. As an example, suppose the random variable A_j denotes the net cash flow occurring at the end of period j, $j = 0, 1, \ldots, N$. Hence, the present worth is given by

$$PW = \sum_{j=0}^{N} (1 + i)^{-j} A_j. \tag{14-1}$$

Since the expected value of a sum of random variables equals the sum of the expected values of the random variables, then the expected present worth is given by[1]

$$E[PW] = \sum_{j=0}^{N} (1 + i)^{-j} E[A_j], \tag{14-2}$$

where $E[\cdot]$ denotes the expected value. Furthermore, since the A_j's are statistically independent, then the variance of present worth is given by[2]

$$V[PW] = \sum_{j=0}^{N} (1 + i)^{-2j} V[A_j], \tag{14-3}$$

where $V[\cdot]$ denotes the variance, and $\sigma[\cdot] = \sqrt{V[\cdot]}$ will denote the standard deviation.

The central limit theorem, from probability theory, establishes that the sum of independently distributed random variables tends to be normally distributed as the number of terms in the summation increases.[3] Hence, as N increases, the PW tends to be normally distributed, with a mean value of $E[PW]$ and a variance of $V[PW]$.

Example 14-1

Consider the following data, where the MARR = 10% per year:

j	$E[A_j]$	$V[A_j]$
0	−$10,000	1×10^6
1–10	$1,800	4×10^4

Determine the probability of a negative present worth if the cash flows are independent.

---

[1] Recall that the expected value of a constant times a random variable equals the constant times the expected value of the random variable.

[2] Recall that the variance of a constant times a random variable equals the constant squared times the variance of the random variable.

[3] An additional condition is that the random variables have finite moments. The condition is usually met in risk analysis.

Solution

$$E[\text{PW}] = -\$10,000 + \sum_{j=1}^{10} (1.10)^{-j}\$1,800$$

$$= -\$10,000 + \$1,800(P/A, 10\%, 10)$$

$$= \$1,059,$$

$$V[\text{PW}] = 1 \times 10^6 + (4 \times 10^4)\sum_{j=1}^{10}(1.10)^{-2j},$$

which reduces to[4]

$$V[\text{PW}] = 1 \times 10^6 + (4 \times 10^4)(P/A, 10\%, 20)/2.10$$

$$= 116.217 \times 10^4,$$

$$\sigma[\text{PW}] = \sqrt{116.217 \times 10^4} = \$1,078.$$

Figure 14-1 depicts the problem. The probability of PW ≤ 0 (where S is the standard normal deviate) is obtained as follows:

$$P_r(\text{PW} \leq 0) = P_r\left(S \leq \frac{0 - E[\text{PW}]}{\sigma[\text{PW}]}\right)$$

$$= P_r\left(S \leq \frac{0 - 1,059}{\$1,078}\right) \tag{14-4}$$

$$= P_r(S \leq -0.98)$$

A table of probabilities for S is given in Appendix E. By symmetry,

$$P_r(\text{PW} \leq 0) = 1 - P_r(S \leq 0.98)$$

$$= 1 - 0.8365 = 0.1635. \qquad \blacksquare$$

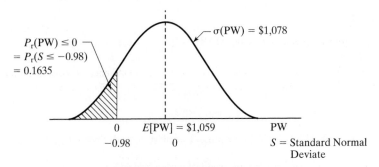

Figure 14-1 Example 14-1 results.

---

[4] The relation $\displaystyle\sum_{j=1}^{N}(1 + i)^{-2j} = (P/A, i\%, 2N)/(2 + i)$ is used.

14.2.3 Comparisons of Alternatives Using Normal Distribution Estimates

Riskiness of one or more alternatives can be described by direct (or calculated) estimates of outcomes in terms of probabilities. If it can be assumed that the outcomes can be estimated in terms of normal distribution(s), the process can be facilitated and probabilistic inferences (and comparisons) can rather easily be calculated.

We will make use of the general relationships that, when considering the difference between two variable outcomes X and Y, the following relationships are true:

(a)   The expected value of the difference is

$$E[X-Y] = E[X] - E[Y]. \tag{14-5}$$

(b)   The standard deviation of the difference
(i) when outcomes are independent is

$$\sigma(X-Y) = \sqrt{\sigma_X^2 + \sigma_Y^2}, \text{ and} \tag{14-6}$$

(ii) when outcomes are perfectly correlated is

$$\sigma(X-Y) = \sigma_X + \sigma_Y. \tag{14-7}$$

Example 14-2

Suppose two alternatives, A and B, are estimated to have normally distributed AW outcomes (in thousands of $) as follows:

Alternatives	A	B
$E(AW)$	900	500
$\sigma(AW)$	750	250

Figure 14-2 shows relevant diagrams for each alternative by itself. We will not show calculations of probabilities of negative AWs herein, but their relative magnitudes are indicated by shaded areas.

The expected value and standard deviations of the difference between the alternatives (A-B) can be calculated from Equations 14-5 through 14-7, respectively, as

$$E[AW_{(A-B)}] = 900 - 500 = 400;$$

$$\sigma[AW_{(A-B)}] = \sqrt{(750)^2 + (250)^2} = 791 \text{ (if independent);}$$

$$\sigma[AW_{(A-B)}] = 750 + 250 = 1,000 \text{ (if perfectly correlated).}$$

Assuming that a normal distribution describes the difference between the AW outcomes for the two alternatives, Figure 14-3 illustrates the results obtained. Calculations of probabilities of decision reversal (i.e., that B will turn out to have a higher AW than A) are as follows:

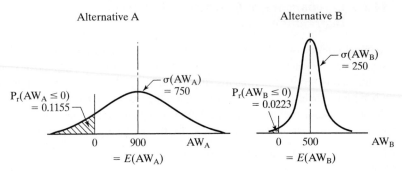

Figure 14-2   Example 14-2—Each alternative alone.

Difference Between Alternatives A and B:

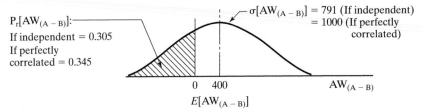

Figure 14-3   Example 14-2—Difference between alternatives.

If independent, then

$$Pr[AW_{(A-B)} \leq 0] = Pr\left[S \leq \frac{0-400}{791}\right] = Pr[S \leq -0.51] = 0.305$$

If perfectly correlated, then

$$Pr[AW_{(A-B)} \leq 0] = Pr\left[S \leq \frac{0-400}{1,000}\right] = Pr[S \leq -0.4] = 0.345$$

■

14.3   Monte Carlo Simulation of Risk

The Monte Carlo simulation technique is an especially useful means of analyzing situations involving risk to obtain approximate answers when a physical experiment or the use of analytical approaches is either too burdensome or not feasible. It has enjoyed widespread acceptance in practice because of the analytical power it makes possible without the necessity for complex mathematics. It is especially adaptable to computation by digital computers. Indeed, computer languages have been developed especially to facilitate Monte Carlo simulation.

The technique is sometimes descriptively called the *method of statistical trials*. It involves, first, the random selection of an outcome for each variable (element) of interest, the combining of these outcomes with any fixed amounts, and calculation, if necessary, to obtain one trial outcome in terms of the desired answer (measure of merit). Done repeatedly, this will result in enough trial outcomes to obtain a sufficiently close approximation of the mean, variance, distribution shape, or other characteristic of the desired answer. Figure 14-4 schematically shows this process applied to investment project analysis.

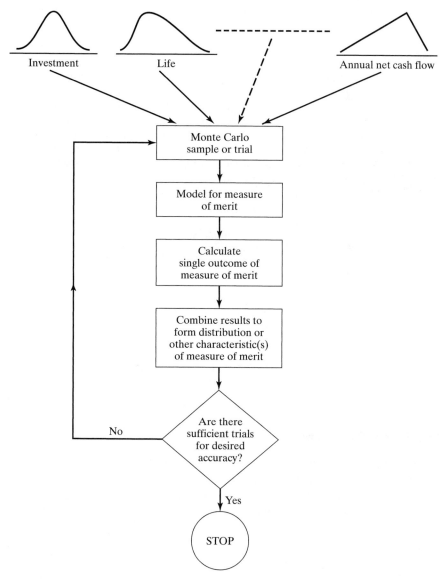

Figure 14-4   Schematic of Monte Carlo technique applied to investment project analysis.

The key requisite of the Monte Carlo technique is that the outcomes of all variables of interest be *randomly* selected (i.e., that the probability of selection of all possible outcomes be in exact accord with their respective probability distributions). This is accomplished through the use of tables of random numbers and relating these numbers to the distributions of the variables. *Random numbers* are numbers that have been generated in such a way that there is an equal probability of any number appearing each time, regardless of what sequence is experienced at any prior time. Appendix C contains one page of these numbers. The next simple example will demonstrate the Monte Carlo technique.

Example 14-3

As an illustration of Monte Carlo simulation applied to one variable or element, suppose the annual net cash flow for a project is estimated to have the distribution shown in Table 14-1.

This random simulation can be accomplished through tabular methods by assigning random numbers to each outcome in proportion to the probability of each outcome. Because two-digit probabilities are given in this case, sets of only two random digits are needed, which are shown in Table 14-2.

Now one can generate net cash flow outcomes by picking random numbers[5] and determining the net cash flow that corresponds to each, according to the list in Table 14-2. Table 14-3 lists 10 two-digit random numbers taken arbitrarily from a table of random numbers, such as that in Appendix C, together with the corresponding net cash flows, taken from Table 14-2.

TABLE 14-1 Example of Frequency Distribution for Net Annual Cash Flow

Net cash flow	P(Net cash flow)
$10,000	0.10
15,000	0.50
20,000	0.25
25,000	0.15

TABLE 14-2 Assignment of Random Numbers for Example in Table 14-1

Net cash flow	Random numbers
$10,000	00–09
15,000	10–59
20,000	60–84
25,000	85–99

■

[5] *Note:* The random numbers should be taken from the table in such a way as to assure randomness or nonrepetitiveness by randomly selecting a point to begin in the table and randomly selecting the direction of movement within the table (such as up, down, to the right, etc.).

TABLE 14-3 Generation of Outcomes for Example in Table 14-1

Random number	Net cash flow outcome
47	$15,000
91	25,000
02	10,000
88	25,000
81	20,000
74	20,000
24	15,000
05	10,000
51	15,000
74	20,000

It may be of interest to note that the mean net cash flow, based on the simulated outcomes from Example 14-3, is $175,000/10 = $17,500. This compares with a mean of $17,250 for the known distribution shown in Table 14-1. Results for 10 simulated outcomes would not always turn out this close. However, in general, the larger the number of Monte Carlo trials, the closer is the approximation to the desired answer(s).

For Monte Carlo simulations in which a computer is not used, it is sometimes helpful to use a graph of the distribution function (cumulative frequency function) instead of a table matching random numbers to the various outcomes. Figure 14-5 shows a graph of the distribution function for the example in Table 14-1. Once a graph such as that in Fig. 14-5 has been constructed, outcomes are generated. First, a random number table, such as that in Appendix C, is used to obtain random values that correspond to the ordinate scale (vertical axis) with the decimal removed. For each random number, a horizontal line is drawn until it meets the curve. Then a vertical line is dropped to the abscissa (horizontal axis), and the outcome is thus determined. The dotted line in Fig. 14-5 illustrates the generation of a sample cash flow outcome.

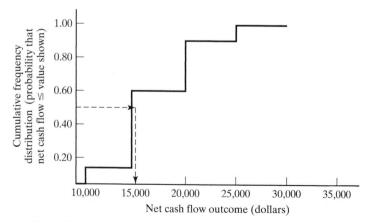

Figure 14-5   Sample cumulative frequency distribution for net cash flow.

The preceding example is for a discrete outcome distribution, but it should be noted that the same principle applies for continuous distributions. For continuous distributions, the tabular method is usually impractical, but the graphical method is readily acceptable, as is shown in the next section.

14.3.1 Generation of Random Normal Values

It is quite common for random phenomena to possess a normal distribution and for element outcomes to be estimated as normally distributed. The Monte Carlo technique can conveniently be used for simulation of random outcomes in such cases.

The basic quantity needed to generate randomly distributed normal outcomes is called a *random normal deviate,* or random normal number. A random normal deviate is merely a random number of standard deviations from the mean of a standard normal distribution. Random normal deviates can be obtained directly from a graph of the cumulative standard normal distribution. Such a graph is shown in Fig. 14-6.

For a normal distribution, the probability of an occurrence near the mean is greater than the probability of an occurrence farther from the mean. This is reflected in Fig. 14-6, since the relative frequency of occurrence at each outcome value is proportional to the slope of the cumulative frequency curve.

To obtain random normal deviates, a table of random numbers is used to select numbers between 0.000 and 0.999 on the ordinate scale of the cumulative frequency distribution. (*Note:* More or less than three decimal places can be used as desired for accuracy.) For each random number, a horizontal line and a vertical line can be drawn

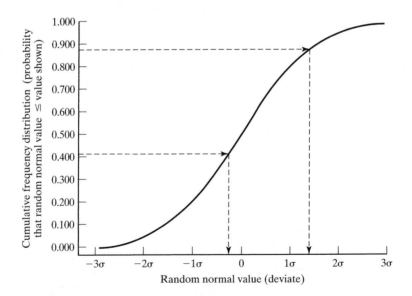

Figure 14-6 Cumulative standard normal frequency distribution for generation of random normal deviates.

to find the corresponding random normal deviate. This is shown for two sample random numbers in Fig. 14-6, and the results are summarized here:

Random number	Random normal deviate
405	−0.24
877	1.16

Tables of random normal deviates can be generated by a procedure such as that just given. Such a table is presented in Appendix D. Tables of random normal deviates save much effort, for they enable us to generate a Monte Carlo sample from a normal distribution merely by using this relation: **outcome value = mean + (RND × standard deviation)**, where RND denotes the random normal deviate. As an example, suppose a project has a mean life of 8 years and a standard deviation of 2 years. Generated lives using the preceding random normal deviates, for example, can then be calculated as

$$8 - 0.24(2) = 7.52 \text{ year}$$

and

$$8 + 1.16(2) = 10.32 \text{ year}.$$

14.3.2 Generation of Uniformly Distributed Values[6]

Whenever the cumulative distribution function of a random variable can be expressed mathematically, random outcomes of that variable can be generated from random numbers by direct mathematical substitution. We next present an example of the development of a mathematical model for the generation of uniformly distributed values.

A uniform continuous distribution, with a minimum value a and a maximum value b, has a density function and cumulative frequency distribution as shown in Fig. 14-7. For this distribution, the mean equals $(a + b)/2$, the variance equals $(b - a)^2/12$, and the range equals $(b - a)$.

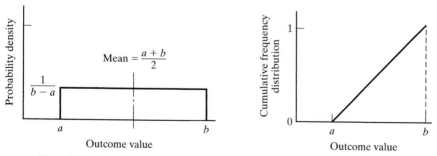

Figure 14-7   Density and distribution function for uniform continuous distribution.

[6] Methods of generating random variates from a variety of different probability distributions are given by R. E. Shannon, *Systems Simulation: The Art and Science* (Englewood Cliffs, NJ: Prentice-Hall, Inc., 1975).

To illustrate the generation of outcomes according to this distribution, let RN denote a random number, and let RN_m denote the highest-valued random number. By similar triangles, it can be seen that

$$\text{outcome value} = a + \frac{RN}{RN_m}(b - a)$$

(14-8)

$$= a + (RN \text{ expressed as a decimal})(b - a).$$

An equivalent statement is

$$\text{outcome value} = \frac{a + b}{2} - \frac{(b - a)}{2} + \frac{RN}{RN_m}(b - a).$$

(14-9)

If an element or variable is uniformly distributed with a mean of 8 and a range of 6, random outcomes can be generated using the relation

$$8 - \frac{6}{2} + \frac{RN}{RN_m}(6) = 5 + (RN \text{ expressed as a decimal})(6).$$

Example 14-4

Example of the use of Monte Carlo simulation for nonindependent elements: One of the valuable features of the Monte Carlo technique is that it provides an analysis tool for cases in which elements are not independent and thus are difficult or impossible to manipulate so as to obtain the desired answers analytically. For example, suppose that the life of a project is described by some distribution whose mean is a function of the annual cash flow of the project. Further, suppose that the annual cash flow itself is described by some distribution, and that it is desired to determine the distribution of the PW (present worth) of those cash flows over the life of the project. The following illustrates the use of the Monte Carlo technique for this type of situation.

The annual net cash flow of a project is estimated to be normally distributed with a mean of $10,000 and a standard deviation of $2,000. The life of the project is estimated to be uniformly distributed with a mean of 0.0005 of the annual net cash flow (rounded to the nearest integer year) and a range (difference between maximum and minimum life) of 6 years. Table 14-4 demonstrates the use of the Monte Carlo technique for 10 trials to obtain an estimate of the mean of the PW of the cash flows, at an interest rate of 10% per year.

The estimated mean PW from the limited number of trials in Table 14-4 is $374,800/10 = $37,480. Repeated trials would doubtless result in a more accurate answer. It is worthy of note that the exact answer for this situation is not the same as $10,000(P/A, 10%, 5) = $37,910, where $10,000 is the mean cash flow and 5 years is the mean life. ∎

Example 14-5

Example of the Monte Carlo technique applied to an economic analysis for a single project: To illustrate the use of the Monte Carlo technique to calculate the measure of merit for a single project, for which the element outcomes are estimated as variables, consider a case in which the estimates are as follows:

Investment:	Normally distributed with mean of $100,000 and standard deviation of $5,000
Life:	Uniformly distributed with minimum of 4 yr and maximum of 16 yr (rounded to nearest integer)
Market value (MV):	$10,000 (single outcome)
Annual net cash flow:	$14,000 with probability 0.4
	$16,000 with probability 0.4
	$20,000 with probability 0.2

TABLE 14-4 Monte Carlo Example with Nonindependent Elements (Example 14-4)

Random normal deviate (RND)	Annual net cash flow (ANCF) [$10,000 + RND($2,000)]	Three random numbers (RN)	Project life N $\left[0.0005(\text{ANCF}) -3 + \dfrac{RN}{999}(6)\right]$	Project life N to nearest integer	$(P/A,10\%,N)$	PW of cash flows [ANCF $\times$ $(P/A,10\%,N)$]
0.944	$ 8,112	443	3.65	4	3.170	$25,700
−1.140	7,720	511	3.64	4	3.170	24,500
1.353	12,706	549	6.54	7	4.868	62,000
0.466	10,912	169	3.48	3	2.487	27,100
0.732	11,464	656	6.65	7	4.868	56,000
−1.853	6,394	955	5.84	6	4.355	27,800
−0.411	9,188	783	6.29	6	4.355	40,100
0.488	10,976	572	5.92	6	4.355	48,000
−0.351	9,298	842	6.75	7	4.868	45,400
−1.336	7,328	372	2.89	3	2.487	18,200
						$\Sigma = \overline{\$374,800}$

Note: Cash flows are re-invested at the MARR so PWs can be added.

It is assumed that all elements subject to variation vary independently of one another, and it is desired to obtain a good estimate of the distribution characteristics for the AW by using an interest rate of 10%. For purposes of the illustration, only five repetitions of the Monte Carlo simulation will be made. However, perhaps several thousand repetitions would be needed to obtain sufficiently accurate AW distribution information. Table 14-5 shows the calculations.

The estimate of the expected AW, based on the very limited sample as calculated in Table 14-5, is $-\$18,032/5 = -\$3,606$.

An estimate of the standard deviation of the AW can be obtained from the relation

$$\sigma[AW] = \sqrt{\frac{\sum_{i=1}^{k}(AW_i - E[AW])^2}{k-1}} \tag{14-10}$$

$$= \sqrt{\frac{\$58,100,000}{4}} = \$3,850.$$

This estimate should also come closer to the true standard deviation with increasing numbers of Monte Carlo trials.   ■

Example 14-6

Example of the Monte Carlo technique applied to an economic comparison of two independent projects: Suppose that two competing projects, A and B, have the following estimated distributions of AW and it is desired to estimate the distribution of the difference in AW between the projects, using Monte Carlo simulation (assume that the outcomes are independent of each other):

Project A		Project B
AW	P(AW)	
-$ 5,000	0.10	AW is normally distributed
10,000	0.30	with mean of $25,000 and
20,000	0.50	standard deviation of $10,000
30,000	0.10	

Sample calculations to obtain the desired answers are shown in Table 14-6. The mean difference in AW, on the basis of the limited simulation in Table 14-6, is $12,680/4 = $3,170. To reemphasize, a much larger number of Monte Carlo trials than the number illustrated is needed before meaningful estimates of the distribution of the difference in the AW for the two projects can be made.   ■

Example 14-7

Example of the Monte Carlo technique applied to an economic comparison of two projects with correlated elements: Two competing projects have the following outcome distribution characteristics:

	Project	
	A	B
Investment (normally distributed)		
Mean:	$50,000	$60,000
Standard deviation:	$20,000	$5,000
Life (uniformly distributed and rounded to the nearest year)		
Minimum:	3 yr	2 yr
Maximum:	7 yr	12 yr

TABLE 14-5 Monte Carlo Example 14-5 for a Single Project

Random normal deviate (RND)	Investment, P [$100,000 + RND($5,000)]	Three random numbers (RN)	Project life, N $\left[4 + \dfrac{RN}{999}\,(16-4)\right]$	Project life N to nearest integer	One random number	Annual receipts, A $\begin{bmatrix}\$14,000 \text{ for } 0\text{--}3 \\ 16,000 \text{ for } 4\text{--}7 \\ 20,000 \text{ for } 8\text{--}9\end{bmatrix}$	AW $[-\{(P - MV)(A/P, 10\%, N) \\ + MV(10\%)\} + A]$
0.30	$101,500	693	4 + 8.32	12	2	$14,000	-$ 432
-0.92	95,400	192	4 + 2.30	6	5	16,000	- 4,700
0.13	100,650	092	4 + 1.10	5	1	14,000	- 9,000
-0.16	99,200	490	4 + 5.87	10	4	16,000	+ 500
0.54	102,700	314	4 + 3.77	8	1	14,000	- 4,400
							$\Sigma = -\$18,032$

Note: MV is the market value ($10,000) at EOY N

TABLE 14-6   Monte Carlo Comparison of Two Independent Projects (Example 14-6)

One random number	AW for project A [−$5,000 for 0 / 10,000 for 1–3 / 20,000 for 4–8 / 30,000 for 9]	Random normal deviate (RND)	AW for project B [= $25,000 + RND($10,000)]	Difference in AW for two projects
4	$20,000	0.636	$31,360	$11,360
1	10,000	−0.179	22,210	12,210
4	20,000	−2.546	− 460	− 20,460
6	20,000	0.457	29,570	9,570
				$\Sigma =$ $12,680

It is thought that the investment outcomes are completely correlated ($\rho = +1$) for the two projects (i.e., when the investment for one project occurs high, the investment for the second project occurs correspondingly high, etc). On the other hand, the lives for the two projects are thought to be independently distributed. Table 14-7 demonstrates the use of the Monte Carlo technique for generating the distribution of the difference in the capital recovery costs for the two projects. For purposes of the illustration, zero market value and 10% interest are assumed and only three trials are shown. Note that the investment amounts for the two projects are generated using the same random normal deviates, thus reflecting complete correlation. On the other hand, the lives for the two projects are generated using independent random digits, reflecting the independence of life outcomes.   ■

For typical practical problems, hundreds or even thousands of Monte Carlo trials are required in order to reduce sampling variation to a level sufficiently low that the desired answers possess the level of accuracy thought necessary. This is often too laborious a task to accomplish by hand, but it can be done very efficiently with digital computers.

TABLE 14-7   Monte Carlo Comparison of Projects with Correlated Elements (Example 14-7)

Random normal deviate (RND)	Investment (P) for A [$50,000 + RND($20,000)]	Investment (P) for B [$60,000 + RND($5,000)]	Two random numbers	Life for A $\left[3 + \dfrac{RN}{99}(4) \right]$ (rounded)
0.178	$53,560	$60,890	95	8
−0.507	49,860	57,460	04	4
0.362	57,240	61,810	08	4

Two random numbers	Life for B $\left[2 + \dfrac{RN}{99}(10) \right]$ (rounded)	Capital recovery cost for A [$P(A/P, 10\%, N)$]	Capital recovery cost for B [$P(A/P, 10\%, N)$]	Difference in capital recovery cost [project B − project A]
16	4	$10,000	$19,150	$ 9,150
98	12	15,550	8,400	− 7,150
00	2	18,030	35,000	16,970

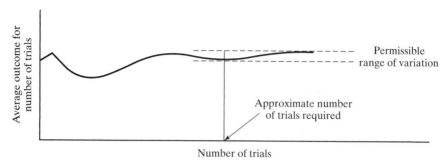

Figure 14-8   Number of trials.

14.4 Method for Determining Required Number of Monte Carlo Trials and Limitations

An easy method for determining the approximate number of Monte Carlo trials required to obtain sufficiently accurate answers is to keep a running tally (plot) on the average answer(s) of interest for increasing numbers of trials and to judge the number of trials at which those answer(s) have become stable enough to be within the accuracy required.

It is to be expected that the average outcome(s) will dampen or stabilize with increasing numbers of trials. This phenomenon is illustrated in Fig. 14-8 by the wavy line. Figure 14-8 also shows how a given typical permissible range of error can result in an indicated approximate number of trials required.

The Monte Carlo technique possesses real limitations that should be recognized. As for any analysis technique, the results can be no more accurate than the model and estimates used. The technique also inherently possesses the same problems of statistical variation and the need for experimental design that are encountered in direct physical experimentation. Finally, it should be emphasized that a sufficiently large number of Monte Carlo trials must be performed to reduce sampling variation (range of error) to a level that is tolerable in view of the accuracy needed and economically justified.

PROBLEMS

14-1. An individual is considering an investment of $10,000 in a venture. It is expected that annual receipts of $2,000 will occur over a 10-year period. However, the annual receipts are not guaranteed to occur at the expected level. The standard deviation for annual receipts is estimated to be $200 each year. A MARR of 10% is to be used to determine the expected value and standard deviation of present worth, assuming that annual receipts are statistically independent. Assuming a normal distribution of PWs, what is the probability that the PW will be less than $1,800?

14-2. The AW of a project is estimated to be normally distributed, with an expected value of $5,000 and a standard deviation of $4,000. What is the probability of the project's having an AW of

a. less than $5,000?

b. less than $0?

c. more than $7,000?

14-3. The PW of Cost (PW-C) outcomes of two mutually exclusive alternatives are each thought to be best described as normally distributed with the following estimates (in $millions):

Alternative	Q	X
E[PW-C]	15	12
σ[PW-C]	2	20

a. Which is best on the basis of lowest (i) expected value? (ii) variance?

b. Determine the probability that each will have a PW-C greater than a perceived "ceiling" of $18 million.

c. Assume that the difference between the two PW-Cs is independent and normally distributed. What is the probability Alternative Q will turn out to be better than X?

14-4. An initial investment of $9,000 results in annual receipts of $2,500 until the project is terminated. The probability distribution for the life of the project is estimated to be as follows:

N	P(N)
4	0.10
5	0.25
6	0.45
7	0.15
8	0.05

Use a MARR of 15% per year to determine the probability the investment is profitable (i.e., its PW ≥ 0).

14-5. The market value of a prospective asset is a random variable that depends on the life of the asset, according to the following table:

	Probability of market value				
Life	$5,000	$10,000	$15,000	$20,000	$25,000
2	—	—	0.20	0.50	0.30
4	—	0.20	0.50	0.20	0.10
6	0.30	0.30	0.30	0.10	—
8	0.30	0.50	0.20	—	—

It is thought that each of the asset lives is equally likely to occur. If the investment in the asset is $50,000 and the interest rate is 15%, show how you can obtain a distribution of the capital recovery cost for the asset by setting up a table and generating five trial outcomes.

14-6. The estimated element outcomes for a key project are as follows:

Investment:	Normally distributed with a mean of $1,000,000 and a variance of 16,000,000
Life:	5 yr with probability 0.2
	7 yr with probability 0.7
	9 yr with probability 0.1
Net annual cash flow:	Uniformly distributed between $120,000 and $340,000 per yr
Market value:	$0

All element outcomes are independent of each other. Demonstrate how to obtain a distribution of the AW by generating five outcomes. From this, obtain estimates of the mean and variance of the AW. Assume a MARR of 10% per year.

15
Decision Criteria and Methods for Risk and Uncertainty

15.1 Introduction

This chapter will be devoted first to numerous criteria for aiding in making decisions for what are classically called *risk problems* (i.e., problems in which probabilities of various possible outcomes can be estimated). These criteria are

1. dominance,
2. aspiration level,
3. most probable future,
4. expected—value,
5. expectation variance,
6. certain monetary equivalence, and
7. expected utility.

The final section illustrates the use of several decision rules and principles that can be used for *uncertainty problems* (i.e., when the probabilities of various possible outcomes cannot be estimated).

15.2 General Model for Risk Problems

Table 12-1 showed a general formulation for decision problems in which there are various possible outcomes (called *states of nature*) for which probabilities can be estimated.

TABLE 15-1   Sample Problem with Probabilities Known—
Payoffs in $M of Net PW

| Alternatives | State of nature (probability of state) | | | |
	S_1 (0.5)	S_2 (0.1)	S_3 (0.1)	S_4 (0.3)
I	3	−1	1	1
II	4	0	−4	6
III	5	−2	0	2
IV	2	−2	0	0
V	5	−4	−1	0

A typical problem involving five alternatives and four states of nature, together with associated probabilities, is shown in Table 15-1. The numbers in the body of the table can be thought of as returns or profits in $M (where $M denotes thousands of dollars) of net PW, such that a negative number is a cost.

The following sections describe and illustrate several criteria for decision making, using the example in Table 15-1. Most of the criteria apply to classical risk problems in which probabilities of various outcomes can be estimated. However, the first two criteria or principles can be applied to decision problems even when the probabilities are not known.

15.3 Dominance Criterion or Elimination Check

The first step in making a decision when the results for all alternatives and states of nature can be quantified is to eliminate from consideration any alternatives that are clearly not to be performed, regardless of the state of nature that occurs. If the result for any alternative X is better than the result for some other alternative Y, for all possible states of nature, then alternative X is said to *dominate* alternative Y, and thus Y can be dropped from further consideration.

Example 15-1
Given the problem depicted in Table 15-1, check for dominance and take appropriate action.

Solution   By systematic visual inspection, you can determine that alternative I dominates IV and that alternative III dominates V. (*Note:* Actually, III and V are equally good for state of nature S_1, but V is never better than III.) Thus, alternatives IV and V can both be eliminated from further consideration on the grounds that no rational decision maker would choose either alternative.   ■

The sample problem in Table 15-1 can now be reduced to the problem shown in Table 15-2.

15.4 Aspiration-Level Criterion

The aspiration-level criterion involves selecting some level of aspiration and then choosing so as to maximize (or minimize) the probability of achieving this level. An

TABLE 15-2 Sample Problem in Table 15-1 After
Dominated Alternatives Have Been Eliminated—
Payoffs in $M of Net PW

| | State of nature (probability of state) | | | |
| | S_1 | S_2 | S_3 | S_4 |
Alternatives	(0.5)	(0.1)	(0.1)	(0.3)
I	3	−1	1	1
II	4	0	−4	6
III	5	−2	0	2

aspiration level is simply some level of achievement (such as profit) the decision maker desires to attain or some level of negative results (such as cost) to be avoided.

Example 15-2
Given the decision problem in Table 15-2, determine which would be the best alternative if the decision maker has each of the following aspiration levels: (a) possible result of at least 5; (b) possible negative result (loss) no worse than −1.

   Solution

   (a) Alternatives II and III have possible results of 5 or greater. The probabilities of these results are 0.3 and 0.5, respectively. Hence, alternative III would be the choice.

   (b) Only alternative I has a possible result that is no more negative than −1. Thus, alternative I would be the choice.   ■

It is commonly thought that some form of the aspiration-level criterion is the most widely used of all principles in management decision making. The following are cases in which use of aspiration levels makes intuitive good sense:

   1. When it is costly or too time consuming to determine all the reasonable alternatives and their prospective results, you may choose to search for alternatives only until an alternative is found that gives a reasonable probability of achieving the aspiration level.

   2. Occasionally, a given alternative is available for only a limited time and action must be taken before information on all the reasonable alternatives and their prospective results can be developed. For example, equipment at a particular price may be available only if an agreement to purchase is made within a matter of hours or days.

   3. Sometimes, it is difficult or impossible to evaluate the results for each alternative, but it may be possible to determine which alternatives do meet the aspiration level of the decision maker. In this case, a reasonable criterion is to choose the alternative that maximizes the probability of achieving the aspiration level.

15.5 Most Probable Future Criterion

The most probable future criterion suggests that, as the decision maker considers the various possible outcomes in a decision, he or she overlooks all except the most probable one and acts as though it were certain.

Many decisions are based on this principle, since, in fact, only the most probable future is seriously considered (thus making the problem virtually one of "assumed certainty").

Example 15-3
Given the decision problem in Table 15-2, determine which would be the best alternative, using the most probable future criterion.

  Solution   The most probable future is state of nature S_1 (probability $= 0.5$). The results for S_1 range from 3 for alternative I to 5 for alternative III. Of these, alternative III has the best result and would thus be the choice.   ■

15.6 Expected Value Criterion

Using the expectation principle, and thereby choosing to optimize the expected payoff or cost (expressed in equivalent terms), simplifies a decision situation by weighting all dollar payoffs or costs by their probabilities. The criterion is often known as the *expected monetary value,* or *EMV.* As long as the dollar consequences of possible outcomes for each alternative are not very large in the eyes of the decision maker, the expectation principle can be expected to be consistent with a decision maker's behavior.

  The general formula for finding the expected outcome (value) of a variable x for any alternative A_i having k discrete outcomes is

$$E[x] = \sum_{j=1}^{k} x_j \cdot P(x_j), \qquad (15\text{-}1)$$

where

$$E[x] = \text{expected value of } x,$$
$$x_j = j\text{th outcome of } x, \text{ and}$$
$$P(x_j) = \text{probability of } x_j \text{ occurring.}$$

If the notation for the general model in Table 12-1 is used, this can be expressed as

$$E[A_i] = \sum_{j=1}^{k} R_{ij} \times P(S_j). \qquad (15\text{-}2)$$

Example 15-4
Given the same decision problem shown in Table 15-2, show which alternative is best by the expected value criterion.
  Solution

Alternative, A_i	$E[A_i]$
I	$3(0.5) - 1(0.1) + 1(0.1) + 1(0.3) = 1.8$
II	$4(0.5) + 0(0.1) - 4(0.1) + 6(0.3) = 3.4$
III	$5(0.5) - 2(0.1) + 0(0.1) + 2(0.3) = 2.9$

Thus, alternative II, having the highest expected result, is best.   ■

15.7 Expectation-Variance Criterion

The *expectation-variance criterion* or procedure, sometimes called the *certainty equivalence method,* involves reducing the economic desirability of a project into a single measure that includes consideration of the expected outcome, as well as variation of that outcome. One simple example is

$$Q = E[x] - A \cdot \sigma[x], \qquad (15\text{-}3)$$

where

$$
\begin{aligned}
Q &= \text{expectation-variance measure,} \\
E[x] &= \text{mean or expected monetary outcome,} \\
\sigma[x] &= \text{standard deviation of monetary outcome, and} \\
A &= \text{coefficient of risk aversion.}^{[1]}
\end{aligned}
$$

The variance of a variable x for any alternative having k discrete outcomes is

$$V[x] = \sum_{j=1}^{k} (x_j - E[x])^2 \, P(x_j), \qquad (15\text{-}4)$$

where $V(x) = $ variance of x, and all other symbols were defined with Eq. 15-1. A more convenient form for calculating $V[x]$, is

$$V[x] = \sum_{j=1}^{k} x_j^2 P(x_j) - (E[x])^2. \qquad (15\text{-}5)$$

Example 15-5
Given the decision problem in Table 15-2, determine which alternative is best by using the criterion in Eq. 15-3, with $A = 0.7$.

    Solution   Using Eq. 15-5 gives these results:

Alternative, A_i	$V[A_i]$
I	$(3)^2(0.5) + (-1)^2(0.1) + (1)^2(0.1) + (1)^2(0.3) - (1.8)^2 = 1.76$
II	$(4)^2(0.5) + (0)^2(0.1) + (-4)^2(0.1) + (6)^2(0.3) - (3.4)^2 = 8.84$
III	$(5)^2(0.5) + (-2)^2(0.1) + (0)^2(0.1) + (2)^2(0.3) - (2.9)^2 = 5.69$

[1] Donald Farrar (*The Investment Decision Under Uncertainty,* Englewood Cliffs, NJ: Prentice-Hall, Inc., 1962) and others have shown that, as long as there is a diminishing marginal utility of money, the correspondence between a firm's coefficient of risk aversion and its utility function of monetary outcome is

$$A = -\frac{U''(E[x])}{2}.$$

    That is, the coefficient of risk aversion is equal to the negative of one-half of the second derivative of the utility function, evaluated at the expected monetary outcome. (Refer to section 15.9)

Using Eq. 15-3 and calculated results for $E[x]$ from the preceding example of the expected value criterion gives the following:

Alternative	$Q = E[x] - 0.7\sqrt{V[x]}$
I	$1.8 - 0.7\sqrt{1.76} = 0.87$
II	$3.4 - 0.7\sqrt{8.84} = 1.32$
III	$2.9 - 0.7\sqrt{5.69} = 1.23$

Thus, alternative II is highest and therefore the best by this particular expectation-variance criterion. ∎

There are innumerable other expectation–variance criteria that can be applied, depending on the risk preferences and sophistication of the decision maker and his or her analyst. For example, Cramer and Smith, in a classic article,[2] recognize that the desirability of an investment project is a function of not only the expected value and variance, but also of the investment amount in the individual project. Hence, they developed an evaluation model of the form

$$Q = E[x] - A\sigma[x]^a I^b, \tag{15-6}$$

where

Q = certainty equivalence or expectation–variance measure,

$E[x]$ = expected monetary outcome,

A = coefficient of risk aversion,

$\sigma[x]$ = standard deviation of monetary outcome,

I = project investment amount, and

a and b = constants.

Cramer and Smith further show detailed examples of how one can empirically obtain all the constants for the use of the model. The next example is a simple application.

Example 15-6

Suppose the outcomes and investments required for two competing projects are estimated as follows:

Alternative	Outcomes in net AW		Investment required
	$E[AW]$	$\sigma[AW]$	
A	$10,000	$25,000	$22,500
B	8,000	4,000	40,000

It is desired to show which alternative would be preferred if each of the following criteria is used:

(a) expected value,

[2] R. H. Cramer and B. E. Smith, "Decision Models for the Selection of Research Projects," *The Engineering Economist* 9, no. 2 (Winter 1964).

(b) expectation–variance, using Eq. 15-3 with a high coefficient of risk aversion A for both projects, and

(c) expectation–variance, using Eq. 15-6 with $A = 0.40$ for project A and 0.75 for project B and constants $a = b = 0.50$.

Solution

(a) Project A is better; $E(\text{AW})_A > E(\text{AW})_B$.

(b) Project B would be better, intuitively. (*Note:* This is true if the coefficient of risk aversion A for both projects were anything higher than 0.095.)

$$A = \frac{\$10,000 - \$8,000}{\$25,000 - \$4,000} = 0.095$$

(c) Project A:

$$Q = \$10,000 - 0.4(\$25,000)^{0.5}(\$22,500)^{0.5} = \$520.$$

Project B:

$$Q = \$8,000 - 0.75(\$4,000)^{0.5}(\$40,000)^{0.5} = \$5,000.$$

Thus, on the basis of the preceding calculations, project B is better.   ■

15.8 Certain Monetary Equivalence Criterion

An offshoot of the expectation–variance criterion is to determine subjectively the *certain monetary equivalence* of any set of results for any alternative. The *certain monetary equivalent,* or *CME,* is merely the monetary amount for certain at which the decision maker would be indifferent between that amount and various possible monetary outcomes. This concept is very useful in practice and can be applied to situations involving gains (payoffs) or losses (costs). While it can be most meaningfully applied to risk situations in which various payoffs or costs and their respective probabilities are known, it can also be used in situations involving uncertainty regarding payoffs or costs, probabilities, or both. Next, we present several examples.

Example 15-7

Suppose a decision maker is faced with either a possible loss of $100,000 with probability 0.01 or no loss with probability 0.99. He desires to decide what is the maximum amount he would be willing to pay in order to avoid the risk of loss.

Solution   The desired quantity is a certain monetary equivalent. It is quite subjective, depending on the decision maker's risk preferences, particularly considering the consequences of the possible monetary outcomes in relation to the total assets at his disposal. As a guide, you might calculate the expected monetary value, EMV, as follows:

$$\text{EMV} = -\$100,000(0.01) + \$0(0.99) = -\$1,000.$$

If the possible $100,000 loss poses little threat in the eyes of the decision maker, so that he is called *risk neutral* in this situation, he could reasonably designate the EMV as his CME. However, most decision makers are at least somewhat *risk averse* and thus will be willing to pay a certain amount of more than $1,000—say, $5,000—in order to avoid the chance of loss of $100,000. This illustrates why most people are willing to purchase liability insurance even though the known cost of the policy is higher than the expected losses to be covered by the policy. On the other hand, occasionally there are decision makers who are *risk seeking* in nature and would pay only something less than $1,000 for certain in order to avoid the risk of loss. A possible, but not necessarily rational, extreme is a decision maker who enjoys the risk of loss of $100,000 so much (for example, he may like to boast about it to his friends) that he is unwilling to pay any amount to avoid the risk.   ■

Example 15-8

Suppose a decision maker is again confronted with the sample problem in Table 15-2 and desires to designate her CME for each alternative so as to choose the best.

   Solution   Again, the specification of CMEs is very subjective, reflecting the decision maker's relative weighting of the consequences of the various possible gains and losses. As a starting point, she might consider the EMVs for alternatives I, II, and III, which were previously calculated to be 1.8, 3.4, and 2.9, respectively. After considering this and the range of possible outcomes involved, a particular decision maker might choose CMEs of, say, 1.0, 2.0, and 2.5, respectively. In this case, alternative III, with the highest CME, would be preferred.

   The CME criterion can also be used for situations in which probabilities are neither known nor estimable. In such cases, any CMEs determined are even more subjective than otherwise; nevertheless, they can express "gut feelings" about the uncertainties and risk preferences involved.   ■

Example 15-9

Suppose a decision maker has a prospective project that is estimated to bring the following possible equivalent returns (for which the respective probabilities are thought to be too nebulous to estimate):

$$\text{Return in \$M: } -50, \text{ or } 150, \text{ or } 250.$$

It is desired to demonstrate the determination of his CME for the project (i.e., to identify the single lump amount he would accept as being just as valuable to him as the variable return).

   Solution   This requires very subjective judgment on the basis of the decision maker's intuitive feelings about the desirability (or nondesirability) of the possible gains or losses and their respective likelihoods. If he very much abhors the possible loss of $50M, he might be willing to *pay* something like $10M or $25M (CME = −$10M or −$25M) or more to avoid the risk. On the other hand, if he is strongly attracted to the possible gain of $150M or $250M, and if he thinks the probabilities of them are quite high, he might specify a CME approaching the $250M gain.   ■

15.9 Expected Utility Criterion

The *expected utility criterion* or method has particular usefulness for analyzing projects in which the potential gain or loss is of significant size compared to the total funds available to the firm. More specifically, if the marginal utility or desirability of each dollar potentially to be gained or lost is not a constant, the utility of dollars rather than just the amount of dollars is relevant, and it may then be worthwhile to use the expected utility method rather than the probabilistic monetary method.

The expected utility method consists of determining the *cardinal utility* (e.g., relative degree of usefulness or desirability to the decision maker) of each of the possible outcomes of a project or group of projects on some numerical scale and then calculating the expected value of the utility to use as the measure of merit.

The application of this method is based on the premise that it is possible to measure the attitudes of an individual or decision maker toward risk. If the decision maker is consistent with herself, then a relation between monetary gain or loss and the utility or relative desirability of that gain or loss can be obtained through the decision maker's answers to a series of questions and resultant computations, as explained next.

15.9.1 Steps in Deriving Utility-of-Money Function

1. Select two possible monetary outcomes within the range of interest. For example, say you pick $0 and $10,000.

2. Assign arbitrary utility indices to these monetary outcomes, the only restriction being that the index for the higher monetary outcome be higher than the index for the lower monetary outcome. For example, say you assign an index of 1 to a $0 outcome and 20 to a $10,000 outcome.

3. The utility value of other monetary outcomes can be found by having the decision maker answer questions based on the following relation: Given any three monetary amounts, $X < $Y < Z, and known utility values for any two of these amounts, the utility of the third amount can be found by the equation

$$U[\$Y] = P \times U[\$X] + (1 - P) \times U[\$Z], \qquad (15\text{-}7)$$

where

$$P = \text{probability},$$
$$U[X] = \text{utility of } \$X, \text{ etc.}$$

   (a) *To obtain utility values for monetary amounts within any two amounts,* X *and* Z, *for which utility values have been assigned or calculated,* ask questions such as, "What monetary amount for certain, Y, would you desire just as highly as a $P\%$ chance of X and a $(1 - P\%)$ chance of Z? (*Note:* It is generally thought that it is easiest for decision makers to think in terms of $P = 0.5$, though P can be any value between 0 and 1.)

   For example, say you let $P = 50\%$, and suppose the decision maker decides he would desire $3,000 for certain just as much as a 50% chance of $0 outcome and a 50% chance of $10,000 outcome. The utility of $3,000 can then be calculated as

$$U[\$3,000] = 0.5 \times U[\$0] + 0.5 \times U[\$10,000]$$
$$= 0.5 \times 1 + 0.5 \cdot 20$$
$$= 10.5.$$

   (b) To obtain utility values for monetary outcomes less than or greater than those for which utility values have been assigned or calculated, ask questions such as, "What relative chances of monetary outcomes of $X versus $Z would be just as desirable as a certain monetary outcome of $Y?" (*Note:* $X or $Z is the amount for which the utility value is to be determined.)

For example, suppose it is desired to find the utility of $20,000 given the utility values just obtained for $0, $3,000, and $10,000. Suppose the question posed is, "What relative chances of monetary outcomes of $3,000 versus $20,000 would be just as desirable as a certain outcome of $10,000?" Suppose further that the considered answer by the decision maker is 40% chance of $3,000 and 60% chance of $20,000. The utility value of $20,000 can then be calculated as

$$0.4 \times U[\$3,000] + 0.6 \times U[\$20,000] = U[\$10,000],$$
$$0.4 \times 10.5 + 0.6 \times U[\$20,000] = 20, \text{ and}$$
$$U[\$20,000] = 26.3.$$

4. Questions and computations in step 3 can be continued as long as utility values are needed. These can, in turn, be graphed to show utility values for the entire range of monetary outcomes of interest. A graph based on the preceding values is shown in Fig. 15-1.

In carrying out the utility derivation procedure, inconsistencies in the decision maker's replies may be discovered (e.g., two or more utility values calculated for the same monetary outcome or an extremely jagged utility-of-money function). If this happens, it becomes necessary to requestion to obtain judgments that are internally consistent.

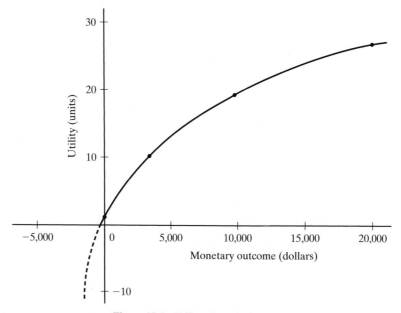

Figure 15-1   Utility-of-money function.

The use of expected utility value as a decision criterion has a real advantage over the expected monetary value, such as expected annual worth or expected present worth. Procedures based on expected monetary values virtually overlook the severe consequences of widely varying possible outcomes and merely take a weighted average of all outcomes. The expected utility procedure overcomes this objection by incorporating these variance influences directly into the computations. A large loss may be assigned a large negative utility by the individual, or he may assign a very great positive utility to a large increment in wealth, thus automatically bringing variance influences into the calculated results. This is demonstrated in the next example.

Example 15-10
As an example of calculation of expected utility, suppose the decision maker having the utility-of-money function in Fig. 15-1 is faced with a project that is expected to have monetary outcomes according to the following probabilities:

Monetary outcome (in net PW)	Probability
$20M	0.05
10M	0.15
0M	0.30
-2M	0.50

It is desired to compare calculated results using expected utility and expected monetary value or outcome.

Solution   The necessary calculations are shown in Table 15-3. Thus, the expected utility method indicates an unfavorable project (expected utility is less than 1.0 when the utility of a monetary outcome of $0 is 1.0). In contrast, the expected monetary outcome (EMV) indicates a favorable project (expected net PW is greater than $0). ■

Again, it should be emphasized that the expected utility method is useful for analyzing projects in which the potential gain or loss is of significant size compared to the monetary resources of the individual or firm for which the analysis is made. If a graph of the utility-of-money function can be closely approximated by a straight line over the range between the maximum and minimum monetary returns under consideration, then the expected monetary value can be used in place of the expected utility without significant error.

TABLE 15-3   Calculation of Expected Monetary Outcome (EMV) and Expected Utility

Monetary outcome (in net PW)	Probability of monetary outcome	Monetary outcome × probability	Utility of outcome	Utility × probability
$20M	0.05	$1.0M	26.3	1.31
10M	0.15	1.5M	20.0	3.00
0M	0.30	0M	1.0	0.30
-2M	0.50	-1.0M	-10.0	-5.00
		Expected net PW: Σ = $1.5M		Expected utility: Σ = -0.39

15.9.2 The Use of Utility Measurements

The method of assigning utilities to outcomes can be quite useful for gaining understanding of the rationale behind decisions made in situations involving risk. Indeed, if a decision maker specifies a utility-of-money function and if the economic analyst can predict monetary outcomes of individual projects that the decision maker accepts or believes, then the analyst can specify the project acceptance or rejection choices that would presumably turn out to be the same as those of the decision maker (neglecting nonmonetary factors). Thus, the problem of the analyst in dealing with decision under risk would be "solved." That is, the analyst could provide the manager with recommendations consistent with the manager's own thinking, thus allowing the the manager to address other problems.

Several limitations to use of the expected utility method should be recognized. First, it is often time consuming, and it is difficult to obtain a consistent utility-of-money function for an individual or organization. Second, responses for determining utility-of-money functions may well change over time; indeed, they may change even from day to day because of changes in the mood or temporary outlook of the person being questioned. Finally, a utility function for a particular set of alternative projects is not necessarily valid for another set of alternatives. Many intangible considerations taint the choice of any specific weighting. A decision maker might indicate a utility function that clearly shows a conservative approach in her attitudes toward corporate actions, but she might have an entirely different set of attitudes for investing with her own personal finances, such as in the stock market.

Under some conditions, it is expedient to employ methods that retain the concepts of utility functions without having to enumerate the full range of utilities and continuing through formal expected utility calculations. These informal uses can serve well to solidify subjective evaluations of risk situations.

15.9.3 Multicriteria Utility Models

Significant developments in the theory and practice of using multicriteria (e.g., multiple objectives, attributes, or factors) utility models have been summarized in a definitive work by Keeney and Raiffa.[3]

In general, the utility, $U(x) = U(x_1, x_2, \ldots, x_n)$, of any combination of outcomes $(x_1, x_2, \ldots, x_n)$ for n criteria $(X_1, X_2, \ldots, X_n)$ can be expressed as either (a) an additive or (b) a multiplicative function of the individual criteria utility functions $U_1(x_1)$, $U_2(x_2), \ldots, U_n(x_n)$, provided that each pair of criteria is

1. preferentially independent of its complement (i.e., the preference order of consequences for any pair of criteria does not depend on the levels at which all other criteria are held) and

2. utility independent of its complement (i.e., the conditional preference order for lotteries involving only changes in the levels of any pair of criteria does not depend on the levels at which all other criteria are held).

---

[3] R. L. Keeney and H. Raiffa, *Decisions with Multiple Objectives: Preferences and Value Tradeoffs* (New York: John Wiley & Sons, Inc., 1976).

The mechanics of building additive and multiplicative utility functions is beyond the scope of this book. The interested reader is referred to Keeney and Raiffa.

15.10 Expectation and Variance Criteria Applied to Investment Projects

The next examples show how expectation and variance criteria may be calculated for a project in which one or more of the variables (elements) are thought to vary according to independent discrete probabilities. The same type of analysis could be used for comparison of two or more alternatives.

Example 15-11

A single project is estimated to have a variable life and other element outcomes as follows:

Investment:	$10,000
Life:	3 yr, with probability = 0.3
	5 yr, with probability = 0.4
	7 yr, with probability = 0.3
Salvage value:	$2,000
Annual receipts:	$5,000
Annual disbursements:	$2,200

Find the expected annual worth and the standard deviation of the annual worth if the MARR is 8%.

Solution

For life = 3 yr, net AW is

$$\$5,000 - \$2,200 - [(\$10,000 - \$2,000)(A/P, 8\%, 3) + \$2,000(8\%)] = -\$460.$$

For life = 5 yr, net AW is

$$\$5,000 - \$2,200 - [(\$10,000 - \$2,000)(A/P, 8\%, 5) + \$2,000(8\%)] = \$630.$$

For life = 7 yr, net AW is

$$\$5,000 - \$2,200 - [(\$10,000 - \$2,000)(A/P, 8\%, 7) + \$2,000(8\%)] = \$1,110;$$

$$E[\text{AW}] = \sum_{\text{yr}} \text{AW} \times P(\text{AW})$$

$$= -\$460(0.3) + \$630(0.4) + \$1,110(0.3) = \$446;$$

$$V[\text{AW}] = \sum_{\text{yr}} (\text{AW})^2 \times P(\text{AW}) - (E[\text{AW}])^2$$

$$= (-\$460)^2 \times 0.3 + (\$630)^2 \times 0.4$$

$$+ (\$1,110)^2 \times 0.3 - (\$446)^2$$

$$= \$401,000$$

$$\sigma[\text{AW}] = \sqrt{\$401,000} = \$631.$$

Note that the expected net AW at an expected life of 5 years, which is $446, is less than the net AW at the assumed-certain life of 5 years, which is $630. Variation of project life can have a very marked effect on the results of an economic evaluation. In general, the greater the life variation, the higher the expected capital recovery cost based on that variation compared to the capital recovery cost at the assumed-certain life equal to the expected life.   ∎

Example 15-12

Assume the same conditions as in Example 15-11, except that annual receipts is also a random variable and is $7,000 with a probability of 0.33 or $4,000 with a probability of 0.67. Further, assume that the variation of project life occurs independently of variation of annual receipts. Show the net AW for all possible occurrences, and compute the expected net AW.

	Project life		
Annual receipts	3 yr $(P = 0.3)$	5 yr $(P = 0.4)$	7 yr $(P = 0.3)$
$7,000 $(P = 0.33)$	$1,540	$2,630	$3,110 ⎫
$4,000 $(P = 0.67)$	−1,460	−370	110 ⎭ Net AW

Solution

The expected net AW is given by

$$E[AW] = \sum_R \sum_N (AW \mid N,R)P(N)P(R),$$

where R denotes annual receipts and N denotes project life. Hence,

$$E[AW] = \$1,540(0.3)(0.33) + \$2,630(0.4)0.33)$$
$$+ \$3,110(0.3)(0.33) - \$1,460(0.3)(0.67)$$
$$- \$370(0.4)(0.67) + \$110(0.3)(0.67)$$
$$= \$437.$$ ∎

In view of the risk and uncertainty regarding numerous variables or elements as typically found in economic analyses, it is reasonable that the measure of merit or desirability for one or more projects can be expressed as one or more continuous distributions. As an illustration, if the net annual worths for projects A and B in Example 15-6 (using the expectation-variance criterion) were distributed normally, the situation could be depicted as in Fig. 15-2. It can be seen in Fig. 15-2 that the probability of a loss (negative net AW) for project A is much higher than for project B; hence, project B might be chosen. If the respective distributions are extremely skewed rather than normally distributed, the indicated decision may differ. For example, suppose that project A is skewed to the right and project B is skewed to the left as shown in Fig. 15-3. On the basis of these conditions, project B might no longer be considered the more desirable.

If a probabilistic monetary model involves simple mathematical functions for the individual elements considered, then it sometimes can be mathematically manipulated so as to obtain directly the desired parameters or characteristics of the measure of merit. Chapter 14 described how Monte Carlo simulation can be utilized to obtain approximations for models of virtually unlimited complexity.

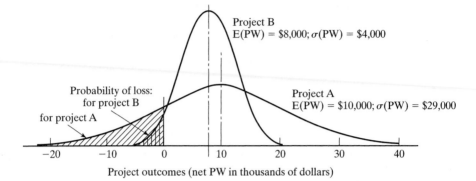

Figure 15-2   Outcome data for alternative projects assuming normal distributions.

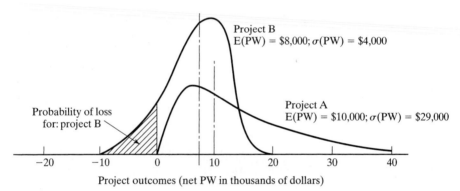

Figure 15-3   Same alternative projects as in Fig. 15-2, except with skewed distributions.

15.11 Miscellaneous Decision Rules for Complete Uncertainty

In this section, we will describe some decision rules or principles for choosing from among alternatives in situations in which there is complete uncertainty about certain probabilities. These decision rules apply to situations in which there are a number of alternatives (*courses of action*) and a number of possible outcomes (*states of nature*), and in which the result (*effect*) of each alternative on each possible outcome is known, but the probability of occurrence of each possible outcome is not known.

The most difficult aspect of using these decision rules is deciding which one to use for making a decision. In effect, these decision rules reflect various degrees of optimism or pessimism and should be chosen according to which reflect certain management views involving intuition and appropriateness for a particular situation. The greatest defense for the use of any of these rules is that their use will promote explicitness and consistency in decision making under complete uncertainty.

TABLE 15-4   Sample Problem Involving Complete
Uncertainty Payoffs—Net PW ($M)

Alternatives	State of nature			
	S_1	S_2	S_3	S_4
I	3	−1	1	1
II	4	0	−4	6
III	5	−2	0	2

A representation of a typical problem is given by the matrix in Table 15-4.

Note that this is the same as the problem in Table 15-2, except that now probabilities are not known. The next sections explain and illustrate each of several decision rules for this type of problem.

15.11.1 Maximin or Minimax Rule

The *maximin rule* suggests that the decision maker determine the minimum profit (payoff) associated with each alternative and then select the alternative that maximizes the minimum profit. Similarly, in the case of costs, the *minimax rule* suggests that the decision maker determine the maximum cost associated with each alternative and then select the alternative that minimizes the maximum cost. These decision rules are conservative and pessimistic, for they direct attention to the worst outcome and then make the worst outcome as desirable as possible.

Example 15-13

Given the payoffs for each of three alternatives and for each of four possible states of nature (chance occurrences) in Table 15-4, determine which alternative would maximize the minimum possible payoff.

Solution   The minimum possible payoff for alternative I is −1, for alternative II is −4, for alternative III is −2. Hence, alternative I would be chosen as maximizing these minimum payoffs. ∎

15.11.2 Maximax or Minimin Rule

The maximax or minimin rules are direct opposites of their counterparts just discussed, and thus reflect extreme optimism. The *maximax rule* suggests that the decision maker determine the maximum profit associated with each alternative and then select the alternative that maximizes the maximum profit. Similarly, in the case of costs, the *minimin rule* indicates that the decision maker should determine the minimum cost associated with each alternative and then select the alternative that minimizes the minimum cost.

Example 15-14

Given the same payoff matrix as in Table 15-4, determine which alternative would maximize the maximum payoff.

Solution   The maximum possible payoff for alternative I is 3. Similarly, for II, the maximum payoff is 6, and for III, it is 5. The highest of these is 6, which occurs with alternative II; so, alternative II is the maximax choice. ∎

15.11.3 Laplace Principle or Rule

The Laplace rule simply assumes that all possible outcomes are equally likely and that one can choose on the basis of expected outcomes as calculated by using equal probabilities for all outcomes. There is a common tendency toward this assumption in situations in which there is no evidence to the contrary, but the assumption (and, therefore, the rule) is of highly questionable merit.

Example 15-15

Given the same payoff matrix as in Table 15-4, determine which alternative is best by using the Laplace rule.

Solution

$$E[\text{alt. I}]: \qquad 3 \times \tfrac{1}{4} - 1 \times \tfrac{1}{4} + 1 \times \tfrac{1}{4} + 1 \times \tfrac{1}{4} = 1.00;$$

$$E[\text{alt. II}]: \qquad 4 \times \tfrac{1}{4} + 0 \times \tfrac{1}{4} - 4 \times \tfrac{1}{4} + 6 \times \tfrac{1}{4} = 1.50;$$

$$E[\text{alt. III}]: \qquad 5 \times \tfrac{1}{4} - 2 \times \tfrac{1}{4} + 0 \times \tfrac{1}{4} + 2 \times \tfrac{1}{4} = 1.25.$$

Thus, alternative II, giving the highest expected payoff, is best.   ∎

15.11.4 Hurwicz Principle or Rule

The Hurwicz rule is intended to reflect any degree of moderation between extreme optimism and extreme pessimism that the decision maker may wish to choose. The rule may be stated explicitly as follows:

Select an index of optimism a such that $0 \le a \le 1$. For each alternative, compute the weighted outcome: $a \times$ (value of profit or cost if most favorable outcome occurs) + $(1 - a) \times$ (value of profit or cost if least favorable outcome occurs). Choose the alternative that maximizes the weighted outcome.

A practical difficulty of the Hurwicz rule is that it is difficult for the decision maker to determine a proper value for a, the weighting factor. The Hurwicz rule also lacks several of the desirable properties of a good decision rule, and it can even lead to results that are obviously counter to intuition.

Example 15-16

Given the same payoff matrix as in Table 15-4, calculate which alternative would be best, using the Hurwicz rule, for an index of optimism of 0.75. Also graph the calculated payoff for each alternative over the entire range of the index of optimism.

Solution

$$\text{alt. I:} \qquad 0.75(3) + 0.25(-1) = 2.0;$$

$$\text{alt. II:} \qquad 0.75(6) + 0.25(-4) = 3.5;$$

$$\text{alt. III:} \qquad 0.75(5) + 0.25(-2) = 3.25.$$

Thus, alternative II, giving the highest payoff, is best. The graph is shown in Fig. 15-4.   ∎

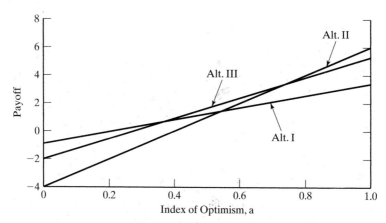

Figure 15-4   Graphed payoffs over range of optimism indices.

15.11.5 Minimax Regret Rule

The minimax regret rule, proposed by L. J. Savage, is similar to the minimax and maximin rules, but is intended to counter some of the ultraconservative results given by those rules. This rule suggests that the decision maker examine the maximum possible *regret* (loss because of not having chosen the best alternative for each possible outcome) associated with each alternative and then select the alternative that minimizes the maximum regret.

Example 15-17

Given the same payoff matrix as in Table 15-4, show which alternative would be chosen on the basis of minimizing the maximum regret. Develop a regret matrix to obtain a solution.

Solution   The result is shown in Table 15-5.

TABLE 15-5   Regret Matrix for Problem of Table 15-4

Alternatives	State of nature				Maximum of states
	S_1	S_2	S_3	S_4	
I	2(= 5 − 3)	1	0	5	5
II	1(= 5 − 4)	0	5	0	5
III	0(= 5 − 5)	2	1	4	4 (min. of all max.)

Thus, it can be seen that the worst (highest) regret for alternative I is 5, for alternative II is 5, and for alternative III is 4. The minimum of these maximum regrets is 4 for alternative III, and thus alternative III is the choice.   ∎

PROBLEMS

15-1. Consider the following matrix of net PWs (in $M) for four alternatives and various possible business conditions for which probabilities can be estimated.

Alternatives	Business condition (and probability)		
	Excellent (0.3)	Fair (0.5)	Poor (0.2)
I	30	25	−15
II	15	15	15
III	35	30	−5
IV	45	10	−5

Show what can be determined, or which alternative is preferred, using each of the following criteria:

a. Does any alternative dominate all the other alternatives? Is any alternative dominated by any other alternative? Which alternative(s) is (are) left for consideration after checking for dominance?

b. if the aspiration level is to minimize the chance of a loss and maximize the chance to make at least 28

c. most probable future

d. expected value

e. expectation–variance, using the function in Eq. 15-3 where $\sigma[x]$ is approximated as (maximum outcome − minimum outcome)/5 and where $A = 1.0$

f. certain monetary equivalence, using your own subjective assessment, assuming you are the top decision maker for a firm with $100 million in assets

15-2. Suppose you were faced with the same matrix of net PWs as in Problem 15-1 but you do not have probability estimates for the various business conditions. Show which alternative should be chosen, using each of the following decision rules or principles:

a. maximin rule

b. maximax rule

c. Laplace principle

d. Hurwicz principle, with 0.5 optimism

e. minimax regret rule

15-3. Consider the following matrix of equivalent annual costs (in $M) for three alternatives and various possible conditions:

Alternatives	State of nature (and probability)		
	Kind (0.3)	Erratic (0.6)	Perverse (0.1)
Alpha	20	28	40
Bravo	25	25	25
Charlie	11	27	45

Show what can be determined, or which alternative is preferred, using each of the following criteria:

a. Is any alternative dominated?

b. if the aspiration level is to have a possible cost no greater than 26

c. the most probable future

d. expected value

e. certain monetary equivalence, using your own subjective assessment, assuming you are a manager who is risk neutral, and thus you base your decisions on long-run averages

15-4. Suppose you are faced with the same matrix of equivalent annual costs as in Problem 15-3, except that you feel the probability estimates are invalid and cannot be used. Show which alternative should be chosen, using each of the following decision rules or principles:

a. minimax rule

b. minimin rule

c. Laplace principle

d. Hurwicz principle, with 0.8 index of optimism

e. minimax regret rule

f. minimize variability, or range (i.e., max. minus min.)

15-5. Given the associated matrix of costs, for various mutually exclusive alternatives, show which is best, using the following decision rules or principles:

a. minimax rule

b. minimin rule

c. Laplace principle

d. Hurwicz principle, with $\frac{2}{3}$ optimism

e. minimax regret rule

	State of nature			
Alternative	A	B	C	D
I	18	18	10	14
II	14	14	14	14
III	5	26	10	14
IV	14	22	10	10
V	10	12	12	10

15-6. Suppose the flip of a fair coin will determine whether you gain X or lose X. What certain amount would you personally be willing to pay (or accept), instead of the random outcome, if X is

a. $0.10?

b. $1.00?

c. $10?

d. $100?

e. $1,000?

f. $10,000?

15-7. Suppose a business opportunity has a 0.25 chance of making a PW of X and a 0.75 chance of making $0. For what certain amount would you be just willing to sell the opportunity if the money is for you personally and X is

a. $1,000?

b. $10,000?

    c. $100,000?

    d. $1,000,000?

15-8. Answer Problem 15-7 if the money belongs to a large corporation for which you are the decision maker. Are you more risk averse if making such decisions for a corporation or for yourself personally?

15-9. Entrepreneur Y has a utility index of 108 for $11,000 and 75 for $0. He is indifferent between a 0.5 chance at $11,000 plus a 0.5 chance at a $20,000 loss and a certainty of $0. What is his utility index for a loss of $20,000?

15-10. Entrepreneur Z has a utility index of 10 for $18,750, 6 for $11,200, and zero for $0. What probability combination of $0 and $18,750 would make her indifferent to $11,200 for certain?

15-11. Two economists, Alfred M. Dismal and J. Maynard Science, are arguing about the relative merits of their respective decision rules. Dismal says he always takes the act with the greatest expected monetary value; Science says she always takes the act with the greatest expected utility, and her utility function for money is $U = 10 + 0.2M$, where M is the monetary payoff. For decisions involving monetary payoffs, who will make the better choices?

15-12. You have a date for the economic analysis ball; the admission is $20, which you do not have. On the day of the dance your psychology instructor offers you either $16 for certain or a 50–50 chance at nothing or $24. Which choice would you make, assuming you had no other sources of funds or credit? Why? If the utility of $16 is 20, and the utility of $0 is zero, what does this imply about the utility of $20?

15-13. Develop a utility function for yourself for the monetary outcomes of −$100,000, −$10,000, +$10,000, +$40,000, and +$200,000. Start with the following monetary outcomes and arbitrarily assigned units:

Monetary outcome	Utility units
$ 1,000	10
15,000	30

Write the questions you ask yourself, and show your calculations. Finally, plot the results with monetary outcome on the x-axis.

15-14. Suppose that the utility-of-money function of a decision maker is described as utility = ln(monetary outcome in thousands of dollars) between the monetary outcome limits of $100 and $1,000,000. The monetary outcomes and associated probabilities for two competing projects are as follows:

Project	Monetary outcome (gain)	Probability
A	$ 1,000	0.33
	10,000	0.33
	19,000	0.33
B	$ 3,000	0.3
	10,000	0.4
	12,000	0.1
	13,000	0.2

Show which project is preferable by (a) the expected monetary method and by (b) the expected utility method. (*Note:* ln means logarithm to the base *e*.)

15-15. A certain project requires an investment of $10,000 and is expected to have net annual receipts minus disbursements of $2,800. The salvage value as a function of life, together with associated probabilities, is as shown:

Life	Salvage value	Probability
3	$4,000	0.25
5	2,000	0.50
7	0	0.25

Find the expected net AW and the standard deviation of net AW if the MARR is 8%.

15-16. Work Problem 15-15 with the change that the net annual receipts minus disbursements is a random variable, independent of the life, and is estimated to be $1,800 with probability 0.2, $2,800 with probability 0.6, and $3,800 with probability 0.2.

15-17. Project Stochastic is estimated to require an investment of $25,000, have a life of 5 years, and $0 salvage value, and have an annual net cash flow of $5,000 with 30% probability, $10,000 with 50% probability, and $12,000 with 20% probability. If the MARR is 15%, calculate the expected value and variance of the net AW for project Stochastic.

15-18. Project Variate is estimated to require an investment of $25,000 and have an annual net cash flow of $16,000 and a $0 salvage value. The life for project Variate is estimated to be 1 year with 10% probability, 5 years with 50% probability, and 10 years with 40% probability. If the MARR is 15%, calculate the expected value and variance of the net AW.

15-19. Plot a frequency histogram for the projects in Problems 15-17 and 15-18, distinguishing between the two by shaded coding. Which project would probably be thought more desirable if the decision maker were (a) conservative, thus not prone to take risks, or (b) a maximizer of expectations, regardless of risk.

15-20. Suppose that the expectation–variance decision function for a given project is equal to the net AW minus a constant times the standard deviation of the net AW.
  a. For the projects in Problems 15-17 and 15-18, determine which appears to be the more desirable if the constant coefficient is 0.6.
  b. At what value of the constant coefficient are the two projects equally desirable?

15-21. The mean and standard deviations of the rate of return for project X are estimated to be 15% and 5%, respectively. Similarly, the mean and standard deviation of the rate of return for a competing project Y are estimated to be 25% and 18%, respectively.
  a. If the expectation–variance function for the decision maker is the expected rate of return minus 0.1 times the variance of the rate of return (in integer amounts), show which project would be more desirable.
  b. For the function in part (a), at what value of the coefficient applied to the variance would the projects be considered equally desirable?

15-22. Suppose that in Problem 15-15 the interest on capital is 4% with probability 0.5 and 12% with probability 0.5, and suppose that interest varies independently of the life of the project.
  a. Calculate the expected net AW.
  b. Plot histograms of outcomes for Problem 15-15 and for this problem to compare variability.

15-23. A specific project requires an investment of $100,000 and is expected to have a salvage value of $20,000. It is thought equally likely that the life will turn out to be 6, 10, or 12 years. The net annual cash inflow is twice as likely to be $30,000 as either $35,000 or $18,000. The minimum required rate of return is 10%.

    a. Develop a table showing the net AW for all combinations of the two variables.

    b. Assuming that the variable outcomes are independent, calculate $E[AW]$ and $V[AW]$.

15-24. Suppose a decision maker is to be given $100. As part of the deal, she must pick one of two bets. In the first bet, there is a 50–50 chance of either winning $200 or losing $50. In the second bet, there is a 50–50 chance of winning $350 or losing the original $100. Determine which bet would (probably) be chosen, and briefly discuss why, if the decision maker is

    a. "risk neutral" for the amounts involved.

    b. you, acting on your own behalf.

    c. a struggling student desperately needing $40 now.

    d. a "risk seeker," who is most motivated by the possibility of a maximum win.

16
Decision Tree Analysis

16.1 Introduction

Decision trees, also commonly called *decision flow networks* and *decision diagrams,* are powerful means for depicting and facilitating the analysis of important problems, especially those that involve sequential decisions and variable outcomes over time. Decision trees have great usefulness in practice because they make it possible to look at a large complicated problem in terms of a series of smaller simpler problems, and they enable objective analysis and decision making that include explicit consideration of the risk and effect of the future.

The name *decision tree* is descriptive of the appearance of a graphical portrayal, for it shows branches for each possible alternative for a given decision and branches for each possible outcome (event) that can result from each alternative. Such networks reduce abstract thinking to a logical visual pattern of cause and effect. When costs and returns are associated with each branch and probabilities are estimated for each possible outcome, then analysis of the decision tree can clarify choices and risks.

16.2 A Deterministic Decision Tree

The most basic form of decision tree occurs when each alternative can be assumed to result in a single outcome—that is, when certainty is assumed. The replacement problem in Fig. 16-1 illustrates this. The problem as shown reflects that the decision on whether to replace the old machine with the new machine is not just a one-time decision, but rather one that recurs periodically. That is, if the decision is made to keep the old machine at decision point 1, then later, at decision point 2, a choice again has to

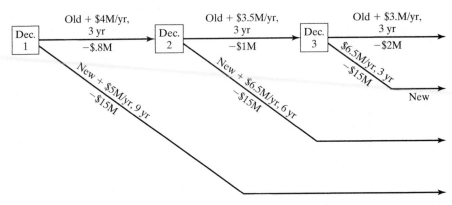

Figure 16-1   Deterministic replacement example.

be made. Similarly, if the old machine is chosen at decision point 2, then a choice again has to be made at decision point 3. For each alternative, the cash inflow is shown above the arrow and the cash investment opportunity cost is shown below the arrow.

For this problem, we are concerned initially with which alternative to choose at decision point 1. But an intelligent choice at decision point 1 should take into account the later alternatives and decisions that stem from it. Hence, the correct procedure in analyzing this type of problem is to *start* at the most distant decision point, determine the best alternative and quantitative result of that alternative, and then "roll back" to each successive decision point, repeating the procedure until finally the choice at the initial or present decision point is determined. By this procedure, we can make a present decision that directly takes into account the alternatives and expected decisions of the future.[1]

For simplicity in this example, timing of the monetary outcomes will first be neglected, which means that a dollar has the same value regardless of the year in which it occurs. Table 16-1 shows the necessary computations and decisions. Note that the monetary outcome of the best alternative at decision point 3 ($7.0M for the "Old")

TABLE 16-1   Monetary Outcomes and Decisions at Each Point—
Deterministic Replacement Example of Fig. 16-1

Decision point	Alternative	Monetary outcome		Choice
3	Old	$3M(3) − $2M	= $ 7.0M	Old
	New	$6.5M(3) − $15M	= $ 4.5M	
2	Old	$7M + $3.5M(3) − $1M	= $16.5M	
	New	$6.5M(6) − $15M	= $24.0M	New
1	Old	$24M + $4M(3) − $0.8M	= $35.2M	Old
	New	$5M(9) − $15M	= $30.0M	

[1] This procedure is a special (and simple) case of dynamic programming (DP). For a discussion of the use of DP in replacement studies, the reader is referred to C. S. Park and G. P. Sharpe-Bette, *Advanced Engineering Economics* (New York: John Wiley & Sons, 1990).

becomes part of the outcome for the "Old" alternative at decision point 2. Similarly, the best alternative at decision point 2 ($24.0M for the "New") becomes part of the outcome for the "Old" alternative at decision point 1.

By following the computations in Table 16-1, we can see that the answer is to keep the "Old" now and plan to replace it with the "New" at the end of 3 years. But this does not mean that the old machine should necessarily be kept for a full 3 years and then a new machine bought without question (refer to Chapter 9). Conditions may change at any time, thus necessitating a fresh analysis—probably a decision tree analysis—based on estimates that are reasonable in light of conditions at that later time.

16.3 A Deterministic Decision Tree Considering Timing

For decision tree analyses, which involve working from the most distant decision point to the nearest decision point, the easiest way to take into account the timing of money is to use the present worth approach and thus discount all monetary outcomes to the decision points in question. To demonstrate, Table 16-2 shows computations for the same replacement problem of Fig. 16-1, using an interest rate of 25% per year.

Note from Table 16-2 that, when taking into account the effect of timing by calculating present worths at each decision point, the indicated choice is not only to keep the "Old" at decision point 1, but also to keep the "Old" at decision points 2 and 3. This result is not surprising since the high interest rate tends to favor the alternatives with lower initial investments, and it also tends to place less weight on long-term returns.

16.4 Consideration of Random Outcomes

The deterministic replacement example of Fig. 16-1 did not include one of the most powerful elements in the use of decision trees: the formal consideration of variable outcomes to which probabilities of occurrence can be assigned. Suppose that, for each

TABLE 16-2 Decisions at Each Point with Interest = 25% per Year for Deterministic Replacement Example of Fig. 16-1

Decision point	Alternative	PW of monetary outcome		Choice
3	Old	$3M(P/A,3) - $2M		
		$3M(1.95) - $2M	= $3.85M	Old
	New	$6.5M(P/A,3) - $15M		
		$6.5M(1.95) - $15M	= -$2.33M	
2	Old	$3.85(P/F,3) + $3.5M(P/A,3) - $1M		
		$3.85(0.512) + $3.5M(1.95) - $1M	= $7.79M	Old
	New	$6.5M(P/A,6) - $15M		
		$6.5M(2.95) - $15M	= $4.18M	
1	Old	$7.79M(P/F,3) + $4M(P/A,3) - $0.8M		
		$7.79M(0.512) + $4M(1.95) - $0.8M	= 10.98M	Old
	New	$5.0M(P/A,9) - $15M		
		$5.0M(3.46) - $15M	= $2.30M	

alternative, there are two possible monetary outcomes, depending on whether the demand is "high" or "low." In such a case, the decision tree problem of Fig. 16-1 would appear as in Fig. 16-2. Note that, for each alternative in Fig. 16-2, there is shown a circle from which are drawn arrows to represent each possible chance event or state of nature which can result, such as demand being either "high" (H) or "low" (L).

     In order to solve this problem—that is, to determine the best alternative for each decision point, etc.—it is necessary to determine the outcome (usually expressed in monetary units) and the probability of occurrence for each possible chance event. Then the criterion (measure of merit) for choice (usually expected PW of monetary outcomes) can be decided and the solution can be computed by the same procedure as before; that is, criterion outcomes and decisions are determined for the most distant decision points first, and then the procedure is successively repeated, moving back in time until the decision for decision point 1 is determined.

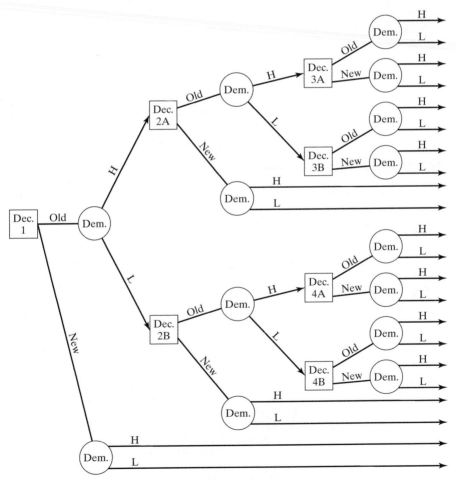

Figure 16-2   Probabilistic replacement example.

16.5 A Classical Decision Tree Problem

Next, we present a brief description of a classical problem for which decision trees are very useful for analysis and solution.[2]

An oil wildcatter must decide whether to drill or not to drill at a given site before his option expires. He is uncertain whether the hole will turn out to be dry, wet, or a gusher. The net payoffs (in present worths) for each state are −$70,000, $50,000, and $200,000, respectively. The initially estimated probabilities that each state will occur are 0.5, 0.3, and 0.2, respectively. Figure 16-3 is a decision flow network depicting this simple situation. Table 16-3 shows calculations to determine that the best choice for the wildcatter is to drill, on the basis of an expected monetary value of $20,000 versus $0 if he does not drill. Nevertheless, this may not be a clear-cut decision, because of the risk of a $70,000 loss and because the wildcatter might reduce the risk by obtaining further information.

Suppose it is possible for the wildcatter to take seismic soundings at a cost of $1,000. The soundings will disclose whether the terrain below has no structure (outcome NS), an open structure (outcome OS), or a closed structure (outcome CS).

Instead of using Bayesian methods for revision of probabilities (to be discussed later), let us assume that the probabilities of the various possible well outcomes, given the various seismic sounding outcomes, are as shown in Fig. 16-4, which is a flow diagram for the entire problem.[3] The solution of the problem, using the expected monetary value (EMV) criterion is shown in Table 16-4. It should be noted that the alternative "seismic soundings" is now best, with an expected monetary outcome of $31,550.

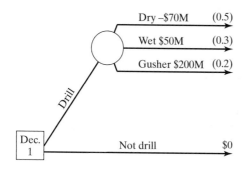

Dry −$70M   (0.5)

Wet $50M   (0.3)

Gusher $200M   (0.2)

Drill

Dec. 1   Not drill $0

Figure 16-3 Oil wildcatter problem diagram, before consideration of seismic soundings.

TABLE 16-3 Expected Monetary Calculations for the Oil Wildcatter Problem Before Consideration of Seismic Soundings

Drill:	−$70,000(0.5) + $50,000(0.3) + $200,000(0.2) = $20,000
Not drill:	= $0

[2] H. Raiffa, *Decision Analysis: Introductory Lectures on Choices under Uncertainty* (Reading, MA: Addison-Wesley, 1968).

[3] Problem 16-11 at the end of this chapter is a statement of the sampling or added-study probabilities which, when combined with the prior probabilities in Fig. 16-3, result in the posterior probabilities in Fig. 16-4.

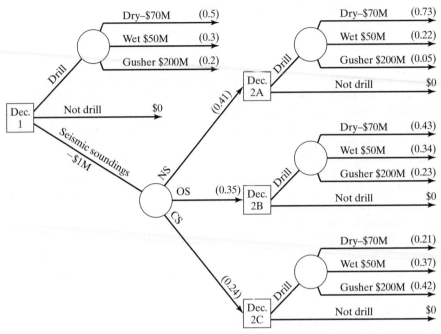

Figure 16-4 Oil wildcatter problem with seismic soundings taken into consideration.

TABLE 16-4 Expected Monetary Calculations for the Oil Wildcatter Problem
with Consideration of Seismic Soundings

Decision point	Alternative	Expected monetary value		Choice
2A	Drill	−$70M(0.73) + $50M(0.22) + $200M(0.05) = −$30M		
	Not drill		$0	Not drill
2B	Drill	−$70M(0.43) + $50M(0.34) + $200M(0.23) =	$32.9M	Drill
	Not drill		$0	
2C	Drill	−$70M(0.21) + $50M(0.37) + $200M(0.42) =	$87.5M	Drill
	Not drill		$0	
1	Drill		$20M	
	Not drill		$0	
	Seismic soundings	$0(0.41) + $32.9M(0.35) + $87.5M(0.24) − $1M	= $31.55M	Seismic soundings

16.6 Constructing a Decision Tree

Now that decision trees (diagrams) have been introduced and the mechanics of using the diagrams to arrive at an initial decision have been illustrated, the steps involved can be summarized as follows:

1. Identify the points of decision and alternatives available at each point.

2. Identify the points of uncertainty and the type or range of possible outcomes at each point.

3. Estimate the values needed to make the analysis, especially the probabilities of different outcomes and the costs/returns for various outcomes and alternative actions.

4. Analyze the alternatives, starting with the most distant decision point(s) and working back, to choose the best initial decision.

The preceding example used the EMV as the decision criterion. However, if outcomes can be expressed in terms of utility units, then the decision maker can use the expected utility as a decision criterion. Alternatively, the decision maker may be willing to express his certain monetary equivalent (CME) for each chance outcome node and use that as his decision criterion.

Because a decision diagram can quickly become discouragingly, if not unmanageably, large, it is generally best to start out by structuring a problem simply by considering only major alternatives and outcomes in order to get an initial understanding or "feel" for the problem. Then you can develop more information on alternatives and outcomes that seem sufficiently important to affect the final decision, until you are satisfied that the study is sufficiently complete (in view of the nature and importance of the problem and the time and study resources available).

The proper diagramming of a decision problem is, in itself, generally very useful to the understanding of the problem, and it is essential to correct subsequent analysis.

The placement of decision points and chance outcome nodes from the initial decision point to the base of any later decision point should give a correct representation of the information that will and will not be available when the decision maker actually has to make the choice represented by the decision point in question. The decision tree diagram should show the following:

1. all initial or immediate alternatives among which the decision maker wishes to choose;

2. all uncertain outcomes and future alternatives that the decision maker wishes to consider because they may directly affect the consequences of initial alternatives;

3. all uncertain outcomes that the decision maker wishes to consider because they may provide information that can affect her future choices among alternatives, and hence indirectly affect the consequences of initial alternatives.

It should also be noted that the alternatives at any decision point and the outcomes at any outcome node must be

1. mutually exclusive (i.e., no more than one can possibly be chosen), and

2. collectively exhaustive (i.e., one must be chosen or something must occur if the decision point or outcome node is reached).

16.7 Use of Bayesian Statistics to Evaluate the Worth of Further Investigative Study

One alternative that frequently exists in an investment decision problem is further research or investigation before deciding on the investment. This means making an intensive objective study, hopefully by a fresh group of people. It may involve such aspects as undertaking additional research and development study, making a new analysis of market demand, or possibly studying anew future operating costs for particular alternatives.

The concepts of Bayesian statistics provide a means for utilizing subsequent information to modify estimates of probabilities and also a means for estimating the value of further economic investigative study. To illustrate how the worth of sample information is obtained, consider the short-term machine replacement decision situation shown in Fig. 16-5. Here, each alternative has two possible chance outcomes: "high" or "low" demand. It is estimated that each outcome is equally likely to occur, and the monetary result expressed as PW is shown above the arrow for each outcome. Again, the amount of investment for each alternative is shown below the respective lines. Based on these amounts, the calculation of the expected monetary outcome (net PW) is shown in Table 16-5, which indicates that the "Old" should be chosen.

To demonstrate the use of Bayesian statistics, suppose that you are considering the advisability of undertaking an independent intensive investigation before deciding upon the "Old" versus the "New." Suppose also that this further study would cost $0.1M. In order to use the Bayesian approach, it is necessary for management to assess the conditional probabilities that the intensive investigation will yield certain results. These probabilities reflect explicit measures of management's confidence in the ability of the investigation to predict the outcome. Sample assessments are shown

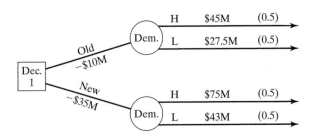

Figure 16-5   One-stage replacement problem.

TABLE 16-5   Expected Monetary Values
for Problem in Fig. 16-5

Old:	$45M(0.5) + $27.5M(0.5) − $10M = $26.25M
New:	$75M(0.5) + $43M(0.5) − $35M = $24.0M

TABLE 16-6   Management's Assessment of Confidence
in Investigation Results

$P(h \mid H) = 0.70$
$P(h \mid D) = 0.20$
$P(d \mid H) = 0.30$
$P(d \mid D) = 0.80$

Key:	Investigation-predicted demand	Actual demand
	h = High	H = High
	d = Low	D = Low

in Table 16-6. As an explanation, $P(h/H)$ means the probability that the predicted demand is "high," given that the actual demand will turn out to be "high."

Appendix 16-A contains a formal statement of Bayes's theorem, as well as a tabular format for ease of calculations in the discrete outcome case. Tables 16-7 and 16-8 use this format for revision of probabilities, on the basis of data in Table 16-6 and the prior probabilities of 0.5 that the demand will be high and 0.5 that the demand will be low.

The probabilities calculated in Tables 16-7 and 16-8 can now be used to assess the alternative of further investigation. Figure 16-6 shows a decision tree diagram for this alternative, as well as the two original alternatives. Note the demand probabilities entered on the branches according to whether the *investigation* indicates "high" (h) or "low" (d) demand.

TABLE 16-7   Computation of Posterior Probabilities Given that
Investigation-Predicted Demand Is High (h)

(1)	(2)	(3)	(4)	(5)
			= (2)(3)	= (4)/Σ(4)
State	Prior	Confidence		Posterior
(actual	probability	assessment	Joint	probability
demand)	P(state)	$P(h \mid$ state)	probability	P(state $\mid h$)
H	0.5	0.70	0.35	0.78
D	0.5	0.20	0.10	0.22
			Σ = 0.45	

TABLE 16-8   Computation of Posterior Probabilities Given that
Investigation-Predicted Demand Is Low (d)

(1)	(2)	(3)	(4)	(5)
			= (2)(3)	= (4)/Σ(4)
State	Prior	Confidence		Posterior
(actual	probability	assessment	Joint	probability
demand)	P(state)	$P(d \mid$ state)	probability	P(state $\mid d$)
H	0.5	0.30	0.15	0.27
D	0.5	0.80	0.40	0.73
			Σ = 0.55	

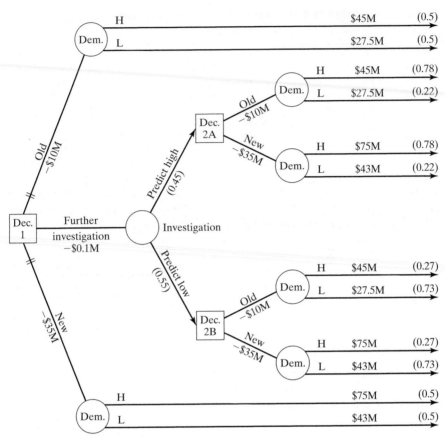

Figure 16-6   Replacement problem with alternative of further investigation.

The expected outcome for the alternative of further investigation can now be calculated. This is done by the standard decision tree principle of determining the decision at the most distant points and working back. This is shown in Table 16-9. It is worthy to note that the 0.45 and 0.55 probabilities that investigation-predicted

TABLE 16-9   Expected Monetary Outcome for Replacement Problem of Fig. 16-6

Decision point	Alternative	Expected monetary outcome		Choice
2a	Old	$45M(0.78) + $27.5M(0.22) − $10M	= $31.15M	
	New	$75M(0.78) + $43M(0.22) − $35M	= $32.96M	New
2b	Old	$45M(0.27) + $27.5M(0.73) − $10M	= $22.23M	Old
	New	$75M(0.27) + $43M(0.73) − $35M	= $16.64M	
1	Further investigation	$32.96M(0.45) + $22.23M(0.55) − $0.1M = $26.96M		Further investigation
	Keep old	(from Table 14-5): $26.25M		
	New	(from Table 14-5): $24.00M		

demand will be "high (h)" and "low (d)," respectively, are obtained from the total in column (4) of the Bayesian revision calculations shown in Tables 16-7 and 16-8.

Thus, from Table 16-9, it can be seen that the alternative of further investigation, with an expected return of $26.96M, is the best present course of action by a slight margin. While the figures used here do not reflect much advantage to the further investigation, the advantage potentially can be great.

It is often thought useful to show results of calculations and choices between alternatives directly on the decision diagrams. For example, the replacement problem in Fig. 16-6 (and the calculated results and choices in Table 16-9) might be shown as in Fig. 16-7. The numbers in small boxes next to each outcome node represent the expected value (or other indicator of desirability) of outcomes beyond that point. The "double slash" marks for all alternatives except one emanating from each decision point indicate alternatives that would *not* be chosen. The number in the small box next to each decision point indicates the value of the best alternative at each point.

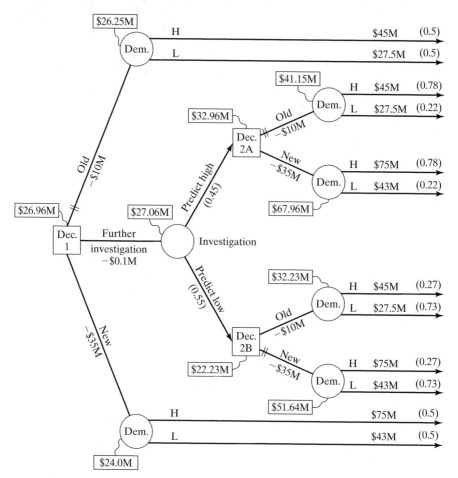

Figure 16-7 Replacement problem with alternative of further investigation (showing useful notation conventions).

16.8 Alternate Method of Analysis

Some analysts and decision makers prefer to show the criterion values (such as expected monetary value) at the end of each possible path through the tree. (*Note:* If the monies along the path are at significantly different points in time, the criterion values should be expressed in terms of equivalent worths, such as PW or AW.) Then, the "roll back" technique can be used to determine the optimal choice at each decision point and obtain the same initial decision as when using the previous method. For example, the problem in Fig. 16-6 and Fig. 16-7 could be shown as in Fig. 16-8, which also includes the "notation conventions" explained in the last section. The criterion values at the end of each branch are placed in oval boxes to distinguish them from the other outcome and investment values emanating from each chance node and decision point. As an example, the criterion value for the third path from the top, $34.9M, in

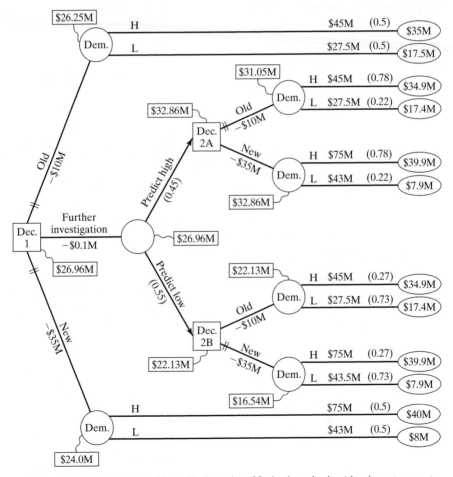

Figure 16-8 Replacement problem with alternative of further investigation (showing outcomes at end of each branch and subsequent analysis).

TABLE 16-10   Expected Monetary Outcomes for Replacement Problem of Figure 16-8

Decision point	Alternative	Expected monetary outcome	Choice
2a	Old	$34.9M(0.78) + $17.4M(0.22) = $31.05M	
	New	$39.9M(0.78) + $7.9M(0.22) = $32.86M	New
2b	Old	$34.9M(0.27) + $17.4M(0.73) = $22.13M	Old
	New	$39.9M(0.27) + $7.9M(0.73) = $16.54M	
1	Further investigation	$32.86M(0.45) + $22.13M(0.55) = $26.96M	Further investigation
	Old	35M(0.5) + 17.5M(0.5) = $26.25M	
	New	40M(0.5) + 8M(0.5) = $24.00M	

Fig. 16-8 is obtained by adding −$0.1M, −10M, and $45M for that path. Calculations of expected monetary outcomes in Fig. 16-8 are shown in Table 16-10.

16.9 Examples of Decision Tree Applications

The decision tree technique can be useful in a very wide range of decision situations. To give some idea of the breadth of potential applications, two examples follow.

16.9.1 Small versus Large Asset

Figure 16-9 shows a situation in which a firm is initially faced with the decision between a small machine and a large machine for a use in which demand for the

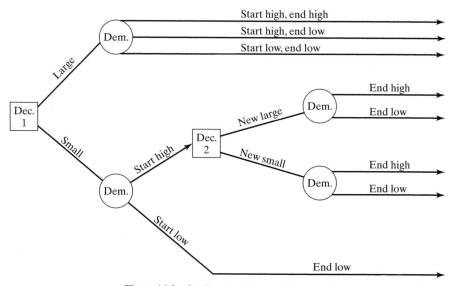

Figure 16-9   Small versus large machine example.

machine is uncertain, but subject to probabilistic estimates. Further, if the firm should invest in a small machine now, it has the future choice of whether to invest in another small machine according to the anticipated demand at the time of that future decision.

16.9.2 Facilities Modernization

Figure 16-10 shows a situation in which a firm is faced with the decision of whether to invest in major automation of the plant's facilities. The new equipment is supposed to result in reduced labor cost, but its technical performance (perf.) is critical and subject to variation. Also, the monetary outcome is influenced strongly by the total market demand (MKT.) for the product and by whether or not competitors (compet.) also automate. The diagram shows two decision stages; but, of course, further stages can be enumerated if that is thought desirable.

16.10 Decision Trees and Real Options

Although we have shown that decision trees are applicable to a wide range of problem situations where uncertainty plays a role, we need to focus on a particular capability that is very important in capital investment decisions—decision trees enable the analyst to model real options that are hidden in classical net present worth analysis. These options, not treated formally in classic single outcome analyses, are often understood by management and factored informally into their assessment of the analysis (and the assessment of the analyst!). This presence of hidden options is most easily understood with an example.

Example—Hytech Industries

Hytech is considering opening an entirely new electronic interface using Radio Frequency Identification with a new coding system that deviates from some standards, but offers advantages to bulk chemical manufacturers. The cost to develop the manufacturing capacity and build the market is estimated at $700 million, and the resulting net cash flows (inflows) after income taxes are estimated at $100 million per year (albeit with much potential variation depending upon acceptance of the product). The after-tax MARR is 15% per year, and the new product will have a life of 20 years.

Solution as Classical Single Outcome Analysis   The net present worth of this project is

$$\text{NPV} = -\$700 + \$100[P/A, 15\%, 20] = -\$74.1 \text{ million,}$$

and the recommendation would be to avoid the investment.

Suppose, however, that the $100 million cash inflow per year is merely an average. If the device catches on, potential sales will exceed the annual capacity for the new facilities and annual inflows will be limited to $200 million. In fact, if this happens, a second plant can be acquired 1 year later for an additional $500 million investment and it would be able to generate an additional $150 million cash inflow annually. On the other hand, if the new product bombs, sales could effectively be zero, but the plant could be salvaged for $150 million.

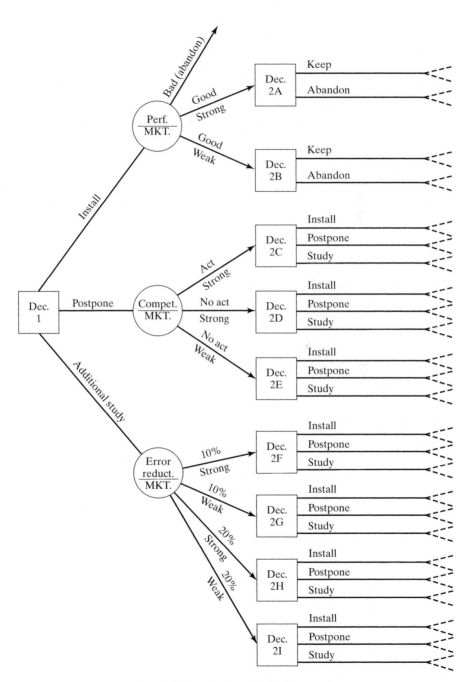

Figure 16-10   Facilities modernization example.

Solution as a Decision Tree with Embedded Options   The decision tree appears as shown in Figure 16-11. If the end-of-path convention is used, then the tip values are as shown in Table 16-11. Note that path 3, Abandon when first year inflows exceed $200 million, is not reasonable and is not evaluated. Similarly, paths 4 and 7 would involve expansion when the market has not developed and are not evaluated.

As shown in the table, when first year net inflows are high ($200 million +), the better of paths 1 and 2 is to expand and realize a present worth of $926 million. If response is tepid ($100 million inflow), then path 5 is better than path 6. The present worth of path 5 is negative, but staying in business recoups some of the investment and is better than abandoning the project. Finally, if the market is negligible, path 9 is better than path 8, because the ability to salvage some of the investment mitigates the loss.

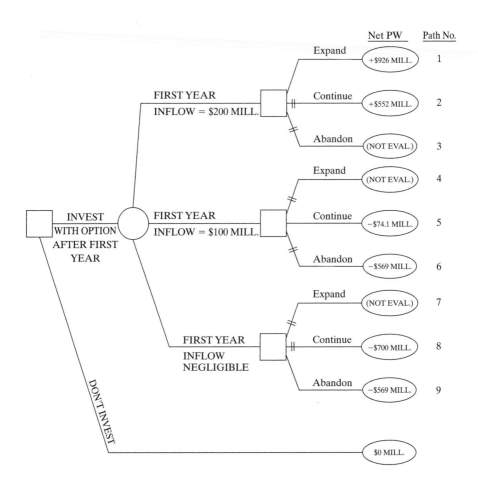

Figure 16-11   Hytech Industries Example.

TABLE 16-11   Decision Tree Endpoint Evaluation

Path	First year inflows	Option	Evaluation	Net P W
1	$200 million +	Expand	$-$700 + $200 [P/A, 15\%, 20]$	+$926 million
			$-$500 [P/F, 15\%, 1]$	
			$+$150[P/A, 15\%, 19] [P/F, 15\%, 1]$	
2	$200 million +	Continue	$-$700 + $200 [P/A, 15\%, 20]$	+$552 million
3	$200 million +	Abandon	Not evaluated	
4	$100 million	Expand	Not evaluated	
5	$100 million	Continue	$-$700 + $100 [P/A, 15\%, 20]$	$-$74.1 million
6	$100 million	Abandon	$-$700 + $150 [P/F, 15\%, 1]$	$-$569 million
7	Negligible	Expand	Not evaluated	
8	Negligible	Continue	$-$700$	$-$700 million
9	Negligible	Abandon	$-$700 + $150 [P/F, 15\%, 1]$	$-$569 million

If each outcome of the first year cash flows were equally likely and management is risk neutral, then the expected monetary value is $94.3(= [+$926 -$74.1-$569] / 3)million, an improvement of $168.4 million over the $74.1 million loss projected in the single outcome analysis. This improvement can be viewed as the value of the combined options to expand, if the market develops, or to salvage some investment if the project is a flop. Chapter 17 develops other approaches for the valuation of real options. They have become an important means of quantifying downstream impacts of projects and a valuable tool in the analysis of capital investment projects. ■

16.11 Advantages and Disadvantages of Decision Tree Analysis

The systematic approach of decision tree analysis has its merits and demerits. Indeed, what is a pro to some analysts and decision makers may well be a con to others. The following is a synthesis of often claimed advantages:

1. *Makes uncertainty explicit.* The uncertainty the analyst feels about estimates or projects is recognized and incorporated in the analysis.

2. *Promotes more reasoned estimating procedures.* Requiring that estimates be given as probability distributions, rather than as single values, and requiring that these estimates be broken into elements, forces more attention on the estimating.

3. *Encourages consideration of the whole problem.* The systematic approach forces the analyst or decision maker to come to quantitative grips with the interactions among various facets of his or her problem.

4. *Helps communication.* It facilitates the provision of inputs, in an unambiguous quantitative manner, from experts and analysts, as needed, and it provides these results to the decision maker in a clear manner.

5. *Helps determine the need for data and study.* The systematic examination of the value of information in a decision context helps suggest the gathering and compilation of data from new sources.

6. *Stimulates generation of new alternatives.* Detailed decision analysis helps the decision maker and his or her staff to think hard about new, viable alternative actions.

7. *Helps "sell" decision.* A hard, thorough analysis can be used to emphasize that a decision has not been made frivolously and to rally support for the decision.

8. *Provides framework for contingency planning.* Decision analysis not only results in an initial decision, but it can be used as a basis for continuous reevaluation of a decision problem that has a distant time horizon.

The following is a summary of frequently claimed disadvantages:

1. *Tends to exclude consideration of intangibles.* Because a well-done decision analysis is thorough, there is a tendency to place too great a reliance on the quantitative results.

2. *The clear basic questions are often the most difficult.* Often, the decision maker would rather take refuge in the fuzziness and complexity of real-life situations than reveal preferences in a number of broken-down or starkly simple decision situations.

3. *Requires expert articulation of the thought process.* A decision maker may be a great synthesizer of interconnected considerations through subconscious thinking, but he may be unable to give a verbal description of his thought process, thus making it appear that he is much more restricted in the complexity of his analysis than is actually true.

4. *Decision analysts tend to lack compassion.* Persons who elect to get into formalized, systematic analysis are so prone to attach numbers to everything that they tend to exclude many human and artistic qualities, thereby inhibiting creativity.

16.12 Summary

The decision tree approach may appear to be complex, but it needs to be no more complex than the decision situation involved. Any investment problem can be examined at many levels of detail. A major difficulty in setting up a decision tree analysis is to strike the appropriate level. In general, the appropriate level allows decision makers to consider major future alternatives commensurate with the consequences of those alternatives without becoming so concerned with detail and refinement that the key factors are obscured.

Decision tree methodology is a basis for investment analysis, evaluation, and decision and is a means of making explicit the process that should be at least intuitively present in good investment decision making. Use of this methodology will help force a consideration of alternatives, define problems for further investigation, and clarify for the decision maker the nature of the risks she faces and the estimates she must make.

PROBLEMS

16-1. Because of shifting rock formations, a community will be in danger of the collapse of an upstream dam for a year starting now. A permanent replacement dam has been started, but it will take a year to complete. If there is a dam collapse, it will destroy the town, but no deaths would be expected because there is an efficient warning system.

One alternative is to make temporary repairs on the existing dam and to construct temporary levees. Such repairs would greatly decrease the probability of the collapse of the dam. If the collapse should occur, the levees might or might not hold the water back from the town.

The decision maker in this problem has decided to call in experts to give opinions. Because the collapse of the old dam depends in part on the underlying geological features of the area, the experts can better assess the likelihood of dam collapse if they conduct some expensive geological tests.

Diagram the decision maker's problem in the form of a decision tree.

16-2. The president of the High Point Carolina Company, Bob Foscue, must decide quickly whether or not to lease a large manufacturing area that has become available adjacent to the firm's present facilities. He is convinced that if he does not lease the extra space now, it will not be feasible for him to move or otherwise expand his plant capacity for at least 2 years. If he does not lease the space and runs out of production capacity during the period, he may be forced next year to make a difficult decision that could be critical to the future of his small business. This decision will be between (a) failure to fill all customer orders by rationing limited supplies among existing accounts and (b) using outside contract furniture builders to produce for demand in excess of the firm's capacity. The consequences of choice (a) would be a significant slowing of the firm's aggressive growth momentum. However, choice (b) has an excellent chance of working out well except for two potentially fatal dangers: the contract suppliers might develop dangerous competition with the firm's designs, or they might fail to meet quality and shipping quantity requirements.

Foscue believes that, if he does not lease the additional space, the question of whether or not he will run out of production capacity and be forced to choose between (a) and (b) depends on two key factors: (1) retailer demand for his furniture during the coming year's buying season and (2) his own decision on whether or not to continue an existing merchandising arrangement with a large direct-mail catalog firm. The direct merchandising arrangement provides an outlet for considerable sales volume, but at modest prices. Foscue is very sure that the merchandising firm will not offer materially improved prices or terms. His own final decision about renewal of the contract will depend on his overall evaluation of supply and demand factors at the time. His current decision about the lease, by limiting the amount of furniture he can supply without going to outside contractors, may have some impact on his subsequent decision on the renewal of the contract. However, Foscue does not consider it feasible or worthwhile to lay out all of the important developments occurring prior to the time of renegotiation.

He is fairly confident, however, that if he does not renew the contract, then total demand will be low enough so that he can meet it fully without either expanding beyond his present manufacturing space or relying upon contract producers.

Draw a decision tree that would be appropriate for a first model of Foscue's problem.

16-3. A purchasing manager is faced with deciding whether or not to stock a large supply of metal. The uncertain variable is the future price of the metal. The following are present worths of consequences and prior probabilities for the various perceived outcomes:

Future price	P(future price)	PW if do stock	PW if do not stock
High	0.3	$100,000	$0
Medium	0.5	– 10,000	0
Low	0.2	–100,000	0

For $6,000, it is possible to hire a consulting firm that would be able to make a fairly accurate forecast in terms of whether the price will go up or down as follows:

If the future is going to be	Then the probabilities the consultant will predict the price will go up or down are as follows:	
	Up	Down
High	0.9	0.1
Medium	0.4	0.6
Low	0.8	0.2

a. Diagram the problem in the form of a decision tree.

b. Determine what would be the better alternative, using the EMV criterion.

16-4. The Norva Company has already spent $80,000 developing a new electronic gauge and is now considering whether or not to market it. Tooling for production would cost $50,000. If the gauge is produced and marketed, the company estimates that there is only one chance in four that the gauge would be successful. If successful, the net cash inflows would be $100,000 per year for 8 years. If not successful, the net cash outflows would be $30,000 per year for 2 years, after which time the venture would be terminated. The MARR is 20% per year.

a. Draw a decision tree and determine the better alternative, using the EMV criterion based on present worths.

b. Suppose the market research group can make a market survey that with probability 0.8 will *predict* a success if the gauge will turn out to be a success and with probability 0.9 will *predict* failure if the gauge will turn out to be unsuccessful. Should the survey be undertaken first? What is the expected value of the survey to the company?

16-5. Suppose, given the alternatives in the small versus large machine example in Fig. 16-9, the demands are assumed to be random variables with present worths of outcomes as follows:

At Decision Point 1:

If "Large," normal distribution with

$$\text{Expected outcome} = \$500M$$
$$\text{Standard deviation} = \$200M$$

If "Small," discrete distribution with

      P(Start high) = 0.70
      P(Start low) = 0.30, with outcome
                 "End low" having uniform distribution between
                 $150M and $450M

At Decision Point 2:
If "New large," normal distribution with

            Expected outcome = $650M
            Standard deviation = $250M
If "New small," normal distribution with

            Expected outcome = $550M
            Standard deviation = $150M

Demonstrate the use of Monte Carlo simulation (see Chapter 14) for developing data to approximate the distribution of the *difference* between the present worths of the "Large" and "Small" alternatives at decision point 1. Set up a table to show your random numbers and random deviates and the subsequent calculations, and demonstrate by generating five full outcomes for the desired distribution.

16-6. Given the two-stage decision situation shown in Fig. 16-12, determine which is the best initial decision. Use the expected PW method and a MARR of 12%. To give the

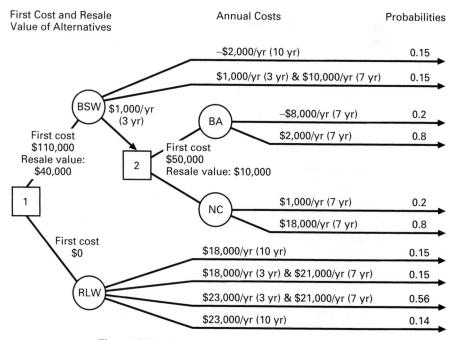

First Cost and Resale   **Annual Costs**   **Probabilities**
Value of Alternatives

                       −$2,000/yr (10 yr)   0.15

          $1,000/yr (3 yr) & $10,000/yr (7 yr)   0.15

BSW $1,000/yr   −$8,000/yr (7 yr)   0.2
     (3 yr)   BA

First cost $2,000/yr (7 yr)   0.8
$110,000   First cost
Resale value: $50,000
$40,000   2 Resale value: $10,000

                   $1,000/yr (7 yr)   0.2
1   NC
                  $18,000/yr (7 yr)   0.8

First cost
$0 $18,000/yr (10 yr)   0.15

        $18,000/yr (3 yr) & $21,000/yr (7 yr)   0.15

RLW $23,000/yr (3 yr) & $21,000/yr (7 yr)   0.56

          $23,000/yr (10 yr)   0.14

Figure 16-12   Two-stage decision situation for Problem 16-6.

problem a physical context, the following letter symbols have been employed for each alternative:

> BSW—Build small warehouse
> RLW—Rent large warehouse
> BA—Build addition
> NC—No change

16-7. A firm must decide between purchasing an automatic (machine that costs $50,000, will last 10 years, and will have $0 salvage value) and purchasing a manual machine that costs $20,000 and will last 5 years and have $0 salvage value. If the manual machine is purchased initially, after 5 years, a decision will have to be made between a manual machine having the same characteristics affecting cost as the first manual machine and a semiautomatic machine costing $40,000 that would have a $20,000 salvage value after 5 years of life. The annual operating costs for the machines are as follows: automatic, $10,000; manual, $14,000; semiautomatic, $11,000.

    a. Graphically construct a decision tree to represent this situation.

    b. Determine which decision would be made at each point, using the PW method and a MARR of 10%.

    c. At what interest rate would the decision between the manual and semiautomatic machine be reversed?

16-8. Suppose you are faced with the same alternatives and dollar outcome consequences as in the replacement problem depicted in Fig. 16-6. However, the initial estimates of probability of demand are high, 0.6; low, 0.4. Furthermore, management's assessment of confidence in further investigation results, using the notation in Table 16-6, are

$$P(h \mid H) = 0.80,$$
$$P(h \mid D) = 0.40,$$
$$P(d \mid H) = 0.20, \text{ and}$$
$$P(d \mid D) = 0.60.$$

Calculate the choice at each decision point to determine the best initial decision. How close is the initial decision with these revised probabilities to the initial decision for the original problem depicted in Fig. 16-6?

16-9. Figure 16-13 is a decision tree portrayal of a building lease-versus-buy problem with input data supplied. Investment requirements are shown as negative numbers; probabilities associated with each outcome are shown in parentheses. The annual cash savings and duration of those savings are shown together at each relevant outcome. Salvage values in the cases of abandonments are assumed to occur at the end of the 25-year study period. Determine the best decision, using the expected net PW method with a MARR of 0%.

16-10. Figure 16-14 is a simplified portrayal of the relevant factors for deciding whether to start an applied research project. Determine the answer, assuming that the decision points are each 1 year apart and the MARR is 20%. Use the expected net PW method.

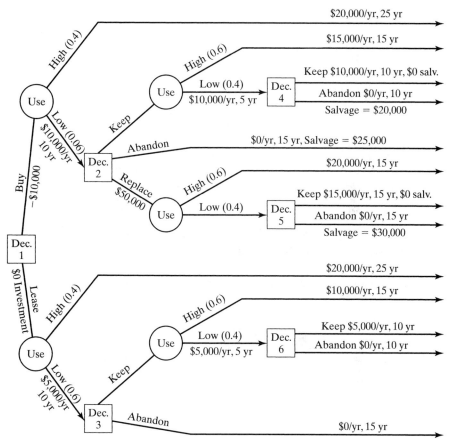

Figure 16-13   Decision tree for Problem 16-9.

16-11. Suppose, for the oil wildcatter problem for which prior probabilities are shown in Fig. 16-3, the probabilities of the possible outcomes for the added study (seismic soundings) are as follows:

Given that the well state will turn out to be:	Then the probability that the seismic soundings will indicate ____ structure is:		
	No	Open	Closed
Dry	0.6	0.3	0.1
Wet	0.3	0.4	0.3
Soaking	0.1	0.4	0.5

Given the structure indicated by the seismic sounding, use Bayes's theorem to calculate the posterior probabilities regarding what the well state will turn out to be. Check your results against the probabilities shown in Fig. 16-4. (*Note:* The tabular structure shown in Appendix 16-A should be helpful.)

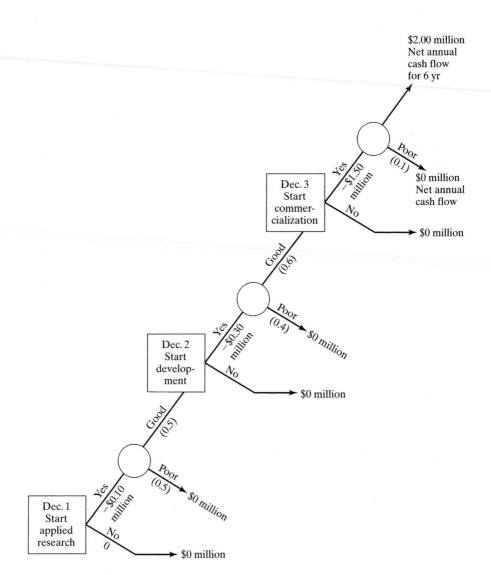

Figure 16-14   Simplified tree, showing only relevant information, for Problem 16-10.

APPENDIX 16-A

Bayes's Theorem and Tabular Format for Calculation of Posterior Probabilities

In general, if there are n mutually exclusive, exhaustive possible outcomes S_1, S_2, S_i, S_n, and the results of additional study, such as sampling or further investigation, are X, such that X is discrete and $P(X) \neq 0$, and prior probabilities $P(S_i)$ have been established, then Bayes's theorem for the discrete case can be written as

$$P(S_i \mid X) = \frac{P(X \mid S_i)P(S_i)}{\sum_i P(X \mid S_i)P(S_i)}. \qquad (16\text{-}A\text{-}1)$$

The posterior probability $P(S_i \mid X)$ is the probability of outcome S_i given that additional study resulted in X. The probability of X and S_i occurring, $P(X \mid S_i)P(S_i)$, is the "joint" probability of X and S_i, or $P(X, S_i)$. The sum of all the joint probabilities is equal to the probability of X. Therefore, Eq. 16-A-1 can be written as

$$P(S_i \mid X) = \frac{P(X \mid S_i)P(S_i)}{P(X)}. \qquad (16\text{-}A\text{-}2)$$

A format for application is presented in Table 16-A-1. The columns are as follows:

COLUMN

(1) S_i: the potential states of nature
(2) $P(S_i)$: the estimated prior probability of S_i (*Note:* This column sums to unity.)

TABLE 16-A-1 Format for Applying Bayes's Theorem in Discrete Outcome Cases

(1) State	(2) Prior probability	(3) Probability of sample outcome, or confidence assessment, X	(4) = (2)(3) Joint probability	(5) = (4)/Σ(4) Posterior probability $P(S_i \mid X)$
S_1	$P(S_1)$	$P(X \mid S_1)$	$P(X \mid S_1)P(S_1)$	$P(X \mid S_1)P(S_1)/P(X)$
S_2	$P(S_2)$	$P(X \mid S_2)$	$P(X \mid S_2)P(S_2)$	$P(X \mid S_2)P(S_2)/P(X)$
⋮	⋮	⋮	⋮	⋮
S_i	$P(S_i)$	$P(X \mid S_i)$	$P(X \mid S_i)P(S_i)$	$P(X \mid S_i)P(S_i)/P(X)$
⋮	⋮	⋮	⋮	⋮
S_n	$P(S_n)$	$P(X \mid S_n)$	$P(X \mid S_n)P(S_n)$	$P(X \mid S_n)P(S_n)/P(X)$
	$\sum_{i=1}^{n} P(S_i) = 1.0$		$\sum_{i=1}^{n} P(X \mid S_i)P(S_i) = P(X)$	$\sum_{i=1}^{n} P(S_i \mid X) = 1.0$

(3) $P(X \mid S_i)$: the conditional probability of getting sample or added study results X, given that S_i is the true state

(4) $P(X \mid S_i)P(S_i)$: the point probability of getting X and S_i; the summation of this column is $P(X)$, which is the probability that the sample or added study results in outcome X

(5) $P(S_i \mid X)$: the posterior probability of S_i, given that the sample outcome resulted in X; numerically, the ith entry is equal to the ith entry of column (4) divided by the sum of column (4) (*Note:* This column sums to unity.)

PART FOUR

Specialized Topics

17
Capital Investment Decisions as Real Options

17.1 Introduction

Companies make capital investments to exploit opportunities for shareholder (owner) wealth creation. These opportunities are *real options,* which allow a firm to invest capital now, or in some situations to postpone all or part of the investment until later. In recent years, engineers and managers have become aware of the need to analyze real options, which are options available to a firm when it invests in real assets, such as plant, equipment, and land.

Managers need a rational framework for deciding whether a project should be implemented now or delayed until later when some of its risk or uncertainty has been resolved. This option to postpone all or part of a capital investment has intrinsic value that is generally not recognized in traditional investment decision studies of project profitability. Accordingly, the purpose of this chapter is to present an expanded framework for valuing real options associated with capital investment decisions. We achieve this purpose by reprinting an excellent article from the *Harvard Business Review,* written by Timothy A. Luehrman and entitled "Investment Opportunities as Real Options: Getting Started on the Numbers" (July–August 1998).

The real options approach to capital investments is based on an interesting analogy about financial options. A company with an opportunity to invest capital actually owns something much like a financial call option—the company has the right but not the obligation to invest in (purchase) an asset at a future time of its choosing. When a firm makes an irreversible capital investment, it exercises its call option, which has value by virtue of the flexibility it gives the firm. The reprinted article provides a framework for exercising the option in an optimal manner.

An example of a postponable investment is coal-fired generating capacity of an electric utility. Anticipated capacity needed for the next 10 years can be added in one large expansion project, or the capacity addition can be more flexibility acquired in staggered stages, which permit the utility to better respond to future demand characteristics and possibly different types of generating capacity, such as natural gas or nuclear. If the utility company decides to go ahead with a single, large, irreversible expansion project, it eliminates the option of waiting for new information that might represent a more valuable phased approach to meeting customers' demands for electricity. The lost option's value is an *opportunity cost* that must be included in the overall evaluation of the investment. This is the essence of the real options approach to capital investment—to fairly value the option of waiting to invest in all or a part of the project and to include this value in today's metric of overall project profitability. Clearly, viewing capital investments as real options forces a greater emphasis to be placed on the value of information in risky situations facing a firm.

17.2 Investment Opportunities as Real Options[1]

The analogy between financial options and corporate investments that create future opportunities is both intuitively appealing and increasingly well accepted. Executives readily see why investing today in R&D, or in a new marketing program, or even in certain capital expenditures (a phased plant expansion, say) can generate the possibility of new products or new markets tomorrow. But for many nonfinance managers, the journey from insight to action, from the puts and calls of financial options to actual investment decisions, is difficult and deeply frustrating.

Experts do a good job of explaining what option pricing captures that conventional discounted-cash-flow (DCF) and net-present-value (NPV) analyses do not. Moreover, simple option pricing for exchange-traded puts and calls is fairly straightforward, and many books present the basics lucidly. But at that point, most executives get stuck. Their interest piqued, they want to know How can I use option pricing on my project? and How can I use this with real numbers rather than with sterilized examples? Unfortunately, how-to advice is scarce on this subject and mostly aimed at specialists, preferably with Ph.D.'s. As a result, corporate analyses that generate real numbers have been rare, expensive, and hard to understand.

The framework presented here bridges the gap between the practicalities of real-world capital projects and the higher mathematics associated with formal option-pricing theory. It produces quantitative output, can be used repeatedly on many projects, and is compatible with the ubiquitous DCF spreadsheets that are at the heart of most corporate capital-budgeting systems. What this framework cannot supply is absolute precision: when a very precise number is required, managers will still have to call on technical experts with specialized financial tools. But for many projects in many companies, a "good enough" number is not only good enough but considerably better than the number

---

[1] Reprinted by permission of the *Harvard Business Review*. From "Investment Opportunities as Real Options: Getting Started on the Numbers," by Timothy A. Luehrman, July–August 1998, pages 51–67. Copyright © 1998 by the Harvard Business School Publishing Corporation; all rights reserved.

a plain DCF analysis would generate. In such cases, forgoing some precision in exchange for simplicity, versatility, and explicability is a worthwhile trade.

We'll begin by examining a generic investment opportunity—a capital budgeting project—to see what makes it similar to a call option. Then we'll compare DCF with the option-pricing approach to evaluating the project. Instead of looking only at the differences between the two approaches, we will also look for points of commonality. Recognizing the differences adds extra insight to the analysis, but exploiting the commonalities is the key to making the framework understandable and compatible with familiar techniques. In fact, most of the data the framework uses come from the DCF spreadsheets that managers routinely prepare to evaluate investment proposals. And for option values, the framework uses the Black–Scholes option-pricing table instead of complex equations. Finally, once we've built the framework, we'll apply it to a typical capital-investment decision.

17.3 Mapping a Project Onto an Option

A corporate investment opportunity is like a call option because the corporation has the right, but not the obligation, to acquire something—let us say, the operating assets of a new business. If we could find a call option sufficiently similar to the investment opportunity, the value of the option would tell us something about the value of the opportunity. Unfortunately, most business opportunities are unique, so the likelihood of finding a similar option is low. The only reliable way to find a similar option is to construct one.

To do so, we need to establish a correspondence between the project's characteristics and the five variables that determine the value of a simple call option on a share of stock. By mapping the characteristics of the business opportunity onto the template of a call option, we can obtain a model of the project that combines its characteristics with the structure of a call option. The option we will use is a *European* call, which is the simplest of all options because it can be exercised on only one date, its expiration date. The option we synthesize in this way is not a perfect substitute for the real opportunity, but because we've designed it to be similar, it is indeed informative. The diagram "Mapping an Investment Opportunity onto a Call Option" shows the correspondences making up the fundamental mapping.

Many projects involve spending money to buy or build a productive asset. Spending money to exploit such a business opportunity is analogous to exercising an option on, for example, a share of stock. The amount of money expended corresponds to the option's *exercise price* (denoted for simplicity as X). The present value of the asset built or acquired corresponds to the *stock price (S)*. The length of time the company can defer the investment decision without losing the opportunity corresponds to the option's *time to expiration (t)*. The uncertainty about the future value of the project's cash flows (that is, the riskiness of the project) corresponds to the *standard deviation of returns* on the stock (σ). Finally, the time value of money is given in both cases by the *risk-free rate of return* (r_f). By pricing an option using values for these variables generated from our project, we learn more about the value of the project than a simple discounted-cash-flow analysis would tell us.

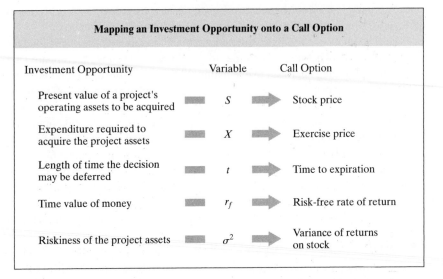

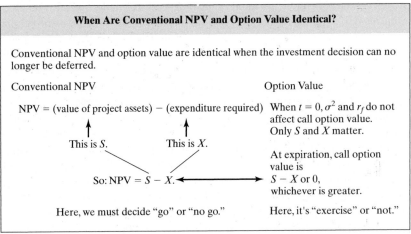

17.4 Linking NPV and Option Value

Traditional DCF methods would assess this opportunity by computing its net present value. NPV is the difference between how much the operating assets are worth (their present value) and how much they cost:

$$\text{NPV} = \text{present value of assets}$$
$$- \text{required capital expenditure.}$$

When NPV is positive, the corporation will increase its own value by making the investment. When NPV is negative, the corporation is better off not making the investment.

When are the project's option value and NPV the same? When a final decision on the project can no longer be deferred; that is, when the company's "option" has reached its expiration date. At that time, either

$$\text{the option value} = S - X$$

or

$$\text{the option value} = 0,$$

whichever is greater. But note that

$$NPV = S - X$$

as well, because we know from our map that S corresponds to the present value of the project assets and X to the required capital expenditure. To reconcile the two completely, we need only observe that when NPV is negative, the corporation will not invest, so the project value is effectively zero (just like the option value) rather than negative. In short, both approaches boil down to the same number and the same decision. (See the diagram "When Are Conventional NPV and Option Value Identical?")

This common ground between NPV and option value has great practical significance. It means that corporate spreadsheets set up to compute conventional NPV are highly relevant for option pricing. Any spreadsheet that computes NPV already contains the information necessary to compute S and X, which are two of the five option-pricing variables. Accordingly, executives who want to begin using option pricing need not discard their current DCF-based systems.

When do NPV and option pricing diverge? When the investment decision may be deferred. The possibility of deferral gives rise to two additional sources of value. First, we would always rather pay later than sooner, all else being equal, because we can earn the time value of money on the deferred expenditure. Second, while we're waiting, the world can change. Specifically, the value of the operating assets we intend to acquire may change. If their value goes up, we haven't missed out; we still can acquire them simply by making the investment (exercising our option). If their value goes down, we might decide not to acquire them. That also is fine (very good, in fact) because, by waiting, we avoid making what would have turned out to be a poor investment. We have preserved the ability to participate in good outcomes and insulated ourselves from some bad ones.

For both of these reasons, being able to defer the investment decision is valuable. Traditional NPV misses the extra value associated with deferral because it assumes the decision cannot be put off. In contrast, option pricing *presumes* the ability to defer and provides a way to quantify the value of deferring. So to value the investment, we need to develop two new metrics that capture these extra sources of value.

17.4.1 Quantifying Extra Value: NPVq

The first source of value is the interest you can earn on the required capital expenditure by investing later rather than sooner. A good way to capture that value is to suppose you put just enough money in the bank now so that when it's time to invest, that money plus the interest it has earned is sufficient to fund the required expenditure. How much money is that? It is the discounted present value of the capital expenditure.

In option notation, it's the present value of the exercise price, or PV(X). To compute PV(X), we discount X for the requisite number of periods (t) at the risk-free rate of return (r_f):

$$PV(X) = X \div (1 + r_f)^t.$$

The extra value is the interest rate (r_f) times X, compounded over however many time periods (t) are involved. Alternatively, it is the difference between X and PV(X).

We know that conventional NPV is missing that extra value, so let's put it in. We have seen that NPV can be expressed in option notation as

$$NPV = S - X.$$

Let's rewrite it using PV(X) instead of X. Thus,

$$\text{"modified" NPV} = S - PV(X).$$

Note that our modified NPV will be greater than or equal to regular NPV because it explicitly includes interest to be earned while we wait. It picks up one of the sources of value we are interested in.

Modified NPV, then, is the difference between S (value) and PV(X) (cost adjusted for the time value of money). Modified NPV can be positive, negative, or zero. However, it will make our calculations a lot easier if we express the relationship between cost and value in such a way that the number can never be negative or zero.

So instead of expressing modified NPV as the *difference* between S and PV(X), let's create a new metric: S *divided* by PV(X). By converting the difference to a ratio, all we are doing, essentially, is converting negative values to decimals between zero and one.[2] We'll call this new metric NPVq, where "q" reminds us that we are expressing the relationship between cost and value as a quotient:

$$NPVq = S \div PV(X).$$

Modified NPV and NPVq are not equivalent; that is, they don't yield the same numeric answer. For example, if $S = 5$ and PV(X) = 7, NPV = -2 but NPVq = 0.714. But the difference in the figures is unimportant because we haven't lost any information about the project by substituting one metric for another. When modified NPV is positive, NPVq will be greater than one; when NPV is negative, NPVq will be less than one. Anytime modified NPV is zero, NPVq will be one. There is a perfect correspondence between them, as the diagram "Substituting NPVq for NPV" shows.

17.4.2 Quantifying Extra Value: Cumulative Volatility

Now let's move on to the second source of additional value, namely that while we're waiting, asset value may change and affect our investment decision for the better. That possibility is very important, but naturally it is more difficult to quantify because we are not actually sure that asset values will change or, if they do, what the future values

---

[2] There are other mathematical advantages, beyond the scope of this article, associated with using the quotient instead of the difference. Also, students of economics will recognize the similarity between NPVq and the famous "Tobin's q," which measures the ratio of the value of an asset to its replacement cost.

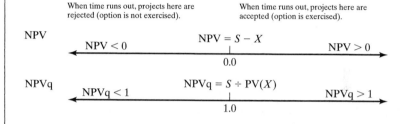

Substituting NPVq for NPV

We can rank projects on a continuum according to values for NPVq, just as we would for NPV. When a decision can no longer be deferred, NPV and NPVq give identical investment decisions, but NPVq has some mathematical advantages.

will be. Fortunately, rather than measuring added value directly, we can measure *uncertainty* instead and let an option-pricing model quantify the value associated with a given amount of uncertainty. Once again, we'll go through two steps. First, we'll identify a sensible way to measure uncertainty. Then we'll express the metric in a mathematical form that will be easier for us to use but will not cause us to lose any practical content.

The only way to measure uncertainty is by assessing probabilities. Imagine that the project's future value is to be drawn from an urn containing all possible future values, weighted according to their likelihood of occurring. That is, if a value of $100 were twice as likely as $75 or $125, there would be twice as many $100 balls in the urn as $75 balls or $125 balls.

How can we quantify this uncertainty? Perhaps the most obvious measure is simply the range of all possible values: the difference between the lowest and the highest possibilities. But we can do better than that by taking into account the relative likelihood of values between those extremes. If, for example, very high and very low values are less likely than "medium" or "average" values, our measure of uncertainty should reflect that. The most common probability-weighted measure of dispersion is *variance*, often denoted as *sigma squared* (σ^2). Variance is a summary measure of the likelihood of drawing a value far away from the average value in the urn. The higher the variance, the more likely it is that the values drawn will be either much higher or much lower than average. In other words, we might say that high-variance assets are riskier than low-variance assets.

Variance is an appealing measure of uncertainty, but it's incomplete. We have to worry about a time dimension as well: how much things can change while we wait depends on how long we can afford to wait. For business projects, things can change a lot more if we wait two years than if we wait only two months. So in option valuation, we speak in terms of variance *per period*. Then our measure of the total amount of uncertainty is variance per period times the number of periods, or $\sigma^2 t$.

This sometimes is called *cumulative variance*. An option expiring in two years has twice the cumulative variance as an otherwise identical option expiring in one year, given the same variance per period. Alternatively, it may help to think of cumulative variance as the amount of variance in the urn times the number of draws you are allowed, which again is $\sigma^2 t$.

Cumulative variance is a good way to measure the uncertainty associated with business investments. Now we'll make two modifications, again for mathematical convenience, that won't affect the ability of the variable to tell us what we want to know about uncertainty. First, instead of using the variance of project *values*, we'll use the variance of project *returns*. In other words, rather than working with the actual dollar value of the project, we'll work with the percentage gained (or lost) per year. There is no loss of content because a project's return is completely determined by the project's value:

$$\text{return} = \frac{(\text{future value} - \text{present value})}{\text{present value}}.$$

The probability distribution of possible values is usually quite asymmetric; value can increase greatly but cannot drop below zero. Returns, in contrast, can be positive or negative, sometimes symmetrically positive or negative, which makes their probability distribution easier to work with.

Second, it helps to express uncertainty in terms of *standard deviation* rather than variance. Standard deviation is simply the square root of variance and is denoted by σ. It tells us just as much about uncertainty as variance does, but it has the advantage of being denominated in the same units as the thing being measured. In our business example, future asset values are denominated in units of currency—say, dollars—and returns are denominated in percentage points. Standard deviation, then, is likewise denominated in dollars or percentage points, whereas variance is denominated in *squared* dollars or *squared* percentage points, which are not intuitive. Since we are going to work with returns instead of values, our units will be percentage points instead of dollars.

To make these refinements to our measure of total uncertainty, we do the following:

First, stipulate that σ^2 denotes the variance of returns per unit of time on our project.

Second, multiply variance per period by the number of periods (t) to get cumulative variance ($\sigma^2 t$).

Third, take the square root of cumulative variance to change units, expressing the metric as standard deviation rather than variance. Let's call this last quantity *cumulative volatility* ($\sigma\sqrt{t}$) to distinguish it from cumulative variance.

17.4.3 Valuing the Option

Together, our two new call-option metrics, NPVq and $\sigma\sqrt{t}$, contain all the information needed to value our project as a European call option using the Black–Scholes model. They capture the extra sources of value associated with opportunities. And they are composed of the five fundamental option-pricing variables onto which we mapped our business opportunity. NPVq is actually a combination of four of the five variables: S, X, r_f, and t. Cumulative volatility combines the fifth, σ, with t. (See the diagram "Linking Our Metrics to the Black–Scholes Model.") By combining variables in this way, we get to work with two metrics instead of five. Not only is that easier for most of us to grasp, it also allows us to plot two-dimensional pictures, which can be helpful substitutes for equations in managers' discussions and presentations. Finally, each of the metrics has a natural business interpretation, which makes option-based analysis less opaque to non-finance executives.

The graph "Locating the Option Value in Two-Dimensional Space" shows how to use NPVq and $\sigma\sqrt{t}$ to obtain a value for the option. NPVq is on the horizontal axis, increasing from left to right. As NPVq rises, so does the value of the call option. What causes higher values of NPVq? Higher project values (S) or lower capital expenditures (X). Note further that NPVq also is higher whenever the present value of X is lower. Higher interest rates (r_f) or longer time to expiration (t) both lead to lower present values of X. Any of these changes (lower X or higher S, r_f, or t) increases the value of a European call.

Cumulative volatility is on the vertical axis of the graph, increasing from top to bottom. As $\sigma\sqrt{t}$ increases, so does call value. What causes higher values of $\sigma\sqrt{t}$? Greater uncertainty about a project's future value and the ability to defer a decision longer. Either of these changes (higher σ or t) likewise increases the value of a European call.

Plotting projects in this two-dimensional space creates a visual representation of their relative option values. No matter where you start in the graph, call value increases when you move down, to the right, or in both directions at once. Projects in

Linking Our Metrics to the Black–Scholes Model

Our two new metrics together contain all five variables in the Black–Scholes model. Combining five variables into two lets us locate opportunities in two-dimensional space.

Investment Opportunity	Call Option	Variable	Option Value Metrics
Present value of a project's operating assets to be acquired	Stock price	S	
Expenditure required to acquire the project assets	Exercise price	X	NPVq
Length of time the decision may be deferred	Time to expiration	t	
Time value of money	Risk-free rate of return	r_f	
Riskiness of the project assets	Variance of returns on stock	σ^2	$\sigma\sqrt{t}$

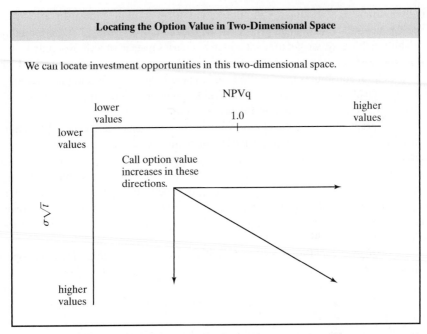

the lower-right corner of the graph are high on both NPVq and $\sigma\sqrt{t}$ metrics and their option value is high compared with projects in the upper-left corner.

Locating various projects in the space reveals their value relative to one another. How do we get absolute values? That is, how can we get a *number*? Having gotten this far, we find that getting a number is easy. Because NPVq and $\sigma\sqrt{t}$ contain all five Black–Scholes variables, we can fill in a table with Black–Scholes call values that correspond to *every pair* of NPVq and $\sigma\sqrt{t}$ coordinates. I call this "pricing the space," and the table does it for us.

The exhibit "Using the Black–Scholes Option-Pricing Model to 'Price the Space' " shows part of the filled-in Black–Scholes table. Each number expresses the value of a specific call option as a percentage of the underlying project's (or asset's) value. For example, for a project whose NPVq equals 1.0 and $\sigma\sqrt{t}$ equals 0.5, the value given in the table is 19.7%. Any European call option for which NPVq is 1.0 and $\sigma\sqrt{t}$ is 0.5 will have a value equal to 0.197 times S. If the assets associated with a particular project have a value (S) of $100, then the project viewed as a call option has a value of $19.70. If S were $10, the call option value would be $1.97, and so forth. Option values in the table are expressed in relative terms, as percentages of S, rather than in absolute dollars, to enable us to use the same table for both big and small projects. It's also convenient not to have to manipulate the Black–Scholes equation every time we want to value a project. The Black–Scholes model is used once, to generate the table itself.[3] After that, we need only locate our project in the table and multiply by a factor of S.

---

[3] This form of option-pricing table is nearly as old as the Black–Scholes model itself. I first ran into it as an M.B.A. student using Richard Brealey and Stewart Myers's text, *Principles of Corporate Finance* (New York: McGraw-Hill, 1981).

Using the Black-Scholes Option-Based Pricing Model to "Price the Space"

Each number in the table gives the value of a European call for specified values of NPVq and $\sigma\sqrt{t}$, as a percentage of s, the value of project assets.

Black–Scholes value of a European call option, expressed as a percentage of underlying asset value.

Example:

Suppose $S = \$100$
$X = \$105$
$t = 1$ year
$r_f = 5\%$
$\sigma = 50\%$ per year

then NPVq = 1.0

and $\sigma\sqrt{t} = 0.50$.

The table gives a value of 19.7%.

Interpretation:

Viewed as a call option, the project has a value of:

Call value = $0.197 \times \$100 = \$19.70$.

Compare this to its conventional NPV:

NPV = $S - X$
= $\$100 - \$105$
= $-\$5$.

NPVq

$\sigma\sqrt{t}$	0.80	0.82	0.84	0.86	0.88	0.90	0.92	0.94	0.96	0.98	**1.00**	1.02	1.04	1.06	1.08
0.05	0.0	0.0	0.0	0.0	0.0	0.0	0.1	0.3	0.6	1.2	**2.0**	3.1	4.5	6.0	7.5
0.10	0.0	0.1	0.2	0.3	0.5	0.8	1.2	1.7	2.3	3.1	**4.0**	5.0	6.1	7.3	8.6
0.15	0.5	0.7	1.0	1.3	1.7	2.2	2.8	3.5	4.2	5.1	**6.0**	7.0	8.0	9.1	10.2
0.20	1.5	1.9	2.3	2.8	3.4	4.0	4.7	5.4	6.2	7.1	**8.0**	8.9	9.9	10.9	11.9
0.25	2.8	3.3	3.9	4.5	5.2	5.9	6.6	7.4	8.2	9.1	**9.9**	10.9	11.8	12.8	13.7
0.30	4.4	5.0	5.7	6.3	7.0	7.8	8.6	9.4	10.2	11.1	**11.9**	12.8	13.7	14.6	15.6
0.35	6.2	6.8	7.5	8.2	9.0	9.8	10.6	11.4	12.2	13.0	**13.9**	14.8	15.6	16.5	17.4
0.40	8.0	8.7	9.4	10.2	11.0	11.7	12.5	13.4	14.2	15.0	**15.9**	16.7	17.5	18.4	19.2
0.45	9.9	10.6	11.4	12.2	12.9	13.7	14.5	15.3	16.2	17.0	**17.8**	18.6	19.4	20.3	21.1
0.50	11.8	12.6	13.4	14.2	14.9	15.7	16.5	17.3	18.1	18.9	**⟨19.7⟩**	20.5	21.3	22.1	22.9
0.55	13.8	14.6	15.4	16.1	16.9	17.7	18.5	19.3	20.1	20.9	**21.7**	22.4	23.2	24.0	24.8
0.60	15.8	16.6	17.4	18.1	18.9	19.7	20.5	21.3	22.0	22.8	**23.6**	24.3	25.1	25.8	26.6
0.65	17.8	18.6	19.3	20.1	20.9	21.7	22.5	23.2	24.0	24.7	**25.3**	26.2	27.0	27.7	28.4
0.70	19.8	20.6	21.3	22.1	22.9	23.6	24.4	25.2	25.9	26.6	**27.4**	28.1	28.8	29.5	30.2
0.75	21.8	22.5	23.3	24.1	24.8	25.6	26.3	27.1	27.8	28.5	**29.2**	29.9	30.6	31.3	32.0
0.80	23.7	24.5	25.3	26.0	26.8	27.5	28.3	29.0	29.7	30.4	**31.1**	31.8	32.4	33.1	33.8
0.85	25.7	26.5	27.2	28.0	28.7	29.4	30.2	30.9	31.6	32.2	**32.9**	33.6	34.2	34.9	35.5
0.90	27.7	28.4	29.2	29.9	30.6	31.3	32.0	32.7	33.4	34.1	**34.7**	35.4	36.0	36.6	37.3

Why is the option value $19.70 less than the asset value of $100? We've been analyzing sources of extra value associated with being able to defer an investment. The key is to remember that *extra* refers to a comparison between option value and *net present value* (NPV), not to a comparison between option value and present value (S). In this example, we are not expecting the option value to be greater than S; we are expecting it to be greater than NPV, which is S minus capital expenditures (X). S equals $100 here, but we didn't say what X was. Since NPVq equals 1.0 in this example, X must in fact be greater than $100; otherwise NPVq would be greater than 1.0. For concreteness, suppose that this is a one-year option—that is, suppose we can defer the decision for one year—and that the risk-free rate of return (r_f) is 5%. Then for NPVq to equal 1.0, X must be $105. Recall that

$$NPVq = S \div PV(X)$$
$$= \$100 \div (\$105 \div 1.05).$$

Thus, the conventional NPV is actually negative:

$$NPV = S - X$$
$$= \$100 - \$105$$
$$= -\$5.$$

And an option value of $19.70 really is substantially greater than conventional NPV.

17.5 Using the Framework: An Example in Seven Steps

To illustrate how to apply the framework, consider this example of a hypothetical, but representative, capital investment. Division managers at a company we'll call Franklin Chemical are proposing a phased expansion of their manufacturing facilities. They plan to build a new, commercial-scale plant immediately to exploit innovations in process technology. And then they anticipate further investments, three years out, to expand the plant's capacity and to enter two new markets. The initial investment is obviously strategic because it creates the opportunity for subsequent growth. Yet executives responsible for the company's capital budget are unimpressed by the project because its NPV is essentially zero. (Cash flow projections and NPV calculations for the projected investment are shown in the table "Franklin Chemical's Initial Calculations for a Proposed Expansion.") In fact, in the annual jockeying for funds, this program may not beat out competing alternatives. Its champions are frustrated and feel sure that the company's conventional NPV approach is missing something.

They are right.

This project has considerable option value because the initial expenditure of $125 million buys the right to expand (or not) three years later. This is important because the expenditures in the third year are large—three times the initial investment.

17.5.1 Step 1 Is to Recognize the Option and Describe It

It takes practice to recognize the options that may be buried in conventional projects. However, there are at least two easy ways to see the option in our example.

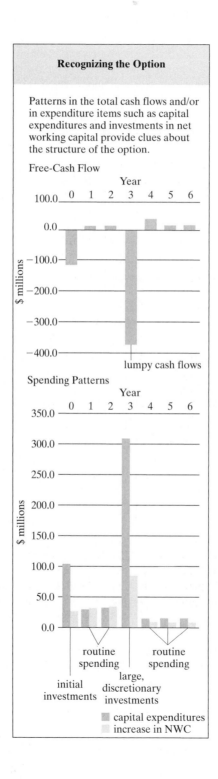

Franklin Chemical's Initial Calculations for a Proposed Expansion

Discounted-cash-flow valuation: conventional NPV is our starting point.

Year	0	1	2	3	4	5	6
Operating projections							
revenues		455.0	551.0	800.0	1080.0	1195.0	1255.0
− cost of good sold		341.3	414.9	596.0	811.1	893.9	941.3
= gross profit		113.8	136.1	204.0	268.9	301.1	313.8
− SG&A expense		110.4	130.0	219.2	251.6	280.3	287.4
= operating profit		3.3	6.1	−15.2	17.3	20.8	26.3
Cash flow calculation							
EBIT (1-tax rate)		2.2	4.0	−10.0	11.5	13.7	17.4
+ depreciation		19.0	21.0	21.0	46.3	48.1	50.0
− capital expenditures	100.0	8.1	9.5	307.0	16.0	16.3	17.0
− increase NWC	25.0	4.1	5.5	75.0	7.1	8.0	9.7
= free cash flow, assets	−125.0	9.0	10.0	−371.0	34.7	37.5	40.7
+ terminal value, assets (perpetuity value with 5% per year growth)							610.3
Discount to present value							
× discount factor (12%)	1.000	0.893	0.797	0.712	0.636	0.567	0.507
= PV (by year)	−125.0	8.0	8.0	−264.1	22.0	21.3	329.8
NPV (sum of all years)	**0.1**						

Figures are in $millions and have been rounded.

The first is simply to look beyond the numbers and examine the project's description. It surely says something about the two-phased nature of the program by way of justifying the large outlays in year 3. The other is to examine the pattern of the project cash flows over time. The cash flows in the chart are very uneven: two figures are an order of magnitude larger than the other five, and both of these are negative. A graph of the capital-expenditures line would clearly show the spike in spending in year 3. Such a large sum is almost surely *discretionary*. That is, the company can choose not to make the investment, based on how things look when the time comes. This is a classic expansion option, sometimes called a *growth option*. (See the graphs "Recognizing the Option.")

Franklin's project has two major parts. The first part is to spend $125 million now to acquire some operating assets. The second part is an option to spend an additional sum, more than $300 million, three years from now to acquire the additional capacity and enter the new markets. The option here is a call option, owned by the company, with three years to expiration, that can be exercised by investing certain amounts in net working capital (NWC) and fixed assets. Viewing the project in this way, we want to evaluate the following: NPV (entire proposal) = NPV (phase 1 assets) + call value (phase 2 assets).

Phase 1 refers to the initial investment and the associated cash flows. It can be valued using NPV as usual. *Phase 2* refers to the opportunity to expand, which may or may not be exploited in year 3. To value phase 2, we will use the framework outlined above to synthesize a comparable call option and then value it.

17.5.2 Step 2 Is to Map the Project's Characteristics onto Call Option Variables

This mapping will create the synthetic option we need and indicate where in the DCF spreadsheet we need to go to obtain values for the variables. The value of the underlying assets (S) will be the present value of the assets acquired when and if the company exercises the option. The exercise price (X) will be the expenditures required to acquire the phase 2 assets. The time to expiration (t) is three years, according to the projections given in the DCF analysis, although we might want to quiz the managers involved to determine whether the decision actually could be made sooner or later. The three-year risk-free rate of interest (r_f) is 5.5% (which is the market rate of interest on a three-year U.S. government bond). Note for comparison that the risk-adjusted discount rate being applied in the spreadsheet is 12%. Finally, the standard deviation of returns on these operating assets (σ) is not given anywhere in the spreadsheet. For now, we'll assume that figure is 40% per year, a value that is neither particularly high nor low. The insert "How to Estimate Cumulative Volatility" explains ways to obtain values for σ in cases like these.

How to Estimate Cumulative Volatility

The variable in our option-pricing model that managers are least accustomed to estimating is variance (σ^2), or standard deviation (σ), which we used to get our metric for cumulative volatility ($\sigma\sqrt{t}$). For a real option, σ cannot be found in a newspaper or in a financial statement, and most people do not have highly developed intuition about, say, the annualized standard deviation of returns on assets associated with entering a new market. In the example given in the text, we assume σ is 40% per year. Is that reasonable? Here are several sound approaches to creating or judging estimates of σ:

Take a(n Educated) Guess

Assets to which you would assign higher hurdle rates because of their higher-than-average systematic risk are also likely to have higher values of σ. How high is "high" for standard deviation? Returns on broad-based U.S. stock indexes had a standard deviation of approximately 20% per year for most of the past 15 years; exceptions (upward spikes) were associated with events like the 1987 stock-market crash and the 1990–1991 Persian Gulf crisis. Individual stocks generally have a higher standard deviation than the market as a whole; returns on General Motors' stock, for example, have a σ of about 25% per year. Individual projects within companies can be expected to have a still higher σ. When I work with manufacturing assets and have no specific information at all about σ, I begin by examining a range: from 30% to 60% per year.

Gather Some Data

For some businesses, we can estimate volatility using historical data on investment returns in the same or related industries. Alternatively, we might compute what is called *implied volatility* using current prices of options traded on organized exchanges. The idea is to observe a market price for an option whose parameters are all known except for σ. We then use a model like Black–Scholes to figure out what σ must be given all the other variables. Today we can get implied volatility for shares of a very large number of companies in many industries. Often it is possible to use implied volatilities from options on stocks to infer σ for assets in corresponding industries. The quality and availability of such data have improved enormously in the past ten years.

Simulate σ

Spreadsheet-based projections of a project's future cash flows, together with Monte Carlo simulation techniques, can be used to synthesize a probability distribution for project returns. Once you have the synthesized distribution, the computer can quickly calculate the corresponding standard deviation. Simulation software packages for desktop computers are commercially available and work with the same popular spreadsheet applications that generate your company's DCF models. These tools also have become far more widely available and much easier to use in recent years.

17.5.3 Step 3 Is to Rearrange the DCF Projections for Two Purposes: to Separate Phase 1 from Phase 2 and to Isolate Values for *S* and *X*

I generally find it easier to work on S and X first. That requires making a judgment about what spending is discretionary versus nondiscretionary or what spending is routine versus extraordinary. It also requires making a similar judgment about which cash inflows are associated with phase 1 as opposed to phase 2.

In this project, expenditures on net working capital and fixed assets obviously are lumpy. The very large sums in year 3 clearly are discretionary and form part of the exercise price (X). The smaller sums in other years are plausibly routine and may be netted against phase 2 cash inflows, ultimately to be discounted and form part of S, the value of the phase 2 assets.

Sometimes it is easy to separate phase 1 cash flows from phase 2 cash flows because whoever prepared the DCF analysis built it up from detailed, phase-specific operating projections. When that's the case, as it is in our example, all we have to do is use the disaggregated detail underlying the summary DCF analysis, and the table "Franklin's Projections Rearranged" presents this information for our example. At other times, we have to allocate cash flows to each phase. A common expedient is simply to break out the phase 1 cash inflows and terminal value. Then, phase 2 cash inflows and terminal value are whatever is left over. Note that when we discount cash flows for the two phases separately, we obtain the same NPV as before.

17.5.4 Step 4 Is to Establish a Benchmark for Phase 2's Option Value Based on the Rearranged DCF Analysis

Having separated phases 1 and 2, we can get a conventional discounted-cash-flow NPV for each, which can be seen in the table showing Franklin Chemical's rearranged calculations. This table shows that phase 1 alone has a positive NPV of $16.3 million while phase 2's NPV is –$16.2 million. The sum of the two is the same NPV, $0.1 million, that we obtained originally. Already, we have a quantitative option-related insight. The value of the whole proposal must be *at least* $16.3 million because the option value of phase 2, whatever it turns out to be, cannot be less than zero. In fact, if the option value of phase 2 is significant, the project as a whole will have a much higher value than $16.3 million, to say nothing of the $0.1 million we started with. This insight is available only when we separate the project's two phases and realize that we will have a choice about whether or not to undertake phase 2.

Our first DCF benchmark for phase 2 is –$16.2 million. Actually, though, phase 2's conventional NPV is worse than that, and it's worth a digression to see why. The DCF valuation contains a common mistake. It discounted the discretionary spending in year 3 at the same 12% risk-adjusted rate that had already been applied to the project's cash flows. That rate is almost certainly too high because such expenditures are rarely subject to the same operating and product-market forces that make the project's cash flows risky. Construction costs, for example, may be uncertain, but they

Franklin's Projections Rearranged

The DCF projections have been rearranged to separate phases 1 and 2 and isolate S and X.

Phase 1

Year	0	1	2	3	4	5	6
cash flow	0.0	9.0	10.0	11.0	11.6	12.1	12.7
+ terminal value							191.0
– investment	–125						
× discount factor (12%)	1.000	0.893	0.797	0.712	0.636	0.567	0.507
= PV (each year)	–125.0	8.0	8.0	7.8	7.3	6.9	103.2
NPV (sum of years)	**16.3**	This is the conventional NPV of phase 1.					

Phase 2

Year	0	1	2	3	4	5	6
cash flow				0.0	23.1	25.4	28.0
+ terminal value							419.3
– investment				–382			
× discount factor (12%)				0.712	0.636	0.567	0.507
= PV (each year)				–271.9	14.7	14.4	226.6
NPV (sum of years)	**–16.2**	This is the conventional NPV of phase 2.					

Phases 1 and 2

Year	0	1	2	3	4	5	6
cash flow	0.0	9.0	10.0	11.0	34.7	37.5	40.7
+ terminal value							610.3
– investment	–125			–382			
× discount factor (12%)	1.000	0.893	0.797	0.712	0.636	0.567	0.507
= PV (each year)	–125.0	8.0	8.0	–264.1	22.0	21.3	329.8
NPV (sum of years)	**0.1**	This is the same conventional NPV we got previously.					

Figures are in $millions and have been rounded.

are usually much more dependent on engineering factors, weather conditions, and contractors' performance than on customers' tastes, competitive conditions, industry capacity utilization, and such. Overdiscounting future discretionary spending leads to an optimistically biased estimate of NPV. To see the magnitude of this effect, discount the year 3 expenditures of $382 million at 5.5% instead of 12% (again, it's as if we were putting investment funds into treasury bonds between now and year 3). Then, phase 2 has a conventional DCF value of –$69.6 million, not –$16.2 million, and the NPV for the whole project goes from $0.1 million to –$53.4 million, a very substantial difference (See the table "Getting the Right Benchmark.")

17.5.5 Step 5 Is to Attach Values to the Option-Pricing Variables

Having reformulated the DCF spreadsheet, we can now pull values for S and X from it. X is the amount the company will have to invest in net working capital and fixed assets (capital expenditures) in year 3 if it wants to proceed with the expansion: $382 million. S is the present value of the new phase 2 operating assets. In other words, it's the DCF value now (at time zero) of the cash flows those assets are expected to produce in the fourth year and beyond. The same table shows that to be $255.7 million. The other option-pricing variables have already been mentioned: t is 3 years, r_f is 5.5%, and σ is 40% per year.

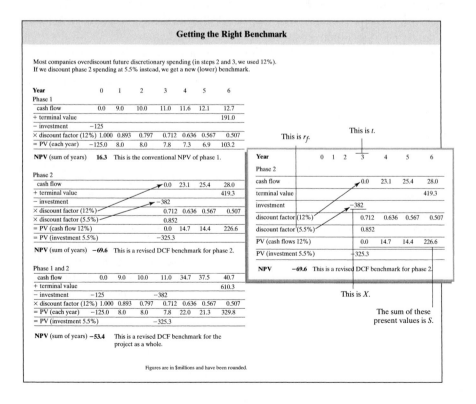

17.5.6 Step 6 Is to Combine the Five Option-Pricing Variables into Our Two Option-Value Metrics: NPVq and $\sigma\sqrt{t}$

In this case:

$$\text{NPVq} = S \div \text{PV}(X) = \frac{\$255.7}{382 \div (1.055)^3} = 0.786$$

And

$$\sigma\sqrt{t} = 0.4 \times \sqrt{3} = 0.693.$$

The exhibit "Deriving the Option-Value Metrics for Franklin's Project" shows how all five variables were derived and how they combine to form our two metrics.

17.5.7 Step 7 Is to Look up call Value as a Percentage of Asset Value in Our Black–Scholes Option-Pricing Table

The table does not show values that correspond exactly to our computed value for NPVq and $\sigma\sqrt{t}$, but by interpolating we can see that the value of our synthesized call option is about 19% of the value of the underlying assets (S). Accordingly, the dollar value of the option is 0.19 times $255.7 million, which equals $48.6 million. Recall the value of the entire proposal is given by: NPV (entire proposal) = NPV (phase 1 assets) + call value (phase 2 assets). Filling in the figures gives: NPV (entire proposal) = $16.3 million + $48.6 million = $64.9 million.

Our final estimate of $64.9 million is a long way from the original figure for NPV of $0.1 million and even further from –$53.4 million. Yet the option-pricing analysis uses the same inputs from the same spreadsheet as the conventional NPV. What looks like a marginal-to-terrible project through a DCF lens is in fact a very attractive one. This seems especially so when we compare $64.9 million with the required initial investment of $125 million; the value associated with the option to expand the plant in year 3 is fully half again as much as the initial investment. Few projects look so good.

What should you do next? All the things you would usually do when evaluating a capital project. Perform sensitivity analyses. Check and update assumptions. Examine particularly interesting or threatening scenarios. Compare and interpret the analysis in light of other historical or contemporary investments and transactions. In addition, now that you've begun pricing synthesized real options, you may want to consider adding a few other items to your list. Some are discussed further in the insert "How Far Can You Extend the Framework?" They include checking for project features that would make it more suitable to use an American rather than a European option. They also include checking for clear disadvantages associated with deferring investment, such as competitive preemption, which would offset some or all of the sources of value associated with waiting. Some of these concerns can be handled by straightforward modifications to the framework. Others require more sophisticated modeling than either this framework alone or conventional NPV generally can provide. Even in those cases, though, a naïvely formulated option value will augment whatever insight may be drawn from a

DCF treatment alone. Remember that simply by recognizing the structure of the problem we gleaned an important insight about the value of the project in our example (that it had to be at least $16.3 million) before we actually priced the option.

Does the framework really work? Yes. Even though we have taken some liberties, we know more about our project after using it than we did before. And if it seemed worthwhile, we could further refine our initial estimate of option value. But the key to getting useful insight from option pricing sooner rather than later is to build on, rather than abandon, the DCF-based NPV analysis your company already uses. Had we set out to value the option from scratch, it would have been more difficult and taken longer. It also would have been hard to tell how well we had done and when to stop working on it. Option pricing should be a complement to existing capital-budgeting systems, not a substitute for them. The framework presented here is a way to start where you are and get somewhere better.

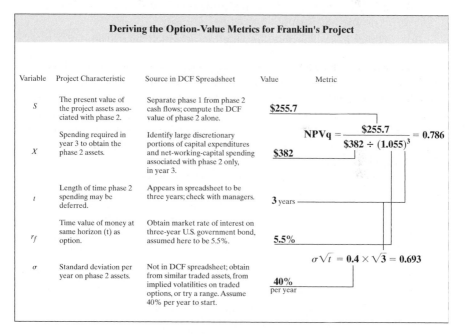

Deriving the Option-Value Metrics for Franklin's Project

Variable	Project Characteristic	Source in DCF Spreadsheet	Value	Metric
S	The present value of the project assets associated with phase 2.	Separate phase 1 from phase 2 cash flows; compute the DCF value of phase 2 alone.	$255.7	
X	Spending required in year 3 to obtain the phase 2 assets.	Identify large discretionary portions of capital expenditures and net-working-capital spending associated with phase 2 only, in year 3.	$382	
t	Length of time phase 2 spending may be deferred.	Appears in spreadsheet to be three years; check with managers.	3 years	
r_f	Time value of money at same horizon (t) as option.	Obtain market rate of interest on three-year U.S. government bond, assumed here to be 5.5%.	5.5%	
σ	Standard deviation per year on phase 2 assets.	Not in DCF spreadsheet; obtain from similar traded assets, from implied volatilities on traded options, or try a range. Assume 40% per year to start.	40% per year	

$$NPVq = \frac{\$255.7}{\$382 \div (1.055)^3} = 0.786$$

$$\sigma\sqrt{t} = 0.4 \times \sqrt{3} = 0.693$$

How Far Can You Extend the Framework?

Real corporate projects will present immediate challenges to some of the simplifications underlying this framework. Can the framework be souped up to handle more complex problems? Or does its very simplicity present insurmountable limitations? When might it generate seriously misleading information?

Bells and whistles can be added to the framework fairly easily, although they require extra data, and the details are beyond the scope of this article. For example, the framework assumes that the amount and timing of a project's capital expenditures (X) are certain. But what if they are not? That's usually the case with business opportunities. The framework can be adapted to handle those circumstances, but the adaptation helps only if we can describe the uncertainty. Specifically, we need to know the probability distribution of X and the joint probability distribution of S and X. That is, it matters whether X tends to be high when S tends to be high; whether the opposite is true (that is, X tends to be high when S is low and vice versa); or whether they are both not only uncertain but also unrelated.

As another example, suppose the uncertainty (or variance) associated with a project changes over time. That's fairly common, and it makes sense that it should affect our estimate of cumulative volatility. Once again, if we know *how* the variance changes over time, or if we can make plausible guesses, the framework can be adapted without much trouble. The insert "Where to Find Additional Help" cites some readings that address how such real-world problems affect option values.

Both of these examples describe situations in which the real limitation is not the framework but rather the data or our knowledge of the project's parameters. Even when we know that we lack necessary data, the framework can help by showing us what the effect on value would be if the data were one thing or another. We might conclude that it's worthwhile to gather or create better data.

Some other real-world complications are thornier. The framework does a good job of capturing the extra value associated with deferring an investment decision. But what if there are particular costs associated with deferral? For example, companies trying to be first to market with the next generation of a hot product will incur large costs if deferral allows a competitor to preempt them. Anytime there are predictable costs to deferring, the option to defer an investment is less valuable, and we would be foolish to ignore those costs. If such additional costs were the only issue, we could easily handle them in our framework as long as we knew when the decision to invest or not to invest finally would be made. But in the real world, we often don't know. Companies may not be compelled to invest at a certain moment, but rather may have discretion to time their investments. So the problem is to decide not only whether to invest but also when.

In effect, many real options are *American* rather than European. American options can be exercised at any time prior to expiration; European options may be exercised only at expiration. The option-pricing table embedded in this framework prices European options. American options are more valuable than European options whenever the costs associated with deferral are predictable. When that's the case, I recommend using the framework to set the lower boundary of the option value and then supplementing that figure with another table or a spreadsheet-based algorithm to convert European option values to the corresponding American option values. That is not quite as easy as merely rewriting the pricing table because it would take a three-dimensional or possibly four-dimensional table to accommodate the extra variables needed. But it can be set up in a spreadsheet. In short, the combination of costs associated with deferral and a company's ability to time investments calls for some added analysis. But the framework needs to be augmented, not scrapped.

Finally, the Black–Scholes option-pricing model that generated the numbers in the table makes some simplifying assumptions of its own. They include assumptions about the form of the probability distribution that characterizes project returns. They also include assumptions about the *tradability* of the underlying project assets; that is, about whether those assets are regularly bought and sold. And they include assumptions about the ability of investors to continually adjust their investment portfolios. When the Black–Scholes assumptions fail to hold, this framework still yields qualitative insights but the numbers become less reliable. Consequently, it may be worthwhile to consult an expert about alternative models to improve the quantitative estimates of option value.

WHERE TO FIND ADDITIONAL HELP

Graduate-level corporate-finance textbooks cover the basics of option pricing, beginning from first principles:

Zvi Bodie and Robert C. Merton, *Finance* (Upper Saddle River, NJ: Prentice Hall, 1998).

Richard A. Brealey and Stewart Myers, *Principles of Corporate Finance*, 5th edition (New York: McGraw-Hill, 1996).

Stephen A. Ross, Randolph W. Westerfield, and Jeffrey Jaffe, *Corporate Finance*, 4th edition (Chicago: Richard D. Irwin, 1996).

Other books go beyond the basics to treat specialized problems and present more advanced models for option pricing. Well-known titles include

John C. Hull, *Options, Futures, and Other Derivatives,* 3rd edition (Upper Saddle River, NJ: Prentice Hall, 1997).

David G. Luenberger, *Investment Science* (New York: Oxford University Press, 1998).

Paul Wilmott, Jeff Dewynne, and Sam Howison, *Option Pricing* (Oxford, UK: Oxford Financial Press, 1993).

A few books focus on real options in particular. Each takes a somewhat different approach to modeling corporate opportunities:

Martha Amram and Nalin Kulatilaka, *Real Options: Managing Strategic Investment in an Uncertain World* (Boston, MA: Harvard Business School Press, 1998).

Avinash K. Dixit and Robert S. Pindyck, *Investment Under Uncertainty* (Princeton, NJ: Princeton University Press, 1994).

Lenos Trigeorgis, *Real Options: Managerial Flexibility and Strategy in Resource Allocation* (Cambridge, MA: MIT Press, 1996).

Shorter readings on selected related topics include

Avinash K. Dixit and Robert S. Pindyck, "The Options Approach to Capital Investment," *Harvard Business Review* May–June 1995.

W. Carl Kester, "Today's Options for Tomorrow's Growth," *Harvard Business Review* March–April 1984.

Timothy A. Luehrman, "What's It Worth? A General Manager's Guide to Valuation," *Harvard Business Review* May–June 1997.

Lenos Trigeorgis, ed., *Real Options in Capital Investment: Models, Strategies, and Applications* (Westport, CT.: Praeger, 1995).

Finally, expanded versions of the option-pricing table published in this article are available in the Brealey and Myers text cited above and in Luehrman, "Capital Projects as Real Options: An Introduction," *Harvard Business Series* case no. 295–074.

PROBLEMS

Use Table 17-1 to determine Options Values for the following problems:

17-1. Refer to the Franklin Chemical example on page 506+. The initial assessment of NPV for this project is $0.1 million, which indicates that the 12% MARR has been attained. (a) Explain why the positive NPV may not be good enough to justify this project, even though the project's IRR exceeds the risk-adjusted MARR. (b) The standard deviation of returns (σ) has been assumed to be 40% per year. Perform a sensitivity analysis for $\pm 50\%$ changes in σ for the Franklin Project to determine whether the decision to invest in additional plant capacity three years later is reversed.

17-2. Refer to the Franklin Chemical example on page 506+. If the three-year risk-free interest rate (r_f) is 8% per year rather than 5.5%, does the recommendation to acquire additional capacity in three years change? How sensitive is this decision to the value of the risk-free rate of return?

17-3. A stock in WGS Enterprises is currently selling at $41 per share. By using Table 17-1, what is the value of a call option to purchase a share for $60 one year from now if the risk free rate is 2% per year and the standard deviation of returns on the stock is 25% per year? What is the value of a call option four years from now? Suppose the standard deviation is 50% and recompute the same values. What are your conclusions about the relationships between option values and these parameters?

17-4. What is the value of the simple option to expand in Section 16-10 if the capacity of the plant is exceeded? Recall that this would be discovered in one year and involve an outlay of $500 million then, with cash inflows of $150 million per year for the following 19 years. The risk-free interest rate is 2% per year. Use Table 17-1, and follow Timothy Luehrman's advice for establishing a value of σ.

17-5. A certain electric utility company has an opportunity to build a new electric power plant. In the first phase of the project, an outlay of $6 million is needed immediately ($t = 0$) to construct the housing for the turbine and generating equipment. In the second phase one year later, the turbine and generating equipment can be acquired if demand warrants this expansion of capacity. The equipment will cost $106 million. In year one ($t = 1$ yr.), the present worth of the completed power plant is a random variable with a mean of $112 million and $\sigma = 0.4$. If the risk-free interest rate is 6%, should the power plant be built?

17-6. (Research Question) NPV_q and $\sigma\sqrt{t}$ contain all the information needed to value a capital investment proposal as a European call option, using the Black–Scholes model.
   a. List the key assumptions that underpin the Black–Scholes model.
   b. How realistic are they for time-phased capital investments?

TABLE 17-1 Option Values

Multiply the percentage in the table by the share price to get the approximate value of the call option.

| | | | | | NPV_q = Share price ÷ PV of exercise price | | | | | | | |
$\sigma\sqrt{t}$	0.5	0.6	0.7	0.8	0.9	1	1.1	1.2	1.3	1.4	1.5	2
0.25	0%	2%	1%	2.80%	5.90%	10%	14.70%	20%	24.70%	29.40%	33.80%	50%
0.50	2.60%	5.10%	8.20%	12%	15.70%	19.70%	23.70%	27.60%	31.30%	34.80%	38%	51.30%
0.75	9.80%	13.70%	17.80%	21.80%	25.60%	29.20%	32.70%	35.90%	39%	41.70%	44.30%	55%
1.00	19%	23.60%	27.70%	31.60%	35%	38.30%	41.20%	44%	46.50%	48.80%	50.90%	59.50%
1.25	29%	33.50%	37.40%	40.90%	44.60%	46.80%	49.30%	51.60%	53.70%	55.70%	57.40%	64.50%
1.50	38.80%	42.90%	46.50%	49.60%	52.30%	54.70%	56.80%	58.80%	60.50%	62.10%	63.60%	69.40%
1.75	48%	51.70%	54.80%	57.50%	60.20%	61.80%	63.60%	65.30%	66.70%	68%	69.20%	74%
2.00	56.50%	59.70%	62.40%	64.60%	66.60%	68.30%	69.80%	71.10%	72.30%	73.40%	74.40%	78.30%

18
Activity-Based Costing and Activity-Based Management[1]

18.1 Introduction

One of the main organizational challenges in this era of increasing economic global-ization is the correct identification of costs at the output unit level, and of the factors that drive such costs (i.e., things that cause expenses to be what they are). It is expected that correct cost identification will give light to production improvement efforts. This is accomplished by focusing efforts on those aspects of the product or the production process that consume more resources and, therefore, account for higher percentages of the total cost. In addition, pricing techniques considering target costs and modern product design techniques are informed by costing information.

Economists and accountants have experimented with different production cost-ing techniques over the years, culminating with a widespread adoption of activity-based costing (ABC) in the early 1990s. The collection of ABC-informed decisions a manager can make when better cost projections are available is what is known as activity-based management, or ABM. We next present an overview, with extensive examples.

18.1.1 Understanding the Problem

Overhead allocation to products has haunted accountants and cost engineers since factories grew big enough to have more than one product line and diversified product presentations (i.e., size, color, packaging, etc.) whose production processes

---

[1] This chapter was written with substantial contributions from Dr. Jerome P. Lavelle, Department of Industrial Engineering, North Carolina State University, Raleigh, NC and Dr. Alexandra Medina-Borja, Grado Department of Industrial and Systems Engineering, Virginia Tech, Falls Church, VA.

shared factory facilities and support services, sometimes with concurrent production schedules. It is being argued that incorrect overhead allocation follows the same logic as when you go out with a couple of friends to have lunch and you get a salad and a soda costing you $9 while your friends order the most expensive dish on the menu, in addition to a glass of imported wine, costing them $30 each. At the end of the meal they suggest you three split the check. Thus, each member of the party will end up paying [$9 + 2 × $30]/3 = $23. We could say that, if product lines could have feelings, low overhead consumption product lines would certainly feel exactly as you would— shortchanged; your friends are under cost, whereas you are over cost.

There is a compelling reason to attempt accurate overhead allocation—to arrive at a monetary figure for what it takes to produce the product, or to deliver the service, that is close enough to reality that the price assigned to the product is appropriate to, or at least greater than, the cost to produce it. This is an important and growing problem. Since the early 1950s, the proportion of the total cost contributed by direct costs is being replaced by overhead costs. Back then, the reported average overhead cost accounted for around 25% of the total cost. Today, average overhead costs amount to over 65%. This is a function of automation and lean manufacturing, which reduce the direct labor and direct material bills, but cause a lot of burden due to machinery, computer systems, and other customer and supplier support activities.

The next simple example illustrates the impact of overhead costs on total cost.

Example 18-1

The Bread Company produces and sells baked bread. Two main products are produced in their Alexandria factory: the $1/2$-pound wheat loaf and the 1-pound French loaf. Company accountants gather from the bakers exactly how much flour, water, eggs, yeast, and oil are used to prepare each loaf of each type of bread. They also know the time it takes the baker to prepare the dough and the oven time for each. All these are easily traceable costs. However, there are additional costs incurred to prepare the loafs that are not directly related to each type of unit of bread, such as the cost of leasing the facilities, oven and tools, depreciation, the salaries and benefits of the sales department staff who dedicate their time to market both types of bread, and other salaries, such as the facilities janitor, night security, and general administration. Transportation costs for delivery and disposal of leftovers are also part of this "difficult-to-allocate" category of expenses.

Furthermore, accountants and managers know it is more difficult to market the daily production of French bread. The sales department personnel dedicate more time with their customers to sell French bread's daily production. There is also a higher probability that, on any given day, the lots of French bread will not be totally allocated to customers, and will have to be disposed of, sometimes selling them to organizations that buy old bread at a price lower than the direct cost (materials and direct labor cost). On occasions, the only option is to dispose of unsold products. Transportation for disposal of waste needs to be arranged. It is obvious that one lot of French bread will incur these indirect costs at a different proportion than one lot of the wheat bread. The activities that generate these expenses are undertaken in a different way, and therefore affect the production of one lot of French bread differently than one lot of wheat bread. However, accountants do not know how to assign indirect costs, other than by allocating the total costs—of the lease, transportation, sales, and other overhead salaries and machine depreciation—to product units by volume of bread output (i.e., by dividing the total overhead cost by the number of units or pounds produced in a period).

To illustrate this point, let's assume the following:

$$\text{Total indirect cost} = \$10,000$$
$$\text{Wheat bread production} = 8,000 \text{ units}$$
$$\text{French bread production} = 2,000 \text{ units}$$

Traditional cost accounting practices will divide $10,000/10,000 loaves, thereby assigning an indirect cost of $1 to each unit of bread, no matter what is the proportion of the activities really undertaken to generate that cost.

Direct labor and materials costs for French bread amount to $1,500, and for wheat bread they amount to $10,000.

Traditional cost accounting techniques will arrive at the following conclusions:

	Wheat bread	French bread
Units produced	8,000	2,000
Direct costs per unit	$10,000/8,000 = $1.25 per loaf	$1,500/2,000 = $0.75 per loaf
Indirect costs per unit	$1 per loaf	$1 per loaf
Total cost per unit	$2.25	$1.75
Selling price per unit	$3	$2.50
% profit	33.3%	42.9%

Important managerial decisions are made on the basis of these cost accounting results. According to the procedure used, both breads are profitable and it is worthwhile to keep producing French loaves. This conclusion is reached by reading the accountant's records, despite the fact that at least three times per week the total production is not sold and needs to be disposed of.

> At the end of the fiscal year exercise, the Bread Company seems to have shrunk its profit margin, it is far from what was projected. Upper management is puzzled.

The problem of getting a *better* (notice we do not say "the right") way to allocate overhead costs can be better approached by applying Activity-Based Costing. ABC seeks to identify expense-generating activities and to then find factors or proxy variables that represent the volume of the resources consumed to perform the activity needed in the production of each type of product. Thus, those drivers should be correlated to that expense account.

> In our Bread Company example, the actual sale promotion that representatives do every day to allocate the lots of production among customers is the activity that the sales department cost center performs. The driver for this activity should represent its intensity, such as the number of calls a sales representative makes to sell one lot of bread. The allocation then will be done based on those "drivers" of overhead expenses. Sales department salaries will be allocated on the basis of the division of effort, as measured by a proxy driver (i.e., the number of calls).

The success of ABC is based on a sensible selection of cost pools, activities, resources, and drivers. Thus, management needs to identify activities that represent all those types of indirect expense. This method will most likely give light to the fact that French bread production generates a higher rate of overhead expenses, therefore becoming less profitable than it seems otherwise.

On the basis of new total costs for each unit of bread, management can make new pricing decisions, as well as probably reprogramming the lot size for each type of bread and changing the policy regarding disposal of leftovers.   ■

18.1.2 What Are Activities and Activity-Based Costing?

Activity-based costing is a newly reemphasized and expanded set of methodologies for enterprisewide management of business costs. It focuses on detailing costs and assigning those costs to the items that cause them to occur. The underlying philosophy of ABC is stated by Liggett, et al.[2], as follows:

> Certain *activities* are carried out in the manufacture of *products*. Those activities consume a firm's *resources,* thereby creating *costs.* The *products,* in turn, consume *activities.* By determining the amount of resource (and the resulting cost) consumed by an *activity* and the amount of an *activity* consumed in manufacturing a *product,* it is possible to directly trace manufacturing *costs* to *products.*

Thus, one can see that *activities* are at the heart of an ABC system. Activities to produce a marketable product have classically been divided into two areas: direct and indirect (or overhead) activities. Direct activities are actions that accomplish a task *directly* associated with the final product or service produced. Examples include component insertion, painting, drilling, backboard wiring, thread tapping, and wave solder. Indirect activities are actions that are necessary to provide a good or service to a customer, but their effect is *indirect.* Examples of indirect activities might include material procurement, kitting, issuing a purchase order, material handling, machine setup, maintenance, and inspection. In addition, *overhead* items such as marketing, sales, utility expenses, machine depreciation, engineering and technical support, and customer warranties can also be classified as indirect activities. As one can see, all of these items, both direct and indirect, are necessary in delivering the product to the customer.

It is important to determine the cost of each activity to provide for a tracing of the true cost components in a majority of tasks. The next example shows the calculation of activity costs in a telephone customer-order processing work cell.

Example 18-2

Five workers in a telephone-order business cell process customer orders for one product line of the firm. The workers carry out two basic tasks, categorized as *telephone* and *nontelephone tasks* by the company. The *telephone tasks* include taking product orders, logging customer complaints and feedback, and providing on-line product inquiry assistance. These tasks are performed 75%, 10%, and 15% of the time, respectively, when a worker is engaged in a *telephone task.* The *nontelephone tasks* include invoicing and filling orders, order shipping preparation, and inventory updating and maintenance. These nontelephone tasks are performed 35%, 25%,

---

[2] H. R. Liggett, J. Trevino, and J. P. Lavelle, "Activity-Based Cost Management Systems in Advanced Manufacturing Environment," *Economic and Financial Justification of Advanced Manufacturing Technologies,* H. R. Parsaei, ed. (New York: Elsevier Science Publishers, 1992).

and 40% of the time, respectively, when the worker is engaged in such *nontelephone tasks*. The five workers include two workers dedicated to 100% telephone tasks, one worker who handles 50% telephone and 50% nontelephone tasks, and two workers who are split 10% telephone and 90% nontelephone tasks. Assuming a 2,000-hour work-year and a labor-plus-benefits rate of $15.00 per hour, the annual costs of the telephone-order business cell are calculated as follows:

One full-time employee annual costs = (2,000 hrs/yr) × ($15/hrs) = $30,000/year.

Telephone Tasks:

Taking product orders:

$30,000 × 2.7 employees × 75% of time in task   = $60,750 per year

Logging customer complaints/feedback:

$30,000 × 2.7 employees × 10% of time in task   = $8,100 per year

Providing on-line product inquiry assistance:

$30,000 × 2.7 employees × 15% of time in task   = $12,150 per year

Subtotal = $81,000 per year

Nontelephone Tasks:

Invoicing/billing orders:

$30,000 × 2.3 employees × 35% of time in task = $24,150 per year

Order shipping preparation:

$30,000 × 2.3 employees × 25% of time in task = $17,250 per year

Inventory updating/maintenance:

$30,000 × 2.3 employees × 40% of time in task = $27,600 per year

■

Subtotal = $69,000 per year

The concept of *resources* is also important in an ABC system. Resources are assets available in the enterprise for use in direct and indirect activities. Examples include production equipment, labor, tooling, computer time, and floor space. Associated with the use of these resources is a *cost*. When the resource is used, a cost can be assigned to the activity (direct or indirect) that caused that cost. In this way, costs, and the activities that *drive* those costs, can be identified and managed. The principle of activity-based costing systems, then, involves identifying activities that consume resources that have costs associated with them.

18.2 Traditional versus Activity-Based Costing Systems

Traditional costing systems are sometimes referred to as absorption-based accounting systems. This designation refers to the manner in which overhead and indirect expenses are assigned to cost centers. As described previously, this assignment is sometimes done on an arbitrary basis and results in cost centers often *absorbing* costs that they do not directly cause.

Activity-based costing systems are different from traditional costing in three important aspects. First, they depart from the way in which manufacturing overhead costs are managed. Second, they focus on process costs and not on product costs—providing the basis for more accurate cost reduction. Last, they provide a more detailed cost structure than that found in typical traditional (absorption-based) costing systems. These differences allow for more accurate and timely management of both direct and indirect costs.

18.2.1 Overhead Allocation

Traditional cost accounting systems allocate overhead (indirect) costs in terms of volume-based production indicators, such as direct labor hours, material cost, and direct labor cost (just as in our Bread Company example). Of these, many firms choose direct labor hours as their basis for allocation. The effect of allocating overhead costs in this manner is to spread an overhead cost item in direct proportion to the chosen volume-based indicator for each costing cell in the enterprise. Just as with our split lunch check example of Section 18.1.1, high-volume products often subsidize lower-volume products when traditional costing systems are used. This happens because lower-volume products are assigned less than their share of overhead and indirect costs. In the case of a high-volume dedicated assembly line, a traditional accounting system approach would allocate more overhead cost to it, in comparison to the flexible assembly cell, based on the fact that there is more direct labor content in the operations in that costing center.

Consider the example of allocating overhead to two production costing centers, one a high-volume dedicated assembly line (with a mix of automation and manual operations, and few product changeovers) and the other a lower-volume flexible production cell (with heavy automation, little direct labor content, and small lot sizes). Because of the fixed nature of the high-volume dedicated line, it will require fewer support costs than will the flexible assembly cell. Table 18-1 lists items for which the low-volume line (small lot size, etc.) will require relatively more resources, due to its flexible, low-volume nature.

Traditional allocations typically cause the low-volume products (those manufactured on the flexible cell) to be assigned a disproportionate share of the overhead, indirect, and support expenses. Sometimes manufacturers are able to recognize this and price the undercosted low-volume products at a premium in order to account for the "spreading" of overhead in the costing system. This causes two effects. First, the reporting system will identify these products as having the highest market margins, and thus sales and marketing forces will target them to increase their commissions (usually at the expense of the more profitable high-volume products). Second, prices on the overcosted higher-volume products will be higher than necessary. This leaves

TABLE 18-1   Additional Resources Consumed
by Low-Volume Flexible Assembly Cell

• Material pick orders	• Manufacturing engineering support	• Routing sheets
• Manufacturing assembly aids	• Machine program downloads	• Material handling requests
• Setup time	• Shipping invoices	• Work orders
• Scheduling, receiving, warehousing, testing, and maintenance resources		

TABLE 18-2   Cost Basis for Traditional versus Activity-Based Costing

Overhead cost item	Traditional costing	ABC costing
Utility expenses	Direct labor hours	Metered utility expense
Maintenance activities	or Direct labor cost	Maintenance hours billed
Order entry		No. of orders processed
Receiving activities		No. of receipts processed
Engineering support		No. of hours billed
Design change orders		No. of orders processed
Inspection activities		No. of items inspected
Production setups		No. of setups required

a manufacturer vulnerable to other firms, which have either a less diverse product offering or a costing system that leads to accurate costing, to offer the same product at a lower cost. Table 18-2 illustrates several examples of allocation bases for traditional versus activity-based costing systems.

In an ABC system, costs are created by, and assigned to, sources: products, suppliers, or customers. Either a *product* activity (for instance, material procurement) is causing the cost, or an activity related to maintaining *supplier relationships* (such as the cost of sponsored quality assurance training) is causing the cost, or a *customer* (such as the number of distribution points required to service a customer) is causing the cost. ABC accounts for all costs and traces them back to one of these three sources. Furthermore, traditional costing systems do not always account for the total cost of operations. Many times they ignore "below-the-line" expenses, such as R&D, distribution, marketing, customer service, and administration, and tend to group these costs into a single expense pool. Many companies treat these expenses as fixed costs and do not attempt to assign them to customers, distribution channels, markets, or even products.[3]

18.2.2 Cost Hierarchies

Horngren et al.[4] present a categorization of costs as a hierarchy. Such categorization allocates costs into different cost pools through different types of cost drivers (cost-allocation bases). It is important to note that there will be occasions on which the difficulty in getting drivers causally related to activity costs will force the allocation into a different level of cost object. The categorization of Horngren et al. becomes useful in those situations. When one faces an extreme degree of difficulty in determining cause and effect relationships between the drivers and the

---

[3] M. C. O'Guin, *The Complete Guide to Activity-Based Costing* (Englewood Cliffs, NJ: Prentice Hall, 1991).

[4] Charles T. Horngren, George Foster, and Srikant M. Datar, *Cost Accounting: A Managerial Emphasis* (Upper Saddle River, NJ, Prentice Hall, 1999).

activity costs related to production, the activity is considered a business sustaining activity.

According to the preceding categorization, ABC systems commonly use a four-part cost hierarchy to identify cost-allocation bases:

1. Output unit-level costs represent resources used on activities performed on each individual unit of product or service, such as maintenance, energy, machine depreciation, repairs, etc.

2. Batch-level costs represent resources used on activities that are related to a group of units of product(s) or service(s), rather than to each individual unit of product or service, such as setup costs and procurement costs.

3. Product-sustaining costs or service-sustaining costs are resources used on activities undertaken to support individual products or services. Design costs, engineering costs, and promotion costs fall in this category.

4. Facility-sustaining costs are resources used on activities that cannot be traced to individual products or services, nor to product lots or batches, but are used to support the organization as a whole. General administration costs, rent of office space, and building security are some examples.[5]

Figure 18-1 depicts the hierarchy and relationships among activities, resources, drivers, and cost objects (products and services).

Example 18-3

To further illustrate several ABC concepts, consider the JPL Company, which is manufacturing two models of pagers with an annual production volume of 500,000 units each. Historically, JPL has used a traditional product-costing system, with direct labor hours as the allocation base for all of its manufacturing overhead costs. The total manufacturing overhead costs, as well as the total direct material and direct labor costs of both products for the current year, are given in Table 18-3.

Recently, the company installed an activity-based costing system and, subsequently, identified six main activities to be responsible for most of the manufacturing overhead costs.

**TABLE 18-3   Total Manufacturing Costs
for Models A and B (current year)**

Category	Model A	Model B
Direct material cost:	$ 280,060	$ 150,000
Direct labor cost:	$ 350,000	$ 250,000
Direct labor hours:	35,000	25,000
Total manufacturing overhead costs:	$12,000,000	

[5] Horngren et al., 1999.

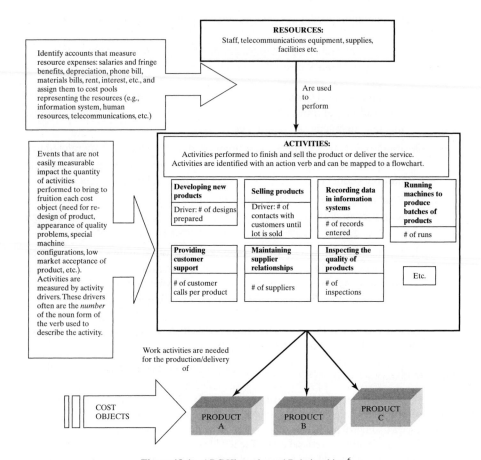

Figure 18-1   ABC Hierarchy and Relationships.[6]

The overhead costs were distributed to these six activities through activity accounting, identifying the resources used in each activity. Also, six corresponding activity cost drivers and their budget consequences for the current year were established. The activities, their cost drivers, resultant overhead costs, and budgeted rates are provided in Table 18-4.

Besides establishing the data in Table 18-3, the ABC system also measured the levels of each activity (or cost driver rates) required by the two products. These data are given in Table 18-5.

For post-audit purposes, the company ran a cost comparison between the two costing systems. To arrive at product costs for models A and B, the next two sections show the computations involved in the traditional (absorption-based) costing system and the ABC system, respectively.

---

[6] Adapted from Charles T. Horngren, George Foster, and Srikant M. Datar, *Cost Accounting: A Managerial Emphasis, (Upper Saddle River, NJ, Prentice Hall,* June 1999) and Gary Cokins, *Activity-Based Cost Management* (John Wiley & Sons, Inc., 2001).

TABLE 18-4   Activity-Based Costing Data for JPL Example

Activity	Cost driver	Costs	Budgeted rate
Production:	No. of machine hours	$8,000,000	200,000 hours
Engineering:	No. of engineering change hours	1,000,000	40,000 hours
Material handling:	No. of material moves	1,000,000	60,000 moves
Receiving:	No. of batches	800,000	500 batches
Quality assurance:	No. of inspections	800,000	20,000 inspections
Packing and shipping:	No. of products	400,000	1,000,000 products

TABLE 18-5   Cost Driver Rates for Models A and B

Activity	Model A	Model B
Production:	50,000 machine-hours	150,000 machine-hours
Engineering:	15,000 orders	25,000 orders
Material handling:	20,000 moves	40,000 moves
Receiving:	150 batches	350 batches
Quality assurance:	6,000 inspections	14,000 inspections
Pack and ship:	500,000 products	500,000 products

Cost Comparison Using Traditional Costing

1. Obtain the overall manufacturing overhead cost rate use the following formula:

$$\text{cost rate} = (\text{total overhead costs}) \div (\text{total direct labor hours})$$
$$= \$12,000,000 / 60,000 \text{ hours}$$
$$= \$200 \text{ per direct labor hour}$$

2. Compute total cost per unit for each model:

Activity	Model A	Model B
Direct material cost:	$ 280,060	$150,000
Direct labor cost:	$ 350,000	$ 250,000
Allocated overhead:	(35,000 hr × $200) = $ 7,000,000	(25,000 hr × $200) = $5,000,000
Total costs:	$7,630,060	$5,400,000
Cost per unit (based on 500,000 units):	$15.26	$10.80

(b) Cost Comparison Using Activity-Based Costing

1. Establish applied activity rate of the six activities:

applied activity rate = activity cost in Table 18-4 ÷ budgeted cost driver rate in Table 18-4

2. Determine the activity cost to be traced to each pager model:

Cost traced to each pager model = Cost driver rate (from Table 18-5)
                                  × applied activity rate (from step 1)

3. Compute total cost per unit for each model:

	Applied activity rate	Model A	Model B
Production:	$40/mach. hr.[*]	$2,000,000[†]	$6,000,000
Engineering:	$25/hr	375,000	625,000
Material handling:	$16.67/move	333,400	666,600
Receiving:	$1,600/batch	240,000	560,000
Quality assurance:	$40/inspection	240,000	560,000
Pack and ship:	$0.40/product	200,000	200,000
		$12,000,000	
Total overhead costs:		$3,388,400	$8,611,600
Direct material cost:		$280,060	$150,000
Direct labor cost:		350,000	250,000
Total costs:		4,018,460	9,011,600
Cost per unit:		$8.04	$18.02

[*]$40/machine-hour = $8,000,000/200,000 machine hours
[†]$2,000,000 = $40/machine-hour × 50,000 machine-hours

Summary of Comparison

The two costing systems produced different cost estimates as summarized in Table 18-6. Using absorption-based (traditional) costing, model A costs more to make than model B. However, on the basis of activity-based costing, model A is less expensive to make than model B. Because of differences in the number of *transactions* that affect indirect (overhead) costs, the reality is that model B actually incurs more overhead cost than model A, as demonstrated through the activity-based analysis.

As acknowledged earlier, companies making decisions based on distorted costs may unknowingly price some of their products out of the market, while selling others at a loss. Likewise, firms may make key decisions on the basis of competitors' market prices that do not consider the true costs of manufacturing their new or revised products. ∎

18.2.3 Process Focus

Activity-based costing systems focus on activities and the cost of those activities, rather than on products. This focus identifies real opportunities for process improvement and cost reduction. The ABC system, then, functions as a data provider to management decisions by providing feedback on resources consumed per operating period. This functionality is beyond the scope of traditional costing systems, which originally were designed primarily for inventory valuation for financial and income tax purposes. By focusing on activity costs, a manager can accurately track and evaluate the activities that, together, constitute the total product cost. Managers can directly see the effects of order entry, equipment setup, maintenance and engineering support, inspection, and many other costs. Traditional costing systems do not provide enough detail to track costs at that level, and they often make no provision for reporting those costs in a timely manner. As indicated in Table 18-7, ABC is useful in product-line management, as well as enterprisewide decisions.

TABLE 18-6 Product Cost Comparison
(Traditional Versus ABC)

	Model A	Model B
Traditional costing system	$15.26	$10.80
Activity-based costing system	8.04	18.02

TABLE 18-7 Strategic Decisions Supported by ABC Systems

Decision type	Examples
Product-line management	Product pricing
	Make-versus-buy decisions
	Product mix
	Facility expansion
	Employee reduction
	Off-shore expansion
Enterprisewide	Advanced technology investment
	TQM implementation
	Life-cycle cost management
	JIT implementation
	Capital investment management

ABC allows a firm to distinguish costs that are customer driven and to explicitly include them as part of the cost of serving specific customers. Because of this, a company may find that some customers are too expensive to service. ABC data provide a basis either to discontinue serving that customer or to negotiate changes in the producer–customer relationship. In such cases, costs may be shared or possibly transferred from the producer to the customer. Traditional costing systems do not provide a level of detail and assignment to be able to identify the costs of such situations.

18.3 ABC as a Two-Stage Procedure

The use of ABC requires that drivers relating activities and products/services first be determined. Time spent with the client discussing each individual service alternative may be an example of a driver. Second, for those activities that do not have a specific driver associated with them, or whenever the driver is difficult to collect, expert estimation can be used. This methodology is made possible through a deep understanding of the business processes. To accomplish this, it is recommended that a process map be created in advance. A process map will give light to the series of activities and their interrelations to create the final product or service.

Therefore, if necessary, a company can use activity cost drivers that are a combination of actual data collected in the company (such as the number of trips or the number of hours per client) and subjective estimations made by experts in the activity (most likely the actual resources used for the activity).

No matter what the size and scope of the ABC system to be implemented, one of the most important challenges is to identify valid and measurable (meaning cost- and time-effective measurable) activity cost drivers. The final selection should be guided by the identification of those events which cause the activity costs to experience significant swings. An important, and sometimes overlooked, requirement is that the activity cost driver should be a proxy measure of the intensity of the activity and, ultimately, positively correlated to the activity cost. Overhead costs incurred are classified on the basis of resources, which in turn are assigned to activities—to the extent possible—on the basis of a cause-and-effect relationship with the cost driver(s).

The ideal situation would be to identify cost drivers that, if increased or reduced, in turn create proportional fluctuations in the activity (increase or decrease in transactions) and in the activity cost. Determining causality is sometimes a difficult problem and, for ABC implementation, we could be satisfied with the presence of a strong positive correlation.

Example 18-4 illustrates this situation.

Example 18-4

PLASTIC, Inc., a polymers' extrusion company, is in the initial phases of ABC implementation. The ABC design team has identified work activities and is now seeking to select activity cost drivers.

The list of activities is given in Table 18-8.

An evaluation of the process map for the manufacturing of customized orders for different extruded products reveals that sales take place in two ways: (1) customers willing to receive a quote for special products contact the sales department, or (2) sales representatives contact former clients to identify potential new projects. As production is customized to the needs of the customer, different moldings need to be cast and tested before putting them in production. A lot of scrap material is left after the tests. Quality control is also not standard; one inspection can last hours, whereas others can last minutes.

Table 18-8   List of Activities and Potential Cost Drivers

Activities	Potential cost drivers—Available data
• Customer contact	• Weight of scrap left after setup and testing
• Quote preparation	• Number of production runs
• Engineering work	• Number of customer contacts
• Production preparation—setup	• Number of quotes
	• Engineering hours
• Production management and supervision	• Product complexity as evaluated by production managers
	• Number of setups
	• Setup hours
• Quality assurance	• Intensity of activities as measured by line workers
	• Sales volume in dollars
	• Number of inspections
	• Time spent in inspections

Table 18-9   Activities and Corresponding Cost Drivers for PLASTIC, Inc.

Activities	Final activity cost drivers
Customer relations ⟶	Number of customer contacts
Quote preparation ⟶	Number of quotes prepared
Engineering work ⟶	Engineering hours
Casting ⟶	Product complexity as evaluated by production managers
Extruder setup ⟶	Setup hours
Production test ⟶	Weight (in pounds) of scrap left after setup and testing
Production management and supervision ⟶	Intensity of activities as measured by management
Quality assurance ⟶	Number of hours of inspection

As cause and effect needs to govern the selection, it was difficult to select the drivers for production preparation and quality assurance. The following are several considerations that guided the final selection:

- After a lot of discussion, the activity "production preparation and testing" was subdivided into Casting, Extruder Setup, and Testing. This benefited from the availability of drivers seemingly correlated with such activity costs.
- Regarding inspection, it was agreed that number of inspections was not correlated with the intensity of the activity. However, time dedicated to inspection was positively correlated with inspection expenses.
- Some employees felt that sales volume would be a good driver for quote preparation. However, some of the employees in that department noted that not all quotes resulted in a sale, while all of them consumed time. Number of quotes was then decided to be a better driver to represent the cost of this activity. Table 18-9 shows the final activities and corresponding drivers. The arrows illustrate the positive correlation with the driver.   ∎

18.4 ABC/ABM Uses

The following sections primarily discuss and illustrate how ABC/ABM systems provide "value-added" information and functionality.

18.4.1 The Future of Activity-Based Management

The key for management is to determine where money is being generated and where money is being wasted. ABC implementation is most worthwhile when management objectives regarding the use of the information for decision making are clear. Organizations often generate excessive information, which overwhelms their people. Thanks to the advances of computers, data warehousing and data mining tools, the excess of data can now be more manageable as it is transformed into decision-making information. But to make the best use of that technology, top management has to have a plan for using the potential information.

Several global implementations of ABC/ABM have demonstrated that ABM can be at the core of a firm's strategic tools. ABC/ABM can also support what Cokins[7] calls "process-based thinking." Cokins presents 10 significant advances in ABC/ABM thinking that he predicts will become standard practices in the 21st century:

1. Evolution of the expenditures and cost assignment views from two stages to multiple stages if necessary.

2. Transformation of general ledger data for decision support with the use of on-line analytical processing (OLAP) technology for reporting purposes (mostly based in data warehouse systems and multidimensional databases used for quantitative analysis). On-line reporting systems are believed to be the key feature enabling the strategic use of ABC in ABM.

3. Reaffirmation of the two views of costs: horizontal cost assignment to processes and vertical cost assignment, based on cost objects, both of them related and using the same activities.

4. Development of a three-level ABC/ABM disaggregation of activities for a more refined system (the three-dimensional view refers to depth and layers of detail in the system).

5. Time phasing of ABC into ABM for a process view, so that "root-cause analysis" problem solving can take place.

6. Business process visualization of costs combining ABC/ABM data with flowcharting software, including 3-D depictions.

7. Integration of ABC/ABM data with strategy and measurement of performance. ABM systems today integrate ABC information with broader management tasks.

8. Emphasis on predictive costing. Organizations are increasingly using ABC/ABM to test their decisions for the future (estimating and predicting future costs).

9. Identification and treatment of organizational and business sustaining costs independent of product costs.

10. Shortened time interval between ABC/ABM recalculations—more frequently than on an annual basis.

18.4.2 Capital Budgeting/Economic Justification Using ABC Data

Capital budgeting can be defined as the decisions, most often made at the upper level of management, that allocate funds to investment alternatives. As discussed in

---

[7] Gary Cokins, *Activity-Based Cost Management: An Executive's Guide* (John Wiley & Sons, Inc. 2001).

Chapter 11, capital budgeting involves issues such as setting an appropriate MARR, identifying and evaluating sources of funding, allowing for dependencies between projects, evaluating riskiness of alternatives, and analyzing issues surrounding budget periods, communication, and the timing of projects. However, much of what engineering practitioners consider capital budgeting is alternative project evaluation, based on discounted cash flow analysis utilizing the "separation principle."

Because estimated effects are so important to the capital budgeting process, an accurate accounting of true costs becomes important in providing realistic and meaningful analyses. Effective cost management, then, is vitally important to competitive enterprises in the context of the capital budgeting process. Activity-based costing systems identify and track the "right" cost items, as well as the levels of those items, that are necessary in capital budgeting analyses.

18.4.3 Product Design Decisions Using ABC Data

In today's manufacturing firms, there is an increased recognition of the effect that product design has on many downstream operations. Decisions made at design time affect nearly every aspect of producing a product, delivering it to the customer, and servicing it over its intended life cycle. Much of the focus on design activities has been part of the total quality management (TQM) movement in the United States. Programs with acronyms such as DFx, QFL, and DOE (Table 18-10) have all focused attention on the *process* of product design.

Activity-based costing data can be very important to the product designer, and some of the *first* ABC systems (such as those at Tektronix and Hewlett-Packard) were designed specifically for this purpose. Activity-based systems identify activities that create costs—both direct and indirect—so that product designers can evaluate the effect that decisions such as component selection, tolerances, surface finish requirements, and material selection have. Such decisions ultimately affect items such as the number of parts required. the machine routing required, and the testing and inspection required. As an example, tighter tolerances mean more inspection resources—inspectors, time on the line, equipment requirements, training—which translates to more costs.

Many company-wide quality efforts focus on, and recognize the importance of, cost data on product designs. Designers who use well-structured ABC data are fully

TABLE 18-10 Quality Programs that Focus Attention at the Design Stage

DFx	Design For x, where x can be any aspect of the product life cycle (e.g., manufacturability, testability, disposability, reworkability, or serviceability).
QFD	Quality Function Deployment uses a house-of-quality matrix that aids in mapping customer requirements into product design features.
DOE	Design of Experiments, part of the philosophies developed by Genichi Taguchi,[8] espouses a focus on design variables and a reduction in variation in those variables through the use of design-of-experiment testing during the design phase of product development.

[8] G. Taguchi, at al. *Quality Engineering in Production Systems* (New York: McGraw-Hill, 1959).

enlightened on *all* of the effects of their design decisions. Often, however, companies are led into inappropriate action by thinking, "Well, we have cost data—let's incorporate its use during product design!" By managing costs by using traditional volume-based indicators, companies are encouraged, in their product designs, to seek inappropriate design objectives such as eliminating (reducing) direct labor hours, machine hours, or material costs, Each of these design objectives may well ultimately increase total product costs by adding to the indirect and overhead cost items.

Another important contemporary design issue involves the time required. Timeliness of product introduction to the marketplace is very important in today's competitive marketplace and, many times, it ultimately affects a product's success or failure. Because of this, design lead times need to be shortened to permit a firm to react to product and market changes. Today's design environment calls for a drive toward short cycle times and rapid introduction of products into the marketplace. Designs with few part types require less manufacturing time and money. Fewer part types translates to fewer engineering drawings, specification sheets, tooling sheets, routing sheets, engineering shop aids, and other types of documentation. Also, fewer parts require less incoming inspection resources, storage space, manufacturing machine types, test sets, and mean reduced setup and changeover expenses at the machine level. All of these items translate into lowered indirect expenses.

Many companies are focusing on getting designs to market more quickly. Hewlett-Packard's "Product Generation Team" is one example of a concerted corporatewide effort to bring cost-effective and streamlined products to market in the shortest interval possible. Use of ABC data encourages a focus on minimizing costs and increasing efficiencies in product designs. Two primary tenets of design-for-manufacturing (DFM) guidelines are to minimize the number of components in product design and to maximize the modularity of product designs. Use of ABC data facilitates these, and many other, salutary effects on product designs.

18.5 Current Issues in the Use of ABC Systems

A few of the current issues in the use of activity-based costing data and in activity-based costing system implementations are discussed next.

18.5.1 ABC System Design Issues

The implementation of an activity-based costing system is not a trivial matter, and several key issues that must be addressed in the system design phase greatly affect any implementation. Primary among these issues are the intended uses, or objectives, of the system. This affects the degree of definition of cost pools, activity centers, and cost drivers. If the system's purpose is for product costing, less detail may be required compared with systems designed to assess specific problems or manage certain costs. Systems whose use is toward enterprisewide continuous improvement systems initiatives require even more detail. Kaplan[9] suggests that cost systems can be built to

[9] R. S. Kaplan, "One Cost System Isn't Enough," *Harvard Business Review* (Jan–Feb. 1988).

address three different problems: (1) inventory valuation, (2) operational control, and (3) individual product cost measurement. Brimson,[10] on the other hand, lists four such potential objectives: (1) product costing, (2) managing cash and inventory, (3) cost control, and (4) decision support.

18.5.2 Hesitancy to Invest in ABC

All new investments in manufacturing and business firms require an impetus for such action. Investments in new insurance carriers, new automated production equipment, new officewide word processing software, or a new cost accounting system are examples of management decisions that affect the way business is done. Often, however, management is reluctant to commit to such investments. U.S. managers are notorious for looking for short-term solutions to problems—be they simple or complex.

An individual championing an ABC implementation may face much resistance. One common obstacle is the "We've got a costing system now and it isn't broke, so why do you want to fix it?" attitude. As with most systemwide information systems, it will require sound data and detailed plans to convince decision makers to invest. As with other "change the way we do business" investments, an ABC implementation may require a "proving" phase as a pilot project.

In the future, more and more firms will be moving toward incorporating the power of activity-based costing as part of their business systems. In considering ABC implementations, companies should consider the wisdom of two quotes, the first attributed to Henry Ford and the second anonymous: "If you need *it,* and don't invest in it, then you'll pay for it without getting it," and "If you do what you've always done, then you'll get what you've always gotten." ABC strikes at the heart of many competitive issues currently facing U.S. firms, which would do well to consider the effects that investing, *or not investing*, will have on them in the future.

18.5.3 ABC System Integration Issues

Most ABC research has focused on software implementations at the PC- or network-based levels. Little has been accomplished in integrating ABC principles with other enterprisewide integrated information systems. Formal inclusion in enterprise-level software architectures is being explored right now.

In ABC implementations, isolated systems suffer from problems of effort duplication, data errors, and update inconsistencies with other business software systems. Duplication costs, system validity, and the reaction of some employees to put little faith in data not coming from "the" companywide business system are other problems that may affect these implementations.

---

[10] J. A. Brimson, *Activity Accounting: An Activity-Based Costing Approach* (New York: Wiley & Sons, Inc., 1991).

18.6 Summary

As Cooper[11] points out, economist John M. Clark questioned the traditional unit-based allocation of manufacturing costs to production units as early as 1923. The precepts of ABC were largely ignored until the latter part of the 20th century, when several important effects increased interest and use. The following can be considered the drivers of ABC in its early inceptions: First, the direct-labor portion of products' costs has decreased to relatively low percentages. Second, intense worldwide competition has forced firms in many sectors to make order-of-magnitude increases in the effectiveness and efficiency of their operations. Third, current technology in micro-computers and personal computers, together with information technology, has greatly increased the collection, inspection, and use of data in manufacturing operations on large-scale and real-time bases.

These effects have combined to reawaken U.S. companies to the benefits of ABC-type data, the use of the data generated by ABC systems to inform decision making, and improvement initiatives such as Business Process Reengineering (BPR), total quality management, marketing strategies, and corporate strategy. Current advances have increased the adoption of ABC/ABM, as difficulties in tracing, storing, and maintaining the required data and making automated calculations and reports have been overcome.

REFERENCES

Anderson, Shannon and Mark Young. *Implementing Management Innovations: Lessons Learned from Activity Based Costing in the U.S. Automotive Industry.* Boston: Kluwer Academic Publishers, 2001.

Beaujon, B. and R. S. Singhal. "Understanding the Activity Costs in an Activity-Based Cost System." *Journal of Cost Management,* Spring 1990.

Brimson, J. A. *Activity Accounting: An Activity-Based Costing Approach.* New York: Wiley & Sons, Inc., 1991.

Canada, J. R. and W. G. Sullivan. *Economic and Multi-Attribute Evaluation of Advanced Manufacturing Systems.* Englewood Cliffs, NJ: Prentice Hall, 1989.

Cokins, Gary. *Activity-Based Cost Management.* NewYork: John Wiley & Sons, Inc., 2001.

Cooper, Robin. "The Two-Stage Procedure in Cost Accounting—Part One." *Journal of Cost Management,* I-2, Summer 1987, pp. 43–51.

Cooper, R. "Cost Classification in Unit-Based and Activity-Based Manufacturing Cost Systems." *Journal of Cost Management,* Fall 1990.

Gurowka, Jim. "ABC Software—Is There a Future for ABC Software?" *FOCUS Magazine: for the performance management professional,* Issue 2, 2000.

Horngren, Charles T., George Foster, and Srikant M Data. *Cost Accounting: A Managerial Emphasis.* Upper Saddle River, NJ: Prentice Hall, 1999.

---

[11]R. Cooper, "Cost Classification in Unit Based and Activity-Based Manufacturing Cost Systems," *Journal of Cost Management* (Fall 1990).

Kaplan, Robert. "The Four Stage Model of Cost Management." *Management Accounting*, February 1990.

Kaplan, R. S. "One Cost System Isn't Enough." *Harvard Business Review*, Jan–Feb 1988.

Kolb, Mathew and Joseph Donelly. "Best Practices in Activity-Based Management." In *Arthur Andersen's Global Lessons in Activity-Based Management*. Steve Player and Roberto Lacerda (editors), New York: John Wiley & Sons Inc., 1999.

Johnson, H. "Managing Costs: An Outmoded Philosophy." *Manufacturing Engineering,* May 1989.

Nachtmann, H. and K. L. Needy, "Methods for Handling Uncertainty in Activity Based Costing Systems." *The Engineering Economist*, Vol. 48, No. 3, 2003, pp. 259–282.

O'Guin, M. C. *The Complete Guide to Activity-Based Costing*. Englewood Cliffs, NJ: Prentice Hall, 1991.

Steve Player and Roberto Lacerda (editors), *Arthur Andersen's Global Lessons in Activity-Based Management*. New York: John Wiley & Sons Inc., 1999.

Taguchi, G. et al. *Quality Engineering in Production Systems*. New York: McGraw-Hill, 1959.

PROBLEMS

18-1. ABC systems are being implemented on a wide-scale basis by the largest manufacturers in the United States. Do you see ABC systems as implementable by small and medium size enterprises? Explain your answer.

18-2. What do you see as the major obstacles that face a company that has decided to "go for it" and implement ABC on an enterprisewide basis?

18-3. What are the characteristics of the firm that you think has the most to gain from an ABC implementation? In what environments will ABC have the most impact?

18-4. Do you think that ABC can be a panacea, or do you see it as just another management fad that eventually will blow away and be forgotten?

18-5. In a small group (or by yourself), brainstorm and make a list of all of the monetary and nonmonetary attributes that could (should) be included in an economic analysis justification of implementing an ABC system.

18-6. Categorize each of the costs given as either direct or indirect:

machine run costs	machine operator wages
insurance costs	machine depreciation
utility costs	cost of product sales force
material handling costs	support (administrative) staff salaries
engineering drawings support	cost of materials
cost to market the product	machine labor overtime expenses
cost of storage	cost of tooling and fixtures

Assume a traditional costing system is in place when you answer.

18-7. In the Bread Company example given in Example 18-1, identify the necessary cost pools to implement ABC.

18-8. In the same Bread Company example, management has identified the potential drivers and activities shown:

Activities	Cost drivers
Oven cleaning	Number of units produced
Oven set up	Number of pounds produced
Oven maintenance	Number of units unsold
Facilities cleaning	Number of pounds unsold
Sales and promotion of daily production	Square footage of machinery, including oven
Sales of leftovers	Square footage of storage space
Transportation and disposal of leftovers	Number of calls to customers
	Number of customers
Transportation of sold units	Number of units sold
	Number of trips required to dispose of unsold units
	Mileage on truck driven to dispose of unsold units
	Number of production lots
	Average time spent by sales representative to sell one order

Select the most appropriate ones, assigning them to each activity, and justify your selection.

18-9. The Bread Company team charged with implementing ABC has come up with the following costs and driver information:

Cost information	Cost drivers' information		
		French bread	Wheat bread
Transportation of sold products: $1,000			
Transportation and disposal of unsold products: $1,500			
Sales department salaries: $6,000	Number of units produced	2,000	8,000
Maintenance: $500	Number of units unsold by regular means (average)	800	50
Facilities cleaning: $400	% of units disposed	90%	1%
Telephone: $300	Number of production lots	200	400
Promotion: $300	Number of delivery trips	1 per lot	1 per lot
	Number of disposal trips	1 per lot	none
	Time spent on the phone by sales representative to sell one lot	3 hours	0.5 hour
	Time spent by sales representative to sell leftovers of one lot	2.25 hours	0.25 hour
	Setups and cleaning (oven maintenance)	1 per lot	1 per lot
	Number of ads/promotion	4	1
	Square footage of storage space	100 Sq Ft	200 Sq. Ft.

To dispose of the unsold units by selling them, the Bread Company accepts a cost recovery price of $0.50 per loaf, independent of the type of bread.

a. Consider this information and prepare an ABC analysis of the profitability of each one of the products. Get the direct cost information from the example given. Compare with the results in Example 18.1. Explain what you think happened.

b. Prepare an alternative strategy for commercializing bread. Include marketing as well as manufacturing strategies on the basis of the results above.

18-10. Depict a process map for the telephone company case in Example 18-2, and identify the operations that generate a cost and where in the general ledger one could find cost center information to feed this map for ABC purposes.

18-11. LABS Laboratories is a pharmaceutical company challenged to arrive at a more realistic cost of their two main products: an antibiotic in suspension (a powder that the pharmacist has to dilute in water before dispensing it to the patient) and an antiinflammatory product in tablets. The table shows the overhead costs associated with the production of both products.

Management is aware that the production of the pressed tablets requires a higher energy consumption and that, in general, the production is much more messy, requiring a lot of cleaning, plant and machine maintenance, and protection equipment for the operators. However, the production volume of the antibiotic is much higher. Accountants have been allocating overhead in relationship to production volumes in Kg.

Overhead costs	
Energy	$200,000
Plant janitor	$ 30,000
Maintenance	$ 40,000
Protection equipment	$ 5,000

	Production in units	Direct costs	Kg of production	Energy per hour of production
Tablets	1,000,000	$400,000	50,000	200KW
Antibiotics	2,000,000	$500,000	120,000	30KW

The main energy consumption machine for the tablets is a compactor that handles up to 200 Kg per hour. Antibiotic powder can be produced at a rate of 100 Kg per hour.

Management has assigned a team to look at the possible cost drivers to begin collecting data. The team is struggling to identify drivers and has come up with only the following "facts." Maintenance of the tablets' equipment needs to be performed every 24 hours of operation (roughly every fourth day), while the powder equipment can be maintained every 200 hours. Protection equipment is replaced every week, except when maintenance is in progress. There are four operators in the tablets' production line that require protection and two in the antibiotics' production line. Each tablet is sold at $0.25 and each vial with antibiotic powder is sold at $2.50. Set up an ABC costing scheme, evaluate the selling prices, and make recommendations for a profit margin of 35%.

18-12. FELCAM, a service company, provides two different types of services. Service A, which requires minimal sales effort and Service B, which is still not very well known in the market. As expected, Service B consumes a lot of time from the sales representatives. Instead of allocating an equal amount of sales expenses to each one of the products, the company can use estimation techniques to map out accurately this cost to the individual services. Sales expenses include the representatives' salaries, phone and Internet time, and some travel to selected potential customers. There is also a great amount of time spent preparing the presentations and materials to show to individual clients, since there is always an attempt to customize them to the client's potential use of the service. These expenses are shown in the following list:

Sales representatives salaries	$140,000
Phone and Internet	4,000
Travel expenses	20,000
Presentations materials and equipment	10,000

The following activity information is also available:

	Service A	Service B
Number of sales appointments	40	75
Number of follow-up calls	20	125
Estimated % of effort	25%	75%
Number of presentations	10	30
Volume of sales	$1,500,000	$300,000
Number of finalized sales	500	100

The company wants to know whether it is pricing Service B correctly.

a. Use ABC to arrive at a recommendation regarding ABC.

b. What is the cost of delivering each of the services?

18-13. The budgeted manufacturing overhead costs of WGS Window Company for 2005 are listed as follows:

Type of costs	Cost pools
Electric power	$ 500,000
Work cells	3,000,000
Materials handling	1,000,000
Quality control inspections	1,000,000
Product runs (machine setups)	500,000
Total budgeted overhead costs	$6,000,000

For the last 5 years, the cost accounting department has been charging overhead production costs on the basis of machine hours. The machine hour is the cost driver in determining the applied manufacturing overhead costs rate. The VP for production, Hal

Jacobs, estimates the budgeted capacity for 2005 to be 1,000,000 machine hours. The predetermined applied factory overhead cost rate will, therefore, be $6.00 per machine hour ($6,000,000 budgeted overhead costs divided by 1,000,000 budgeted machine hours).

Phil Stolzer, the president, recently attended a one-day seminar on an Activity-Based Cost system (ABC), which allocates overhead costs on the basis of activity-based cost drivers. After attending the seminar, he believes that the ABC method results in more reliable cost data. This reliability will lead to better and more accurate pricing policies that will give the company an edge over its competitors. With this system, manufacturing overhead costs are correlated to various activity-cost bases, rather than to one allocation base, such as machine hours, direct labor hours, units of production, etc.

Stoizer plans to implement this application of overhead costs, on the basis of ABC techniques that require the use of cost pools and cost drivers. Peter Brock, production manager, upon the president's request, has provided the following data regarding the expected total 2005 activity of the activity-based cost drivers for those budgeted overhead costs previously listed under the cost pool section:

Type of cost	Activity-based cost drivers
Electric power	100,000 kilowatt hours
Work cells	600,000 square feet
Materials handling	200,000 materials moves
Quality control inspections	100,000 # of inspections
Product runs (setups)	50,000 product runs

Larry Ryan, the VP of Marketing, received an offer to sell 5,000 windows to a local construction company that is currently building a high-rise office building adjacent to a shopping mall. Larry Ryan asks Sue Pretora, head of the cost accounting department, to prepare cost estimates for manufacturing the 5,000 windows. With this information, a comparison can be made between the present system of applying overhead cost on machine hours and the new ABC system, on the basis of various cost drivers. Pretora provided the following data concerning the production of 5,000 windows:

Direct materials cost	$ 100,000
Direct labor costs	$ 300,000
Machine hours	10,000
Direct labor hours	15,000
Electric powers kilowatt hours	1,000
Work cells square feet	8,000
No. of materials handling moves	100
No. of quality control inspections	50
Number of product runs (setups)	25

a. What are the manufacturing costs per window unit under the present cost accounting system that applies manufacturing overhead costs on the basis of machine hours?

b. What is the manufacturing cost per window unit if the activity-based costing system is implemented?

c. If the preceding two cost accounting systems will result in different cost estimates, which cost accounting system is preferable for pricing policy?

19
Multiple-Attribute Decision Making and the Analytic Hierarchy Process[1]

19.1 Introduction

Although it is very useful to use cost or profit as a measure of desirability, many decisions between alternatives cannot be measured only in these terms. Most firms have other objectives, such as customer service, goodwill, community reputation, job satisfaction, safety, employment stability, etc. These factors, which cannot be expressed directly in cost or profit terms, often are called *intangibles, irreducibles,* or *nonmonetary attributes.*

Even though a decision maker may have readily definable objectives, he or she might still have a significant problem defining the attributes (sometimes also called criteria) by which the attainment of objectives can be measured. For example, the attainment of a "safety" objective in an automobile purchase decision might be measured by such criteria as weight, maximum possible speed, and interior air bags.

The ultimate aim of the analyst, or decision maker, with respect to multiple objectives and criteria, should be to use rational methods of evaluating them, so that a single measure of value may be associated with each alternative in a decision problem.

To provide perspective and motivation for the problems associated with multiple-attribute decision making, consider an example involving the choice of a computer-aided design (CAD) workstation. Typical data are summarized in Table 19-1. Three vendors and "do nothing" constitute the list of feasible alternatives (choices) in the decision problem, and a total of seven attributes is judged sufficient for purposes of

---

[1] Some of this chapter is based on J. R. Canada and W. G. Sullivan, *Economic and Multiattribute Evaluation of Advanced Manufacturing Systems* (Englewood Cliffs, NJ: Prentice Hall, 1989).

TABLE 19-1   CAD Workstation Selection Problem

	Alternative			
Attribute	Vendor A	Vendor B	Vendor C	Reference ("Do nothing")
Cost of purchasing the system	$115,000	$338,950	$32,000	$0
Reduction in design time	60%	67%	50%	0
Flexibility	Excellent	Excellent	Good	Poor
Inventory control	Excellent	Excellent	Excellent	Poor
Quality	Excellent	Excellent	Good	Fair
Market share	Excellent	Excellent	Good	Fair
Machine utilization	Excellent	Excellent	Good	Poor

discriminating among the alternatives. Aside from the question of which workstation to select, other significant questions come to mind in multiattribute decision making:

1. How are the criteria (attributes) chosen in the first place?
2. Who makes the subjective judgments regarding nonmonetary criteria such as "quality"?
3. What response is required—a partitioning of alternatives or a rank ordering of alternatives, for instance?

Several simple, though workable and credible, models for selecting among alternatives such as those in Table 19-1 are described in this chapter.

19.2 Choice of Criteria

The choice of attributes (criteria) by which to judge alternative designs, systems, products, processes, etc., is one of the most important tasks in multiple-attribute decision making. (The most important task, of course, is to identify feasible alternatives from which to select.) It has been observed that the articulation of attributes for a particular decision can, in some cases, shed enough light on the problem to make the final choice obvious to all involved!

Consider again the data in Table 19-1. These general observations regarding the attributes used to discriminate among alternatives can immediately be made:

1. Each attribute distinguishes at least two alternatives—in no case should identical values for an attribute apply to all alternatives.
2. Each attribute captures a unique dimension or facet of the decision problem (i.e., attributes are independent and nonredundant).
3. All attributes, in a collective sense, are assumed to be sufficient for purposes of selecting the best alternative.
4. Differences in values assigned to each attribute are presumed to be meaningful in distinguishing among feasible alternatives.

In practice, selection of a set of attributes is usually the result of group consensus, and it is clearly a subjective process. The final list of attributes, both monetary and nonmonetary, is therefore heavily influenced by the decision problem at hand, as well as by an intuitive feel for which criteria will or will not pinpoint relevant differences among feasible alternatives. If too many attributes are chosen, the analysis will become unwieldy and difficult to manage. Too few attributes, on the other hand, will limit discrimination among alternatives. Again, judgment is required to decide what number is "too few" or "too many." If some attributes in the final list lack specificity or cannot be quantified, it will be necessary to subdivide them into lower-level attributes that can be measured.

To illustrate the preceding points, we might consider adding an attribute called "cost of operating and maintaining the system" in Table 19-1 to capture a vital dimension of the CAD system's life-cycle cost. The attribute "flexibility" should perhaps be subdivided into two other more specific criteria such as "ability to interface with computer-aided manufacturing equipment" (e.g., numerically controlled machine tools) and "capability to create and analyze solid-geometry representations of engineering design concepts." Finally, it might be constructive to aggregate two attributes in Table 19-1, namely, "quality" and "market share." Because there is no difference in values assigned to these two attributes across the four alternatives, they could be combined into a single attribute, perhaps named "achievement of greater market share through quality improvements."

19.3 Selection of a Measurement Scale

Identifying feasible alternatives and appropriate attributes represents a large portion of the work associated with multiple-attribute decision making. The next task is to develop metrics (measured scales or descriptors) that permit various states of each attribute to be represented. For example, in Table 19-1, "dollars" was the obvious choice for the metric of purchase price. A subjective assessment of flexibility was made on a metric having five gradations that ranged from "poor" to "excellent." The gradations were poor, fair, good, very good, and excellent. In many problems, the metric is simply the scale upon which a physical measurement is made. For instance, anticipated noise pollution for various routings of an urban highway project might be a relevant attribute whose metric is "decibels."

19.4 Dimensionality of the Decision Problem

This first way of dealing with data, such as those shown in Table 19-1, is called *single-dimensioned analysis*. (The dimension corresponds to the number of metrics used to represent the attributes that discriminate among alternatives.) Collapsing all information to a single dimension is popular in practice, because many analysts and decision makers believe that a complex problem can be made tractable in this manner. In fact, several useful models presented later are single dimensioned. Such models are termed *compensatory*, because changes in values of a particular attribute can be offset by, or traded off against, opposing changes in another attribute.

A second basic way to process the information in Table 19-1 is to retain the individuality of the attributes as the best alternative is being determined. That is, there is no attempt to collapse attributes to a common scale. This is referred to as *full-dimensioned analysis* of the multiple attribute decision problem. For example, if r^* attributes have been chosen to characterize the alternatives under consideration, the predicted values for all r^* attributes are considered in the choice.

19.5 Selected Analysis Techniques

Numerous techniques have been developed over the past many years to facilitate analysis of multiple-attribute decision problems. To give the readers an introduction to the diversity and application of available methods, we will concentrate on the following, which are given in approximate order of increasing complexity and power:

1. Alternatives–Attributes Score Card;
2. Ordinal Scaling;
3. Weighting Factors;
4. Weighted Evaluation of Alternatives;
5. Analytic Hierarchy Process.

19.5.1 Alternatives–Attributes Score Card

Graphical techniques are very powerful, because they allow us to describe the multiple attribute decision problem so that decision makers can readily understand and absorb large amounts of information. Graphical techniques often do not include specific "weighting" of criteria, which may be advantageous, because attribute weights are often nebulous or differ greatly among various decision makers.

A scorecard is a matrix of alternatives versus attributes, together with numbers or other symbols to represent the outcome expected for each alternative with respect to each attribute. Table 19-2 is an example in which qualitative and quantitative estimates are provided for the performance of five attributes used to judge the "value" or "worth" of four alternatives for modernizing a manufacturing company. Ease of interpretation of the scorecard (to facilitate decision making) can be obtained by such devices as symbols and colors for "best" and "worst" alternatives for each attribute.

Figure 19-1 shows the use of shaded circles to portray visually the relative evaluations of alternatives with respect to the five criteria given in Table 19-2. As shown in the key, the evaluations for each attribute, except net present worth, were categorized in five ways, from "exceptional" (full shading) down to "poor" (no shading). Net present worth is shown to be "exceptional" (with full shading) for $500M, with proportionately less shading down to $0M. While the use of shaded circles does cause loss of the specific language and quantitative information given in Table 19-2, it is easy to scan for relative comparisons.

In examining the scorecard information in Table 19-2 and Fig. 19-1, you might be led to conclude that alternative P-4 and perhaps P-3 are definitely not as desirable as alternatives P-1 and P-2. Often a final decision can be made using this method, but it should be recognized that any relative weights or importances of the attributes have

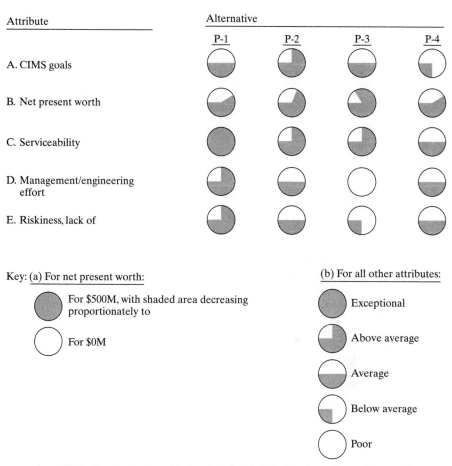

Figure 19-1   Graphical portrayal (using shaded circles) for four alternatives and five attributes.

TABLE 19-2   Sample Scorecard for Four Alternatives and Five Attributes

Attribute	Alternative			
	P-1	P-2	P-3	P-4
A. CIMS goals	Average	Above average	Average	Below average
B. Net present worth	$300M	$350M	$400M	$290M
C. Serviceability	<1 hr	1–1.5 hr	1–1.5 hr	1.5–2 hr
D. Management/ engineering effort	500 hr	700 hr	1,100 hr	700 hr
E. Riskiness, lack of	Above average	Average	Below average	Average

☐ Best alternative for attribute

⌐ ⌐ Worst alternative for attribute

not been assigned formally. *In future examples, we have elected to drop alternatives P-3 and P-4 from the analysis to simplify the ensuing discussion.*

19.5.2 Ordinal Scaling

An ordinal scaling is simply a ranking of criteria in order of preference. Before attributes are weighted (or alternatives are evaluated), it is often desired to rank them in order of decreasing preference. This might be done by presenting the decision maker with a list of attributes (or alternatives) and asking him or her to rank them in order of preference. There is, however, a procedure that may make this task easier and provides a check on the internal consistency of the value judgments obtained. This is called the *method of paired comparisons* and is illustrated next.

Consider the five attributes in Table 19-2, which we now want to rank in order. They are designated as follows:

A. CIMS goals;

B. net present worth;

C. serviceability;

D. management/engineering effort;

E. riskiness, lack of.

The method of paired comparisons submits attributes to the decision maker two at a time for preference judgments. In general, if there are N factors, $N(N-1)2$ pairs must be judged. Assume that the results of this process are indicated by the following list of preference statements (as usual, the symbol > means "is preferred to" and = means "is equal to"):

1. A < B	6. B > D
2. A > C	7. B > E
3. A > D	8. C > D
4. A > E	9. C = E
5. B > C	10. D < E

A good way to depict the preceding pairwise comparisons and then determine rankings is shown by the matrix in Table 19-3. In that matrix, "P" is shown for each pair in which the row factor (attribute) is preferred to the column factor. Note that the diagonal of the matrix is empty (since a given factor cannot be preferred to itself). A good way to make sure that all pairs of factors have been considered is to recognize that there should be a P either above *or* below the diagonal for all pairs, or in case of equal preferences, an "=" is shown both above *and* below the diagonal. Note that on the right-hand side of the matrix is shown the number of times the factor in each row is preferred. Thus, for this example, it is found that the rank order of attributes is B > A > C = E > D.

The foregoing scheme of deducing rankings assumes transitivity of preferences; that is, if A > C and B > A, then B must > C. If the number of times a given factor is preferred to another is equal for two or more factors (except in the case of ties),

TABLE 19-3   Illustration of Preference Comparisons[a]

	A	B	C	D	E	Number of times preferred
A	—		P	P	P	3
B	P	—	P	P	P	4
C			—	P	=	1 1/2
D				—		0
E			=	P	—	1 1/2

[a] "P" if row factor is preferred to column factor, or "=" if they are
equally preferred.

there is evidence of lack of consistency (i.e., intransitivity), which suggests the need
for questioning preference judgments for the attributes involved.

19.5.3 Weighting Factors

Many numerical formula methods for assigning weights exist that are easy to
use, but are generally less defensible than direct assignment of weights on the basis of
preference comparisons among criteria. Several of these formula-based methods are
described next, and sample calculations are shown in Table 19-4 for the same five criteria (attributes) given in Table 19-2. The weights are expressed as percentages.

1. *Uniform or equal weights.* Given N attributes, the weight for each is

$$W_i = 1/N \times 100\%.$$

2. *Rank sum weights.* If R_i is the rank position of attribute i (with 1 the highest
   rank, etc.) and there are N attributes, then rank sum weights W_i for each
   attribute, may be calculated as

$$W_i = \frac{N - R_i + 1}{\sum\limits_{i=1}^{N} (N - R_i + 1)} \times 100\%.$$

3. *Rank reciprocal weights.* Rank reciprocal weights, using the same notation
   as before, may be calculated as

$$W_i = \frac{1/R_i}{\sum\limits_{i=1}^{N} (1/R_i)} \times 100\%.$$

When comparing the methods in Table 19-4, note that the rank reciprocal
method gives the highest weight for the first-ranked attribute. We might choose
among the preceding weighting methods according to which provides the closest
approximation to the independently judged weight for the highest ranked attribute. If
we take that independently judged weight to be the 40%, it is seen to be closest to the
44% for the rank reciprocal method.

TABLE 19-4 Sample Calculation of Weights by Several Formulas

	Uniform	Rank sum			Rank reciprocal	
		(A)	(B)	(C)	(D) = 1/(A)	(E)
Attribute	$W_i = 100\%/N$	R_i^a	$N - R_i + 1$	$W_i = \dfrac{(B) \times 100\%}{\Sigma(B)}$	$1/R_i$	$W_i = \dfrac{(D) \times 100\%}{\Sigma(D)}$
A. Net present worth	20	1	5	33	1	44
B. CIMS goals	20	2	4	27	0.5	22
C. Serviceability	20	3	3	20	0.33	14
D. Management/ engineering effort	20	5	1	7	0.20	9
E. Riskiness, lack of	20	4	2	13	0.25	11
	$\Sigma = \overline{100}$		$\Sigma = \overline{15}$	$\Sigma = \overline{100}$	$\Sigma = \overline{2.28}$	$\Sigma = \overline{100}$

[a] Ranks shown with 1 = highest (best), etc.

19.5.4 Weighted Evaluation of Alternatives

Once weights have been assigned to attributes, the next step is to assign numerical values regarding the degree to which each alternative satisfies each attribute. This is generally a difficult judgment task, using an arbitrary scale of, say, between 0 to 10 or 0 to 1,000 to reflect relative evaluations for each alternative and each attribute.

Example 19-1

Suppose that we are comparing two alternatives on the basis of how well they satisfy the five attributes having rank reciprocal weights developed in Table 19-4. The attributes, together with the subjective evaluation of how well a particular alternative meets each on the basis of a scale of 0 to 10, are shown in Table 19-5. [*Note:* These evaluation ratings roughly correspond to the graphical portrayals of alternatives P-1 and P-2 in Fig. 19-1, with a fully shaded circle representing an evaluation rating of 10, a half-shaded circle representing an evaluation rating of 5, etc.]

Once the evaluations have been made, the results can be calculated, as in Table 19-6, to arrive at weighted evaluations of attributes for each alternative. Thus, the summed weighted evaluation is 72.9 for alternative P-1 and 73.1 for alternative P-2, as calculated using the following equation:

$$\text{weighted evaluation} = \sum_i (\text{normalized attribute weight}) \times \text{evaluation rating})/10$$

This indicates alternative P-2 is marginally better, even though it happened to have lower evaluation ratings for three out of five attributes.   ■

TABLE 19-5 Sample Evaluation Ratings of How Well Each Alternative Satisfies Each Attribute

	Alternative	
Attribute	P-1	P-2
A. CIMS goals	7.5	9
B. Net present worth	6	7
C. Serviceability	10	7.5
D. Management/engineering effort	8	6
E. Riskiness, lack of	8	6

TABLE 19-6 Calculation of Weighted Evaluations of Alternatives

Attribute	Normalized attribute weight (from table 19-4)	Alternative P-1		Alternative P-2	
		Evaluation rating	Weighted evaluation	Evaluation rating	Weighted evaluation
A. CIMS goals	22	7.5	16.5	9	19.8
B. Net present worth	44	6	26.4	7	30.8
C. Serviceability	14	10	14.0	7.5	10.5
D. Management/ engineering effort	9	8	7.2	6	5.4
E. Riskiness, lack of	11	8	8.8	6	6.6
			$\Sigma = 72.9$		$\Sigma = 73.1$

19.6 The Analytic Hierarchy Process—Introduction

The analytic hierarchy process (AHP) was developed and documented primarily by Thomas Saaty.[2] Applications of this methodology have been reported in numerous fields, such as transportation planning, portfolio selection, corporate planning, marketing, and others.

The strength of the AHP method lies in its ability to structure a complex, multiperson, multiattribute, and multiperiod problem hierarchically. Pairwise comparisons of the elements (usually, alternatives and attributes) can be established by using a scale indicating the strength with which one element dominates another with respect to a higher-level element. This scaling process can then be translated into priority weights (scores) for the comparison of alternatives.

The use of the AHP to solve a decision problem consists of five stages:

1. construction of a decision hierarchy by breaking down the decision problem into a hierarchy of decision elements and identifying decision alternatives;
2. determination of the relative importance of attributes and subattributes (if any);
3. determination of the relative standing (weight) of each alternative with respect to each next-higher level attribute or subattribute;
4. determination of indicator(s) of consistency in making pairwise comparisons;
5. determination of the overall priority weight (score) of each alternative.

The process of computing the priority vectors and consistency ratios is quite laborious and is best accomplished with a computer program. The AHP methodology has been incorporated in at least two software packages: Expert Choice[3] and Auto-Man.[4] Expert Choice is a generic decision problem software package, whereas Auto-Man (a public-domain package) was designed specifically for the evaluation of manufacturing alternatives. *To demonstrate manually the AHP stages, we will evaluate the automation alternatives P-1, P-2, and P-3 previously considered in Section 19.5.*

19.6.1 Construction of the Decision Hierarchy

The AHP begins by decomposing a complex decision problem into a hierarchy of subproblems. We will concentrate on what Saaty calls "functional" hierarchies as applied to multiattribute decision problems. He uses the term "element" to apply to the overall objective, attributes, subattributes, sub-subattributes, and so on, and alternatives of a problem as follows:

> The top level, called the *focus,* consists of only one element—the broad, overall objective. Subsequent levels may each have several elements, although their number is very

---

[2] Thomas L. Saaty, *The Analytic Hierarchy Process* (New York: McGraw-Hill Book Company, 1980); Thomas L. Saaty, *Decision Making for Leaders* (Belmont, CA: Wadsworth Publishing Company, Inc., 1982).

[3] E. H. Forman, T. L. Saaty, M. A. Selly, and R. Waldron, *Expert Choice* (McLean, VA: Decision Support Software, Inc., 1983).

[4] Stephen F. Weber, *AutoMan: Decision Support Software for Automated Manufacturing Investments* (U.S. Department of Commerce, National Institute of Standards and Technology, NISTIR 89-4166, 1989).

small—between 5 and 9. Because the elements in one level are to be compared with one another against a criterion in the next higher level, the elements in each level must be of the same order of magnitude.[5]

Figure 19-2 shows the standard form of the AHP hierarchy. The focus, or objective, of the decision problem is shown at the top level of the hierarchy. The second level consists of attributes considered important in achieving the overall objective. Subsequent levels are created by dividing attributes into subattributes, subattributes into sub-subattributes, and so on. The alternatives are shown in the bottom level. Figure 19-3 shows a typical four-level hierarchy applied to a career choice problem.

It is important to note that the selected attributes and subattributes should be *independent*. The mathematics of the AHP (as presented in this chapter) are based on the principle of hierarchic composition. This principle states that the elements on a single level of the hierarchy are independent and their relative importance (priority weights) does not depend on the elements at the next lower level of the hierarchy. Consider the Level II attributes shown in Fig. 19-3. According to the principle of hierarchic composition, the attributes *money, job security, family life,* and *work environment* are independent: A change in the value of one of these attributes will not change the value of any other attribute. Furthermore, the importance of these attributes with respect to career choice satisfaction is not affected by the set of alternatives under consideration.

The following guidelines from Arbel and Seidmann should be considered when constructing hierarchies:[6]

1. The number of levels used should be chosen to represent effectively the problem at hand.
2. The order of the levels should reflect a logical causal relationship between adjacent levels.
3. The number of members in a particular level should be chosen to describe the level in adequate detail, but should not cause unnecessary complexity.

Let us apply the AHP method to an adaptation of the problem first introduced in Table 19-2. We will limit our consideration to three alternatives (P-1, P-2, and P-3) for modernizing a manufacturing company. The focus of the decision problem is selecting the "best overall automated system," and the "best" system can be defined by five reasonably independent attributes: A. CIMS goals, B. Net present worth, C. Serviceability, D. Management/engineering effort, and E. Riskiness, lack of. In this example, the attributes capture the required level of detail so that subattributes are not necessary. The resulting three-level hierarchy is shown in Fig. 19-4.

---

[5] Thomas L. Saaty, *Decision Making for Leaders* (Belmont, CA: Wadsworth Publishing Company, Inc., 1982), p. 28.

[6] A. Arbel and A. Seidmann, "Performance Evaluation of Flexible Manufacturing Systems," *IEEE Transactions on Systems, Man, and Cybernetics* SMC-14, no. 4 (July–August. 1984):606–617.

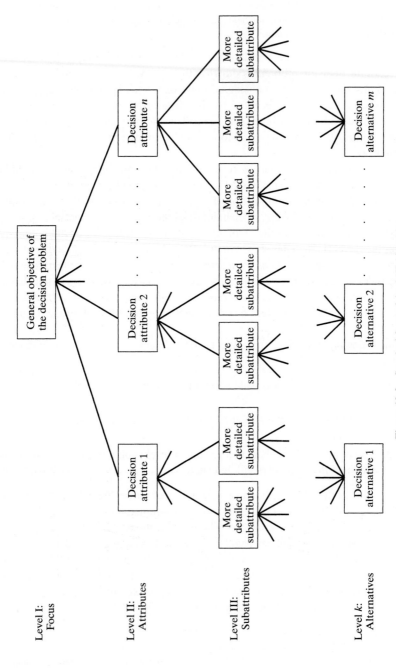

Figure 19-2 Standard form of the AHP hierarchy.

Level I:
Focus

Level II:
Attributes

Level III:
Subattributes

Level k:
Alternatives

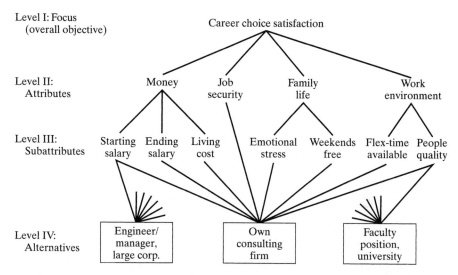

Figure 19-3   Complete hierarchy for sample career choice problem. (*Note:* For clarity, full lines indicate relationships between Level IV, and levels above it are shown only for alternative "Own Consulting Firm." Similar and equal numbers of lines applicable for each of the other two alternatives are shown only partially.) Source: J. R. Canada, et al., "How to Make a Career Choice; the Use of the Analytic Hierarchy Process," *Industrial Management* 27, no. 5 (1985).

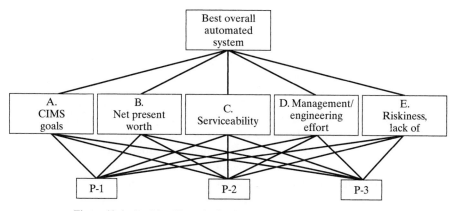

Figure 19-4   Decision hierarchy for the automation alternatives example.

19.6.2 Determining the Importance of Attributes and Subattributes

Once the hierarchy is established, priorities (relative importance weights) must be established for each set of elements at every level of the hierarchy. Priority data are obtained by asking various decision makers to evaluate a set of elements at one hierarchical level in a pairwise fashion regarding their relative importance with respect to an element at the next higher level of the hierarchy. This is accomplished by having decision makers respond to questions of the following form: "With respect to the best overall manufacturing system, how much more important is net present worth than lack of riskiness?" and, at the level of the alternatives, "With respect to serviceability, how strongly do you prefer P-1 to P-2?"

The response to each question takes the form of a value from one to nine and its reciprocal. The magnitude of the response indicates the strength of preference of one decision element to another. Saaty's suggested numbers to express degrees of preference between the two elements x and y are as follows:

If x is … as (than) y,	then the preference number to assign is
Equally important/preferred	1
Weakly more important/preferred	3
Strongly more important/preferred	5
Very strongly more important/preferred	7
Absolutely more important/preferred	9

Even numbers (2, 4, 6, 8) can be used to represent compromises among the preferences above. For inverse comparisons such as y to x, the reciprocal of the preference number for x to y is used. The major assumptions of the AHP comparison procedure are

1. the relative weights of attributes within a level, conditional on each attribute in the immediately preceding level, are unidimensional;
2. pairwise judgments of attributes encompass all relevant aspects of importance;
3. decision makers can evaluate subproblems in an accurate and relatively consistent manner.

The results of the pairwise comparisons are placed in a matrix. Table 19-7 shows a matrix of preference numbers for the pairwise comparisons between attributes considered important to the selection of a manufacturing system. For example, Table 19-7 shows that, with respect to the overall focus, Serviceability (C) is weakly more important than Management/engineering effort (D) and is equally as important as Riskiness, lack of (E), and so on. Similar statements could be made about other rows. Note that the entries in each column are the reciprocals of the entries in the corresponding row, indicating the inverse relation of relative strength when attribute y is compared to attribute x versus x being compared to y. The only combinations that really require thoughtful judgment are those above *or* below the diagonal, because the "mirror-image" counterpart of each number is the reciprocal of that number and the diagonal entries are all 1 (since any given element must surely be equally preferred with that element). Note that the right-hand side of Table 19-7 shows the same paired comparison results, both in decimal form and summed by columns to facilitate later calculations.

TABLE 19-7   Matrix of Paired Comparisons (Including Decimal Equivalents) for Attributes

With respect to the "best overall automated system"	A	B	C	D	E	Decimal equivalents				
						A	B	C	D	E
A. CIMS goals	1	1/3	5	6	5	1	0.333	5	6	5
B. Net present worth	3	1	6	7	6	3	1	6	7	6
C. Serviceability	1/5	1/6	1	3	1	0.20	0.167	1	3	1
D. Management/ engineering effort	1/6	1/7	1/3	1	1/4	0.167	0.143	0.333	1	0.25
E. Riskiness, lack of	1/5	1/6	1	4	1	0.20	0.167	1	4	1
					$\Sigma =$	4.567	1.810	13.333	21	13.25

When the decision maker inputs the preferences on a pairwise basis, it is assumed that he or she does not know the vector of weights that characterizes the relative strength of each element with respect to a specific element at the next higher level in the hierarchy. Thus, after obtaining the pairwise judgments as in Table 19-7, the next step is the computation of a vector of priorities (or weighting of the elements in the matrix). In terms of matrix algebra, this consists of calculating the "principal vector" (eigenvector) of the matrix, and then normalizing it to sum to 1.0 or 100%. Standard programs are available for computing the principal vector of a matrix.

> We next present an approximation method for determining the principal vector, which is thought to provide sufficiently close results for most applications:[7] Divide the elements of each column by the sum of the column (i.e., normalize the column), then add the elements in each resulting row, and divide this sum by the number of elements in the row.

Table 19-8 shows the normalized matrix obtained by dividing each element in Table 19-7 by the sum of its respective column. Finally, row entries in the last two columns of Table 19-8 consist of the sum of the five elements in the row and the average of those row elements (principal vector), respectively. The results (principal vector) are that the attributes have the following approximate priority weights:

A.	CIMS goals	0.288
B.	Net present worth	0.489
C.	Serviceability	0.086
D.	Management/engineering effort	0.041
E.	Riskiness, lack of	0.096
	$\Sigma =$	1.000

[7] The approximation method discussed in this chapter can be in error by as much as 25% when the comparison matrix is inconsistent. A better approximation is to raise the matrix to a larger power until it converges.

TABLE 19-8 Normalized Matrix of Paired Comparisons and Calculations of Priority Weights (Approximate Attribute Weights)

	A	B	C	D	E	Row Σ	Approx. Principal Vector (Average = $\Sigma/5$)
A	0.219	0.184	0.375	0.286	0.377	1.441	0.288
B	0.657	0.553	0.450	0.333	0.453	2.446	0.489
C	0.044	0.092	0.075	0.143	0.076	0.430	0.086
D	0.036	0.079	0.025	0.048	0.019	0.207	0.041
E	0.044	0.092	0.075	0.190	0.074	0.477	0.096
$\Sigma =$	1.000	1.000	1.000	1.000	1.000		1.000

19.6.3 Determining the Relative Standing of Alternatives with Respect to Attributes

The next stage of the AHP is to determine the principal vector (priorities) of each of the alternatives with respect to each of the attributes to which they relate in the next higher level in Fig. 19-4. Typically, this is accomplished using the pairwise comparison process, as was demonstrated in the previous section. However, it is also feasible to use quantified performance data (when available) to compute the priority weights (i.e., principal vector) of alternatives with respect to attributes. We will demonstrate the use of both methods in this section.

19.6.3.1 Using Subjective Judgments to Prioritize Alternatives with Respect to Attributes. We will use the pairwise comparison process to determine the priority weights of alternatives with respect to the following attributes: A. CIMS goals, C. Serviceability, and E. Riskiness, lack of. We illustrate this only with respect to the first attribute, *CIMS goals*. Table 19-9 shows illustrative pairwise comparisons, and Table 19-10 shows subsequent calculations paralleling those in Table 19-8, resulting in the approximate principal vector (priority weights that are often descriptively called "evaluation ratings") with respect to CIMS goals. The results are as follows:

Alternative	Priority weight (principal vector)
P-1	0.21
P-2	0.55
P-3	0.24
	$\Sigma = 1.00$

Table 19-11 summarizes the results of evaluating the alternatives with respect to attributes A, C, and E. Note that the uppermost results are those for which calculations were discussed previously and shown in Tables 19-9 and 19-10. Results for attributes C and E are presented without showing the computational details.

TABLE 19-9 Matrix of Paired Comparison Results for Alternatives (with Respect to CIMS Goals)

	P-1	P-2	P-3	P-1	P-2	P-3
P-1	1	1/3	1	1	0.333	1
P-2	3	1	2	3	1	2
P-3	1	1/2	1	1	0.500	1
				$\Sigma = 5$	1.833	$\frac{1}{4}$

TABLE 19-10 Normalized Matrix and Priority Weights for Alternatives (with Respect to CIMS Goals)

	P-1	P-2	P-3	Σ	Approx. principal vector $(\Sigma/3)$
P-1	0.20	0.18	0.25	0.63	0.21
P-2	0.60	0.55	0.50	1.65	0.55
P-3	0.20	0.27	0.25	0.72	0.24
	$\Sigma = 1.00$	1.00	1.00		1.00

TABLE 19-11 Summary of Paired Comparisons and Resulting Priority Weights for Alternatives with Respect to Selected Attributes

		P-1	P-2	P-3	Priority weights	Consistency index (CI)*	Consistency ratio (C.R.)*
A. CIMS goals	P-1	1	1/3	1	0.21		
	P-2	3	1	2	0.55	0.01	0.02
	P-3	1	1/2	1	0.24		
					$\Sigma = 1.00$		
C. Serviceability	P-1	1	2	2	0.50		
	P-2	1/2	1	1	0.25		
	P-3	1/2	1	1	0.25	0.00	0.00
					$\Sigma = 1.00$		
E. Riskiness,	P-1	1	3	4	0.62		
lack of	P-2	1/3	1	2	0.24		
	P-3	1/4	1/2	1	0.14	0.01	0.02
					$\Sigma = 1.00$		

* The relevance and computation of the consistency index and the consistency ratio will be discussed in Section 19.6.4.

19.6.3.2 Using Performance Data to Prioritize Alternatives with Respect to Attributes.
When performance data are available, it is possible to rate alternatives with respect to attributes by using these data instead of the pairwise comparison process. Consider the attributes B. Net present worth and D. Management/ engineering effort. In this example, the numerical performance of each alternative with

respect to these attributes has been estimated. Table 19-12 summarizes the performance data (these data were originally presented in Table 19-2). For each alternative, the net present worth has been estimated in dollars and management/engineering effort has been estimated in terms of "hours required." Notice that, in the case of Net present worth, "higher" values are better, whereas "lower" values are better for Management/engineering effort.

Let us use the performance data to rate the three alternatives with respect to net present worth. When higher values of a performance measure are "better," a one-step normalization process of the data is utilized. To determine the priority weights for alternatives with respect to net present worth, we simply divide each alternative's net present worth by the sum of the net present worth of all alternatives. Doing so results in the following priority weights for alternatives with respect to positive-valued net present worth:

$$\text{P-1} \quad \$300M/(\$300M + \$350M + \$400M) = 0.29;$$
$$\text{P-2} \quad \$350M/(\$300M + \$350M + \$400M) = 0.33;$$
$$\text{P-3} \quad \$400M/(\$300M + \$350M + \$400M) = \underline{0.38};$$
$$\Sigma = 1.00.$$

Note that this normalization process results in P-3 having the highest weight with respect to net present worth and P-1 having the lowest weight. This is consistent with the fact that P-3 had the highest estimated net present worth while P-1 had the lowest.

We will also use performance data to obtain the vector of weights for the attribute Management/engineering effort. The performance measure used was "hours required," thus a low estimated value is desirable. When a lower performance value is "better," a two-step normalization process is utilized. The first step is to compute the ratio of the best (smallest) performance value to each alternative's performance value. The second step is to normalize these ratios such that they sum to one. This process results in the following alternative priority weights with respect to management/engineering effort:

	Ratio	Normalized
P-1	500/500 = 1	0.46
P-2	500/700 = 0.7143	0.33
P-3	500/1100 = 0.4545	0.21
		$\Sigma = 1.00$

TABLE 19-12 Estimated Alternative Performance Data for Selected Alternatives

		Alternative		
Attribute	Unit of measure	P-1	P-2	P-3
B. Net present worth	Dollars	300M	350M	400M
D. Management/engineering effort	Hours required	500	700	1100

TABLE 19-13 Priority Weights for Alternatives
with Respect to Quantified Attributes

	B. Net present worth	D. Management/ engineering effort
P-1	0.29	0.46
P-2	0.33	0.33
P-3	0.38	0.21
	$\Sigma = 1.00$	1.00

Note that, once again, the rank order of alternative performance is retained in the normalized weights (i.e., P-1 has the most weight while P-3 has the least weight). The priority weights for alternatives with respect to attributes B and D are summarized in Table 19-13.

The use of performance data to obtain priority weights is promoted by the Auto-Man software package.[8] The advantage of this approach is its objectivity (subjective judgments are minimized). However, it is important to note that this method assumes that a linear relationship exists between a performance value and its relative weight (e.g., $50 is twice as good as $25). If a linear relationship cannot be assumed for a given attribute, then a pairwise comparison process using the nine-point scale should be used.

19.6.4 Determining the Consistency of Judgments

One of the strengths of the AHP is its ability to measure the degree of consistency present in the subjective judgments made by the decision maker. Judgmental consistency is concerned with the transitivity of preference in the pairwise comparison matrices. Consider the case in which attribute A is judged to be twice as important as attribute B and attribute B is judged to be twice as important as attribute C. Perfect cardinal consistency would then require that attribute A be judged four times as important as attribute C. Saaty developed a method (the AHP) by which one can measure the magnitude of departure from perfect consistency. The AHP includes both a local measure of consistency for individual comparison matrices and a global measure of consistency for the entire decision problem. We will demonstrate the calculation of the local measure of consistency for the automation alternatives example. The computation of the global measure will be discussed in Section 19.9.

The local consistency ratio (C.R.) is an approximate mathematical indicator, or guide, of the consistency of pairwise comparisons. It is a function of what is called the "maximum eigenvalue" and size of the matrix (called a "consistency index"), which is then compared against similar values if the pairwise comparisons had been merely random (called a "random index"). If the ratio of the consistency index to the random index (called a "consistency ratio") is no greater than 0.1, Saaty suggests the consistency is generally quite acceptable for pragmatic purposes. We will demonstrate the computation of a consistency ratio by using the matrix of comparisons given in Table 19-7.

[8] Stephen F. Weber, *AutoMan: Decision Support Software for Automated Manufacturing Investments* (U.S. Department of Commerce, National Institute of Standards and Technology, NISTIR 89-4166, 1989).

The first step is to multiply the matrix of comparisons, call it matrix $[A]$, by the principal vector or priority weights (right-hand column of Table 19-8) $[B]$ to get a new vector $[C]$:

$$
\begin{array}{ccc}
[A] & [B] & [C] \\
\begin{vmatrix}
1 & 0.33 & 5 & 6 & 5 \\
3 & 1 & 6 & 7 & 6 \\
0.2 & 0.167 & 1 & 3 & 1 \\
0.167 & 0.143 & 0.333 & 1 & 0.25 \\
0.2 & 0.167 & 1 & 4 & 1
\end{vmatrix}
\times
\begin{vmatrix}
0.288 \\
0.489 \\
0.086 \\
0.041 \\
0.096
\end{vmatrix}
=
\begin{vmatrix}
1.607 \\
2.732 \\
0.444 \\
0.212 \\
0.485
\end{vmatrix}
\end{array}
$$

Next, divide each element in vector $[C]$ by its corresponding element in vector $[B]$ to find a new vector $[D]$:

$$
[D] = \begin{vmatrix} \dfrac{1.607}{0.288} & \dfrac{2.732}{0.489} & \dfrac{0.444}{0.086} & \dfrac{0.212}{0.041} & \dfrac{0.485}{0.096} \end{vmatrix}
$$

$$
= \begin{vmatrix} 5.58 & 5.59 & 5.16 & 5.17 & 5.05 \end{vmatrix}.
$$

Now, average the numbers in vector $[D]$. This is an approximation of what is called the "maximum eigenvalue," denoted by λ_{max}:

$$
\lambda_{max} = \frac{5.58 + 5.59 + 5.16 + 5.17 + 5.05}{5} = 5.31.
$$

The consistency index (CI) for a matrix of size N is given by the formula

$$
CI = \frac{\lambda_{max} - N}{N - 1} = \frac{5.31 - 5}{5 - 1} = 0.08.
$$

Random indexes (RI) for various matrix sizes, N, have been approximated by Saaty (based on large numbers of simulation runs) as follows:

N	1	2	3	4	5	6	7	8	9	10	11	...
RI	0.00	0.00	0.58	0.90	1.12	1.24	1.32	1.41	1.45	1.49	1.51	...

For the preceding example, the RI = 1.12. The consistency ratio (C.R.) can now be calculated with the relationship

$$
C.R. = \frac{CI}{RI} = \frac{0.08}{1.12} = 0.07.
$$

On the basis of Saaty's empirical suggestion that a C.R. = 0.10 is acceptable, we would conclude that the foregoing pairwise comparisons to obtain priority vectors are reasonably consistent.

19.6.4.1 Calculation for Alternative Comparisons with Respect to CIMS Goals (Table 19-9)   The calculation of a consistency ratio for alternatives is directly parallel to that for attributes as just given. For the pairwise comparisons in

Table 19-9 (alternatives with respect to CIMS goals), the following calculations are shown without explanation:

$$
\begin{array}{ccc}
[A] & [B] & [C]
\end{array}
$$

$$
\begin{vmatrix} 1 & 0.333 & 1 \\ 3 & 1 & 2 \\ 1 & 0.5 & 1 \end{vmatrix} \times \begin{vmatrix} 0.21 \\ 0.55 \\ 0.24 \end{vmatrix} = \begin{vmatrix} 0.633 \\ 1.660 \\ 0.725 \end{vmatrix};
$$

$$
[D] = \begin{vmatrix} \dfrac{0.633}{0.21} & \dfrac{1.660}{0.55} & \dfrac{0.725}{0.24} \end{vmatrix} = \begin{vmatrix} 3.01 & 3.02 & 3.02 \end{vmatrix};
$$

$$
\lambda_{max} = \frac{3.01 + 3.02 + 3.02}{3} = 3.02;
$$

$$
CI = \frac{3.02 - 3}{3 - 1} = 0.01;
$$

$$
C.R. = \frac{0.01}{0.58} = 0.02 \quad \text{(058 is from the short RI table in the previous section).}
$$

The consistency ratios for the comparison matrices of alternatives with respect to attributes A, C, and E are displayed in Table 19-11. Recall that performance data were used to compute the priority vectors of alternatives with respect to attributes B and D. When performance data are used, the consistency index and the consistency ratio are equal to zero.

19.7 Determining the Overall Priority Weights of Alternatives

Table 19-14 summarizes all priority weights in a form that is convenient for calculation of the final result, which is the vector of overall priority weights of alternatives. The following summarizes the content in "weighted evaluation" terminology (from Section 19.5.4):

- The attribute weights (from Table 19-8) are given in the top row in the body of Table 19-14.
- The evaluation ratings regarding how well each alternative meets each attribute (from Table 19-11 and Table 19-13) are given in the last three rows of the body of Table 19-14.
- The weighted evaluation calculated results for each alternative are given in the right-hand column of Table 19-14.

The weighted evaluation for each alternative can be obtained by multiplying the matrix of evaluation ratings by the vector of priority weights and summing over all attributes. Expressed in conventional mathematical notation, we have

weighted evaluation for alternative k =
$$
\Sigma_{\text{all } i \text{ attributes}} (\textbf{priority weight}_i \times \textbf{evaluation rating}_{ik}).
$$

TABLE 19-14 Summary of Priority Weights Labeled as Attribute Weights, Evaluation Ratings,[a] and Weighted Evaluations

| | Attribute | | | | | Alternative priority weights[b] |
	A: CIMS goals	B: Net present worth	C: Serviceability	D: Management/ engineering effort	E: Riskiness, lack of	
Attribute weights	0.288	0.489	0.086	0.041	0.096	
Alternative						
P-1	0.21	0.29	0.50	0.46	0.62	0.324
P-2	0.55	0.33	0.25	0.33	0.24	0.378
P-3	0.24	0.38	0.25	0.21	0.14	0.298

[a] Evaluation ratings are in body of matrix.
[b] Also called alternative "weighted evaluations" in Section 19.5.4

Thus, the priority weight for alternative P-1 is given by

$$0.288(0.21) + 0.489(0.29) + 0.086(0.5) + 0.041(0.46) + 0.096(0.62) = 0.324,$$

which is shown in the right-hand column of Table 19-14. Thus, alternative P-2 (with a priority weight or weighted evaluation of 0.378) is indicated to be more desirable than either alternative P-1 or P-3 (with priority weights of 0.324 and 0.298, respectively).

19.8 Added Explanation Regarding Example

Another way to show the structure of the automation alternatives sample problem, and the results of all priority weights (from Tables 19-8, 19-11, and 19-13) is given in Fig. 19-5. Using the results as displayed in Fig. 19-5, we can calculate the priority weight (weighted evaluation) for any alternative merely by summing the products of weights for all pathways leading to that alternative. Thus, for alternative P-1, the weighted evaluation would be $0.288(0.21) + 0.489(0.29) + 0.086(0.5) + 0.041(0.46) + 0.096(0.62) = 0.324$.

19.9 The Global Consistency Ratio: C.R.H.

The measurement of consistency can be applied to the entire decision hierarchy. The global consistency ratio of the hierarchy (C.R.H.) is obtained by taking the ratio of an

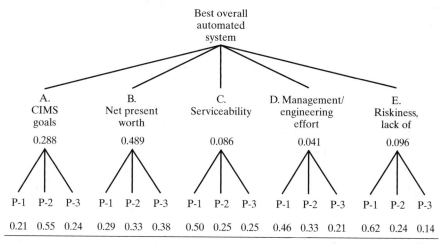

Figure 19-5 Decision hierarchy and priority weight results for the automation alternatives example.

aggregate consistency index M for the entire three-level hierarchy to an aggregate random index, $\overline{M}$. These quantities are computed in the following manner:

$$M = \text{second-level CI} + \begin{vmatrix} \text{vector of} \\ \text{second-level} \\ \text{priority weights} \end{vmatrix} \times \begin{vmatrix} \text{vector of} \\ \text{third-level} \\ \text{CIs} \end{vmatrix} ;$$

$$\overline{M} = \text{second-level RI} + \begin{vmatrix} \text{vector of} \\ \text{second-level} \\ \text{priority weights} \end{vmatrix} \times \begin{vmatrix} \text{vector of} \\ \text{third-level} \\ \text{RIs} \end{vmatrix} .$$

If the ratio of these values (C.R.H.) is no greater than 0.10, then the consistency of the hierarchy is generally acceptable. We will demonstrate the computation of the global consistency ratio for the automation alternatives example.

To compute M, we multiply the priority vector of the second-level attributes by the consistency indices of the third-level comparison matrices and add this result to the consistency index of the second-level comparison matrix. Recall that the consistency index for the attribute comparisons was equal to 0.08 (just before the short RI table). The consistency indices for the subjective comparisons were given in Table 19-11. The consistency index for both attributes B and D (where performance data were used) is equal to zero:

$$M = 0.08 + \begin{vmatrix} 0.288 & 0.489 & 0.086 & 0.041 & 0.096 \end{vmatrix} \times \begin{vmatrix} 0.01 \\ 0.00 \\ 0.00 \\ 0.00 \\ 0.01 \end{vmatrix} = 0.08.$$

To compute $\overline{M}$, we multiply the vector of attribute weights by the vector of RIs corresponding to the size of the alternative comparison matrices and add to the RI corresponding to the size of the attribute comparison matrix:

$$\overline{M} = 1.12 + \begin{vmatrix} 0.288 & 0.489 & 0.086 & 0.041 & 0.096 \end{vmatrix} \times \begin{vmatrix} 0.58 \\ 0.58 \\ 0.58 \\ 0.58 \\ 0.58 \end{vmatrix} = 1.7.$$

The global consistency ratio (C.R.H.) can now be calculated by using the relationship

$$\text{C.R.H.} = M / \overline{M} = \frac{0.08}{1.7} = 0.05.$$

On the basis of Saaty's empirical suggestion that a C.R.H. ≤ 0.10 is acceptable, we would conclude that the foregoing pairwise comparisons to obtain the overall alternative weights are reasonably consistent. This result is expected, because we have already determined that each individual comparison matrix is consistent.

19.9.1 What to Do in the Presence of Inconsistency

If the consistency ratio of an individual matrix or the entire hierarchy is found to be unacceptable, the decision maker should review the judgments made and look for intransitivity. Often, the process of assigning ratings on the nine-point scale is facilitated by, first, rank ordering the elements in the matrix from most important to least important. Rank ordering the elements in the matrix also highlights intransitivities (ratings in the upper diagonal of the matrix should then be greater than or equal to 1 and increase in value across the rows of the matrix). If the source of inconsistency is not apparent or cannot be resolved satisfactorily through the reestimation of preferences, it is possible that there is a problem with the hierarchical formulation of the problem. It may be that the attributes or alternatives being compared are not independent or that they are not directly comparable on a one-to-nine scale.

19.10 Hierarchies with More than Three Levels

In complex decision problems, it is likely that the decision hierarchy will consist of more than three levels. The purpose of this section is to illustrate the AHP computations for a four-level hierarchy. We will once again consider the three automation alternatives: P-1, P-2, and P-3. Figure 19-6 shows the decision elements for this problem, arranged in a four-level hierarchy. The fourth level was created by associating three subattributes with the second-level attribute *CIMS goals.* In this example, the attribute CIMS goals can be fully described by the subattributes Improve product quality, Reduce inventory, and Improve manufacturing flexibility.

19.10.1 Determining the Relative Weights of Attributes and Alternatives

The basic steps of the AHP methodology remain the same regardless of the number of hierarchical levels. Pairwise comparisons must be made (or performance data used) to determine the relative weights of decision elements at one level of the hierarchy with respect to each decision element at the next-higher level of the hierarchy. In the previous three-level example, we first determined the weights of the attributes with respect to the focus and then determined the relative weights of the alternatives with respect to each attribute. The same general procedure applies to the four-level hierarchy, *except* we must perform the intermediate determination of subattribute weights with respect to the higher-level attribute CIMS goals.

The comparison of attributes with respect to the overall goal remains unchanged from the previous example. The results of this process are repeated in Table 19-15. The next step is to determine the relative weights of the third-level subattributes with respect to CIMS goals. The pairwise comparison matrix and resulting priority vector

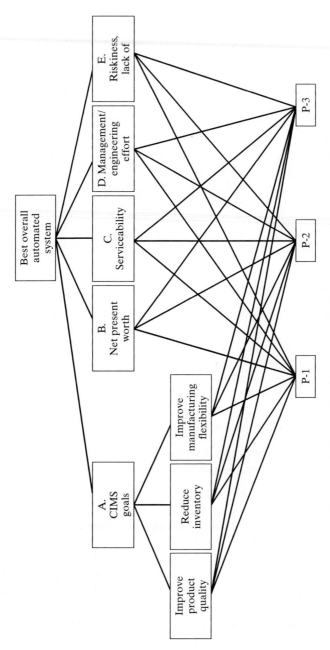

Figure 19-6 Four-level decision hierarchy for the automation alternatives example.

TABLE 19-15   Matrix of Paired Comparisons and
Priority Weights for Attributes

With respect to the "best overall automated system"	A	B	C	D	E	Priority weight	Consistency
A. CIMS goals	1	1/3	5	6	5	0.288	
B. Net present worth	3	1	6	7	6	0.489	CI = 0.08
C. Serviceability	1/5	1/6	1	3	1	0.086	
D. Management/	1/6	1/7	1/3	1	1/4	0.041	C.R. = 0.07
engineering effort	1/5	1/6	1	4	1	0.096	
E. Riskiness, lack of						Σ = 1.000	

TABLE 19-16   Matrix of Paired Comparisons and
Priority Weights for SubAttributes

With respect to "CIMS goals"	A1	A2	A3	Priority weight	Consistency
A1. Improve product quality	1	2	5	0.59	CI = 0.003
A2. Reduce inventory	1/2	1	2	0.28	
A3. Improve manufacturing flexibility	1/5	1/2	1	0.13	C.R. = 0.005
				Σ = 1.00	

TABLE 19-17   Priority Weights of Alternatives with
Respect to Selected Attributes

	Attribute			
	B	C	D	E
P-1	0.29	0.50	0.46	0.62
P-2	0.33	0.25	0.33	0.24
P-3	0.38	0.25	0.21	0.14
CI	0.00	0.00	0.00	0.01
C.R.	0.00	0.00	0.00	0.02

of weights are shown in Table 19-16. The subattribute Improve product quality has been judged to be the most important of the three goals.

The last step is to rate the alternatives with respect to the higher-level decision elements. The comparison matrices with respect to attributes C and E were previously shown in Table 19-11. Performance data were used to rate the alternatives with respect to attributes B and D, and the results were given in Table 19-13. The vectors of alternative weights with respect to these attributes are summarized in Table 19-17. The relative weights of alternatives with respect to CIMS goals are indirectly determined by determining the weights of the alternatives with respect to each of the three subattributes that together define CIMS goals. Table 19-18 shows the resulting comparison matrices and priority weights.

TABLE 19-18 Summary of Paired Comparisons and Resulting Priority Weights for Alternatives with Respect to Subattributes

		P-1	P-2	P-3	Priority weights	Consistency index (CI)	Consistency ratio (C.R.)
A1. Improve	P-1	1	1/2	1	0.26		
product quality	P-2	2	1	1/9	0.21	0.550	0.948
	P-3	1	9	1	0.53		
					$\Sigma = 1.00$		
A2. Reduce	P-1	1	1/2	1	0.24		
inventory	P-2	2	1	3	0.55	0.009	0.016
	P-3	1	1/3	1	0.21		
					$\Sigma = 1.00$		
A3. Improve	P-1	1	1/8	1/3	0.08		
manufacturing	P-2	8	1	3	0.68	0.001	0.002
flexibility	P-3	3	1/3	1	0.24		
					$\Sigma = 1.00$		

Before these weights are used to compute the overall priority of alternatives, we need to check the consistency of the judgments made. As we will see in the next section, the weights shown in Table 19-18 are *not* the weights we will use in the final analysis.

19.10.2 Determining the Consistency of the Comparisons

The consistency ratios of individual comparison matrices are computed as described in Section 19.6.4 and are displayed next to each comparison matrix in Table 19-18. Note that the consistency ratio for the comparison of alternatives with respect to the subattribute Improve product quality was calculated to be 0.948. This value exceeds the empirical upper limit of 0.10. Thus, we would conclude that the judgments made were inconsistent. Before completing the AHP analysis, it is important to resolve any detected inconsistencies.

To resolve the inconsistency problem, let us look more closely at the comparison matrix with respect to Improve product quality (shown in Table 19-18). In the first row and second column of the comparison matrix, the value 1/2 indicates that alternative P-2 is twice as preferred as P-1. The 1 in the third column indicates that P-1 is equally preferred to P-3 with respect to improving product quality. On the basis of these judgments, it would be reasonable to conclude that alternative P-2 should also be twice as preferred as P-3. However, looking at the comparison matrix, P-3 has been judged to be nine times more preferred than P-2. This is the source of the inconsistency detected by the consistency ratio. To resolve the problem, we must decide if P-2 is twice as preferred as P-1 (and therefore P-3) or if P-3 (and therefore P-1) is nine times as preferred as P-2.

For this example, we will assume that P-2 is twice as preferred as both P-1 and P-3. The revised comparison matrix and resulting priority weights are given in

TABLE 19-19 Revised Paired Comparisons and Resulting Priority
Weights for Alternatives with Respect to Subattributes

		P-1	P-2	P-3	Priority weights	Consistency index (CI)	Consistency ratio (C.R.)
A1. Improve	P-1	1	1/2	1	0.25		
product quality	P-2	2	1	2	0.50	0.000	0.000
	P-3	1	1/2	1	0.25		
					$\Sigma = 1.00$		
A2. Reduce	P-1	1	1/2	1	0.24		
inventory	P-2	2	1	3	0.55	0.009	0.016
	P-3	1	1/3	1	0.21		
					$\Sigma = 1.00$		
A3. Improve	P-1	1	1/8	1/3	0.08		
manufacturing	P-2	8	1	3	0.68	0.001	0.002
flexibility	P-3	3	1/3	1	0.24		
					$\Sigma = 1.00$		

Table 19-19. The impact of not resolving the inconsistency problem on the overall
ranking of alternatives will be discussed in the next section.

19.10.3 Determining the Overall Weights of Alternatives

Figure 19-7 summarizes the results of the analysis in a format analogous to
Table 19-14. Using the results as displayed in this format, the overall priority weight
of an alternative is computed by summing the product of weights for all branches that
include the alternative. Thus, for alternative P-1, the priority weight would be

$$P\text{-}1 \text{ weight} = 0.25(0.59)(0.288) + 0.24(0.28)(0.288) + 0.08(0.13)(0.288)$$
$$+ 0.29(0.489) + 0.50(0.086) + 0.46(0.041) + 0.62(0.096)$$
$$= 0.328.$$

The overall priority weights for the three alternatives are given in Fig. 19-7. The
results indicate that alternative P-2 is more desirable than either P-1 or P-3. Note that
similar results were obtained when we used a three-level hierarchy.

Different results would have been obtained if the inconsistency discussed in
Section 19.10.2 had been left unresolved. If we had used the priority weights shown
in Table 19-18 (instead of the revised weights in Table 19-19), the overall weights for
the alternatives would have been

$$P\text{-}1 = 0.330,$$
$$P\text{-}2 = 0.325, \text{ and}$$
$$P\text{-}3 = 0.345.$$

These results erroneously indicate that alternative P-3 is slightly more desirable than
either P-1 or P-2.

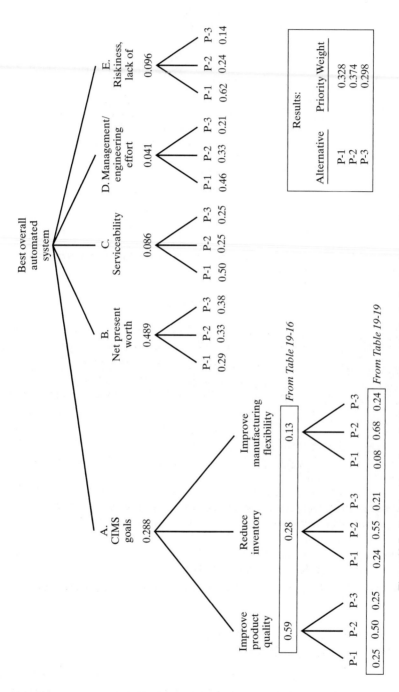

Figure 19-7 Decision hierarchy and priority weight results for the four-level automation alternatives example.

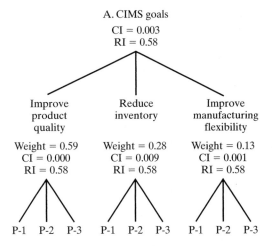

Figure 19-8 Three-level subset of decision hierarchy for calculation of an aggregate consistency index for CIMS goals.

19.10.4 The Global Consistency Measure

The method described in Section 19.9 for computing the global consistency measure was for use with a three-level hierarchy. However, we can generalize this procedure and apply it to a four-level (or more) hierarchy as described next.

The first step of this procedure is to isolate the portion of the hierarchy related to the second-level attribute CIMS goals (see Fig. 19-8). This attribute, its subattributes, and the alternatives make up a three-level hierarchy. We can apply the procedure described in Section 19.9 to compute an aggregate consistency index M for the CIMS goals attribute. Doing so yields $M = 0.006$. This value can now be used in conjunction with the CIs of the other second-level attributes to obtain the C.R.H. for the entire decision problem (see Fig. 19-9).

Now that we have a consistency index associated with each second-level attribute, we can again use the procedure described in Section 19.9 to compute the C.R.H. for the entire hierarchy. The results are

$$M = 0.083, \qquad \overline{M} = 1.7, \text{ and}$$
$$\text{C.R.H.} = M / \overline{M} = 0.049.$$

On the basis of the C.R.H. value, we conclude that our overall consistency is acceptable.

19.11 Problems of Which to Be Aware When Applying the AHP

Though widely used to solve decision problems, the AHP is not without its critics. A number of theoretical complaints have surfaced in the literature, especially when the AHP is applied to capital investment decision problems. The intent of this

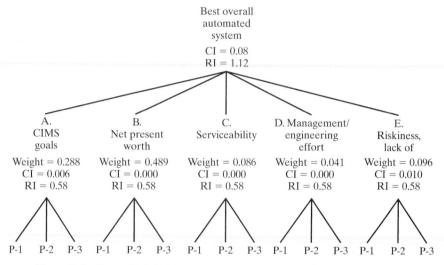

Figure 19-9   Composite three-level hierarchy for computation of the global consistency measure (C.R.H.).

section is to make the reader aware of these issues when applying the AHP to a decision problem.

19.11.1 Vagueness of the Questioning Procedure

One of the major criticisms of the AHP has to do with the manner in which the attribute weights are elicited and assessed. The decision maker is asked questions such as, "Which is more important, good gas mileage or low maintenance, and by how much?" The crucial observation is that such questions are meaningless. A clear interpretation of the question would require knowing what value or range of gas mileage is being compared with what level or range of maintenance.

The conclusion that has been drawn is that the decision maker must be thinking of some average quantities when making such judgments. Otherwise, it would not be reasonable to make such a judgment. Boucher and MacStravic[9] demonstrate that, unless the average quantity interpretation is assumed, the AHP cannot be relied on to give results that are consistent with a simple monetary interpretation of the priority weights. In their paper, the implied dollar values of the alternatives are compared, and it is shown that the ranking given by the AHP is not consistent with the ranking obtained by using the dollar values. Thus, it is important that the decision maker be aware of what is being compared when making judgments of relative importance.

---

[9] Thomas O. Boucher and Elin L. MacStravic, "Multiattribute Evaluation Within a Present Worth Framework and Its Relation to the Analytic Hierarchy Process," *The Engineering Economist* 37, no. 1 (1991):1–32.

19.11.2 Rank Reversal

Another well-documented problem with the AHP is the problem of *rank reversal*.[10] Rank reversal is when the introduction of a new alternative reverses the rankings of previously evaluated alternatives. This phenomenon is attributed to the fact that attribute weights are assessed independently of the alternatives under consideration. Thus, there is no relationship between the weights of the alternatives with respect to attributes and the overall attribute weights. If the anomaly of rank reversal occurs, it is because the principle of hierarchic composition (Section 19.6.1) is being violated. To avoid the problem of rank reversal, it is important that great care be exercised when constructing the decision hierarchy. Not only should the decision attributes be independent of each other, but the attributes should be independent of the alternatives being considered.

19.11.3 Aggregating Benefits and Costs

In many applications of the AHP, two hierarchies are used to evaluate alternatives, a benefits hierarchy and a cost hierarchy. The use of two hierarchies results in two measures associated with each alternative, a benefits priority vector and a cost priority vector. To aggregate these measures, Saaty[11] recommends computing the benefit–cost ratio for each alternative. The "best" alternative is then identified as the one having the highest benefit–cost ratio.

This procedure has been criticized by Bernhard and Canada[12] for failing to yield optimal results for the general case. A correct analysis needs to consider incremental benefits and costs. In addition, we need to consider the relative preference of benefits versus costs. Bernhard and Canada state that an analysis not including such information can lead a decision maker to the wrong conclusion, and they illustrate this point with a numerical example.

A second problem to be aware of is the general arbitrariness inherent in any labeling procedure with regard to whether a negative feature is called a cost or a "disbenefit," or whether a positive feature is labeled a benefit or "negative cost." Bernhard and Canada show that the results obtained using Saaty's benefit–cost ratio method are sensitive to how the benefits and cost hierarchies are defined. However, this problem may be alleviated through the use of an incremental benefit–cost analysis.

19.12 Other AHP-Based Tools for Multicriteria Decision Problems

The success of the AHP has prompted researchers to develop new multicriteria decision-making methodologies that incorporate some of the features of the AHP. The use of a

---

[10] Valerie Belton and Tony Gear, "On a Short-Coming of Saaty's Method of Analytic Hierarchies," *Omega* 11, no. 3 (1983):228–230; James S. Dyer, "Remarks on the Analytic Hierarchy Process," *Management Science* 36, no. 3 (1990):249–258; S. R. Watson and A. N. S. Freeling, "Assessing Attribute Weights," *Omega* 10, no. 6 (1982):582–583.

[11] Thomas L. Saaty, *The Analytic Hierarchy Process* (New York: McGraw-Hill Book Company, 1980).

[12] Richard H. Bernhard and John R. Canada, "Some Problems in Using Benefit/Cost Ratios with the Analytic Hierarchy Process," *The Engineering Economist* 36, no. 1 (1990):56–65.

hierarchy to structure the decision problem and a pairwise comparison process to weight decision elements has appeared in newer methodologies. What makes these methodologies unique are their interpretation of the weights and how the weights are used to select the "best" alternative.

Putrus[13] developed a model specifically for the justification of CIM systems. A specific hierarchy was developed that identifies the important attributes of CIM systems, and pairwise comparisons are used to rate the elements of the hierarchy. These weights are then aggregated to measure total benefits and total risk. The decision to invest in CIM is based on the ratio of total benefits to total risk.

MacStravic and Boucher[14] developed a multicriteria capital investment justification technique that incorporates the finer features of the AHP, but overcomes many of its perceived deficiencies. A hierarchical framework was developed for manufacturing investment decisions. A pairwise comparison process is used to weight decision elements. Unlike the AHP, the weights are then converted into dollar amounts. This allows the "best" alternative to be identified via traditional economic analysis techniques (e.g., present worth).

19.13 Summary

This chapter first described several basic (even simplistic) methodologies for describing and analyzing alternatives for general solutions in which multiple attributes are important to the decision. Then it concentrated on the Analytic Hierarchy Process (AHP) as a formal methodology for structuring even complex multi-period or multi-person problems hierarchically and using a scaling and computational process to determine priority weights (scores) for the comparison of alternatives.

There are numerous other techniques, often very complex or restrictive, which are beyond the scope of this book. The determination of what methodology(ies) to use depends on the nature of the decision problem and the preferences of the decision maker(s). While this determination is a very inexact science, it can be shown that the use of widely differing models (subjective probability distributions, if applicable) will often have less effect on the probable quality of the solution than does the unintended deletion of possible outcomes, alternatives, or important attributes. Good use of multiattribute methodologies should make the decision maker more confident and ultimately make his or her job easier.

PROBLEMS

19-1. Select a class or type of significant decision problem in your work or personal life. Name the three or more significant attributes to be considered for that class of problem. Assume that the weighting of these attributes is nebulous, so you think it desirable to develop only a matrix of alternatives versus attributes, with entries to show how well each alternative

---

[13] Robert Putrus, "Non-Traditional Approach in Justifying Computer Integrated Manufacturing Systems," *AUTOFACT '89* (Oct. 30–Nov. 2, 1989), Detroit, MI.

[14] Elin L. MacStravic and Thomas O. Boucher, "Users Manual: NCIC Decision Support Software for Investment in Advanced Manufacturing Technology," Rutgers University Center for Advanced Food Technology (Aug. 1991).

meets each attribute in whatever measures are appropriate (such as in dollars, time, rank, and so on). Then use colors or other symbol codings to facilitate ease of understanding of the differences by the persons making the final selection among the alternatives.

19-2. Select a class or type of decision problem in your work or personal life involving multiple objectives (not necessarily the same as for Problem 19-1). Name the three to five most important objectives for that class of problem, as you see them, and do the following:

    a. Using the method of paired comparisons, rank the objectives in order of decreasing importance.

    b. Weight them and check for consistency and then normalize the weight to sum to 100.

    c. Identify two to four alternatives for the decision problem and evaluate how well each alternative meets each objective on a scale of 0 to 10. Then multiply the evaluation ratings by the objective weights and sum these for each objective to compute a weighted evaluation for each alternative.

19-3. a. Weight the relative importance of the three or four most important attributes you would consider in selecting a job. Assume that these are the only attributes you will quantitatively consider. Show how you make comparisons for internal consistency and normalize the factor weights to sum to 100.

    b. Using the attribute weights developed in (a), obtain a weighted evaluation of two alternative jobs in which you might be interested by evaluating how well each job satisfies each attribute, using a scale from 0 to 10.

19-4. a. Weight the relative importance of the three to six most important attributes you would consider in selecting a personal car. Assume that these are the only attributes you will quantitatively consider. Show how you make comparisons for internal consistency and normalize the attribute weights to sum to 1,000.

    b. Using the attribute weights developed in (a), obtain a weighted evaluation of three alternative makes of cars you might consider for purchase by evaluating how well each make of car satisfies each attribute. Use a scale from 0 to 20.

19-5. Describe a significant multiattribute decision problem of existing or potential meaningful interest to you. Examples are housing, car, job, travel, and a computer. Identify at least four most relevant attributes that are as independent of each other as possible and at least two or three mutually exclusive alternatives. (*Note:* In the process of further analysis, you might well add or delete alternatives and/or attributes.)

    a. Describe the problem as meaningfully as possible by using two or more graphical techniques. What conclusion (choices), if any, can you make from these?

    b. Rank order the attributes. Can you combine this with one or more of the graphical techniques in part (a) to make choices? If so, describe.

    c. Weight the attributes by the method or formula(s) of your choice and show at least two checks and/or revisions of weights for consistency between initial judgments and preferences.

    d. For each attribute, show a graph of attribute outcome (x-axis) versus evaluation rating (y-axis).

    e. Using parts (c) and (d), show calculations of the weighted evaluations for each alternative and display the results graphically for ease of comparison.

19-6. Two highway alignments have been proposed for access to a new manufacturing plant. On the basis of the data and information given in the associated table, comparison can be made of the two alignments. The better one must be chosen as the proposed highway alignment that will be the connector between an interstate highway and the proposed site. The route length for alignment A is 4.7 miles and for alignment B, 5.1 miles. The monetary costs are as follows:

Item	Alignment A	Alignment B
Land	$ 4,044,662	$ 4,390,000
Bridges	10,134,000	8,701,000
Pavement	4,112,500	4,462,500
Grade drainage	7,050,000	7,650,000
Erosion control	470,000	510,000
Clearing and grading	188,000	204,000
Total	$25,999,162	$25,917,500

Nonmonetary criteria are as follows:

	Alignment A	Alignment B
Maintenance	Moderate	High
Noise pollution	Very good	Good
Cost savings (on gas)	Excellent	Poor
Accessibility to another major roadway	U.S. Highway 41	None
Impact on wildlife	Little	Little
Relocation of residences	2	3
Road condition	Flat	Hilly

Develop and quantify your own weighted evaluation model to determine which highway alignment you would recommend. Show all work.

19-7. Use the weighted evaluation model to make your own selection of one of the three used automobiles for which some data are given in the associated table. State your assumptions regarding miles driven each year, life of the automobile (how long *you* would keep it), market (resale) value at the end of life, interest cost, price of fuel, cost of annual maintenance, and attribute weights and other subjectively based determinations.

	Alternative		
Attribute	Domestic 1	Domestic 2	Foreign
Price	$8,400	$10,000	$9,300
Gas mileage	25 mpg	30 mpg	35 mpg
Type of fuel	Gasoline	Gasoline	Diesel
Comfort	Very good	Excellent	Excellent
Aesthetic appeal	5 out of 10	7 out of 10	9 out of 10
Passengers	4	6	4
Ease of servicing	Excellent	Very good	Good
Performance on road	Fair	Very good	Very good
Stereo system	Poor	Good	Excellent
Ease of cleaning upholstery	Excellent	Very good	Poor
Storage space	Very good	Excellent	Poor

19-8. Utilize the weighted evaluation technique to analyze the following important situation:

	Alternative (firm)		
Attribute	A	B	C
Starting salary	$40,000	$33,500	$38,000
Opportunity for advancement	Excellent	Excellent	Fair
Management attitude	Very good	Excellent	Good
Location	Good	Fair	Poor
Type of work	Excellent	Poor	Very good
Opportunity for continuing education	Very good	Good	Fair

You need to determine which job to take, given acceptable offers from firms A, B, and C. You first determine a set of attributes that are most important in the evaluation and then construct a decision matrix, with alternatives as columns and attributes as rows, and fill it in. (Results are given in the associated matrix.) Analyze the alternatives with the technique according to your perceptions. (Be complete—it took years to get to this point!)

19-9. The table shows evaluation ratings (on a scale of 0 to 20) of how well each of two alternatives satisfies each of four attributes. Show which alternative would be best, using

	Alternative	
Attribute	1	2
A	15	12
B	8	20
C	14	14
D	13	12

the weighted evaluation methodology and each of the following attribute weighting formulas:

a. Uniform.

b. Rank sum, with $A > B > C > D$.

c. Rank reciprocal, with $A > B > C > D$.

(*Note:* Normalize the summed weighted evaluations to equal 100)

19-10. You have just been transferred! Your company has offered two alternatives for a new position with them, and you decide to stay with your present employer:

	Alternative	
Attribute	Dry Gulch	Rapid Creek
Salary	$40,000	$35,000
Proximity to relatives	2,000 miles	500 miles
Promotion potential	Fair	Good
Commuting time (per day)	40 minutes	90 minutes

    a. Determine your weighted evaluation for each alternative by using your own perceptions of relative importances and desirabilities.

    b. Use AHP to determine priority weights by using your own perceptions.

19-11. Consider the following data:

| | AMS Alternative | | | | |
Attribute	A	B	C	Do nothing	Minimum acceptable
P: Purchase price	$115,000	$339,400	$32,000	$0[a]	(Open)
D: Reduced design time	60%	67%[a]	50%	0	25%
L: Impact on manufacturing lead time	Excellent[a]	Excellent	Good	Poor	Good
F: Flexibility	Excellent[a]	Very good	Good	Poor	Good
Q: Quality improvement	Very good	Excellent[a]	Fair	Poor	Good
M: Market share	Up 2%[a]	Up 2%	Up 1%	Down 2%	(Open)

[a] Best outcome.

    If $M > Q > F > P > D > L$, which alternative would you recommend using the weighted evaluation technique? Carefully develop a rationale by which nonmonetary attributes are rated in your analysis.

19-12. Three alternatives are under consideration to improve certain manufacturing operations. The following attributes are considered important to the selection of the best alternative: Reduce product lead time, Improve product quality, and Maximize net present worth.

    Given the following matrix of comparisons, compute the priority weights for attributes with respect to the focus, using AHP methodology:

	A	B	C
A: Reduce product lead time	1	1/5	1/9
B: Improve product quality	5	1	1/2
C: Maximize net present worth	9	2	1

19-13. The AHP methodology presented in this chapter for computing priority weights is an approximation. The priority vector shown in the table was computed by using a better approximation method (the matrix was raised to successively larger powers until it converged). Use the methodology discussed in this chapter to approximate the priority vector. Comment on any differences.

	A	B	C	D	E	Priority weights
A	1	7	8	7	9	0.6143
B	1/7	1	7	7	7	0.2402
C	1/8	1/7	1	2	3	0.0657
D	1/7	1/7	1/2	1	2	0.0476
E	1/9	1/7	1/3	1/2	1	0.0322

19-14. Compute the consistency ratio for the following matrix of comparisons [comment on the source(s) of inconsistency (if any)]:

	A	B	C	D	E	Priority weights
A	1	7	8	7	9	0.6143
B	1/7	1	7	7	7	0.2402
C	1/8	1/7	1	2	3	0.0657
D	1/7	1/7	1/2	1	2	0.0476
E	1/9	1/7	1/3	1/2	1	0.0322

19-15. An engineer expresses his preferences among major attributes with respect to his focus objective of career satisfaction as follows:

	(M)	(W)	(F)
Money (M)	1	2	1/3
Work (W)	1/2	1	1
Family (F)	3	1	1

His evaluation of each of job types A, B, and C are expressed as follows with respect to M, W, and F:

(M)	A	B	C
A	1	4	7
B	1/4	1	2
C	1/7	1/2	1

(W)	A	B	C
A	1	1/2	1
B	2	1	1/3
C	1	3	1

(F)	A	B	C
A	1	2	5
B	1/2	1	2
C	1/5	1/2	1

Using AHP, find the priority weights for all matrices, and for the job types.

19-16. Select a personal decision problem to which you can relate fairly readily (such as alternative jobs, cars, housing, travel, etc.) and for which you have not previously calculated attribute weights, using the AHP methodology. Identify two or three alternatives that might be compared (for reference). Then identify the four most important attributes that are, in your mind, reasonably distinct and independent. Rank order the alternatives, placing the most important first, and so on.

   a. Weight the attributes and normalize the total to 100 points, as exemplified by the *traditional* weighted evaluation method. Do a few checks for consistency until you are satisfied that those weights reasonably reflect your approximate weights. Do not alter on the basis of any later thinking or results.

   b. Weight the attributes by use of the AHP method of pairwise comparisons on the basis of Saaty's ratio preference/importance factors of 1 to 9. Calculate your C.R. (consistency ratio). If it is below 0.10, great—otherwise, go back and reask yourself some or all pairwise comparison questions and recalculate your weights until the C.R. ≤ 0.10.

   c. Repeat part (b), except use Canada's ratio preference/importance factors of 1 to 3. The following reflects scales for parts (b) and (c):

If x(row) is …	Saaty's factors		Canada's factors	
as y(column)	No.	1/No.	No.	1/No.
Equally important/preferred	1	1	1	1
Weakly more important/preferred	3	1/3 = 0.33	1.5	1/1.5 = 0.67
Strongly more important/preferred	5	1/5 = 0.20	2	1/2 = 0.50
Very strongly more important/preferred	7	1/7 = 0.14	2.5	1/2.5 = 0.40
Absolutely more important/preferred	9	1/9 = 0.11	3	1/3 = 0.33

   d. What can you conclude about weighting results, using Saaty's versus Canada's factors regarding closeness to "traditional" [in part (a)]?

19-17. Refer to Table 19-18 and confirm that the consistency ratio for Improve product quality is correct (0.948). Explain in your own words where the inconsistency lies in the matrix.

19-18. Research Question. Investigate the properties of these four types of measurement scales: nominal, ordinal, interval, and ratio. Which measurement scale is implied by the AHP? What can you say about the results of the AHP if these scale properties are not present in the data being used in the analysis?

19-19. NOTE TO INSTRUCTORS: The Apex Inc. case study is presented below. Divide the class into teams of two to three students each and have each team provide a recommendation to the plant manager concerning the proposed investment by Apex in automation. There is no single "correct" answer to the Apex situation, so be creative in preparing a recommendation to the plant manager. Try to balance your quantitative analysis with various intangible factors in play at Apex.

   APEX Inc. is a large company in the Midwestern United States. While the particular product manufactured by APEX is classified, their yearly sales total is estimated at approximately 1 billion dollars. APEX is currently trying to make some improvements to its receiving department's inspection methods. The receiving manager, Joe Smith, is a bit reluctant to change and is openly opposed to any new system for his department. Currently, the receiving department at APEX gets 1 million dollars of material per day. It takes APEX 11 days to completely process this material. Processing the material includes inspection and acceptance or rejection on the basis of this inspection. It can be assumed that the company is using a combination of visual and metrological means of inspecting the incoming material. It takes a combined amount of 11 days to complete both of these tasks. This inefficiency is causing high levels of inventory and uneven flow, resulting in bottlenecks.

   The efforts to reduce the overall inspection time are centered on techniques such as Six Sigma, Just in Time, bar coding, and computer aided inspection, including hard and soft technology. A mechanical engineer named Linda Brown, whose brother is a major stockholder in APEX, has estimated the improvements that the new system will yield. She has calculated that, by implementing the new system, APEX can complete inspection in 7 days versus the current 11 days. Linda claims that the reduced inspection time translates into a one-time reduction of $4 million ($11 million – $7 million = $4 million) in tied up capital. The inventory carrying cost has been determined to be 18% per year. Charles Smith, head of accounting, therefore puts the savings at $720,000 per year. Charles, who is the brother of the receiving manager, Joe Smith, performs his calculation as follows: 4,000,000(18%) = $720,000 per year. The accounting department states the following: "If the investment is worthwhile (a "go" situation), we'll write a mortgage wherein annual payments are made to repay the investment through budget reductions equal to the savings of $720,000". Accounting then computes the amount that can be justified for the investment to be

Initial Investment: $720,000 $(P/A, 25\%, 30 \text{ years}) = \$ 2,876,000.$

Accounting does not indicate where the 25% hurdle rate came from or where the 30-year study period came from. Perhaps the project was considered to be a conventional 30-year mortgage situation. Because the cost of modernizing the receiving area is estimated to be $3 million, the proposal was rejected.

Linda Brown and Charles Smith do not agree on the estimated savings associated with implementing the proposed system. Linda puts the savings at a one-time inventory reduction of $4 million while accounting has computed the savings at $2,876,000, using a mortgage. Linda's method justifies the implementation, since $4 million in savings is greater than the $3 million investment. Also, Linda is not considering the savings from fewer bottlenecks, more floor space, and greater inventory control associated with the modernization of the receiving area, which would further justify the $3 million investment.

What recommendation would you make to management? Be careful to state your assumptions and present any calculations you may make.

APPENDIX A

Tables of Discrete Compounding Interest Factors (for Various Common Values of i from 1/2% to 25%)

TABLE A-1 Discrete Compounding; $i = 1/2\%$

	Single payment		Uniform series				
	Compound amount factor	Present worth factor	Compound amount factor	Present worth factor	Sinking fund factor	Capital recovery factor	
N	To find F given P F/P	To find P given F P/F	To find F given A F/A	To find P given A P/A	To find A given F A/F	To find A given P A/P	N
1	1.0050	0.9950	1.0000	0.9950	1.0000	1.0050	1
2	1.0100	0.9901	2.0050	1.9851	0.4988	0.5038	2
3	1.0151	0.9851	3.0150	2.9702	0.3317	0.3367	3
4	1.0202	0.9802	4.0301	3.9505	0.2481	0.2531	4
5	1.0253	0.9754	5.0502	4.9259	0.1980	0.2030	5
6	1.0304	0.9705	6.0755	5.8964	0.1646	0.1696	6
7	1.0355	0.9657	7.1059	6.8621	0.1407	0.1457	7
8	1.0407	0.9609	8.1414	7.8229	0.1228	0.1278	8
9	1.0459	0.9561	9.1821	8.7790	0.1089	0.1139	9
10	1.0511	0.9513	10.2280	9.7304	0.0978	0.1028	10
11	1.0564	0.9466	11.2791	10.6770	0.0887	0.0937	11
12	1.0617	0.9419	12.3355	11.6189	0.0811	0.0861	12
13	1.0670	0.9372	13.3972	12.5561	0.0746	0.0796	13
14	1.0723	0.9326	14.4642	13.4887	0.0691	0.0741	14
15	1.0777	0.9279	15.5365	14.4166	0.0644	0.0694	15
16	1.0831	0.9233	16.6142	15.3399	0.0602	0.0652	16
17	1.0885	0.9187	17.6973	16.2586	0.0565	0.0615	17
18	1.0939	0.9141	18.7857	17.1727	0.0532	0.0582	18
19	1.0994	0.9096	19.8797	18.0823	0.0503	0.0553	19
20	1.1049	0.9051	20.9791	18.9874	0.0477	0.0527	20
21	1.1104	0.9006	22.0839	19.8879	0.0453	0.0503	21
22	1.1160	0.8961	23.1944	20.7840	0.0431	0.0481	22
23	1.1216	0.8916	24.3103	21.6756	0.0411	0.0461	23
24	1.1272	0.8872	25.4319	22.5628	0.0393	0.0443	24
25	1.1328	0.8828	26.5590	23.4456	0.0377	0.0427	25
26	1.1385	0.8784	27.6918	24.3240	0.0361	0.0411	26
27	1.1442	0.8740	28.8303	25.1980	0.0347	0.0397	27
28	1.1499	0.8697	29.9744	26.0676	0.0334	0.0384	28
29	1.1556	0.8653	31.1243	26.9330	0.0321	0.0371	29
30	1.1614	0.8610	32.2799	27.7940	0.0310	0.0360	30
35	1.1907	0.8398	38.1453	32.0353	0.0262	0.0312	35
40	1.2208	0.8191	44.1587	36.1721	0.0226	0.0276	40
45	1.2516	0.7990	50.3240	40.2071	0.0199	0.0249	45
50	1.2832	0.7793	56.6450	44.1427	0.0177	0.0227	50
55	1.3156	0.7601	63.1256	47.9813	0.0158	0.0208	55
60	1.3488	0.7414	69.7698	51.7254	0.0143	0.0193	60
65	1.3829	0.7231	76.5818	55.3773	0.0131	0.0181	65
70	1.4178	0.7053	83.5658	58.9393	0.0120	0.0170	70
75	1.4536	0.6879	90.7262	62.4135	0.0110	0.0160	75
80	1.4903	0.6710	98.0674	65.8022	0.0102	0.0152	80
85	1.5280	0.6545	105.594	69.107	0.0095	0.0145	85
90	1.5666	0.6383	113.311	72.331	0.0088	0.0138	90
95	1.6061	0.6226	121.22	75.475	0.0082	0.0132	95
100	1.6467	0.6073	129.33	78.542	0.0077	0.0127	100
∞				200.0		0.0050	∞

	Single payment		Uniform series				
	Compound amount factor	Present worth factor	Compound amount factor	Present worth factor	Sinking fund factor	Capital recovery factor	
N	To find F given P F/P	To find P given F P/F	To find F given A F/A	To find P given A P/A	To find A given F A/F	To find A given P A/P	N
1	1.0100	0.9901	1.0000	0.9901	1.0000	1.0100	1
2	1.0201	0.9803	2.0100	1.9704	0.4975	0.5075	2
3	1.0303	0.9706	3.0301	2.9410	0.3300	0.3400	3
4	1.0406	0.9610	4.0604	3.9020	0.2463	0.2563	4
5	1.0510	0.9515	5.1010	4.8534	0.1960	0.2060	5
6	1.0615	0.9420	6.1520	5.7955	0.1625	0.1725	6
7	1.0721	0.9327	7.2135	6.7282	0.1386	0.1486	7
8	1.0829	0.9235	8.2857	7.6517	0.1207	0.1307	8
9	1.0937	0.9143	9.3685	8.5660	0.1067	0.1167	9
10	1.1046	0.9053	10.4622	9.4713	0.0956	0.1056	10
11	1.1157	0.8963	11.5668	10.3676	0.0865	0.0965	11
12	1.1268	0.8874	12.6825	11.2551	0.0788	0.0888	12
13	1.1381	0.8787	13.8093	12.1337	0.0724	0.0824	13
14	1.1495	0.8700	14.9474	13.0037	0.0669	0.0769	14
15	1.1610	0.8613	16.0969	13.8650	0.0621	0.0721	15
16	1.1726	0.8528	17.2578	14.7178	0.0579	0.0679	16
17	1.1843	0.8444	18.4304	15.5622	0.0543	0.0643	17
18	1.1961	0.8360	19.6147	16.3982	0.0510	0.0610	18
19	1.2081	0.8277	20.8109	17.2260	0.0481	0.0581	19
20	1.2202	0.8195	22.0190	18.0455	0.0454	0.0554	20
21	1.2324	0.8114	23.2391	18.8570	0.0430	0.0530	21
22	1.2447	0.8034	24.4715	19.6603	0.0409	0.0509	22
23	1.2572	0.7954	25.7162	20.4558	0.0389	0.0489	23
24	1.2697	0.7876	26.9734	21.2434	0.0371	0.0471	24
25	1.2824	0.7798	28.2431	22.0231	0.0354	0.0454	25
26	1.2953	0.7720	29.5256	22.7952	0.0339	0.0439	26
27	1.3082	0.7644	30.8208	23.5596	0.0324	0.0424	27
28	1.3213	0.7568	32.1290	24.3164	0.0311	0.0411	28
29	1.3345	0.7493	33.4503	25.0657	0.0299	0.0399	29
30	1.3478	0.7419	34.7848	25.8077	0.0287	0.0387	30
35	1.4166	0.7059	41.6602	29.4085	0.0240	0.0340	35
40	1.4889	0.6717	48.8863	32.8346	0.0205	0.0305	40
45	1.5648	0.6391	56.4809	36.0945	0.0177	0.0277	45
50	1.6446	0.6080	64.4630	39.1961	0.0155	0.0255	50
55	1.7285	0.5785	72.8523	42.1471	0.0137	0.0237	55
60	1.8167	0.5505	81.6695	44.9550	0.0122	0.0222	60
65	1.9094	0.5237	90.9364	47.6265	0.0110	0.0210	65
70	2.0068	0.4983	100.676	50.1684	0.0099	0.0199	70
75	2.1091	0.4741	110.912	52.5870	0.0090	0.0190	75
80	2.2167	0.4511	121.671	54.8881	0.0082	0.0182	80
85	2.3298	0.4292	132.979	57.0776	0.0075	0.0175	85
90	2.4486	0.4084	144.86	59.161	0.0069	0.0169	90
95	2.5735	0.3886	157.35	61.143	0.0064	0.0164	95
100	2.7048	0.3697	170.48	63.029	0.0059	0.0159	100
∞				100.000		0.0100	∞

TABLE A-3 Discrete Compounding; $i = 3\%$

	Single payment		Uniform series				
	Compound amount factor	Present worth factor	Compound amount factor	Present worth factor	Sinking fund factor	Capital recovery factor	
N	To find F given P F/P	To find P given F P/F	To find F given A F/A	To find P given A P/A	To find A given F A/F	To find A given P A/P	N
1	1.0300	0.9709	1.0000	0.9709	1.0000	1.0300	1
2	1.0609	0.9426	2.0300	1.9135	0.4926	0.5226	2
3	1.0927	0.9151	3.0909	2.8286	0.3235	0.3535	3
4	1.1255	0.8885	4.1836	3.7171	0.2390	0.2690	4
5	1.1593	0.8626	5.3091	4.5797	0.1884	0.2184	5
6	1.1941	0.8375	6.4684	5.4172	0.1546	0.1846	6
7	1.2299	0.8131	7.6625	6.2303	0.1305	0.1605	7
8	1.2668	0.7894	8.8923	7.0197	0.1125	0.1425	8
9	1.3048	0.7664	10.1591	7.7861	0.0984	0.1284	9
10	1.3439	0.7441	11.4639	8.5302	0.0872	0.1172	10
11	1.3842	0.7224	12.8078	9.2526	0.0781	0.1081	11
12	1.4258	0.7014	14.1920	9.9540	0.0705	0.1005	12
13	1.4685	0.6810	15.6178	10.6349	0.0640	0.0940	13
14	1.5126	0.6611	17.0863	11.2961	0.0585	0.0885	14
15	1.5580	0.6419	18.5989	11.9379	0.0538	0.0838	15
16	1.6047	0.6232	20.1569	12.5611	0.0496	0.0796	16
17	1.6528	0.6050	21.7616	13.1661	0.0460	0.0760	17
18	1.7024	0.5874	23.4144	13.7535	0.0427	0.0727	18
19	1.7535	0.5703	25.1168	14.3238	0.0398	0.0698	19
20	1.8061	0.5537	26.8703	14.8775	0.0372	0.0672	20
21	1.8603	0.5375	28.6765	15.4150	0.0349	0.0649	21
22	1.9161	0.5219	30.5367	15.9369	0.0327	0.0627	22
23	1.9736	0.5067	32.4528	16.4436	0.0308	0.0608	23
24	2.0328	0.4919	34.4264	16.9355	0.0290	0.0590	24
25	2.0938	0.4776	36.4592	17.4131	0.0274	0.0574	25
26	2.1566	0.4637	38.5530	17.8768	0.0259	0.0559	26
27	2.2213	0.4502	40.7096	18.3270	0.0246	0.0546	27
28	2.2879	0.4371	42.9309	18.7641	0.0233	0.0533	28
29	2.3566	0.4243	45.2188	19.1884	0.0221	0.0521	29
30	2.4273	0.4120	47.5754	19.6004	0.0210	0.0510	30
35	2.8139	0.3554	60.4620	21.4872	0.0165	0.0465	35
40	3.2620	0.3066	75.4012	23.1148	0.0133	0.0433	40
45	3.7816	0.2644	92.7197	24.5187	0.0108	0.0408	45
50	4.3839	0.2281	112.797	25.7298	0.0089	0.0389	50
55	5.0821	0.1968	136.071	26.7744	0.0073	0.0373	55
60	5.8916	0.1697	163.053	27.6756	0.0061	0.0361	60
65	6.8300	0.1464	194.332	28.4529	0.0051	0.0351	65
70	7.9178	0.1263	230.594	29.1234	0.0043	0.0343	70
75	9.1789	0.1089	272.630	29.7018	0.0037	0.0337	75
80	10.6409	0.0940	321.362	30.2008	0.0031	0.0331	80
85	12.3357	0.0811	377.856	30.6311	0.0026	0.0326	85
90	14.3004	0.0699	443.35	31.0024	0.0023	0.0323	90
95	16.5781	0.0603	519.27	31.3227	0.0019	0.0319	95
100	19.2186	0.0520	607.29	31.5989	0.0016	0.0316	100
∞				33.3333		0.0300	∞

	Single payment		Uniform series				
	Compound amount factor	Present worth factor	Compound amount factor	Present worth factor	Sinking fund factor	Capital recovery factor	
N	To find F given P F/P	To find P given F P/F	To find F given A F/A	To find P given A P/A	To find A given F A/F	To find A given P A/P	N
1	1.0500	0.9524	1.0000	0.9524	1.0000	1.0500	1
2	1.1025	0.9070	2.0500	1.8594	0.4878	0.5378	2
3	1.1576	0.8638	3.1525	2.7232	0.3172	0.3672	3
4	1.2155	0.8227	4.3101	3.5460	0.2320	0.2820	4
5	1.2763	0.7835	5.5256	4.3295	0.1810	0.2310	5
6	1.3401	0.7462	6.8019	5.0757	0.1470	0.1970	6
7	1.4071	0.7107	8.1420	5.7864	0.1228	0.1728	7
8	1.4775	0.6768	9.5491	6.4632	0.1047	0.1547	8
9	1.5513	0.6446	11.0266	7.1078	0.0907	0.1407	9
10	1.6289	0.6139	12.5779	7.7217	0.0795	0.1295	10
11	1.7103	0.5847	14.2068	8.3064	0.0704	0.1204	11
12	1.7959	0.5568	15.9171	8.8633	0.0628	0.1128	12
13	1.8856	0.5303	17.7130	9.3936	0.0565	0.1065	13
14	1.9799	0.5051	19.5986	9.8986	0.0510	0.1010	14
15	2.0789	0.4810	21.5786	10.3797	0.0463	0.0963	15
16	2.1829	0.4581	23.6575	10.8378	0.0423	0.0923	16
17	2.2920	0.4363	25.8404	11.2741	0.0387	0.0887	17
18	2.4066	0.4155	28.1324	11.6896	0.0355	0.0855	18
19	2.5269	0.3957	30.5390	12.0853	0.0327	0.0827	19
20	2.6533	0.3769	33.0659	12.4622	0.0302	0.0802	20
21	2.7860	0.3589	35.7192	12.8212	0.0280	0.0780	21
22	2.9253	0.3418	38.5052	13.1630	0.0260	0.0760	22
23	3.0715	0.3256	41.4305	13.4886	0.0241	0.0741	23
24	3.2251	0.3101	44.5020	13.7986	0.0225	0.0725	24
25	3.3864	0.2953	47.7271	14.0939	0.0210	0.0710	25
26	3.5557	0.2812	51.1134	14.3752	0.0196	0.0696	26
27	3.7335	0.2678	54.6691	14.6430	0.0183	0.0683	27
28	3.9201	0.2551	58.4026	14.8981	0.0171	0.0671	28
29	4.1161	0.2429	62.3227	15.1411	0.0160	0.0660	29
30	4.3219	0.2314	66.4388	15.3725	0.0151	0.0651	30
35	5.5160	0.1813	90.3203	16.3742	0.0111	0.0611	35
40	7.0400	0.1420	120.800	17.1591	0.0083	0.0583	40
45	8.9850	0.1113	159.700	17.7741	0.0063	0.0563	45
50	11.4674	0.0872	209.348	18.2559	0.0048	0.0548	50
55	14.6356	0.0683	272.713	18.6335	0.0037	0.0537	55
60	18.6792	0.0535	353.584	18.9293	0.0028	0.0528	60
65	23.8399	0.0419	456.798	19.1611	0.0022	0.0522	65
70	30.4264	0.0329	588.528	19.3427	0.0017	0.0517	70
75	38.8327	0.0258	756.653	19.4850	0.0013	0.0513	75
80	49.5614	0.0202	971.228	19.5965	0.0010	0.0510	80
85	63.2543	0.0158	1245.09	19.6838	0.0008	0.0508	85
90	80.7303	0.0124	1594.61	19.7523	0.0006	0.0506	90
95	103.035	0.0097	2040.69	19.8059	0.0005	0.0505	95
100	131.501	0.0076	2610.02	19.8479	0.0004	0.0504	100
∞				20.0000		0.0500	∞

TABLE A-5 Discrete Compounding; $i = 6\%$

	Single payment		Uniform series				
	Compound amount factor	Present worth factor	Compound amount factor	Present worth factor	Sinking fund factor	Capital recovery factor	
	To find F given P	To find P given F	To find F given A	To find P given A	To find A given F	To find A given P	
N	F/P	P/F	F/A	P/A	A/F	A/P	N
1	1.0600	0.9434	1.0000	0.9434	1.0000	1.0600	1
2	1.1236	0.8900	2.0600	1.8334	0.4854	0.5454	2
3	1.1910	0.8396	3.1836	2.6730	0.3141	0.3741	3
4	1.2625	0.7921	4.3746	3.4651	0.2286	0.2886	4
5	1.3382	0.7473	5.6371	4.2124	0.1774	0.2374	5
6	1.4185	0.7050	6.9753	4.9173	0.1434	0.2034	6
7	1.5036	0.6651	8.3938	5.5824	0.1191	0.1791	7
8	1.5938	0.6274	9.8975	6.2098	0.1010	0.1610	8
9	1.6895	0.5919	11.4913	6.8017	0.0870	0.1470	9
10	1.7908	0.5584	13.1808	7.3601	0.0759	0.1359	10
11	1.8983	0.5268	14.9716	7.8869	0.0668	0.1268	11
12	2.0122	0.4970	16.8699	8.3838	0.0593	0.1193	12
13	2.1329	0.4688	18.8821	8.8527	0.0530	0.1130	13
14	2.2609	0.4423	21.0151	9.2950	0.0476	0.1076	14
15	2.3966	0.4173	23.2760	9.7122	0.0430	0.1030	15
16	2.5404	0.3936	25.6725	10.1059	0.0390	0.0990	16
17	2.6928	0.3714	28.2129	10.4773	0.0354	0.0954	17
18	2.8543	0.3503	30.9056	10.8276	0.0324	0.0924	18
19	3.0256	0.3305	33.7600	11.1581	0.0296	0.0896	19
20	3.2071	0.3118	36.7856	11.4699	0.0272	0.0872	20
21	3.3996	0.2942	39.9927	11.7641	0.0250	0.0850	21
22	3.6035	0.2775	43.3923	12.0416	0.0230	0.0830	22
23	3.8197	0.2618	46.9958	12.3034	0.0213	0.0813	23
24	4.0489	0.2470	50.8155	12.5504	0.0197	0.0797	24
25	4.2919	0.2330	54.8645	12.7834	0.0182	0.0782	25
26	4.5494	0.2198	59.1563	13.0032	0.0169	0.0769	26
27	4.8223	0.2074	63.7057	13.2105	0.0157	0.0757	27
28	5.1117	0.1956	68.5281	13.4062	0.0146	0.0746	28
29	5.4184	0.1846	73.6397	13.5907	0.0136	0.0736	29
30	5.7435	0.1741	79.0581	13.7648	0.0126	0.0726	30
35	7.6861	0.1301	111.435	14.4982	0.0090	0.0690	35
40	10.2857	0.0972	154.762	15.0463	0.0065	0.0665	40
45	13.7646	0.0727	212.743	15.4558	0.0047	0.0647	45
50	18.4201	0.0543	290.336	15.7619	0.0034	0.0634	50
55	24.6503	0.0406	394.172	15.9905	0.0025	0.0625	55
60	32.9876	0.0303	533.128	16.1614	0.0019	0.0619	60
65	44.1449	0.0227	719.082	16.2891	0.0014	0.0614	65
70	59.0758	0.0169	967.931	16.3845	0.0010	0.0610	70
75	79.0568	0.0126	1300.95	16.4558	0.0008	0.0608	75
80	105.796	0.0095	1746.60	16.5091	0.0006	0.0606	80
85	141.579	0.0071	2342.98	16.5489	0.0004	0.0604	85
90	189.464	0.0053	3141.07	16.5787	0.0003	0.0603	90
95	253.546	0.0039	4209.10	16.6009	0.0002	0.0602	95
100	339.301	0.0029	5638.36	16.6175	0.0002	0.0602	100
∞				16.6667		0.0600	∞

TABLE A-6 Discrete Compounding; $i = 8\%$

	Single payment		Uniform series				
	Compound amount factor	Present worth factor	Compound amount factor	Present worth factor	Sinking fund factor	Capital recovery factor	
N	To find F given P F/P	To find P given F P/F	To find F given A F/A	To find P given A P/A	To find A given F A/F	To find A given P A/P	N
1	1.0800	0.9259	1.0000	0.9259	1.0000	1.0800	1
2	1.1664	0.8573	2.0800	1.7833	0.4808	0.5608	2
3	1.2597	0.7938	3.2464	2.5771	0.3080	0.3880	3
4	1.3605	0.7350	4.5061	3.3121	0.2219	0.3019	4
5	1.4693	0.6806	5.8666	3.9927	0.1705	0.2505	5
6	1.5869	0.6302	7.3359	4.6229	0.1363	0.2163	6
7	1.7138	0.5835	8.9228	5.2064	0.1121	0.1921	7
8	1.8509	0.5403	10.6366	5.7466	0.0940	0.1740	8
9	1.9990	0.5002	12.4876	6.2469	0.0801	0.1601	9
10	2.1589	0.4632	14.4866	6.7101	0.0690	0.1490	10
11	2.3316	0.4289	16.6455	7.1390	0.0601	0.1401	11
12	2.5182	0.3971	18.9771	7.5361	0.0527	0.1327	12
13	2.7196	0.3677	21.4953	7.9038	0.0465	0.1265	13
14	2.9372	0.3405	24.2149	8.2442	0.0413	0.1213	14
15	3.1722	0.3152	27.1521	8.5595	0.0368	0.1168	15
16	3.4259	0.2919	30.3243	8.8514	0.0330	0.1130	16
17	3.7000	0.2703	33.7502	9.1216	0.0296	0.1096	17
18	3.9960	0.2502	37.4502	9.3719	0.0267	0.1067	18
19	4.3157	0.2317	41.4463	9.6036	0.0241	0.1041	19
20	4.6610	0.2145	45.7620	9.8181	0.0219	0.1019	20
21	5.0338	0.1987	50.4229	10.0168	0.0198	0.0998	21
22	5.4365	0.1839	55.4567	10.2007	0.0180	0.0980	22
23	5.8715	0.1703	60.8933	10.3711	0.0164	0.0964	23
24	6.3412	0.1577	66.7647	10.5288	0.0150	0.0950	24
25	6.8485	0.1460	73.1059	10.6748	0.0137	0.0937	25
26	7.3964	0.1352	79.9544	10.8100	0.0125	0.0925	26
27	7.9881	0.1252	87.3507	10.9352	0.0114	0.0914	27
28	8.6271	0.1159	95.3388	11.0511	0.0105	0.0905	28
29	9.3173	0.1073	103.966	11.1584	0.0096	0.0896	29
30	10.0627	0.0994	113.283	11.2578	0.0088	0.0888	30
35	14.7853	0.0676	172.317	11.6546	0.0058	0.0858	35
40	21.7245	0.0460	259.056	11.9246	0.0039	0.0839	40
45	31.9204	0.0313	386.506	12.1084	0.0026	0.0826	45
50	46.9016	0.0213	573.770	12.2335	0.0017	0.0817	50
55	68.9138	0.0145	848.923	12.3186	0.0012	0.0812	55
60	101.257	0.0099	1253.21	12.3766	0.0008	0.0808	60
65	148.780	0.0067	1847.25	12.4160	0.0005	0.0805	65
70	218.606	0.0046	2720.08	12.4428	0.0004	0.0804	70
75	321.204	0.0031	4002.55	12.4611	0.0002	0.0802	75
80	471.955	0.0021	5886.93	12.4735	0.0002	0.0802	80
85	693.456	0.0014	8655.71	12.4820	0.0001	0.0801	85
90	1018.92	0.0010	12723.9	12.4877	[a]	0.0801	90
95	1497.12	0.0007	18071.5	12.4917	[a]	0.0801	95
100	2199.76	0.0005	27484.5	12.4943	[a]	0.0800	100
∞				12.5000		0.0800	∞

[a] Less than 0.0001.

TABLE A-7 Discrete Compounding; i = 9%

	Single payment		Uniform series				
	Compound amount factor	Present worth factor	Compound amount factor	Present worth factor	Sinking fund factor	Capital recovery factor	
	To find F given P	To find P given F	To find F given A	To find P given A	To find A given F	To find A given P	
N	F/P	P/F	F/A	P/A	A/F	A/P	N
1	1.0900	0.9174	1.000	0.917	1.00000	1.09000	1
2	1.1881	0.8417	2.090	1.759	0.47847	1.56847	2
3	1.2950	0.7722	3.278	2.531	0.30505	0.39505	3
4	1.4116	0.7084	4.573	3.240	0.21867	0.30867	4
5	1.5386	0.6499	5.985	3.890	0.16709	0.25709	5
6	1.6771	0.5963	7.523	4.486	0.13292	0.22292	6
7	1.8280	0.5470	9.200	5.033	0.10869	0.19869	7
8	1.9926	0.5019	11.028	5.535	0.09067	0.18067	8
9	2.1719	0.4604	13.021	5.995	0.07680	0.16680	9
10	2.3674	0.4224	15.193	6.418	0.06582	0.15582	10
11	2.5804	0.3875	17.560	6.805	0.05695	0.14695	11
12	2.8127	0.3555	20.141	7.161	0.04965	0.13965	12
13	3.0658	0.3262	22.953	7.487	0.04357	0.13357	13
14	3.3417	0.2992	26.019	7.786	0.03843	0.12843	14
15	3.6425	0.2745	29.361	8.061	0.03406	0.12406	15
16	3.9703	0.2519	33.003	8.313	0.03030	0.12030	16
17	4.3276	0.2311	36.974	8.544	0.02705	0.11705	17
18	4.7171	0.2120	41.301	8.756	0.02421	0.11421	18
19	5.1417	0.1945	46.018	8.950	0.02173	0.11173	19
20	5.6044	0.1784	51.160	9.129	0.01955	0.10955	20
21	6.1088	0.1637	56.765	9.292	0.01762	0.10762	21
22	6.6586	0.1502	62.873	9.442	0.01590	0.10590	22
23	7.2579	0.1378	69.532	9.580	0.01438	0.10438	23
24	7.9111	0.1264	76.790	9.707	0.01302	0.10302	24
25	8.6231	0.1160	84.701	9.823	0.01181	0.10181	25
26	9.3992	0.1064	93.324	9.929	0.01072	0.10072	26
27	10.2451	0.0976	102.723	10.027	0.00973	0.09973	27
28	11.1671	0.0895	112.968	10.116	0.00885	0.09885	28
29	12.1722	0.0822	124.135	10.198	0.00806	0.09806	29
30	13.2677	0.0753	136.308	10.274	0.00734	0.09734	30
35	20.4140	0.0490	215.711	10.567	0.00464	0.09464	35
40	31.4094	0.0318	337.882	10.757	0.00296	0.09296	40
45	48.3273	0.0207	525.859	10.881	0.00190	0.09190	45
50	74.3575	0.0134	815.084	10.962	0.00123	0.09123	50
55	114.4083	0.0087	1260.092	11.014	0.00079	0.09079	55
60	176.0313	0.0057	1944.792	11.048	0.00051	0.09051	60
65	270.8460	0.0037	2998.288	11.070	0.00033	0.09033	65
70	416.7301	0.0024	4619.223	11.084	0.00022	0.09022	70
75	641.1909	0.0016	7113.232	11.094	0.00014	0.09014	75
80	986.5517	0.0010	10950.574	11.100	0.00009	0.09009	80
85	1517.9320	0.0007	16854.800	11.104	0.00006	0.09006	85
90	2235.5266	0.0004	25939.184	11.106	0.00004	0.09004	90
95	3593.4971	0.0003	39916.635	11.108	0.00003	0.09003	95
100	5529.0408	0.0002	61422.675	11.109	0.00002	0.09002	100

TABLE A-8 Discrete Compounding; $i = 10\%$

	Single payment		Uniform series				
	Compound amount factor	Present worth factor	Compound amount factor	Present worth factor	Sinking fund factor	Capital recovery factor	
N	To find F given P F/P	To find P given F P/F	To find F given A F/A	To find P given A P/A	To find A given F A/F	To find A given P A/P	N
1	1.1000	0.9091	1.0000	0.9091	1.0000	1.1000	1
2	1.2100	0.8264	2.1000	1.7355	0.4762	0.5762	2
3	1.3310	0.7513	3.3100	2.4869	0.3021	0.4021	3
4	1.4641	0.6830	4.6410	3.1699	0.2155	0.3155	4
5	1.6105	0.6209	6.1051	3.7908	0.1638	0.2638	5
6	1.7716	0.5645	7.7156	4.3553	0.1296	0.2296	6
7	1.9487	0.5132	9.4872	4.8684	0.1054	0.2054	7
8	2.1436	0.4665	11.4359	5.3349	0.0874	0.1874	8
9	2.3579	0.4241	13.5795	5.7590	0.0736	0.1736	9
10	2.5937	0.3855	15.9374	6.1446	0.0627	0.1627	10
11	2.8531	0.3505	18.5312	6.4951	0.0540	0.1540	11
12	3.1384	0.3186	21.3843	6.8137	0.0468	0.1468	12
13	3.4523	0.2897	24.5227	7.1034	0.0408	0.1408	13
14	3.7975	0.2633	27.9750	7.3667	0.0357	0.1357	14
15	4.1772	0.2394	31.7725	7.6061	0.0315	0.1315	15
16	4.5950	0.2176	35.9497	7.8237	0.0278	0.1278	16
17	5.0545	0.1978	40.5447	8.0216	0.0247	0.1247	17
18	5.5599	0.1799	45.5992	8.2014	0.0219	0.1219	18
19	6.1159	0.1635	51.1591	8.3649	0.0195	0.1195	19
20	6.7275	0.1486	57.2750	8.5136	0.0175	0.1175	20
21	7.4002	0.1351	64.0025	8.6487	0.0156	0.1156	21
22	8.1403	0.1228	71.4027	8.7715	0.0140	0.1140	22
23	8.9543	0.1117	79.5430	8.8832	0.0126	0.1126	23
24	9.8497	0.1015	88.4973	8.9847	0.0113	0.1113	24
25	10.8347	0.0923	98.3470	9.0770	0.0102	0.1102	25
26	11.9182	0.0839	109.182	9.1609	0.0092	0.1092	26
27	13.1100	0.0763	121.100	9.2372	0.0083	0.1083	27
28	14.4210	0.0693	134.210	9.3066	0.0075	0.1075	28
29	15.8631	0.0630	148.631	9.3696	0.0067	0.1067	29
30	17.4494	0.0573	164.494	9.4269	0.0061	0.1061	30
35	28.1024	0.0356	271.024	9.6442	0.0037	0.1037	35
40	45.2592	0.0221	442.592	9.7791	0.0023	0.1023	40
45	72.8904	0.0137	718.905	9.8628	0.0014	0.1014	45
50	117.391	0.0085	1163.91	9.9148	0.0009	0.1009	50
55	189.059	0.0053	1880.59	9.9471	0.0005	0.1005	55
60	304.481	0.0033	3034.81	9.9672	0.0003	0.1003	60
65	490.370	0.0020	4893.71	9.9796	0.0002	0.1002	65
70	789.746	0.0013	7887.47	9.9873	0.0001	0.1001	70
75	1271.89	0.0008	12708.9	9.9921	a	0.1001	75
80	2048.40	0.0005	20474.0	9.9951	a	0.1000	80
85	3298.97	0.0003	32979.7	9.9970	a	0.1000	85
90	5313.02	0.0002	53120.2	9.9981	a	0.1000	90
95	8556.67	0.0001	85556.7	9.9988	a	0.1000	95
100	13780.6	a	137796	9.9993	a	0.1000	100
∞				10.0000		0.1000	∞

a Less than 0.0001.

TABLE A-9 Discrete Compounding; $i = 12\%$

	Single payment		Uniform series				
	Compound amount factor	Present worth factor	Compound amount factor	Present worth factor	Sinking fund factor	Capital recovery factor	
	To find F given P	To find P given F	To find F given A	To find P given A	To find A given F	To find A given P	
N	F/P	P/F	F/A	P/A	A/F	A/P	N
1	1.1200	0.8929	1.0000	0.8929	1.0000	1.1200	1
2	1.2544	0.7972	2.1200	1.6901	0.4717	0.5917	2
3	1.4049	0.7118	3.3744	2.4018	0.2963	0.4163	3
4	1.5735	0.6355	4.7793	3.0373	0.2092	0.3292	4
5	1.7623	0.5674	6.3528	3.6048	0.1574	0.2774	5
6	1.9738	0.5066	8.1152	4.1114	0.1232	0.2432	6
7	2.2107	0.4523	10.0890	4.5638	0.0991	0.2191	7
8	2.4760	0.4039	12.2997	4.9676	0.0813	0.2013	8
9	2.7731	0.3606	14.7757	5.3282	0.0677	0.1877	9
10	3.1058	0.3220	17.5487	5.6502	0.0570	0.1770	10
11	3.4785	0.2875	20.6546	5.9377	0.0484	0.1684	11
12	3.8960	0.2567	24.1331	6.1944	0.0414	0.1614	12
13	4.3635	0.2292	28.0291	6.4235	0.0357	0.1557	13
14	4.8871	0.2046	32.3926	6.6282	0.0309	0.1509	14
15	5.4736	0.1827	37.2797	6.8109	0.0268	0.1468	15
16	6.1304	0.1631	42.7533	6.9740	0.0234	0.1434	16
17	6.8660	0.1456	48.8837	7.1196	0.0205	0.1405	17
18	7.6900	0.1300	55.7497	7.2497	0.0179	0.1379	18
19	8.6128	0.1161	63.4397	7.3658	0.0158	0.1358	19
20	9.6463	0.1037	72.0524	7.4694	0.0139	0.1339	20
21	10.8038	0.0926	81.6987	7.5620	0.0122	0.1322	21
22	12.1003	0.0826	92.5026	7.6446	0.0108	0.1308	22
23	13.5523	0.0738	104.603	7.7184	0.0096	0.1296	23
24	15.1786	0.0659	118.155	7.7843	0.0085	0.1285	24
25	17.0001	0.0588	133.334	7.8431	0.0075	0.1275	25
26	19.0401	0.0525	150.334	7.8957	0.0067	0.1267	26
27	21.3249	0.0469	169.374	7.9426	0.0059	0.1259	27
28	23.8839	0.0419	190.699	7.9844	0.0052	0.1252	28
29	26.7499	0.0374	214.583	8.0218	0.0047	0.1247	29
30	29.9599	0.0334	241.333	8.0552	0.0041	0.1241	30
35	52.7996	0.0189	431.663	8.1755	0.0023	0.1223	35
40	93.0509	0.0107	767.091	8.2438	0.0013	0.1213	40
45	163.988	0.0061	1358.23	8.2825	0.0007	0.1207	45
50	289.002	0.0035	2400.02	8.3045	0.0004	0.1204	50
55	509.320	0.0020	4236.00	8.3170	0.0002	0.1202	55
60	897.596	0.0011	7471.63	8.3240	0.0001	0.1201	60
65	1581.87	0.0006	13173.9	8.3281	[a]	0.1201	65
70	2787.80	0.0004	23223.3	8.3303	[a]	0.1200	70
75	4913.05	0.0002	40933.8	8.3316	[a]	0.1200	75
80	8658.47	0.0001	72145.6	8.3324	[a]	0.1200	80
∞				8.333		0.1200	∞

[a] Less than 0.0001.

TABLE A-10 Discrete Compounding; $i = 15\%$

	Single payment		Uniform series				
	Compound amount factor	Present worth factor	Compound amount factor	Present worth factor	Sinking fund factor	Capital recovery factor	
N	To find F given P F/P	To find P given F P/F	To find F given A F/A	To find P given A P/A	To find A given F A/F	To find A given P A/P	N
1	1.1500	0.8696	1.0000	0.8696	1.0000	1.1500	1
2	1.3225	0.7561	2.1500	1.6257	0.4651	0.6151	2
3	1.5209	0.6575	3.4725	2.2832	0.2880	0.4380	3
4	1.7490	0.5718	4.9934	2.8550	0.2003	0.3503	4
5	2.0114	0.4972	6.7424	3.3522	0.1483	0.2983	5
6	2.3131	0.4323	8.7537	3.7845	0.1142	0.2642	6
7	2.6600	0.3759	11.0668	4.1604	0.0904	0.2404	7
8	3.0590	0.3269	13.7268	4.4873	0.0729	0.2229	8
9	3.5179	0.2843	16.7858	4.7716	0.0596	0.2096	9
10	4.0456	0.2472	20.3037	5.0188	0.0493	0.1993	10
11	4.6524	0.2149	24.3493	5.2337	0.0411	0.1911	11
12	5.3502	0.1869	29.0017	5.4206	0.0345	0.1845	12
13	6.1528	0.1625	34.3519	5.5831	0.0291	0.1791	13
14	7.0757	0.1413	40.5047	5.7245	0.0247	0.1747	14
15	8.1371	0.1229	47.5804	5.8474	0.0210	0.1710	15
16	9.3576	0.1069	55.7175	5.9542	0.0179	0.1679	16
17	10.7613	0.0929	65.0751	6.0472	0.0154	0.1654	17
18	12.3755	0.0808	75.8363	6.1280	0.0132	0.1632	18
19	14.2318	0.0703	88.2118	6.1982	0.0113	0.1613	19
20	16.3665	0.0611	102.444	6.2593	0.0098	0.1598	20
21	18.8215	0.0531	118.810	6.3125	0.0084	0.1584	21
22	21.6447	0.0462	137.632	6.3587	0.0073	0.1573	22
23	24.8915	0.0402	159.276	6.3988	0.0063	0.1563	23
24	28.6252	0.0349	184.168	6.4338	0.0054	0.1554	24
25	32.9189	0.0304	212.793	6.4641	0.0047	0.1547	25
26	37.8568	0.0264	245.712	6.4906	0.0041	0.1541	26
27	43.5353	0.0230	283.569	6.5135	0.0035	0.1535	27
28	50.0656	0.0200	327.104	6.5335	0.0031	0.1531	28
29	57.5754	0.0174	377.170	6.5509	0.0027	0.1527	29
30	66.2118	0.0151	434.745	6.5660	0.0023	0.1523	30
35	133.176	0.0075	881.170	6.6166	0.0011	0.1511	35
40	267.863	0.0037	1779.09	6.6418	0.0006	0.1506	40
45	538.769	0.0019	3585.13	6.6543	0.0003	0.1503	45
50	1083.66	0.0009	7217.71	6.6605	0.0001	0.1501	50
55	2179.62	0.0005	14524.1	6.6636	a	0.1501	55
60	4384.00	0.0002	29220.0	6.6651	a	0.1500	60
65	8817.78	0.0001	58778.5	6.6659	a	0.1500	65
70	17735.7	a	118231	6.6663	a	0.1500	70
75	35672.8	a	237812	6.6665	a	0.1500	75
80	71750.8	a	478332	6.6666	a	0.1500	80
∞				6.667		0.1500	∞

a Less than 0.0001.

TABLE A-11   Discrete Compounding; $i = 18\%$

	Single payment		Uniform series				
	Compound amount factor	Present worth factor	Compound amount factor	Present worth factor	Sinking fund factor	Capital recovery factor	
N	To find F given P F/P	To find P given F P/F	To find F given A F/A	To find P given A P/A	To find A given F A/F	To find A given P A/P	N
1	1.1800	0.8475	1.000	0.847	1.00000	1.18000	1
2	1.3924	0.7182	2.180	1.566	0.45872	0.63872	2
3	1.6430	0.6086	3.572	2.174	0.27992	0.45992	3
4	1.9388	0.5158	5.215	2.690	0.19174	0.37174	4
5	2.2878	0.4371	7.154	3.127	0.13978	0.31978	5
6	2.6996	0.3704	9.442	3.498	0.10591	0.28591	6
7	3.1855	0.3139	12.142	3.812	0.08236	0.26236	7
8	3.7589	0.2660	15.327	4.078	0.06524	0.24524	8
9	4.4355	0.2255	19.086	4.303	0.05239	0.23239	9
10	5.2338	0.1911	23.521	4.494	0.04251	0.22251	10
11	6.1759	0.1619	28.755	4.656	0.03478	0.21478	11
12	7.2876	0.1372	34.931	4.793	0.02863	0.20863	12
13	8.5994	0.1163	42.219	4.910	0.02369	0.20369	13
14	10.1472	0.0985	50.818	5.008	0.01968	0.19968	14
15	11.9737	0.0835	60.965	5.092	0.01640	0.19640	15
16	14.1290	0.0708	72.939	5.162	0.01371	0.19371	16
17	16.6722	0.0600	87.068	5.222	0.01149	0.19149	17
18	19.6733	0.0508	103.740	5.273	0.00964	0.18964	18
19	23.2144	0.0431	123.414	5.316	0.00810	0.18810	19
20	27.3930	0.0365	146.628	5.353	0.00682	0.18682	20
21	32.3238	0.0309	174.021	5.384	0.00575	0.18575	21
22	38.1421	0.0262	206.345	5.410	0.00485	0.18485	22
23	45.0076	0.0222	244.487	5.432	0.00409	0.18409	23
24	53.1090	0.0188	289.494	5.451	0.00345	0.18345	24
25	62.6686	0.0160	342.603	5.467	0.00292	0.18292	25
26	73.9490	0.0135	405.272	5.480	0.00247	0.18247	26
27	87.2598	0.0115	479.221	5.492	0.00209	0.18209	27
28	102.9665	0.0097	566.481	5.502	0.00177	0.18177	28
29	121.5005	0.0082	669.447	5.510	0.00149	0.18149	29
30	143.3706	0.0070	790.948	5.517	0.00126	0.18126	30
31	169.1774	0.0059	934.319	5.523	0.00107	0.18107	31
32	199.6293	0.0050	1103.496	5.528	0.00091	0.18091	32
33	235.5625	0.0042	1303.125	5.532	0.00077	0.18077	33
34	277.9638	0.0036	1538.688	5.536	0.00065	0.18065	34
35	327.9973	0.0030	1816.652	5.539	0.00055	0.18055	35
40	750.3783	0.0013	4163.213	5.548	0.00024	0.18024	40
45	1716.6839	0.0006	9531.577	5.552	0.00010	0.18010	45
50	3927.3569	0.0003	21813.094	5.554	a	0.18005	50
∞				5.556	a	0.18000	∞

a Less than 0.0001.

TABLE A-12   Discrete Compounding; $i = 20\%$

	Single payment		Uniform series				
	Compound amount factor	Present worth factor	Compound amount factor	Present worth factor	Sinking fund factor	Capital recovery factor	
	To find F given P	To find P given F	To find F given A	To find P given A	To find A given F	To find A given P	
N	F/P	P/F	F/A	P/A	A/F	A/P	N
1	1.2000	0.8333	1.0000	0.8333	1.0000	1.2000	1
2	1.4400	0.6944	2.2000	1.5278	0.4545	0.6545	2
3	1.7280	0.5787	3.6400	2.1065	0.2747	0.4747	3
4	2.0736	0.4823	5.3680	2.5887	0.1863	0.3863	4
5	2.4883	0.4019	7.4416	2.9906	0.1344	0.3344	5
6	2.9860	0.3349	9.9299	3.3255	0.1007	0.3007	6
7	3.5832	0.2791	12.9159	3.6046	0.0774	0.2774	7
8	4.2998	0.2326	16.4991	3.8372	0.0606	0.2606	8
9	5.1598	0.1938	20.7989	4.0310	0.0481	0.2481	9
10	6.1917	0.1615	25.9587	4.1925	0.0385	0.2385	10
11	7.4301	0.1346	32.1504	4.3271	0.0311	0.2311	11
12	8.9161	0.1122	39.5805	4.4392	0.0253	0.2253	12
13	10.6993	0.0935	48.4966	4.5327	0.0206	0.2206	13
14	12.8392	0.0779	59.1959	4.6106	0.0169	0.2169	14
15	15.4070	0.0649	72.0351	4.6755	0.0139	0.2139	15
16	18.4884	0.0541	87.4421	4.7296	0.0114	0.2114	16
17	22.1861	0.0451	105.931	4.7746	0.0094	0.2094	17
18	26.6233	0.0376	128.117	4.8122	0.0078	0.2078	18
19	31.9480	0.0313	154.740	4.8435	0.0065	0.2065	19
20	38.3376	0.0261	186.688	4.8696	0.0054	0.2054	20
21	46.0051	0.0217	225.026	4.8913	0.0044	0.2044	21
22	55.2061	0.0181	271.031	4.9094	0.0037	0.2037	22
23	66.2474	0.0151	326.237	4.9245	0.0031	0.2031	23
24	79.4968	0.0126	392.484	4.9371	0.0025	0.2025	24
25	95.3962	0.0105	471.981	4.9476	0.0021	0.2021	25
26	114.475	0.0087	567.377	4.9563	0.0018	0.2018	26
27	137.371	0.0073	681.853	4.9636	0.0015	0.2015	27
28	164.845	0.0061	819.223	4.9697	0.0012	0.2012	28
29	197.814	0.0051	984.068	4.9747	0.0010	0.2010	29
30	237.376	0.0042	1181.88	4.9789	0.0008	0.2008	30
35	590.668	0.0017	2948.34	4.9915	0.0003	0.2003	35
40	1469.77	0.0007	7343.85	4.9966	0.0001	0.2001	40
45	3657.26	0.0003	18281.3	4.9986	a	0.2001	45
50	9100.43	0.0001	45497.2	4.9995	a	0.2000	50
55	22644.8	a	113219	4.9998	a	0.2000	55
60	56347.5	a	281732	4.9999	a	0.2000	60
∞				5.0000		0.2000	∞

a Less than 0.0001.

TABLE A-13 Discrete Compounding; $i = 25\%$

	Single payment		Uniform series				
	Compound amount factor	Present worth factor	Compound amount factor	Present worth factor	Sinking fund factor	Capital recovery factor	
	To find F given P	To find P given F	To find F given A	To find P given A	To find A given F	To find A given P	
N	F/P	P/F	F/A	P/A	A/F	A/P	N
1	1.2500	0.8000	1.0000	0.8000	1.0000	1.2500	1
2	1.5625	0.6400	2.2500	1.4400	0.4444	0.6944	2
3	1.9531	0.5120	3.8125	1.9520	0.2623	0.5123	3
4	2.4414	0.4096	5.7656	2.3616	0.1734	0.4234	4
5	3.0518	0.3277	8.2070	2.6893	0.1218	0.3718	5
6	3.8147	0.2621	11.2588	2.9514	0.0888	0.3388	6
7	4.7684	0.2097	15.0735	3.1611	0.0663	0.3163	7
8	5.9605	0.1678	19.8419	3.3289	0.0504	0.3004	8
9	7.4506	0.1342	25.8023	3.4631	0.0388	0.2888	9
10	9.3132	0.1074	33.2529	3.5705	0.0301	0.2801	10
11	11.6415	0.0859	42.5661	3.6564	0.0235	0.2735	11
12	14.5519	0.0687	54.2077	3.7251	0.0184	0.2684	12
13	18.1899	0.0550	68.7596	3.7801	0.0145	0.2645	13
14	22.7374	0.0440	86.9495	3.8241	0.0115	0.2615	14
15	28.4217	0.0352	109.687	3.8593	0.0091	0.2591	15
16	35.5271	0.0281	138.109	3.8874	0.0072	0.2572	16
17	44.4089	0.0225	173.636	3.9099	0.0058	0.2558	17
18	55.5112	0.0180	218.045	3.9279	0.0046	0.2546	18
19	69.3889	0.0144	273.556	3.9424	0.0037	0.2537	19
20	86.7362	0.0115	342.945	3.9539	0.0029	0.2529	20
21	108.420	0.0092	429.681	3.9631	0.0023	0.2523	21
22	135.525	0.0074	538.101	3.9705	0.0019	0.2519	22
23	169.407	0.0059	673.626	3.9764	0.0015	0.2515	23
24	211.758	0.0047	843.033	3.9811	0.0012	0.2512	24
25	264.698	0.0038	1054.79	3.9849	0.0009	0.2509	25
26	330.872	0.0030	1319.49	3.9879	0.0008	0.2508	26
27	413.590	0.0024	1650.36	3.9903	0.0006	0.2506	27
28	516.988	0.0019	2063.95	3.9923	0.0005	0.2505	28
29	646.235	0.0015	2580.94	3.9938	0.0004	0.2504	29
30	807.794	0.0012	3227.17	3.9950	0.0003	0.2503	30
35	2465.19	0.0004	9856.76	3.9984	0.0001	0.2501	35
40	7523.16	0.0001	30088.7	3.9995	a	0.2500	40
45	22958.9	a	91831.5	3.9998	a	0.2500	45
50	70064.9	a	280256	3.9999	a	0.2500	50
∞				4.0000		0.2500	∞

a Less than 0.0001.

TABLE A-14 Gradient to Present Worth Conversion Factor for Discrete Compounding (to Find P, Given G)

$$(P/G, i\%, N) = \frac{1}{i}\left[\frac{(1+i)^N - 1}{i(1+i)^N} - \frac{N}{(1+i)^N}\right]$$

n	1%	2%	4%	6%	8%	10%	12%	15%	20%	25%	n
1	0.00	0.00	0.00	0.00	0.00	0.00	0.00	0.00	0.00	0.00	1
2	0.98	0.96	0.92	0.89	0.86	0.83	0.80	0.76	0.69	0.64	2
3	2.92	2.85	2.70	2.57	2.45	2.33	2.22	2.07	1.85	1.66	3
4	5.80	5.62	5.27	4.95	4.65	4.38	4.13	3.79	3.30	2.89	4
5	9.61	9.24	8.55	7.93	7.37	6.86	6.40	5.78	4.91	4.20	5
6	14.32	13.68	12.50	11.46	10.52	9.68	8.93	7.94	6.58	5.51	6
7	19.92	18.90	17.07	15.45	14.02	12.76	11.64	10.19	8.26	6.77	7
8	26.38	24.88	22.18	19.84	17.81	16.03	14.47	12.48	9.88	7.95	8
9	33.69	31.57	27.80	24.58	21.81	19.42	17.36	14.75	11.43	9.02	9
10	41.84	38.95	33.88	29.60	25.98	22.89	20.25	16.98	12.89	9.99	10
11	50.80	47.00	40.38	34.87	30.27	26.40	23.13	19.13	14.23	10.85	11
12	60.57	55.67	47.25	40.34	34.63	29.90	25.95	21.18	15.47	11.60	12
15	94.48	85.20	69.74	57.55	47.89	40.15	33.92	26.69	18.51	13.33	15
20	165.46	144.60	111.56	87.23	69.09	55.41	44.97	33.58	21.74	14.89	20
25	252.89	214.26	156.10	115.97	87.80	67.70	53.10	38.03	23.43	15.56	25
30	355.00	291.72	201.06	142.36	103.46	77.08	58.78	40.75	24.26	15.83	30
35	470.15	374.88	244.88	165.74	116.09	83.99	62.61	42.36	24.66	15.94	35
40	596.85	461.99	286.53	185.96	126.04	88.95	65.12	43.28	24.85	15.98	40
45	733.70	551.56	325.40	203.11	133.73	92.45	66.73	43.81	24.93	15.99	45
50	879.41	642.36	361.16	217.46	139.59	94.89	67.76	44.10	24.97	16.00	50
60	1192.80	823.70	423.00	239.04	147.30	97.70	68.81	44.34	24.99	—	60
70	1528.64	999.83	472.48	253.33	151.53	98.99	69.21	44.42	—	—	70
80	1879.87	1166.79	511.12	262.55	153.80	99.56	69.36	44.47	—	—	80
90	2240.55	1322.17	540.77	268.39	154.99	99.81	—	—	—	—	90
100	2605.76	1464.75	563.12	272.05	155.61	99.92	—	—	—	—	100

TABLE A-15 Gradient to Uniform Series Conversion Factor for Discrete Compounding (to Find A, Given G)

$$(A/G, i\%, N) = \left[\frac{1}{i} - \frac{N}{(1+i)^N - 1}\right]$$

n	1%	2%	4%	6%	8%	10%	12%	15%	20%	25%	n
1	0.0001	0.0000	0.0000	0.0000	0.0000	0.0000	0.0000	0.0000	0.0000	0.0000	1
2	0.4974	0.4950	0.4902	0.4854	0.4808	0.4762	0.4717	0.4651	0.4545	0.4444	2
3	0.9932	0.9868	0.9739	0.9612	0.9487	0.9366	0.9246	0.9071	0.8791	0.8525	3
4	1.4874	1.4752	1.4510	1.4272	1.4040	1.3812	1.3589	1.3263	1.2742	1.2249	4
5	1.9799	1.9604	1.9216	1.8836	1.8465	1.8101	1.7746	1.7228	1.6405	1.5631	5
6	2.4708	2.4422	2.3857	2.3304	2.2763	2.2236	2.1720	2.0972	1.9788	1.8683	6
7	2.9600	2.9208	2.8433	2.7676	2.6937	2.6216	2.5515	2.4498	2.2902	2.1424	7
8	3.4476	3.3961	3.2944	3.1952	3.0985	3.0045	2.9131	2.7813	2.5756	2.3872	8
9	3.9335	3.8680	3.7391	3.6133	3.4910	3.3724	3.2574	3.0922	2.8364	2.6048	9
10	4.4177	4.3367	4.1773	4.0220	3.8713	3.7255	3.5847	3.3832	3.0739	2.7971	10
11	4.9003	4.8021	4.6090	4.4213	4.2395	4.0641	3.8953	3.6549	3.2893	2.9663	11
12	5.3813	5.2642	5.0343	4.8113	4.5957	4.3884	4.1897	3.9082	3.4841	3.1145	12
15	6.8141	6.6309	6.2721	5.9260	5.5945	5.2789	4.9803	4.5650	3.9588	3.4530	15
20	9.1692	8.8433	8.2091	7.6051	7.0369	6.5081	6.0202	5.3651	4.4643	3.7667	20
25	11.4829	10.9744	9.9925	9.0722	8.2254	7.4580	6.7708	5.8834	4.7352	3.9052	25
30	13.7555	13.0251	11.6274	10.3422	9.1897	8.1762	7.2974	6.2066	4.8731	3.9628	30
35	15.9869	14.9961	13.1198	11.4319	9.9611	8.7086	7.6577	6.4019	4.9406	3.9858	35
40	18.1774	16.8885	14.4765	12.3590	10.5699	9.0962	7.8988	6.5168	4.9728	3.9947	40
45	20.3271	18.7033	15.7047	13.1413	11.0447	9.3740	8.0572	6.5830	4.9877	3.9980	45
50	22.4362	20.4420	16.8122	13.7964	11.4107	9.5704	8.1597	6.6205	4.9945	3.9993	50
60	26.5331	23.6961	18.6972	14.7909	11.9015	9.8023	8.2664	6.6530	4.9989	—	60
70	30.4701	26.6632	20.1961	15.4613	12.1783	9.9113	8.3082	6.6627	—	—	70
80	34.2490	29.3572	21.3718	15.9033	12.3301	9.9609	8.3241	6.6656	—	—	80
90	37.8723	31.7929	22.2826	16.1891	12.4116	9.9831	—	—	—	—	90
100	41.3424	33.9863	22.9800	16.3711	12.4545	9.9927	—	—	—	—	100

APPENDIX B

Tables of Continuous Compounding Interest Factors (for r = 10% and 20%)

TABLE B-1 Continuous Compounding; $r = 10\%$

	Discrete flows				Continuous flows		
	Single payment		Uniform series		Uniform series		
	Compound amount factor	Present worth factor	Compound amount factor	Present worth factor	Compound amount factor	Present worth factor	
	To find F given P	To find P given F	To find F given A	To find P given A	To find F given $\overline{A}$	To find P given $\overline{A}$	
N	F/P	P/F	F/A	P/A	$F/\overline{A}$	$P/\overline{A}$	N
1	1.1052	0.9048	1.0000	0.9048	1.0517	0.9516	1
2	1.2214	0.8187	2.1052	1.7236	2.2140	1.8127	2
3	1.3499	0.7408	3.3266	2.4644	3.4986	2.5918	3
4	1.4918	0.6703	4.6764	3.1347	4.9182	3.2968	4
5	1.6487	0.6065	6.1683	3.7412	6.4872	3.9347	5
6	1.8221	0.5488	7.8170	4.2900	8.2212	4.5119	6
7	2.0138	0.4966	9.6391	4.7866	10.1375	5.0341	7
8	2.2255	0.4493	11.6528	5.2360	12.2554	5.5067	8
9	2.4596	0.4066	13.8784	5.6425	14.5960	5.9343	9
10	2.7183	0.3679	16.3380	6.0104	17.1828	6.3212	10
11	3.0042	0.3329	19.0563	6.3433	20.0417	6.6713	11
12	3.3201	0.3012	22.0604	6.6445	23.2012	6.9881	12
13	3.6693	0.2725	25.3806	6.9170	26.6930	7.2747	13
14	4.0552	0.2466	29.0499	7.1636	30.5520	7.5340	14
15	4.4817	0.2231	33.1051	7.3867	34.8169	7.7687	15
16	4.9530	0.2019	37.5867	7.5886	39.5303	7.9810	16
17	5.4739	0.1827	42.5398	7.7713	44.7395	8.1732	17
18	6.0496	0.1653	48.0137	7.9366	50.4965	8.3470	18
19	6.6859	0.1496	54.0634	8.0862	56.8589	8.5043	19
20	7.3891	0.1353	60.7493	8.2215	63.8906	8.6466	20
21	8.1662	0.1225	68.1383	8.3440	71.6617	8.7754	21
22	9.0250	0.1108	76.3045	8.4548	80.2501	8.8920	22
23	9.9742	0.1003	85.3295	8.5550	89.7418	8.9974	23
24	11.0232	0.0907	95.3037	8.6458	100.232	9.0928	24
25	12.1825	0.0821	106.327	8.7278	111.825	9.1791	25
26	13.4637	0.0743	118.509	8.8021	124.637	9.2573	26
27	14.8797	0.0672	131.973	8.8693	138.797	9.3279	27
28	16.4446	0.0608	146.853	8.9301	154.446	9.3919	28
29	18.1741	0.0550	163.298	8.9852	171.741	9.4498	29
30	20.0855	0.0498	181.472	9.0349	190.855	9.5021	30
35	33.1155	0.0302	305.364	9.2212	321.154	9.6980	35
40	54.5981	0.0183	509.629	9.3342	535.982	9.8168	40
45	90.0171	0.0111	846.404	9.4027	890.171	9.8889	45
50	148.413	0.0067	1401.65	9.4443	1474.13	9.9326	50
55	244.692	0.0041	2317.10	9.4695	2436.92	9.9591	55
60	403.429	0.0025	3826.43	9.4848	4024.29	9.9752	60
65	665.142	0.0015	6314.88	9.4940	6641.42	9.9850	65
70	1096.63	0.0009	10417.6	9.4997	10956.3	9.9909	70
75	1808.04	0.0006	17182.0	9.5031	18070.7	9.9945	75
80	2980.96	0.0003	28334.4	9.5051	29799.6	9.9966	80
85	4914.77	0.0002	46721.7	9.5064	49137.7	9.9980	85
90	8103.08	0.0001	77037.3	9.5072	81020.8	9.9988	90
95	13359.7	[a]	127019	9.5076	133587	9.9993	95
100	22026.5	[a]	209425	9.5079	220255	9.9995	100

[a] Less than 0.0001.

	Discrete flows				Continuous flows		
	Single payment		Uniform series		Uniform series		
	Compound amount factor	Present worth factor	Compound amount factor	Present worth factor	Compound amount factor	Present worth factor	
	To find F given P	To find P given F	To find F given A	To find P given A	To find F given $\overline{A}$	To find P given $\overline{A}$	
N	F/P	P/F	F/A	P/A	$F/\overline{A}$	$P/\overline{A}$	N
1	1.2214	0.8187	1.0000	0.8187	1.1070	0.9063	1
2	1.4918	0.6703	2.2214	1.4891	2.4591	1.6484	2
3	1.8221	0.5488	3.7132	2.0379	4.1106	2.2559	3
4	2.2255	0.4493	5.5353	2.4872	6.1277	2.7534	4
5	2.7183	0.3679	7.7609	2.8551	8.5914	3.1606	5
6	3.3201	0.3012	10.4792	3.1563	11.6006	3.4940	6
7	4.0552	0.2466	13.7993	3.4029	15.2760	3.7670	7
8	4.9530	0.2019	17.8545	3.6048	19.7652	3.9905	8
9	6.0496	0.1653	22.8075	3.7701	25.2482	4.1735	9
10	7.3891	0.1353	28.8572	3.9054	31.9453	4.3233	10
11	9.0250	0.1108	36.2462	4.0162	40.1251	4.4460	11
12	11.0232	0.0907	45.2712	4.1069	50.1159	4.5464	12
13	13.4637	0.0743	56.2944	4.1812	62.3187	4.6286	13
14	16.4446	0.0608	69.7581	4.2420	77.2232	4.6959	14
15	20.0855	0.0498	86.2028	4.2918	95.4277	4.7511	15
16	24.5325	0.0408	106.288	4.3325	117.663	4.7962	16
17	29.9641	0.0334	130.821	4.3659	144.820	4.8331	17
18	36.5982	0.0273	160.785	4.3932	177.991	4.8634	18
19	44.7012	0.0224	197.383	4.4156	218.506	4.8881	19
20	54.5981	0.0183	242.084	4.4339	267.991	4.9084	20
21	66.6863	0.0150	296.682	4.4489	328.432	4.9250	21
22	81.4509	0.0123	363.369	4.4612	402.254	4.9386	22
23	99.4843	0.0101	444.820	4.4713	492.422	4.9497	23
24	121.510	0.0082	544.304	4.4795	602.552	4.9589	24
25	148.413	0.0067	665.814	4.4862	737.066	4.9663	25
26	181.272	0.0055	814.227	4.4917	901.361	4.9724	26
27	221.406	0.0045	995.500	4.4963	1102.03	4.9774	27
28	270.426	0.0037	1216.91	4.5000	1347.13	4.9815	28
29	330.299	0.0030	1487.33	4.5030	1646.50	4.9849	29
30	403.429	0.0025	1817.63	4.5055	2012.14	4.9876	30
35	1096.63	0.0009	4948.60	4.5125	5478.17	4.9954	35
40	2980.96	0.0003	13459.4	4.5151	14899.8	4.9983	40
45	8103.08	0.0001	36594.3	4.5161	40510.4	4.9994	45
50	22026.5	a	99481.4	4.5165	110127	4.9998	50
55	59874.1	a	270426	4.5166	299366	4.9999	55
60	162755	a	735103	4.5166	813769	5.0000	60

a Less than 0.0001.

APPENDIX C

Table of Random Numbers[*]

48867	33971	29678	13151	56644	49193	93469	43252	14006	47173
32267	69746	00113	51336	36551	56310	85793	53453	09744	64346
27345	03196	33877	35032	98054	48358	21788	98862	67491	42221
55753	05256	51557	90419	40716	64589	90398	37070	78318	02918
93124	50675	04507	44001	06365	77897	84566	99600	67985	49133
98658	86583	97433	10733	80495	62709	61357	66903	76730	79355
68216	94830	41248	50712	46878	87317	80545	31484	03195	14755
17901	30815	78360	78260	67866	42304	07293	61290	61301	04815
88124	21868	14942	25893	72695	56231	18918	72534	86737	77792
83464	36749	22336	50443	83576	19238	91730	39507	22717	94719
91310	99003	25704	55581	00729	22024	61319	66162	20933	67713
32739	38352	91256	77744	75080	01492	90984	63090	53087	41301
07751	66724	03290	56386	06070	67105	64219	48192	70478	84722
55228	64156	90480	97774	08055	04435	26999	42039	16589	06757
89013	51781	81116	24383	95569	97247	44437	36293	29967	16088
51828	81819	81038	89146	39192	89470	76331	56420	14527	34828
59783	85454	93327	06078	64924	07271	77563	92710	42183	12380
80267	47103	90556	16128	41490	07996	78454	47929	81586	67024
82919	44210	61607	93001	26314	26865	26714	43793	94937	28439
77019	77417	19466	14967	75521	49967	74065	09746	27881	01070
66225	61832	06242	40093	40800	76849	29929	18988	10888	40344
98534	12777	84601	56336	00034	85939	32438	09549	01855	40550
63175	70789	51345	43723	06995	11186	38615	56646	54320	39632
92362	73011	09115	78303	38901	58107	95366	17226	74626	78208
61831	44794	65079	97130	94289	73502	04857	68855	47045	06309
42502	01646	88493	48207	01283	16474	08864	68322	92454	19287
89733	86230	04903	55015	11811	98185	32014	84761	80926	14509
01336	66633	26015	66768	24846	00321	73118	15082	13549	41335
72623	56083	65799	88934	87274	19417	84897	90877	76472	52145
74004	68388	04090	35239	49379	04456	07642	68642	01026	43810
09388	54633	27684	47117	67583	42496	20703	68579	65883	10729
51771	92019	39791	60400	08585	60680	28841	09921	00520	73135
69796	30304	79836	20631	10743	00246	24979	35707	75283	39211
98417	33403	63448	90462	91645	24919	73609	26663	09380	30515
56150	18324	43011	02660	86574	86097	49399	21249	90380	94375
76199	75692	09063	72999	94672	69128	39046	15379	98450	09159
74978	98693	21433	34676	97603	48534	59205	66265	03561	83075
85769	92530	04407	53725	96963	19395	16193	51018	70333	12094
63819	65669	38960	74631	39650	39419	93707	61365	46302	26134
18892	43143	19619	43200	49613	50904	73502	19519	11667	53294
32855	17190	61587	80411	22827	38852	51952	47785	34952	93574
29435	96277	53583	92804	05027	19736	54918	66396	96547	00351
36211	67263	82064	41624	49826	17566	02476	79368	28831	02805
73514	00176	41638	01420	31850	41380	11643	06787	09011	88924
90895	93099	27850	29423	98693	71762	39928	35268	59359	20674
69719	90656	62186	50435	77015	29661	94698	56057	04388	33381
94982	81453	87162	28248	37921	21143	62673	81224	38972	92988
84136	04221	72790	04719	34914	95609	88695	60180	58790	12802
58515	80581	88442	65727	72121	40481	06001	13159	55324	93591
20681	59164	75797	08928	68381	12616	97487	84803	92457	88847

[*] Reproduced with permission from the Rand Corporation, *A Million Random Numbers*. (New York: The Free Press, 1955).

APPENDIX D

Table of Random Normal Deviates[*]

1.102	− .944	.401	.226	1.396	−1.030	−1.723	− .368	2.170	.393
.148	−1.140	.492	−1.210	− .998	.573	.893	− .855	−2.209	− .267
2.372	1.353	− .900	− .554	− .343	.470	−1.033	−1.026	2.172	.195
− .145	.466	.854	− .282	−1.504	.431	− .060	.952	− .343	.735
.104	.732	.604	− .016	− .266	1.372	− .925	−1.594	−2.004	1.925
1.419	−1.853	− .347	.155	−1.078	.623	− .024	.498	.466	.049
.069	− .411	− .661	− .037	.703	.532	− .177	.395	− .278	.240
.797	.488	−1.070	− .721	−1.412	− .976	−1.953	− .206	1.848	.632
− .393	− .351	.222	.557	−1.094	1.403	.173	− .113	.806	.939
− .874	−1.336	.523	.848	.304	− .202	−1.279	.501	.396	.859
.125	−1.170	− .192	1.387	2.291	− .959	.090	1.031	.180	−1.389
−1.091	− .649	− .514	− .232	−1.198	.822	.240	.951	−1.736	.270
2.304	.481	− .987	−1.222	.549	−1.056	.277	− .919	.148	1.517
− .961	2.057	− .546	− .896	.165	− .343	.696	.628	− .929	− .965
− .783	.854	− .139	1.087	.515	− .876	− .448	.485	.589	− .804
.487	.557	.327	1.280	−1.731	− .339	.295	− .724	.720	.331
− .299	.979	− .924	− .649	.574	1.407	− .292	− .775	− .511	.026
1.831	− .937	−1.321	−1.734	1.677	−1.393	−1.187	− .079	− .181	− .844
.243	.466	−1.330	1.078	−1.102	1.123	− .421	− .674	2.951	− .743
−2.181	−1.854	−1.059	− .478	−1.119	.272	− .800	.841	− .061	2.261
.154	− .333	1.011	−1.565	1.261	.776	1.130	1.552	− .563	.558
−1.065	1.610	.463	.062	− .086	.021	1.633	1.788	.480	2.824
1.083	− .760	− .012	.183	.155	.676	−1.315	.067	.213	2.380
.615	− .594	− .028	− .506	− .054	3.173	.817	.210	1.699	1.950
.178	− .500	1.100	1.613	1.048	2.323	− .174	− .033	2.220	− .661
− .507	−1.273	.596	.690	−1.724	−1.689	.163	− .199	− .450	.244
.362	− .588	−1.386	.072	.778	− .591	.365	.465	2.472	1.049
.775	1.546	.217	−1.012	.778	.246	1.055	1.071	.447	− .585
.818	.561	−1.024	2.105	− .868	.060	− .385	1.089	.017	− .873
.014	.240	− .632	− .225	− .844	.448	1.651	1.423	.425	.252
−1.236	−1.045	−1.628	.687	.983	− .840	−1.835	−1.864	1.327	− .408
− .567	−1.161	.010	− .853	.111	1.145	1.015	.056	.141	1.471
.278	−1.783	.170	− .358	.705	− .054	1.098	.707	− .585	− .305
− .959	− .497	.688	− .268	−1.431	− .791	− .727	.958	.237	.092
1.249	.037	.497	.579	− .227	.860	.349	2.355	2.184	−1.744
− .915	− .164	−1.166	1.529	.008	.636	−1.080	− .688	2.444	−1.316
.132	2.809	−1.918	−1.083	− .642	− .179	.339	.637	.063	− .079
− .156	−1.664	1.140	.295	1.086	−2.546	− .002	− .672	.205	− .039
.538	−1.143	− .390	.165	− .160	.457	−1.307	.273	− .670	− .988
.027	− .057	.742	− .149	− .801	1.702	− .346	− .053	.892	−1.181
.023	.423	1.051	− .831	− .325	− .795	−1.129	− .287	.172	− .793
− .196	−1.457	1.060	.557	− .190	− .891	− .768	.282	−1.432	− .447
.133	.577	− .332	−1.932	.220	.189	−1.521	.896	− .781	− .899
.020	− .217	− .856	.605	.072	.520	1.222	− .181	− .266	−1.222
1.405	1.065	1.350	1.353	−2.289	−1.003	.375	−1.621	−1.126	.937
.178	−1.237	− .520	− .603	−1.615	− .358	.605	− .407	−2.579	−1.811
−1.438	.104	−1.821	− .390	− .630	1.294	1.470	.991	− .355	−1.285
1.768	− .175	− .450	.915	− .221	− .019	1.864	.038	.058	1.212
.099	1.076	2.348	−1.550	.458	.147	−1.223	.994	−1.657	1.264
.951	.252	−1.261	− .963	.221	− .036	− .395	− .252	−1.379	1.885

[*] Reproduced with permission from the Rand Corporation, *A Million Random Numbers*. (New York: The Free Press, 1955).

The Standardized Normal Distribution Function,[*] F(S)

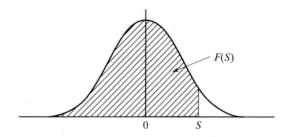

s	0.00	0.01	0.02	0.03	0.04	0.05	0.06	0.07	0.08	0.09
0.0	0.5000	0.5040	0.5080	0.5120	0.5160	0.5199	0.5239	0.5279	0.5319	0.5359
0.1	0.5398	0.5438	0.5478	0.5517	0.5557	0.5596	0.5636	0.5675	0.5714	0.5753
0.2	0.5793	0.5832	0.5871	0.5910	0.5948	0.5987	0.6026	0.6064	0.6103	0.6141
0.3	0.6179	0.6217	0.6255	0.6293	0.6331	0.6368	0.6406	0.6443	0.6480	0.6517
0.4	0.6554	0.6591	0.6628	0.6664	0.6700	0.6736	0.6772	0.6808	0.6844	0.6879
0.5	0.6915	0.6950	0.6985	0.7019	0.7054	0.7088	0.7123	0.7157	0.7190	0.7224
0.6	0.7257	0.7291	0.7324	0.7357	0.7389	0.7422	0.7454	0.7486	0.7517	0.7549
0.7	0.7580	0.7611	0.7642	0.7673	0.7703	0.7734	0.7764	0.7794	0.7823	0.7852
0.8	0.7881	0.7910	0.7939	0.7967	0.7995	0.8023	0.8051	0.8078	0.8106	0.8133
0.9	0.8159	0.8186	0.8212	0.8238	0.8264	0.8289	0.8315	0.8340	0.8365	0.8389
1.0	0.8413	0.8438	0.8461	0.8485	0.8508	0.8531	0.8554	0.8577	0.8599	0.8621
1.1	0.8643	0.8665	0.8686	0.8708	0.8729	0.8749	0.8770	0.8790	0.8810	0.8830
1.2	0.8849	0.8869	0.8888	0.8907	0.8925	0.8944	0.8962	0.8980	0.8997	0.9^20147
1.3	0.90320	0.90490	0.90658	0.90824	0.90988	0.91149	0.91309	0.91466	0.91621	0.91774
1.4	0.91924	0.92073	0.92220	0.92364	0.92507	0.92647	0.92785	0.92922	0.93056	0.93189
1.5	0.93319	0.93448	0.93574	0.93699	0.93822	0.93943	0.94062	0.94179	0.94295	0.94408
1.6	0.94520	0.94630	0.94738	0.94845	0.94950	0.95053	0.95154	0.95254	0.95352	0.95449
1.7	0.95543	0.95637	0.95728	0.95818	0.95907	0.95994	0.96080	0.96164	0.96246	0.96327
1.8	0.96407	0.96485	0.96562	0.96638	0.96712	0.96784	0.96856	0.96926	0.96995	0.97062
1.9	0.97128	0.97193	0.97257	0.97320	0.97381	0.97441	0.97500	0.97558	0.97615	0.97670
2.0	0.97725	0.97778	0.97831	0.97882	0.97932	0.97982	0.98030	0.98077	0.98124	0.98169
2.1	0.98214	0.98257	0.98300	0.98341	0.98382	0.98422	0.98461	0.98500	0.98537	0.98574
2.2	0.98610	0.98645	0.98679	0.98713	0.98745	0.98778	0.98809	0.98840	0.98870	0.98899
2.3	0.98928	0.98956	0.98983	0.9^20097	0.9^20358	0.9^20613	0.9^20863	0.9^21106	0.9^21344	0.9^21576
2.4	0.9^21802	0.9^22024	0.9^22240	0.9^22451	0.9^22656	0.9^22857	0.9^23053	0.9^23244	0.9^23431	0.9^23613
2.5	0.9^23790	0.9^23963	0.9^24132	0.9^24297	0.9^24457	0.9^24614	0.9^24766	0.9^24915	0.9^25060	0.9^25201
3.0	0.9^38650	0.9^38649	0.9^38736	0.9^38777	0.9^38817	0.9^38856	0.9^38893	0.9^38930	0.9^38965	0.9^38999
3.5	0.9^37674	0.9^37759	0.9^37842	0.9^37922	0.9^37999	0.9^38074	0.9^38146	0.9^38215	0.9^38282	0.9^38347
4.0	0.9^46833	0.9^46964	0.9^47090	0.9^47211	0.9^47327	0.9^47439	0.9^47546	0.9^47649	0.9^47748	0.9^47843

For example: $F(2.41) = 0.9^22024 = 0.992024$

[*] Reprinted from A. Hald, *Statistical Tables and Formulas* (New York: John Wiley & Sons, Inc., 1952), by permission of the publisher.

INDEX

TABLE 4-1 Summarization of Discrete Compound Interest Factors and Symbols

To find	Given	Multiply "Given" by factor below	Factor name	Factor functional symbol	Example (answer for $i = 5\%$) (Note: All uniform series problems assume end-of-period payments.)
F	P	$(1+i)^N$	Single sum compound amount	$(F/P, i\%, N)$	A firm borrows \$1,000 for 5 years. How much must it repay in a lump sum at the end of the fifth year? *Ans.*: \$1,276.30
P	F	$\dfrac{1}{(1+i)^N}$	Single sum present worth	$(P/F, i\%, N)$	A company desires to have \$1,000 8 years from now. What amount is needed now to provide for it? *Ans.*: \$676.80
P	A	$\dfrac{(1+i)^N - 1}{i(1+i)^N}$	Uniform series present worth	$(P/A, i\%, N)$	How much should be deposited in a fund to provide for 5 annual withdrawals of \$100 each? First withdrawal 1 year after deposit. *Ans.*: \$432.95
A	P	$\dfrac{i(1+i)^N}{(1+i)^N - 1}$	Capital recovery	$(A/P, i\%, N)$	What is the size of 10 equal annual payments to repay a loan of \$1,000? First payment 1 year after receiving loan. *Ans.*: \$129.50
F	A	$\dfrac{(1+i)^N - 1}{i}$	Uniform series compound amount	$(F/A, i\%, N)$	If 4 annual deposits of \$2,000 each are placed in an account, how much money has accumulated immediately after the last deposit? *Ans.*: \$8,620.20
A	F	$\dfrac{i}{(1+i)^N - 1}$	Sinking fund	$(A/F, i\%, N)$	How much should be deposited each year in an account in order to accumulate \$10,000 at the time of the fifth annual deposit? *Ans.*: \$1,810

Key: i = Interest rate per interest period A = Uniform series amount P = Present worth
     N = Number of interest periods F = Future worth